LTE and the Evolution to 4G Wireless

Design and Measurement Challenges, Second Edition

LTE and the Evolution to 4G Wireless

Design and Measurement Challenges, Second Edition

Moray Rumney

Agilent Technologies, UK Limited

A John Wiley & Sons, Ltd., Publication

This edition first published 2013

© 2013 Agilent Technologies, Inc. Published by John Wiley & Sons, Ltd.

Registered office

John Wiley & Sons Ltd, The Atrium, Southern Gate, Chichester, West Sussex, PO19 8SQ, United Kingdom

For details of our global editorial offices, for customer services and for information about how to apply for permission to reuse the copyright material in this book please see our website at www.wiley.com.

Library of Congress Cataloging-in-Publication Data

LTE and the evolution to 4G wireless: design and measurement challenges/[edited by] Moray Rumney.—Second edition.

     pages cm

  Includes bibliographical references and index.

    ISBN: 978-1-119-96257-1 (cloth: alk. paper) 1. Wireless communication systems. 2. Wireless Internet. I. Rumney, Moray, editor of compilation. II. Long term evolution and the evolution to 4G wireless.

TK5103.2.L798 2013

621.382—dc23

2013000332

A catalogue record for this book is available from the British Library.

Publisher: John Wiley & Sons Limited
Editor-in-Chief: Moray Rumney
Managing Editor : Mary Jane Pahls
Program Manager: Jan Whitacre
Set in 10/13 Agilent Condensed by Eikonal Communications, USA

Foreword to the Second Edition

Since the first edition of this book a number of LTE networks have become operational and customer take-up has been enthusiastic. What is remarkable is that LTE is being launched not on a single band but now on over 16 bands worldwide—and yet my impression is that there has been much less fuss over the launches than was experienced at the introduction of UMTS networks around 2003. At times of launch, bad news tends to hang around longer than good news, so no fuss is generally preferred. And if I am right about the relatively smooth launch of LTE, it is perhaps because the technology represents evolution and not revolution. Maybe operators have learned from the 3G experience to manage their customers' expectations more effectively. Another factor, quite possibly, is that the industry-wide effort to deliver conformance tests for both infrastructure and devices—available almost as soon as the core specifications matured—has paid dividends.

As the chairman of 3GPP RAN Working Group 5, I have been privileged to witness tremendous goodwill and cooperation by the test industry in particular to make conformance testing based on network emulators a reality even before the first LTE network was launched. With the availability of operational handsets thereafter, test development work continues apace. And yet the delivery of conformance tests has not been the only test story. In mid-2010, network operators demanded the ability to measure application layer throughput and other performance factors. Consequently, RAN WG5 strayed from its traditional charter of providing conformance tests for performance requirements defined in the core specifications to deliver a technical report that provides the methodology for user-centric tests that go well beyond the scope of the original performance requirements. This report is already in use by the GCF's Performance Agreement Group that was formed in early 2011; other similar market-driven requests for new types of testing are expected to follow.

These events seem to indicate that reliance on traditional "bit exact" conformance tests is no longer sufficient to ensure that devices will deliver the necessary end-user experience in a real network. A new, end-user driven approach to testing has required thinking beyond the status quo; indeed, it has led to much closer cooperation between industry partners than ever before. This partnership approach is likely to be in further demand as LTE networks and services continue to grow and become more complex.

I believe that the successful early launches of LTE networks are the result of learning from the UMTS experience. The process of test case selection and development has been optimized, but with the continued evolution of LTE through Releases 11 and 12 many technical challenges remain. Features such as carrier aggregation, uplink MIMO, and "over-the-air" MIMO testing, which are covered in the second edition of this book, are not so far away. Keeping up will be as challenging as ever.

Phil Brown
Chairman, TSG T WG1 2003–2005 and TSG RAN WG5 2005–present
December 2012

Foreward to the First Edition

The introduction of any new access technology into the current mobile telecommunication network is going to be filled with challenges, and not all of them can be anticipated. For network operators, convincing tech-savvy customers that new is better will need more than a marketing effort; it will need demonstrable results. The one thing that connects a customer to the network operator is the mobile device or handset. If that device does not appear to work in a variety of situations, then inevitably good reputations will be lost, customers may leave, and ultimately revenue and profits will suffer. No matter how big the investment in the new technology, mastering the physics of the mobile device and its supporting network is the foundation of success.

Expectations are high for Long Term Evolution, or LTE as it commonly called. But these expectations are based on the premise of fault-free performance. Therefore, LTE's initial success will be determined by the ability of handset and radio infrastructure manufacturers to deliver products that conform to 3GPP standards and are robust enough to allow operators to introduce improved services without disruption. Product testing is essential, but it can also be expensive and time-consuming. Manufacturers of new LTE products will have to make difficult decisions regarding what elements require rigorous procedures such as conformance testing and what can be safely left to testing in the field. The bottom line is that compromised quality or outright failure is unacceptable, costing more in the long run if it affects thousands or even millions of users.

Whatever the strategy a manufacturer adopts, the telecommunications test industry will once again play the important (but often unenviable) role of providing the equipment necessary to protect the huge investments made by LTE equipment manufacturers so that they, in turn, can fulfill the demands of the operators. The reality is that operators, manufacturers and the test industry need to work together closely as they try to establish competitive advantage. The scale of investment required to introduce LTE is likely to be the factor that brings about its success, as all players realize the common goal of making the new technology work. In this respect I have seen cooperation between competitors at all levels, whether it is in a 3GPP Standards meeting room with the sharing of work or consensus building, or even at the commercial level where compromises often have to be made to adapt to shifting moods in the market place.

Standards for new technology are essential, but proving adherence to those standards is a serious and expensive task. One approach to managing risk during LTE introduction can be found in the work of the Global Certification Forum (GCF). In March 2008, GCF initiated a process to define the criteria against which the first LTE devices can be certified; this will ensure GCF is well placed to assume market requirements prior to full scale development of conformance test cases. Here is an example of cross-industry commitment and cooperation. To start with, much consensus building was required between operators and manufacturers to select and prioritize the conformance tests being developed by 3GPP TSG RAN WG5. With that selection came the increased confidence needed to invest resources into developing the most relevant test specifications and test platforms in a timely manner. The GCF processes, along with the benefits to be gained by the industry, are explained in this book.

Leading test companies such as Agilent understand the technical challenges ahead. This working knowledge, coupled with an appreciation of market requirements such as device certification and a sensitivity to the needs of their manufacturer customers, qualifies Agilent to weigh in on the complexities and test challenges of LTE. Although no one person or one company can fully comprehend all aspects of LTE, this book sets out to give practical information to practicing engineers who are or will be working with LTE technology. In so doing, the authors are helping prepare the way for successful LTE deployment.

Phil Brown
Chair of 3GPP TSG RAN WG5; Chair of the Global Certification Forum Steering Group 2008
April 2009

Preface

The next generation of cellular technology is dramatically altering the communications landscape, changing the way people access information and interact with one another. At the forefront of the new technology is LTE, the long-term evolution of UMTS, developed by the Third Generation Partnership Project (3GPP). LTE is offering consumers a new level of mobile broadband access while ensuring that network operators achieve greater operational efficiencies and reduced cost of service. The further evolution of LTE through the LTE-Advanced project is now an approved part of the ITU-R IMT-Advanced (4G) program.

Turning LTE and LTE-Advanced into reality takes thousands of hours of engineering development. This book is intended to help make the task easier. The engineers at Agilent have contributed their insights to provide readers with an understanding of LTE that comes from working with the technology on a daily basis, both in the lab and on the committees that are defining the LTE standards. As Agilent engineers, they also have a unique measurement perspective to offer. The authors have shared the best, most current information to help ensure the success of LTE and future generations of cellular technology. The book is not intended, however, to substitute for reading the standards.

In this second edition, the authors have updated the content to the latest standards and included new material on LTE-Advanced where relevant; notably in the discussions of the air interface and physical layer. In addition much has been added on design, verification, and manufacturing test to help engineers meet their design and development goals and schedules.

Chapter 1 introduces LTE with an overview of the technology's objectives, services, and architecture. This chapter covers the standards process and status of the LTE/SAE Trial Initiative (LSTI).

Chapter 2 discusses air interface concepts, including the OFDMA modulation scheme used for the downlink, the SC-FDMA uplink modulation scheme, and multiple antenna techniques such as MIMO.

Chapter 3 gives more detail about the physical layer design. The chapter describes the downlink and uplink frame structures and defines the physical signals and channels. Multiplexing, channel coding, and physical layer measurements are covered. The procedures for the physical layer and radio access are explained along with radio resource management.

Chapter 4 covers upper layer signaling including the specifications for access stratum signaling and non-access stratum protocol states and transitions.

Chapter 5 introduces the System Architecture Evolution (SAE), which is the evolution of the core network to an all-IP system. The SAE is being developed concurrently with the LTE air interface.

Chapter 6 turns to the challenges of system design and testing. The chapter begins by exploring how to use LTE simulation tools to design a system and translate simulations into real signals for testing. This leads to a discussion of testing the new DigRF standard that provides a digital interface between the baseband and RF integrated circuits. Considerable time is spent discussing likely design issues and the challenges of testing LTE receivers and transmitters at the physical and transport channel layers. MIMO test challenges are also addressed in detail including beamforming and UE over-the-air testing. Other sections cover signaling protocol development and test, UE functional test, battery drain test, drive test, and manufacturing test of UEs.

Chapter 7 covers RF and signaling conformance testing and the role of the Global Certification Forum (GCF) and PCS Type Certification Review Board (PTCRB). The role of network operator acceptance testing is also discussed.

Chapter 8 concludes with a summary of the main features in Releases 9 through 12 including more detail on Release 10 LTE-Advanced.

Acknowledgements

The management team of Agilent's Global LTE Initiative would like to recognize the valuable insights and hard work of the many contributors to this book.

Moray Rumney has done an outstanding job of defining the book's focus and the content. He spent many hours writing, working with other authors, and editing to give the book's technical content its final shape.

Our special thanks go to Mary Jane Pahls for without her expertise and dedication to the first and second editions, they would never have happened. Mary Jane has a talent to turn very technical material including the graphics into meaningful and accessible content to provide a consistent end result across a wide range of authors.

Many experts provided valuable input and comments on the content. In addition to the chapter authors listed later, we would like to recognize Russell Barbour, Mike Hurst, Yuqin Shen, Dan Aubertin, Bob Cutler, Phil Lorch, Jing Ya, Hongwei Kong, Dan Jaeger, and Juergen Placht for their contributions. We thank Agilent and our managers for providing their support and encouragement in the second book's creation.

Dr. Michael Leung's enthusiasm was instrumental in getting the first edition done.

The 3GPP standardization process involves people from all over the world. We wish to acknowledge the work of our colleagues in creating the LTE standards documents. Without these documents, this book would not be possible.

Finally, we are grateful for our families who, once again, provided understanding and support during our long hours to complete this work.

Jan Whitacre
LTE Program Lead and Project Director
Agilent Technologies

February 2013

Author Biographies

Ying Bai

Before joining Agilent Technologies, Ying Bai received his Masters in Communication and Information Systems from the Institute of Electronics, Chinese Academy of Sciences, in 2006. Starting at Agilent as an application support engineer for the Signal Sources Division, Bai Ying was initially responsible for offering in-depth technical support to application engineers and customers worldwide, including GSM, W-CDMA, cdma2000® and TD-SCDMA products. In 2008, as an application expert, Bai Ying took on the responsibility of planning the LTE TDD version of Agilent's signal source and signal analysis solutions. He now has product planning, product marketing, and technical support roles for TD-SCDMA and LTE TDD applications.

Randy Becker

Randy Becker obtained his BSE with a major in Electrical Engineering from Walla Walla College in 1997 and an MSEE from the University in Nebraska in 1999. He then joined Hewlett-Packard/Agilent Technologies where he has worked for 14 years in a variety of technical marketing roles. Randy started as a marketing engineer in the Spectrum Analysis Division; two years later he moved to the Signal Sources Division where he stayed for twelve years. Randy is currently a senior application engineer supporting various cellular technologies with a focus in W-CDMA and LTE.

Ed Brorein

Ed Brorein received his BSEE from Villanova University in 1979 and an MSEE from the New Jersey Institute of Technology in 1987. Ed joined Agilent Technologies (at the time Hewlett-Packard) in 1979 and worked as an R&D engineer, manufacturing engineer, and marketing engineer in many various roles, presently as a marketing application engineer. All along Ed has been actively and deeply involved with the design, engineering, and application of DC power products and components. One area of focus for Ed has been extensively working with mobile device developers, helping with the testing of battery life and power management systems.

Peter Cain

Peter Cain is a wireless solution planner working for Agilent's Technical Leadership Organization in Edinburgh, Scotland. Since joining Hewlett-Packard/Agilent Technologies in 1985, Peter has had a variety of roles as an RF engineer, project manager, and marketing specialist. Over the last decade he has directed solution plans and written application notes for *Bluetooth*®, wireless LAN, ultra wideband, and Mobile WiMAX™. Most recently he has applied his knowledge of MIMO to LTE and helped deliver 89600 Wireless Link Analysis. Peter obtained a first class degree in Electronic Engineering at Southampton University in 1981.

Steve Charlton

Steve Charlton contributed to the 1st edition as an employee of Anite Telecoms Ltd. and member of the Anite LTE Layer 2 development team. An engineer for more than 30 years, Steve has been primarily involved in real-time embedded systems, mostly in the telecoms arena.

Niranjan Das

Niranjan Kumar Das contributed to the 1st edition as an employee of Anite Telecoms Ltd. with primary responsibility for the development of the MME protocols for Anite's test system. Since graduating in 1999 with a BE in Computer Science from Dibrugarh University, India, Niranjan has been mainly involved with 3GPP Layer 3 protocol development in UMTS and LTE.

Allison Douglas

Allison Douglas is a product manager in the Microwave and Communications Division at Agilent Technologies in Santa Rosa, CA. Allison joined Agilent Technologies in 2004 as an R&D engineer and since then has been in several different roles, including application engineer supporting cellular technologies and product marketing engineer. She is currently a product manager for the X-Series signal generators. She holds a BSE in Electrical Engineering and Biomedical Engineering from Duke University and an MBA from Gonzaga University.

Jeff Dralla

Jeff Dralla is a product planner and program manager in Agilent's Mobile Broadband Organization focusing on cellular and wireless connectivity manufacturing test solutions. Jeff joined Agilent Technologies in 2004 while completing a B.S. and M.S. in Electrical Engineering from the University of Southern California's Viterbi School of Engineering. Since then Jeff has held various roles within Agilent spanning sales to applications engineering, product planning and marketing, and most recently strategic program management. Jeff's main focus currently is planning next generation test equipment for wireless manufacturing as well as managing Agilent's strategic partnerships with leading wireless chipset companies.

Sandy Fraser

Sandy Fraser is a 25-year veteran of the RF and microwave industry with expertise spanning DC to 100 GHz applied to such diverse technologies as space and military products and infrared. Sandy's career includes over 20 years' experience with a cellular radio focus. During the last twelve years with Agilent Technologies, Sandy has focused on base station emulators for manufacturing test instruments, including the 8922 and the E5515B/C. Today he is the product manager for the E6621A PXT instrument and a leader in LTE technology awareness and training, specializing in LTE protocol and signaling. Sandy is a well-published author and his papers and presentations are appreciated by a global audience. He holds a BSc in Mechanical Engineering from Glasgow University.

Peter Goldsack

Peter Goldsack is an R&D engineer at Agilent Technologies. He has a BS in Mathematics from Edinburgh University, Scotland, and an MS in Electronics from Napier University, also in Edinburgh. Peter has worked at Hewlett-Packard/Agilent Technologies for the past 17 years in a variety of roles within R&D and marketing. He has developed protocol stacks for GSM, GPRS, EGPRS, and LTE and worked in a technical marketing role on GSM, GPRS, EGPRS, W-CDMA, HSDPA, HSUPA, and WiMAX™. Currently his primary responsibility is developing LTE solutions for cellular R&D customers.

Jean-Philippe Gregoire

Jean-Philippe Gregoire received a Masters in Electrical Engineering (microelectronics) from the Université de Liège, Belgium, in 2001. He joined Agilent Technologies the same year. Since then Jean-Philippe has contributed significantly to various projects—from specification to implementation—focusing on baseband digital with a specific interest in MIMO. As member of Agilent Labs, he has lead a research program on closed-loop MIMO and multi-channel fading in a European collaborative framework. Jean-Philippe is the author of several patent applications and technical papers in the field of signal processing, MIMO, and OFDM systems.

Craig Grimley

Craig Grimley joined Hewlett-Packard/Agilent Technologies in 1993 after completing a BEng (Hons) in Electrical and Electronic Engineering from Edinburgh's Heriot-Watt University. Craig initially spent a few years in manufacturing engineering before moving to his current product development research and design role in Agilent's Signal Analysis Division. Over the years Craig has gained measurement experience in many wireless communication signal formats including GSM, EDGE, W-CDMA, DVB-T/C, *Bluetooth*, WLAN, cdma2000, and 1xEVDO, as well as other general-purpose measurement applications including AM/FM and noise figure. His current technology focus is the development of signal analysis measurement capabilities for LTE within the Agilent 89600 VSA product.

Pankaj Gupta

Pankaj Gupta obtained his Bachelor of Technology in Electronics and Communications from Cochin University, India, in 2000. He has worked with SASKEN/Anite for more than eight years. During this time Pankaj has been involved with design and development of 3G conformance test cases. He started attending the 3GPP testing standardization group RAN5 (formerly known as T1) in 2004. Since then he has contributed to the development of R99, HSDPA, HSUPA and LTE conformance signaling test cases. In addition to working on the standards, Pankaj is the test case lead manager within Anite's Conformance Business Unit and is responsible for the Anite Conformance Test product for 3G and LTE test cases.

Ken Horne

Ken Horne is a product planner with the Mobile Broadband Operation at Agilent. Ken Horne graduated in 1985 with an M.Eng in Electronics and Electrical Engineering from Heriot-Watt University, Edinburgh. A member of IET and a Chartered Engineer, Ken has worked in a variety of roles through R&D and marketing in RF engineering and test equipment, joining Hewlett Packard in 1994, where he has had responsibility for many cellular and wireless connectivity test products.

Bob Irvine

Bob Irvine is a senior product manager with JDSU. He graduated from the University of Glasgow in 1991 with a Masters degree in Electronic Engineering. After spending 19 years with Hewlett-Packard/ Agilent Technologies he transferred to JDSU in 2010. During his early career, Bob worked on the launch and support of test equipment for mobile cellular R&D and manufacturing. For the last 15 years he has been working on leading-edge RF test products for the deployment and optimization of wireless cellular networks including GSM, GPRS, UMTS, HSPA, and LTE.

Moto Itagaki

Moto Itagaki brings more than 15 years of wireless technology experience to his role as senior application product planner for cellular and wireless signal analysis solutions at Agilent Technologies. Moto joined Agilent as a firmware research and development engineer for mobile communication test sets. As a product planner over the last decade, Moto has led and influenced signal analysis test application product requirements and designs for GSM, UMTS, IS95/cdma2000/1xEV-DO, 802.16-OFDMA, LTE, and multi-standard radio. Based in the Kobe office, Moto holds an MS in Electrical and Communication Engineering from Tohoku University in Japan.

Naoya Izuchi

Naoya Izuchi is a product marketing engineer for Microwave and Communications Division of Agilent Technologies. He received a BS in Electronic Engineering from Tottori University in Japan and joined Hewlett-Packard/Agilent as a marketing engineer in 1990. Naoya has worked on various RF and microwave test products including the impedance analyzer and network analyzers with Agilent's Component Test Division. After spending 3 years as a product line manager in Agilent's European Marketing Organization, he moved to the Signal Analysis Division and worked as a sales development engineer involved in 3G and 3.5G application support. He is now responsible for the product planning of the LTE and LTE-Advanced Signal Studio software.

Greg Jue

Greg Jue is an application development engineer and scientist working on aerospace/defense applications as a marketing program lead. Greg has worked in Agilent's High Performance Scopes team as well as Agilent EEsof, specializing in WLAN 802.11ac, LTE, WiMAX, aerospace/defense, and SDR applications. He has authored numerous articles, presentations, and application notes, including Agilent's LTE algorithm reference white paper and cognitive radio white paper. Greg pioneered combining design simulation and test solutions at Agilent Technologies, and authored the popular application notes 1394 and 1471 on combining simulation and test. Before joining Agilent in 1995, he worked on system design for the Deep Space Network at the Jet Propulsion Laboratory, Caltech University.

Per Kangru

Per Kangru was the world wide Business Development Manager for Agilent's Networks Solutions Division in the areas of LTE and SAE before joining JDSU in 2010. Per joined Agilent in 2001 working mainly in the area of mobile protocol testing but with MPLS and IP routing conformance testing as well. Over the years at Agilent Per contributed to more than 20 patent applications and he is the sole inventor of several pending patents. Per was Agilent's lead representative in the LTE & SAE Trial Initiative (LSTI) and has been an invited speaker at several industry conferences. Per has a background in basic research in atomic and laser physics from Uppsala University in Sweden.

Eng Wei Koo

Eng Wei Koo has extensive experience in 3GPP and cdma2000 wireless system technologies both as a cdma2000 BTS developer at Motorola and as a lead engineer at Agilent Technologies for protocol test and monitoring solutions. Eng Wei joined Agilent in 2002 and worked on UTRAN and E-UTRAN protocol monitoring solutions, pioneering the development of new and innovative approaches to data analysis. Eng Wei was a lead engineer for LTE network protocol test at Agilent before joining JDSU in 2010. Eng Wei received a Bachelor of Engineering from the University of Queensland, Australia, in 1999.

Gim-Seng Lau

Gim-Seng Lau is an R&D engineer for the Agilent wireless system development. He has worked in the field of RF and wireless engineering for the past 10 years. Currently, he is developing RF test conformance solutions. He also represents Agilent at 3GPP RAN5 and significantly contributes to the development of the standard specifications for LTE, WCDMA, and TD-SCDMA

Michael Lawton

Michael Lawton is a product planning engineer within the Mobile Broadband Operation of Agilent Technologies' Microwave Communications Division. He has been with Agilent for 20 years, spending the majority of his career in product planning working on a variety of different wireless technologies. Michael holds both a BEng and a PhD in Electrical Engineering from the University of Bristol, UK. He has been awarded 4 patents in the areas of wireless networking and fiber optic communications. He has also represented Agilent and served as chair for external groups developing both industry standards and multi-sourcing agreements.

Andrea Leonardi

Andrea Leonardi is an R&D engineer at Agilent Technologies. He has a BS and MS in Computer Science from Midwestern State University, USA. Andrea has worked at Hewlett-Packard/Agilent Technologies since February 1998 in a variety of roles within R&D. He has developed protocol stacks for cdma2000, W-CDMA, HSDPA, HSUPA and LTE one-box-testers. Currently his primary responsibility is developing LTE solutions for cellular R&D customers and he represents Agilent at 3GPP RAN WG5

Dr. Michael Leung

Dr. Michael Leung is a market development manager for Agilent's Asia Electronic Measurement Group. Michael plays a significant role in developing PXI and AXIe modular technology applications and wireless protocol research and testing for 3G (W-CDMA), 3.5G (HSPA), 4G (LTE/LTE-Advanced), and WiFi 802.11ac in Asia. During his 15 years at Agilent, Michael has received five Agilent technical invention awards, contributed more than 25 Agilent technical conference papers, and authored 10 research papers for various international journals and conferences. He is an editor of ICACT Transactions on the Advanced Communications Technology. He received a Master of Science and Doctor of Engineering from Hong Kong Polytechnic University in 1998 and 2005 respectively. Michael is a chartered engineer, a member of IET, and a senior member of IEEE.

Bill McKinley

Bill McKinley has held a number of positions in marketing, R&D, and manufacturing for Hewlett-Packard/Agilent Technologies spanning more than 25 years. He has worked predominantly in RF and microwave technology during this time, initially working with spectrum analyzers and signal generators before moving to mobile communication technologies. His current role is product planner focusing on the requirements of network operators. Bill holds a BSc in Electrical & Electronic Engineering from the University of Abertay in Dundee.

Masatoshi Obara

Masatoshi Obara obtained his BSc in Acoustics Design Engineering from Kyushu Institute of Design (a part of Kyushu University today), Japan, in 1979. He then joined Matsushita Intertechno in Tokyo, which represented test equipment companies such as B&K and DISA in Denmark and Leuven Measurements Systems in Belgium. Starting as a sales engineer, he later became a system engineer to design and develop automated acoustic and vibration test systems. Obara moved to Hewlett-Packard/Agilent Technologies in 1985. During the past 27 years he has developed many RF and microwave test systems for RF and microwave component test, radar and antenna test, satellite receiver test, and analog and digital cellular mobile and base station test. His roles have included system engineer, project manager, and engineering manager.

Mary Jane Pahls

Mary Jane Pahls is the owner of Eikonal Communications, a firm serving the engineering community. She has worked as a writer and editor in the test and measurement and telecommunication industries for more than 20 years, including eight years at Hewlett-Packard Company in Santa Rosa, California. Mary Jane's recent projects, in addition to those for Agilent Technologies, include work for engineering standards groups and for companies in the semiconductor industry. She holds BA and MA degrees from Kent State University.

Venkata Ratnakar Rayavarapu

Venkata Ratnakar Rayavarapu contributed to the 1st edition as an employee of Anite Telecoms Ltd. and member of the Anite Layer 3 protocol development team for LTE. He holds a Masters in Telecommunication Systems from IIT, Kharagpur, India. Over the past eight years Venkata has worked on various mobile technologies including GSM, GPRS, W-CDMA, HSDPA, and HSUPA. He was previously with Hellosoft and Samsung.

Ian Reading

Ian Reading is a strategic program manager with Agilent Technologies Mobile Broadband Organization. Graduating with a BEng in Electronic Engineering from the University of Sheffield, Ian joined Agilent in the UK as a manufacturing engineer in 1984. His subsequent career includes RF design, ETSI GSM committee attendance, and R&D and marketing management for products serving markets that include aerospace, 2G/3G/4G, and WLAN/WPAN/WiMAX.

Moray Rumney

Moray Rumney joined Hewlett-Packard/Agilent Technologies in 1984, after completing a BSc in Electronics from Edinburgh's Heriot-Watt University. Since then, Moray has enjoyed a varied career path, spanning manufacturing engineering, product development, applications engineering, and most recently technical marketing. His main focus has been the development and system design of base station emulators used in the development and testing of cellular phones. Moray joined ETSI in 1991 and 3GPP in 1999 where he was a significant contributor to the development of type approval tests for GSM and UMTS. He currently represents Agilent at 3GPP RAN WG4, where the air interface for HSPA+ and LTE-Advanced is being developed. Moray's current focus is in MIMO "over-the-air" test methods. Moray has published many technical articles in the field of cellular communications and is a regular speaker and chairman at industry conferences. He is a member of IET and a chartered engineer.

Darshpreet Sabharwal

Darshpreet Sabharwal contributed to the 1st edition as an employee of Anite Telecoms Ltd. and a senior software engineer responsible for the development of the medium access control layer for LTE. He has developed protocol stacks for GPRS, EGPRS, and LTE and has worked with various generations of ETSI and 3GPP technologies including 2G, 2.5G, 3G, and LTE. Previously he was a technical leader with Aricent, India. He received his BE in Computer Science with distinction from Guru Nanak Dev Engineering College, India, and holds a postgraduate diploma in Business Administration from Symbiosis, India.

Roland Scherzinger

Roland Scherzinger is a technical marketing engineer for Electronic Test Division Roland joined Hewlett-Packard/Agilent Technologies in 1980 as a test engineer. After holding various positions as a process and manufacturing engineer, he joined the technical marketing team in 1995. Roland's main focus was on test applications for computer buses such as PCI Express®, PCI, PCI-X, and InfiniBand. His responsibilities included application consulting for bring-up and debug, validation, performance, and compliance testing. He has been actively involved in and contributed to a number of PCI-SIG® related activities such as plugfests and developer's conferences. Since 2009, Roland has been the MIPI™ application expert for Agilent in Boeblingen, Germany, and is leading ETD's Digital Test Standards program for the MIPI standards. In this position he is actively participating with MIPI workgroups, plugfests, seminars, and workshops.

Sarabjit Singh

Sarabjit Singh contributed to the 1st edition as a technical architect at Anite Telecoms Ltd., UK, where his main responsibility involved shaping Anite's LTE solutions for cellular R&D customers. Previously, Sarabjit worked with Tata Consultancy Services Ltd as a technical consultant and was a subject matter expert on various mobile technologies for both 3GPP and 3GPP2. He is involved in LTE and HSPA+, and he previously worked on UMTS (Release 99, HSDPA, HSUPA) and CDMA (IS 95 B, cdma2000, 1xEV-DO). He received a BE in Computer Science with distinction from Guru Nanak Dev Engineering College, India, in 1996.

Mark Stambaugh

Mark Stambaugh has a BS in Electrical Engineering from the University of Cincinnati, an MEE from Rice University, and an MS in Computer Science from National Technological University. He joined Hewlett-Packard/Agilent Technologies in 1987 and began his career developing signal generators. Throughout most of his 21 years with Agilent, Mark was part of Agilent's R&D team developing base station emulator products and implementing all the major digital cellular protocols that these instruments have supported since GSM. Concentrating on the physical layer, his roles spanned implementation, system engineering, and technical leadership. Mark has nine patents.

Dr. K. F. Tsang

Dr. K. F. Tsang obtained a PhD from the University of Wales, College of Cardiff, UK. Dr. Tsang is now the chairman and managing director of Citycom Technology Ltd. as well as an associate professor in the Department of Electronic Engineering, City University of Hong Kong. Dr. Tsang has published more than 80 technical papers and is a reviewer for the IEEE Transaction on Circuits and Systems Part I, the Journal of Solid-State Circuits, and the IEEE Transaction on Vehicular Technology. His achievements include receiving the City University of Hong Kong's Applied Research Excellence Award and the Certificate of Merit in both the first Hong Kong Science & Product Innovation Competition in 1998 and the World Chinese Invention Exposition'98. In addition, he has won the EDN Asia Innovator Award, the Ericsson Super-Wireless Application Award, and the Freescale Semiconductor Ltd. Best Award.

Chris Van Woerkom

Chris Van Woerkom obtained his BSEE from the University of California at Davis in 1977 and his MBA from the University of Colorado in 1996. He worked at Hewlett-Packard/Agilent Technologies for 33 years in a wide variety of positions in marketing and technical marketing.

Jinbiao Xu

Jinbiao Xu received the Bachelor's degree in Mathematics and the Master and Ph.D. degrees in Information Engineering from Xidian University at Xi'an, China, in 1991, 1994, and 1997. From 1997 to 1998, he was a post-doctoral researcher on low speech codecs at the Institute of Acoustics, Chinese Academy of Science. Since joining Agilent EEsof EDA in 1999, he has been responsible for the OFDM series wireless library development (including products for DVB-T, ISDB-T, IEEE802.11a, WiMedia, mobile WiMAX, and 3GPP LTE). His current responsibility is to implement digital predistortion, MIMO channel model, and custom OFDM. Jinbiao's research interests includes MIMO, OFDM, pre-distortion, and satellite communication.

Hiroshi Yanagawa

Hiroshi Yanagawa obtained his BSc in Communication Engineering from Shibaura Institute of Technology, Japan, in 1985. He then joined Hewlett-Packard/Agilent Technologies and has worked in various engineering positions over the last 27 years. Hiroshi worked as a marketing engineer for impedance measuring instruments for five years and then moved to custom solution engineering. During this time he developed analog and digital cellular mobile and base station test systems as a system engineer.

Mitsuru Yokoyama

Mitsuru Yokoyama is a lead technologist for wireless applications at Agilent's Microwave and Communications Division in Kobe, Japan. He joined Hewlett-Packard/Agilent Technologies in 1982 after graduating with a BE and ME in Electrical Engineering from Kyoto University. Mitsuru worked initially on software design and development for Agilent's semiconductor test systems, moving to project management of PDC/PHS test products in 1992. Mitsuru represented Agilent at 3GPP committee T1 from its formation in 1999, and he was the first chair of T1's RF sub-group where the UE RF conformance tests were developed. Mitsuru is now engaged in application design and specification development for HSPA, EDGE Evolution, LTE, and other wireless applications.

Ryo Yonezawa

Ryo Yonezawa is currently an R&D engineer at Agilent's Microwave and Communications Division in Kobe, Japan. He has a BS in Electrical Engineering from Hosei University and joined Hewlett-Packard/Agilent Technologies in 1997. He started as a system development engineer and since then has developed RF verification and conformance test systems that support GSM, GPRS, EGPRS, W-CDMA, HSDPA, cdma2000, and EVDO. He currently develops LTE signal generation software.

AUTHOR BIOGRAPHIES

Ben Zarlingo Ben Zarlingo is a product manager for communications test with Agilent's Microwave and Communications Division. He received a BS in Electrical Engineering from Colorado State University in 1980 and has worked for Hewlett-Packard/Agilent Technologies in the areas of spectrum, network, and vector signal analysis with a primary focus on techniques for the design and troubleshooting of emerging communications technologies.

Contents

Chapter 1

LTE Introduction

1.1 Introduction

The challenge for any book tackling a subject as broad and deep as a completely new cellular radio standard is one of focus. The process of just creating the Long Term Evolution (LTE) specifications alone has taken several years and involved tens of thousands of temporary documents, thousands of hours of meetings, and hundreds of engineers. The result is several thousand pages of specifications. Now the hard work is underway, turning those specifications into real products that deliver real services to real people willing to pay real money. A single book of this length must therefore choose its subject wisely if it is to do more than just scratch the surface of such a complex problem.

The focus that Agilent has chosen for this book is a practical one: to explain design and measurement tools and techniques that engineering teams can use to accelerate turning the LTE specifications into a working system. The first half of the book provides an overview of the specifications starting in Chapter 2 with RF aspects and moving through the physical layer and upper layer signaling to the System Architecture Evolution (SAE) in Chapter 5. Due to limited space, the material in Chapters 2 through 5 should be viewed as an introduction to the technology rather than a deep exposition. For many, this level of detail will be sufficient but anyone tasked with designing or testing parts of the system will always need to refer directly to the specifications. The emphasis in the opening chapters is often on visual rather than mathematical explanations of the concepts. The latter can always be found in the specifications and should be considered sufficient information to build the system. However, the former approach of providing an alternative, more accessible explanation is often helpful prior to gaining a more detailed understanding directly from the specifications.

Having set the context for LTE in the opening chapters, the bulk of the remainder of the book provides a more detailed study of the extensive range of design and measurement techniques and tools that are available to help bring LTE from theory to deployment.

1.2 LTE System Overview

Before describing the LTE system it is useful to explain some of the terminology surrounding LTE since the history and naming of the technology is not intuitive. Some guidance can be found in the Vocabulary of 3GPP Specifications 21.905 [1], although this document is not comprehensive. The term LTE is actually a project name of the Third Generation Partnership Project (3GPP). The goal of the project, which started in November 2004, was to determine the long-term evolution of 3GPP's universal

LTE and the Evolution to 4G Wireless: Design and Measurement Challenges, Second Edition, Edited by Moray Rumney.
Copyright Agilent Technologies, Inc. 2013. Published by John Wiley & Sons, Ltd.

mobile telephone system (UMTS). UMTS was also a 3GPP project that studied several candidate technologies before choosing wideband code division multiple access (W-CDMA) for the radio access network (RAN). The terms UMTS and W-CDMA are now interchangeable, although that was not the case before the technology was selected.

In a similar way, the project name LTE is now inextricably linked with the underlying technology, which is described as an evolution of UMTS although LTE and UMTS actually have very little in common. The UMTS RAN has two major components: (1) the universal terrestrial radio access (UTRA), which is the air interface including the user equipment (UE) or mobile phone, and (2) the universal terrestrial radio access network (UTRAN), which includes the radio network controller (RNC) and the base station, which is also known as the node B (NB).

Because LTE is the evolution of UMTS, LTE's equivalent components are thus named evolved UTRA (E-UTRA) and evolved UTRAN (E-UTRAN). These are the formal terms used to describe the RAN. The system, however, is more than just the RAN since there is also the parallel 3GPP project called System Architecture Evolution that is defining a new all internet protocol (IP) packet-only core network known as the evolved packet core (EPC). The combination of the EPC and the evolved RAN (E-UTRA plus E-UTRAN) is the evolved packet system (EPS). Depending on the context, any of the terms LTE, E-UTRA, E-UTRAN, SAE, EPC, and EPS may get used to describe some or all of the system. Although EPS is the only correct term for the overall system, the name of the system will often be written as LTE/SAE or even simply LTE, as in the title of this book.

Figure 1.2-1 shows a high level view of how the evolved RAN and EPC interact with legacy radio access technologies.

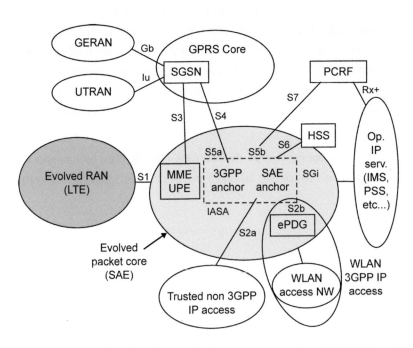

Figure 1.2-1. Logical high-level architecture for the evolved system (from 23.882 [2] Figure 4.2-1)

The 3GPP drive to simplify the existing hybrid circuit-switched/packet-switched core network is behind the SAE project to define an all-IP core network. This new architecture is a flatter, packet-only core network that is an essential part of delivering the higher throughput, lower cost, and lower latency that is the goal of the LTE evolved RAN. The EPC is also designed to provide seamless interworking with existing 3GPP and non-3GPP radio access technologies. The overall requirements for the System Architecture Evolution are summarized in 22.278 [3]. A more detailed description of the EPC is given in Chapter 5.

1.3 The Evolution from UMTS to LTE

The LTE specifications are written by 3GPP, which is a partnership of standards development organizations (SDOs). The work of 3GPP is public and, as will be described in Section 1.6, it is possible to gain access to all meeting reports, working documents, and published specifications from the 3GPP website: www.3gpp.org. The organizational partners that make up 3GPP are the Japanese Association of Radio Industries and Businesses (ARIB), the USA Alliance for Telecommunications Industry Solutions (ATIS), the China Communications Standards Association (CCSA), the European Telecommunications Standards Institute (ETSI), the Korean Telecommunications Technology Association (TTA), and the Japanese Telecommunications Technology Committee (TTC).

Table 1.3-1. Evolution of the UMTS specifications

Release	Functional freeze	Main UMTS feature of release
Rel-99	March 2000	Basic 3.84 Mcps W-CDMA (FDD & TDD)
Rel-4	March 2001	1.28 Mcps TDD (TD-SCDMA)
Rel-5	June 2002	HSDPA
Rel-6	March 2005	HSUPA (E-DCH)
Rel-7	Dec 2007	HSPA+ (64QAM downlink, MIMO, 16QAM uplink) LTE and SAE feasibility study
Rel-8	Dec 2008	LTE work item—OFDMA/SC-FDMA air interface, SAE work item—new IP core network, Dual-carrier HSDPA
Rel-9	December 2009	Home BS, MBMS, multi-standard radio, dual-carrier HSUPA, dual-carrier HSDPA with MIMO, dual-cell HSDPA
Rel-10	March 2011 (protocols 3 months later)	LTE-Advanced (carrier aggregation, 8x DL MIMO, 4x UL MIMO, relaying, enhanced inter-cell interference coordination (eICIC)), 4-carrier HSDPA
Rel-11	September 2012 (protocols 3 months later)	Further eICIC, coordinated multi-point transmission (CoMP), carrier aggregation scenarios, 8-carrier HSDPA
Rel-12	TBD—2014? (Stage 1 March 2013)	Further interference coordination, inter-site carrier aggregation, others TBD including dynamic TDD and LTE-D

Table 1.3-1 summarizes the evolution of the 3GPP UMTS specifications towards LTE. Each release of the 3GPP specifications represents a defined set of features. A summary of the contents of any release can be found at www.3gpp.org/releases.

The date given for the functional freeze relates to the date when no further new items can be added to the release. After this point any further changes to the specifications are restricted to essential corrections. The commercial launch date of a release depends on the period of time following the functional freeze before the specifications are considered stable and then implemented into commercial systems. For the first release of UMTS the delay between functional freeze and commercial launch was several years, although the delay for subsequent releases was progressively shorter. The delay between functional freeze and the first commercial launch for LTE/SAE was remarkably short, being less than a year, although it was two years before significant numbers of networks started operation. This period included the time taken to develop and implement the conformance test cases, which required significant work that could not begin until the feature set of the release was frozen and UEs had been implemented.

After Release 99, 3GPP stopped naming releases with the year and opted for a new scheme starting with Release 4. This choice was driven by the document version numbering scheme explained in Section 1.6. Release 4 introduced the 1.28 Mcps narrow band version of W-CDMA, also known as time division synchronous code division multiple access (TD-SCDMA). Following this was Release 5, in which high speed downlink packet access (HSDPA) introduced packet-based data services to UMTS in the same way that the general packet radio service (GPRS) did for GSM in Release 97 (1998). The completion of packet data for UMTS was achieved in Release 6 with the addition of high speed uplink packet access (HSUPA), although the official term for this technology is enhanced dedicated channel (E-DCH). HSDPA and HSUPA are now known collectively as high speed packet access (HSPA). Release 7 contained the first work on LTE/SAE with the completion of feasibility studies, and further improvements were made to HSPA such as downlink multiple input-multiple output (MIMO), 64QAM on the downlink, and 16QAM on the uplink. In Release 8, HSPA continued to evolve with the addition of numerous smaller features such as dual-carrier HSDPA and 64QAM with MIMO. Dual-carrier HSUPA was introduced in Release 9, four-carrier HSDPA in Release 10, and eight-carrier HSDPA in Release 11.

The main work in Release 8 was the specification of LTE and SAE, which is the main focus of this book. Work beyond Release 8 up to Release 12 is summarized in Chapter 8, although there are many references to features from these later releases throughout this second edition. Within 3GPP there are additional standardization activities not shown in Table 1.3-1 such as those for the GSM enhanced RAN (GERAN) and the IP multimedia subsystem (IMS).

1.4 LTE/SAE Requirements

The high level requirements for LTE/SAE include reduced cost per bit, better service provisioning, flexible use of new and existing frequency bands, simplified network architecture with open interfaces, and an allowance for reasonable power consumption by terminals. These are detailed in the LTE feasibility study 25.912 [4] and in the LTE requirements document 25.913 [5]. To meet the requirements for LTE outlined in 25.913 [5], LTE/SAE has been specified to achieve the following:

- Increased downlink and uplink peak data rates, as shown in Table 1.4-1. Note that the downlink is specified for single input single output (SISO) and MIMO antenna configurations at a fixed 64QAM modulation depth, whereas the uplink is specified only for SISO but at different modulation depths. These figures represent the physical limitation of the FDD air interface in ideal radio conditions with allowance for signaling overheads. Lower peak rates are specified for specific UE categories, and performance requirements under non-ideal radio conditions have also been developed. Comparable figures exist in [4] for TDD operation.

- Scalable channel bandwidths of 1.4 MHz, 3.0 MHz, 5 MHz, 10 MHz, 15 MHz, and 20 MHz in both the uplink and the downlink.

- Spectral efficiency improvements over Release 6 HSPA of 3 to 4 times in the downlink and 2 to 3 times in the uplink.

- Sub-5 ms latency for small IP packets.

- Performance optimized for low mobile speeds from 0 to 15 km/h supported with high performance from 15 to 120 km/h; functional support from 120 to 350 km/h. Support for 350 to 500 km/h is under consideration.

- Co-existence with legacy standards while evolving toward an all-IP network.

Table 1.4-1. LTE (FDD) downlink and uplink peak data rates (from 25.912 [4] Tables 13.1 & 13.1a)

FDD downlink peak data rates (64QAM)			
Antenna configuration	SISO	2x2 MIMO	4x4 MIMO
Peak data rate (Mbps)	100	172.8	326.4

FDD uplink peak data rates (single antenna)			
Modulation depth	QPSK	16QAM	64QAM
Peak data rate (Mbps)	50	57.6	86.4

The headline data rates in Table 1.4-1 represent the corner case of what can be achieved with the LTE RAN in perfect radio conditions; however, it is necessary for practical reasons to introduce lower levels of performance to enable a range of implementation choices for system deployment. This is achieved through the introduction of UE categories as specified in 36.306 [6] and shown in Table 1.4-2. These are similar in concept to the categories used to specify different levels of performance for HSPA.

Table 1.4-2. Peak data rates for UE categories (derived from 36.306 [6] Tables 4.1-1 and 4.1-2)

UE category	Peak downlink data rate (Mbps)	Number of downlink spatial layers	Peak uplink data rate (Mbps)	Number of uplink spatial layers	Support for 64QAM in uplink
Category 1	10.296	1	5.16	1	No
Category 2	51.024	2	25.456	1	No
Category 3	102.048	2	51.024	1	No
Category 4	150.752	2	51.024	1	No
Category 5	302.752	4	75.376	1	Yes
Category 6	301.504	2 or 4	51.024	1, 2, or 4	No
Category 7	301.504	2 or 4	10.2048	1, 2, or 4	No
Category 8	2998.56	8	149.776	8	Yes

Categories 6, 7, and 8 were added in Release 10 for the support of LTE-Advanced (see Section 8.3). There are other attributes associated with UE categories, but the peak data rates, downlink antenna configuration, and uplink 64QAM support are the categories most commonly referenced.

The emphasis so far has been on the peak data rates but what really matters for the performance of a new system is the improvement that can be achieved in average and cell-edge data rates. The reference configuration against which LTE/SAE performance targets have been set is defined in 25.913 [5] as being Release 6 UMTS. For the downlink the reference is HSDPA Type 1 (receive diversity but no equalizer or interference cancellation). For the uplink the reference configuration is single transmitter with diversity reception at the Node B. Table 1.4-3 shows the simulated downlink performance of UMTS versus the design targets for LTE. This is taken from the work of 3GPP during the LTE feasibility study [7]. Table 1.4-4 shows a similar set of results for the uplink taken from [8].

Table 1.4-3. Comparison of UMTS Release 6 and LTE downlink performance requirements

Case 1 500m inter-site distance	Spectrum efficiency		Mean user throughput		Cell-edge user throughput	
	(bps/Hz/cell)	x UTRA	[bps/Hz/user]	x UTRA	(bps/Hz/user)	x UTRA
UTRA baseline 1x2	0.53	x1.0	0.05	x1.0	0.02	x1.0
E-UTRA 2x2 SU-MIMO	1.69	x3.2	0.17	x3.2	0.05	x2.7
E-UTRA 4x2 SU-MIMO	1.87	x3.5	0.19	x3.5	0.06	x3.0
E-UTRA 4x4 SU-MIMO	2.67	x5.0	0.27	x5.0	0.08	x4.4

Table 1.4-4. Comparison of UMTS Release 6 and LTE uplink performance requirements

Case 1 500m inter-site distance	Spectrum efficiency		Mean user throughput		Cell-edge user throughput	
	(bps/Hz/cell)	x UTRA	(bps/Hz/user)	x UTRA	[bps/Hz/user]	x UTRA
UTRA baseline	0.332	x1.0	0.033	x1.0	0.009	x1.0
E-UTRA 1x2	0.735	x2.2	0.073	x2.2	0.024	x2.5
E-UTRA 1x2 MU-MIMO	0.675	x2.0	0.067	x2.0	0.023	x2.4
E-UTRA 1x4	1.103	x3.3	0.110	x3.3	0.052	x5.5
E-UTRA 2x2 SU-MIMO	0.776	x2.3	0.078	x2.3	0.010	x1.1

From these tables the LTE design targets of 2x to 4x improvement over UMTS Release 6 can be seen. Note, however, that UMTS did not stand still and there were Release 7 and Release 8 UMTS enhancements that significantly narrow the gap between UMTS and LTE. The evolution of UMTS continues through Release 12. Although the figures in Tables 1.4-3 and 1.4-4 are meaningful and user-centric, they were derived from system-level simulations and are not typical of the methods used to specify minimum performance. The simulations involved calculation of throughput by repeatedly dropping ten users randomly into the cell. From this data a distribution of performance was developed and the mean user throughput calculated. The cell-edge throughput was defined as the 5th percentile of the throughput cumulative distribution. For this reason the cell-edge figures are quoted per user assuming 10 users per cell, whereas the mean user throughput is independent of the number of users.

When it comes to defining minimum performance requirements for individual UE, the simulation methods used to derive the figures in Tables 1.4-3 and 1.4-4 cannot be used. Instead, the minimum requirements for UMTS and LTE involve spot measurement of throughput at specific high and low interference conditions, and for additional simplicity, this is done without the use of closed-loop adaptive modulation and coding. This approach to defining performance is pragmatic but it means that there is no direct correlation between the results from the conformance tests and the simulated system performance in Tables 1.4-3 and 1.4-4.

1.5 LTE/SAE Timeline

The timeline of LTE/SAE development is shown in Figure 1.5-1. This includes the work of 3GPP in drafting the specifications as well as the conformance test activities of the Global Certification Forum (GCF) and the trials carried out by the LTE/SAE Trial Initiative (LSTI). The work of GCF towards the certification of UE against the 3GPP conformance specifications is covered in some detail in Section 7.4. The LSTI, whose work was completed in 2011, was an industry forum and complimentary group that worked in parallel with 3GPP and GCF with the intent of accelerating the acceptance and deployment of LTE/SAE as the logical choice of the industry for next-generation networks. The work of LSTI was split into four phases. The first phase was proof of concept of the basic principles of LTE/SAE, using early prototypes not necessarily compliant with the specifications. The second phase was interoperability development testing (IODT), which was a more detailed phase of testing using standards-compliant equipment but not necessarily commercial platforms. The third stage was interoperability testing (IOT), similar in scope to IODT but using platforms intended for commercial deployment. The final phase was Friendly Customer Trials, which ran through 2010. GCF certified the first UE against the 3GPP conformance tests in April 2011. By November 2012 there were 102 FDD and 11 TDD commercial networks launched in 51 countries according to the Global Suppliers Association.

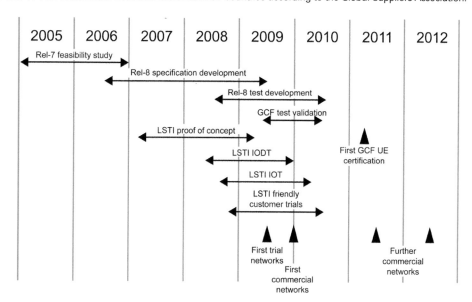

Figure 1.5-1. Projected LTE/SAE timeline

1.6 Introduction to the 3GPP LTE/SAE Specification Documents

The final section in this introductory chapter provides a summary of the LTE/SAE specification documents and where to find them.

1.6.1 Finding 3GPP Documents

A good place to start looking for documents is www.3gpp.org/specifications. From there it is possible to access the specification documents in a number of different ways, including by release number, publication date, or specification number. A comprehensive list of all 3GPP specifications giving the latest versions for all releases can be found at www.3gpp.org/ftp/Specs/html-info/SpecReleaseMatrix.htm. Each document has a version number from which the status of the document can be determined. For instance with 36.101 Vx.y.z, x represents the stability of the document, y the major update, and z an editorial update. If x is 1, then the document is an early draft for information only. If x is 2, then the document has been presented for approval. If x is greater than 2, then the document has been approved and is under change control. Once under change control, the value of x also indicates the release. Therefore a 3 is Release 1999, a 4 is Release 4, a 5 is Release 5, and so on. Most documents in an active release will get updated quarterly, which is indicated by an increment of the y digit. The document will also contain the date when it was approved by the technical specification group (TSG) responsible for drafting it. This date is often one month earlier than the official quarterly publication date.

To avoid confusion, individual documents should be referenced only by the version number. Groups of documents can be usefully referenced by the publication date—e.g., 2008–12—but note that the version numbers of the latest documents for that date will vary depending on how frequently each document has been updated. For example, at 2008–12, most of the physical layer specifications were at version 8.5.0 but most of the radio specifications were at version 8.4.0. It is therefore meaningless to refer to "version 8.x.y" of the specifications unless only one particular document is being referenced.

The set of specifications valid on any publication date will contain the latest version of every document regardless of whether the document was actually updated since the previous publication date. To access the specifications by publication date, go to ftp://ftp.3gpp.org/specs/. Within each date there will be a list of all the Releases and from there each series of specifications can be accessed. If only the latest documents for a Release are required, go to ftp://ftp.3gpp.org/specs/latest/. Newer, less stable, unpublished documents can often be found at ftp://ftp.3gpp.org/specs/Latest-drafts/, although care must be taken when making use of this type of information.

All versions of the releases of any particular document number can be accessed from ftp://ftp.3gpp.org/specs/archive/. This information can also be obtained from ftp://ftp.3gpp.org/Specs/html-info/, which provides the most comprehensive information. From this link the easiest way to proceed is to select a series of documents; e.g., ftp://ftp.3gpp.org/Specs/html-info/36-series.htm. This location will list all 36-series documents with the document numbers and titles. Selecting a document number will access a page with the full history of the document for all releases, including a named rapporteur and the working group (WG) responsible for drafting the document. At the bottom of the page will be a link to the change request (CR) history, which brings up yet another page listing all the changes made to the document and linked back to the TSG that approved the changes.

By tracing back through the CR history for a document it is possible to access the minutes and temporary documents of the TSG in which the change was finally approved. For instance, tracing back through a CR to 36.101 V8.5.0 (2008-12) would lead to a temporary document of the TSG RAN meeting that approved it stored under ftp://ftp.3gpp.org/tsg_ran/TSG_RAN/TSGR_42/. The change history of a document can also be found in the final annex of the document, but linking to the CR documents themselves has to be done via the website. The lowest level of detail is found by accessing the WG documents from a specific TSG meeting. An example for TSG RAN WG4, who develop the LTE 36.100-series radio specifications, can be found at ftp://ftp.3gpp.org/tsg_ran/WG4_Radio/TSGR4_50/. The link to this WG from the document can also be made from the html-info link given above.

The final way to gain insight into the work of the standards development process is to read the email exploders of the various committees. This capability is hosted by ETSI at http://list.etsi.org/.

1.6.2 LTE/SAE Document Structure

The feasibility study for LTE/SAE took place in Release 7, resulting in several Technical Reports of which [1] and [2] are the most significant.

The LTE RAN specifications are contained in the 36-series of Release 8 and are divided into the following categories:

- 36.100 series, covering radio specifications and eNB conformance testing
- 36.200 series, covering layer 1 (physical layer) specifications
- 36.300 series, covering layer 2 and 3 (air interface signalling) specifications
- 36.400 series, covering network signaling specifications
- 36.500 series, covering user equipment conformance testing
- 36.800 and 36.900 series, which are technical reports containing background information.

The latest versions of the 36 series documents can be found at www.3gpp.org/ftp/Specs/latest/Rel-11/36_series/.

The SAE specifications for the EPC are more scattered than those for the RAN and are found in the 22-series, 23-series, 24-series, 29-series, and 33-series of Release 8, with work happening in parallel in Release 9. A more comprehensive list of relevant EPC documents can be found in Chapter 5.

1.7 References

[1]   3GPP TR 21.905 V11.2.0 (2012-09) Vocabulary for 3GPP Specifications

[2]   3GPP TR 23.882 V8.0.0 (2008-09) 3GPP System Architecture Evolution: Report on Technical Options and Conclusions

[3]   3GPP TS 22.278 V12.1.0 (2012-06) Service requirements for the Evolved Packet System (EPS)

[4]   3GPP TR 25.912 V11.0.0 (2012-09) Feasibility study for evolved Universal Terrestrial Radio Access (UTRA) and evolved Universal Terrestrial Radio Access Network (UTRAN)

[5]   3GPP TR 25.913 V9.0.0 (2009-12) Requirements for Evolved Universal Terrestrial Radio Access (E-UTRA) and Evolved Universal Terrestrial Radio Access Network (E-UTRAN)

[6]   3GPP TR 36.306 V11.1.0 (2012-09) Evolved Universal terrestrial Radio Access Network (E-UTRA); User Equipment (UE) radio access capabilities

[7]   3GPP TSG RAN WG1 Tdoc R1-072578 "Summary of downlink performance evaluation," Ericsson, May 2007

[8]   3GPP TSG RAN WG1 Tdoc R1-072261 "LTE performance evaluation — uplink summary," Nokia, May 2007

Links to all reference documents can be found at www.agilent.com/find/ltebook.

Chapter 2

Air Interface Concepts

This chapter covers the radio aspects of LTE, starting in Section 2.1 with an overview of the radio frequency (RF) specifications. Sections 2.2 and 2.3 describe the downlink and uplink modulation schemes in some detail and, finally, Section 2.4 examines the way in which LTE uses multi-antenna methods to improve performance.

2.1 Radio Frequency Aspects

The RF specifications for LTE are covered in two 3GPP technical specification documents: 36.101 [1] for the user equipment (UE) and 36.104 [2] for the base station (BS), which is known in LTE as the evolved node B (eNB) although the more generic term BS is more commonly used. One of the first things to note about LTE is the integration between the frequency division duplex (FDD) and time division duplex (TDD) radio access modes. In the previous Universal Mobile Telephone System (UMTS) specifications, which also supported FDD and TDD, the RF specifications for the UE FDD, UE TDD, base station FDD, and base station TDD modes were covered in separate documents. However, the early decision by 3GPP to fully integrate FDD and TDD modes for LTE has resulted in only one RF specification document each for the UE and the BS. With the higher level of integration between the two modes, the effort required to support them should be less than it was in the past.

The structure of 36.101 [1] for the UE follows the UMTS pattern of minimum requirements for the transmitter and receiver followed by performance requirements for the receiver under fading channel conditions. The final section covers the performance of the channel quality feedback mechanisms. The structure of 36.104 [2] for the BS follows the same pattern as UMTS with transmitter, receiver, and performance requirements.

The purpose of this section is to highlight those aspects of the LTE RF requirements that will be new compared to UMTS. These include issues relating to LTE's support of multiple bands and channel bandwidths as well as those RF specifications peculiar to the use of orthogonal frequency division multiple access (OFDMA) modulation on the downlink and single-carrier frequency division multiple access (SC-FDMA) on the uplink. The RF performance aspects will be covered in Sections 6.5, 6.7, and 7.2.

The first edition of this book was based on the December 2008 Release 8 specifications. Since then there have been substantial additions. In particular, the September 2012 version of 36.101 [1] for Release 11 has more than tripled in length from the Release 8 version in December 2008 as more than 450 change requests have been approved since then. Much of this additional content is related to the addition of 15 new frequency bands, which brings the total to 40 (28 for FDD and 12 for TDD).

LTE and the Evolution to 4G Wireless: Design and Measurement Challenges, Second Edition, Edited by Moray Rumney.
Copyright Agilent Technologies, Inc. 2013. Published by John Wiley & Sons, Ltd..

The other major contributor to the increase in length is the introduction in Release 10 of carrier aggregation (CA), uplink multiple input multiple output (UL-MIMO), and enhanced downlink MIMO (eDL-MIMO). The background to these developments is described in Technical Report 36.807 [3], which like other "800 series" reports is not published. However, it does contain a wealth of technical background information used to develop the specifications.

The incorporation of CA, UL-MIMO, and eDL-MIMO into 36.101 [1] is quite daunting and so a notation system has been derived whereby the new subclauses specific to the new features are named as follows:

- Suffix A, additional requirements need to support CA
- Suffix B, additional requirements need to support UL-MIMO
- Suffix D, additional requirements need to support eDL-MIMO.

The C suffix is reserved for future use. This naming convention makes it easier to determine which subclauses in Sections 5, 6, and 7 of 36.101 [1] apply to UE supporting the new features.

2.1.1 Frequency Bands

Table 2.1-1 shows the IMT-2000 (3G) frequency bands defined by the European Telecommunications Standards Institute (ETSI) and 3GPP. Most of the frequency bands are defined in 36.101 [1] Table 5.5-1, meaning they are recognized by all three International Telecommunications Union (ITU) regions, although it should be noted that definition of a band does not imply its availability for deployment. The exceptions in Table 2.1-1 that are not defined in 36.101 [1] are Bands 15 and 16. These have been defined by ETSI in TS 102 735 [4] for ITU Region 1 (Europe, Middle East, and Africa) only. These bands have not been adopted at this time by ITU Region 2 (Americas) or Region 3 (Asia), which is why they do not appear in 36.101 [1]. Figure 2.1-1 shows how the terms bandwidth, duplex spacing, and gap are used for FDD in Table 2.1-1. The concepts of gap and duplex spacing don't exist for TDD.

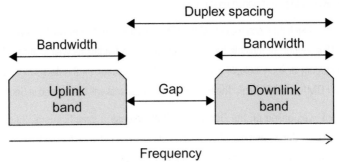

Figure 2.1-1. Explanation of frequency band terms

Table 2.1-1. Defined frequency bands for IMT-2000 (MHz)

Band number	Uplink		Downlink		Bandwidth	Duplex spacing	Gap	Duplex mode
	Low	High	Low	High				
1	1920	1980	2110	2170	60	190	130	FDD
2	1850	1910	1930	1990	60	80	20	FDD
3	1710	1785	1805	1880	75	95	20	FDD
4	1710	1755	2110	2155	45	400	355	FDD
5	824	849	869	894	25	45	20	FDD
6	830	840	875	885	10	35	35	FDD
7	2500	2570	2620	2690	70	120	50	FDD
8	880	915	925	960	35	45	10	FDD
9	1749.9	1784.9	1844.9	1879.9	35	95	60	FDD
10	1710	1770	2110	2170	60	400	340	FDD
11	1427.9	1452.9	1475.9	1500.9	20	48	23	FDD
12	698	716	728	746	18	30	12	FDD
13[1]	777	787	746	756	10	−31	21	FDD
14[1]	788	798	758	768	10	−30	20	FDD
15[2]	1900	1920	2600	2620	20	700	680	FDD
16[2]	2010	2025	2585	2600	15	575	560	FDD
17	704	716	734	746	12	30	18	FDD
18	815	830	860	875	15	45	30	FDD
19	830	845	875	890	15	45	30	FDD
20	832	862	791	821	30	−41	11	FDD
21	1447.9	1462.9	1495.9	1510.9	15	48	33	FDD
22	3410	3490	3510	3590	80	100	20	FDD
23	2000	2020	2180	2200	20	180	160	FDD
24	1626.5	1660.5	1525	1559	34	−101.5	67.5	FDD
25	1850	1915	1930	1995	65	80	15	FDD
26	814	849	859	894	35	45	10	FDD
27	807	824	852	869	17	45	28	FDD
28	703	748	758	803	45	55	10	FDD
33	1900	1920	1900	1920	20	0	0	TDD
34	2010	2025	2010	2025	15	0	0	TDD
35	1850	1910	1850	1910	60	0	0	TDD
36	1930	1990	1930	1990	60	0	0	TDD
37	1910	1930	1910	1930	20	0	0	TDD
38	2570	2620	2570	2620	50	0	0	TDD
39	1880	1920	1880	1920	40	0	0	TDD

40	2300	2400	2300	2400	100	0	0	TDD
41	2496	2690	2496	2690	194	0	0	TDD
42	3400	3600	3400	3600	200	0	0	TDD
43	3600	3800	3600	3800	200	0	0	TDD
44	703	803	703	803	100	0	0	TDD

Note 1: Uplink frequency is higher than downlink frequency.
Note 2: Defined by ETSI for ITU Region 1 only.

Table 2.1-1 shows the large number of options that exist for IMT technologies, which now include LTE. When UMTS was first specified in 1999, only one frequency band was defined. This band became a point around which the industry could focus its efforts in developing specifications and products. In the years since then, bands have been gradually added, and when LTE was specified in 2008, it inherited all the existing UMTS bands plus some new ones added in Release 8. Moreover, with the integration of TDD into the LTE specifications, another eight bands were added to the list.

It is clear from Table 2.1-1 that many of the bands are overlapping or subsets of one another, so the actual RF coverage may not seem to present a problem for power amplifiers and receivers. Where the difficulty lies, however, is in handling the many combinations of filtering that are required to implement the different bands. The bandwidth, duplex spacing, and gap are not constant, which adds to the challenge of designing the specific band filters required for each implemented band. There are also issues in designing efficient antennas to cover the wide range of possible supported bands.

The possibility of variable duplex spacing is not precluded but as of Release 11 has not been developed. For a specified FDD band, variable duplex spacing would mean that the currently fixed relationship between the uplink and downlink channels could become variable. This would increase deployment flexibility but also increase the complexity of the specifications, the equipment design, and network operation.

2.1.2 Channel Bandwidths

A trend in recent years has been for radio systems to be ported to new frequency bands, although typically these systems support only one channel bandwidth. The first release of UMTS, which supported both FDD and TDD modes, used a common CDMA chip rate of 3.84 Mcps and a channel spacing of 5 MHz. Release 4 of UMTS introduced a low chip rate (LCR) TDD option (also known as TD-SCDMA) that used the lower 1.28 Mcps with a correspondingly narrower channel spacing of 1.6 MHz. This was followed in Release 7 by the 7.68 Mcps option with its 10 MHz channel spacing. In Release 8 a dual-carrier version of HSDPA was introduced; however, the wider bandwidth comes from two separate 5 MHz channels. Four-carrier HSDPA was introduced in Release 10 and eight-carrier in Release 11.

The situation for LTE is very different. The OFDMA modulation scheme upon which UMTS is based has as one of its properties the ability to scale its channel bandwidth linearly without changing the underlying properties of the physical layer—these being the subcarrier spacing and the symbol length. The details of the LTE modulation schemes are discussed fully in Sections 2.2 and 2.3. It is sufficient to say at this point that LTE was designed from the start to support six different channel bandwidths. These are 1.4 MHz, 3 MHz, 5 MHz, 10 MHz, 15 MHz, and 20 MHz. Earlier versions of the specifications also supported 1.6 MHz and

3.2 MHz for interworking with LCR TDD, but these were removed when the LTE TDD frame structure was aligned with the FDD frame structure rather than the TD-SCDMA frame structure from UMTS.

The choice of many channel bandwidths means that LTE has more deployment flexibility than previous systems. The wide channel bandwidths of 10, 15, and 20 MHz are intended for new spectrum, with the 2.6 GHz and 3.5 GHz bands in mind. These wider channels offer more efficient scheduling opportunities, which can increase overall system performance. With the potential of having a much wider channel available, individual users might perceive that they have a high bandwidth connection when in fact they are sharing the bandwidth with many other users. An individual user's perception of "owning" the channel comes from the fact that the demand typically is variable and what matters is the peak rate available at the time of the demand. This perception is known as the trunking effect, wherein the wider the channel, the greater the gains. Narrowband systems such as GSM with a channel bandwidth of only 200 kHz are not in a position to instantaneously offer more capacity, even if other users are not making full use of their channel.

The other benefit of a wider channel is the possibility of scheduling users as a function of the channel conditions specific to them. This topic is discussed in more detail in Section 3.4, but the essence is that OFDMA has the ability to schedule traffic over a subset of the channel and thus, with appropriate feedback of the instantaneous channel conditions, can target transmissions at frequencies exhibiting the best propagation conditions and lowest interference.

Table 2.1-2. Combinations of channel bandwidth and frequency band for which RF requirements are defined (36.101 [1] Table 5.6.1-1)

E-UTRA band	Channel bandwidth					
	1.4 MHz	3 MHz	5 MHz	10 MHz	15 MHz	20 MHz
1			Yes	Yes	Yes	Yes
2	Yes	Yes	Yes	Yes	Yes[1]	Yes[1]
3	Yes	Yes	Yes	Yes	Yes[1]	Yes[1]
4	Yes	Yes	Yes	Yes	Yes	Yes
5	Yes	Yes	Yes	Yes[1]		
6			Yes	Yes[1]		
7			Yes	Yes	Yes	Yes[1]
8	Yes	Yes	Yes	Yes[1]		
9			Yes	Yes	Yes[1]	Yes[1]
10			Yes	Yes	Yes	Yes
11			Yes	Yes[1]		
12	Yes	Yes	Yes[1]	Yes[1]		
13			Yes[1]	Yes[1]		
14			Yes[1]	Yes[1]		
17			Yes[1]	Yes[1]		
18			Yes	Yes[1]	Yes[1]	
19			Yes	Yes[1]	Yes[1]	
20			Yes	Yes[1]	Yes[1]	Yes[1]
21			Yes	Yes[1]	Yes[1]	

22			Yes	Yes	Yes[1]	Yes1
23	Yes		Yes	Yes		
24			Yes	Yes		
25	Yes	Yes	Yes	Yes	Yes[1]	Yes[1]
26	Yes	Yes	Yes	Yes[1]	Yes[1]	
27	Yes	Yes	Yes	Yes[1]		
28		Yes	Yes	Yes[1]	Yes[1]	Yes[1,2]
33			Yes	Yes	Yes	Yes
34			Yes	Yes	Yes	
35	Yes	Yes	Yes	Yes	Yes	Yes
36	Yes	Yes	Yes	Yes	Yes	Yes
37			Yes	Yes	Yes	Yes
38			Yes	Yes		
39			Yes	Yes	Yes	Yes
40				Yes	Yes	Yes
41			Yes	Yes	Yes	Yes
42			Yes	Yes	Yes	Yes
43			Yes	Yes	Yes	Yes
44			Yes	Yes	Yes	Yes

Note 1: Bandwidth for which a relaxation of the specified UE receiver sensitivity requirement (36.101 Clause 7.3) is allowed.

The 5 MHz channel bandwidth option for LTE is an obvious choice for re-farming of existing UMTS spectrum. This re-farming will not benefit from trunking gains over UMTS but still has the possibility of gains through frequency-selective scheduling. The 1.4 MHz and 3 MHz options are targeted at re-farming of narrowband systems such as GSM and cdma2000®. Even the 1.4 MHz option will have significant trunking gains over 200 kHz GSM as well as the ability to do some frequency-selective scheduling. The consequence of a system that has so much flexibility in terms of frequency bands and channel bandwidths is the complexity that is created. Several of the LTE RF requirements described in this section reflect this growth in complexity: requirements that in UMTS were expressed as single-valued figures are now represented by multi-dimensional tables.

Although the LTE system could be operated in any of the defined bands at any channel bandwidth, certain combinations are not expected in real deployment, and for such cases no RF performance requirements are defined. Table 2.1-2 shows the combination of channel bandwidths for which performance requirements exist (or do not exist) for the different frequency bands. Table 2.1-2 shows, for example, that no requirements are defined for the 1.4 MHz and 3 MHz bandwidths for several bands including E-UTRA Band 1 (the primary UMTS operating band at 2.1 GHz) as these deployment combinations are not likely. The table also shows that for some combinations there are relaxations in the requirements. For example, there are several bands for which the receiver sensitivity requirements are relaxed when the system operates at 15 MHz and 20 MHz channel bandwidths. At the time of this writing these relaxations are limited to reference sensitivity although the list of affected requirements may grow over time.

2.1.3 Reference Measurement Channels

Before describing the UE and BS RF requirements it is useful to introduce the concept of reference measurement channels (RMCs). These exist for both the downlink and uplink and are used throughout the RF specifications to precisely describe the configuration of signals used to test the UE and BS transmitters and also their receivers.

2.1.3.1 Uplink Reference Measurement Channels

The flexible nature of the uplink transmissions makes it important that the signal definition be explicit when performance targets are specified. As a result, many of the UE transmitter requirements in 36.101 [1] Subclause 6 and some of the receiver requirements in Subclause 7 are defined relative to specific uplink configurations. These are known as uplink RMCs. A similar principle was used in UMTS and the main difference for LTE is the use of SC-FDMA rather than W-CDMA for the air interface.

Since the uplink RMCs are primarily used for testing UE transmitter performance, many of the variables that will be used in real operation are disabled. These include "no incremental redundancy" (1 HARQ transmission), "normal cyclic prefix only," "no physical uplink shared channel (PUSCH) hopping," "no link adaptation," and "for partial allocation the resource blocks (RBs) are contiguous starting at the channel edge." Table 2.1-3 shows an example quadrature phase shift keying (QPSK) uplink reference measurement channel (RMC) for various allocations of up to 75 RBs (75%).

Table 2.1-3. Reference channels for 20 MHz QPSK with partial RB allocation
(36.101 [1] Table A.2.2.2.1-6b)

Parameter	Unit	Value	Value	Value
Channel bandwidth	MHz	20	20	20
Allocated resource blocks		50	54	75
DFT-OFDM symbols per subframe		12	12	12
Modulation		QPSK	QPSK	QPSK
Target coding rate		1/3	1/3	1/5
Payload size	Bits	5160	4776	4392
Transport block CRC	Bits	24	24	24
Number of code blocks per subframe (Note 1)		1	1	1
Total number of bits per subframe	Bits	14400	15552	21600
Total symbols per subframe		7200	7776	10800
UE category		≥ 1	≥ 1	≥ 1

Note 1: If more than one code block is present, an additional CRC sequence of $L = 24$ bits is attached to each code block (otherwise $L = 0$ bit).

For the purposes of testing the BS receiver, further uplink RMCs are defined in 36.104 [2] Annex A. These are referred to as fixed reference channels (FRCs).

2.1.3.2 Downlink Reference Measurement Channels

An example of a single antenna downlink RMC for use with 64QAM PUSCH and common (cell-specific rather than UE-specific) demodulation reference symbols (DMRS) is given in Table 2.1-4. This RMC will be used for performance testing under faded channel conditions.

Table 2.1-4. Fixed reference channel 64QAM R = 3/4 (36.101 [1] Table A.3.3.1-3)

Parameter	Unit		Value				
Reference channel			[R.5 FDD]	[R.6 FDD]	[R.7 FDD]	[R.8 FDD]	[R.9 FDD]
Channel bandwidth	MHz	1.4	3	5	10	15	20
Allocated resource blocks			15	25	50	75	100
Allocated subframes per radio frame			10	10	10	10	10
Modulation		64QAM	64QAM	64QAM	64QAM	64QAM	64QAM
Target coding rate		3/4	3/4	3/4	3/4	3/4	3/4
Information bit payload							
For subframes 1,2,3,4,6,7,8,9	Bits	8504	14112	30576	46888	61664	For subframe 5
Bits		7992	13536	30576	45352	61664	For subframe 0
Bits		6456	12576	28336	45352	61664	Number of code blocks per subframe
		2	3	5	8 1	1	Binary channel bits per subframe
							For subframes 1,2,3,4,6,7,8,9
Bits		11340	18900	41400	62100	82800	For subframe 5
Bits		10476	18036	40536	61236	81936	For subframe 0
Bits		8820	16380	38880	59580	80280	Maximum throughput averaged over 1 frame
Mbps		8.25	13.9	30.4	46.6	61.7	

Note 1: Two symbols allocated to PDCCH for 20 MHz, 15 MHz, and 10 MHz channel BW; three symbols allocated to PDCCH for 5 MHz and 3 MHz; four symbols allocated to PDCCH for 1.4 MHz.
Note 2: Reference signal, synchronization signals, and PBCH allocated per TS 36.211.

It can be seen from Table 2.1-4 that these RMCs are for a fully allocated downlink, and the maximum throughput represented reaches a maximum of 61.7 Mbps for the 20 MHz channel bandwidth case. Note that this peak figure represents the maximum transmitted data rate and is in no way intended to indicate the performance of the downlink in real radio conditions. This peak figure is the reference used for specifying the expected performance, which will be specified relative to the maximum figures.

2.1.4 Transmit Power

This section covers UE and BS transmit power requirements.

2.1.4.1 UE Transmit Power

Over time the transmit power requirements for the UE have become more complex. In earlier standards such as GSM and UMTS Release 99, the transmit power specification was a simple one-to-one relationship between the power class of the UE and the maximum output power. This relationship has become more complicated over time with relaxations that take into account the crest factor of higher-order modulation formats. The trend to specify power back-off started in Release 5 for UMTS with the introduction of a fixed back-off for 16QAM. In Release 6 the fixed back-off was superseded by a more advanced "cubic" metric that related the allowed back-off to a formula that included the cube of the voltage waveform relative to a standard QPSK waveform.

There are four power classes defined for the LTE UE. At the time of this writing, a maximum power requirement is defined only for Class 3 and is specified as 23 dBm ±2 dB for almost all bands. However, the flexibility of the LTE air interface requires consideration of additional dimensions including the channel bandwidth and the size of the power allocation within that bandwidth. The introduction of carrier aggregation and uplink MIMO in Release 10 further complicates the specifications, which now require 15 pages to define all the exceptions to the default UE maximum output power of 23 dBm.

Table 2.1-5 shows the maximum power reduction (MPR) that applies for Power Class 3 depending on the modulation being used and the number of resource blocks transmitted in each channel bandwidth. An RB is the minimum unit of transmission and is 180 kHz wide and 0.5 ms in duration.

Table 2.1-5. Maximum power reduction for power class 3 (36.101 [1] Table 6.2.3-1)

Modulation	Channel bandwidth/transmission bandwidth configuration [RB]						MPR (dB)
	1.4 MHz	3.0 MHz	5 MHz	10 MHz	15 MHz	20 MHz	
QPSK	> 5	> 4	> 8	> 12	> 16	> 18	≤ 1
16QAM	≤ 5	≤ 4	≤ 8	≤ 12	≤ 16	≤ 18	≤ 1
64QAM	> 5	> 4	> 8	> 12	> 16	> 18	≤ 2

The trend shown in Table 2.1-5 is that with increasing modulation depth (meaning higher signal peaks) and increasing transmitted bandwidth, the maximum power is reduced.

The introduction of carrier aggregation (see Section 2.1.11) has resulted in further MPR allowances as shown in Table 2.1-6.

Table 2.1-6. Maximum power reduction for Power Class 3 (36.101 [1] Table 6.2.3A-1)

Modulation	CA bandwidth Class C				MPR (dB)
	50 RB + 100 RB	75 RB + 75 RB	75 RB /100 RB	100 RB + 100 RB	
QPSK	> 12 and ≤ 50	> 16 and ≤ 75	> 16 and ≤ 75	> 18 and ≤ 100	≤ 1
QPSK	> 50	> 75	> 75	> 100	≤ 2
16QAM	≤ 12	≤ 16	≤ 16	≤ 18	≤ 1
16QAM	> 12 and ≤ 50	> 16 and ≤ 75	> 16 and ≤ 75	> 18 and ≤ 100	≤ 2
16QAM	> 50	> 75	> 75	> 100	≤ 3

In addition to the maximum power reductions, which are specified for all operating conditions, there is another class of dynamic MPR known as additional MPR (A-MPR), which comes into play when the network signals the UE. Fourteen different network signaling values have been defined, as shown in Table 2.1-7. The behavior of the UE depends on which band it is using, which channel bandwidth, the number of resource blocks allocated, the modulation depth, the allowed A-MPR, and the specific spurious emission requirements that have to be met under these conditions

Table 2.1-7. A-MPR/spectrum emission requirements (36.101 [1] Table 6.2.4-1)

Network signaling value	Requirements (36.101 subclause)	E-UTRA band	Channel bandwidth (MHz)	Resources blocks (N_{RB})	A-MPR (dB)
NS_01	6.6.2.1.1	Table 5.5-1	1.4, 3, 5, 10, 15, 20	Table 5.6-1	N/A
NS_03	6.6.2.2.1	2, 4, 10, 23, 25, 35, 36	3	> 5	≤ 1
			5	> 6	≤ 1
			10	> 6	≤ 1
			15	> 8	≤ 1
			20	> 10	≤ 1
NS_04	6.6.2.2.2	41	5	> 6	≤ 1
			10, 15, 20	Table 6.2.4-4	
NS_05	6.6.3.3.1	1	10, 15, 20	≥ 50	≤ 1
NS_06	6.6.2.2.3	12, 13, 14, 17	1.4, 3, 5, 10	Table 5.6-1	N/A
NS_07	6.6.2.2.3 6.6.3.3.2	13	10	Table 6.2.4-2	Table 6.2.4-2
NS_08	6.6.3.3.3	19	10, 15	> 44	≤ 3
NS_09	6.6.3.3.4	21	10,15	> 40	≤ 1
				> 55	≤ 2
NS_10		20	15, 20	Table 6.2.4-3	Table 6.2.4-3
NS_11	6.6.2.2.1	23	1.4, 3, 5, 10, 15, 20	Table 6.2.4-5	Table 6.2.4-5
NS_12	6.6.3.3.5	26	1.4, 3, 5	Table 6.2.4-6	Table 6.2.4-6
NS_13	6.6.3.3.6	26	5	Table 6.2.4-7	Table 6.2.4-7
NS_14	6.6.3.3.7	26	10, 15	Table 6.2.4-8	Table 6.2.4-8
NS_15	6.6.3.3.9	26	1.4, 3, 5, 10, 15	Table 6.2.4-9, Table 6.2.4-10	Table 6.2.4-9, Table 6.2.4-10
NS_16	6.6.3.3.9	27	3, 5, 10	Table 6.2.4-11, Table 6.2.4-12, Table 6.2.4-13	
NS_17	6.6.3.3.10	28	5, 10	Table 5.6-1	N/A
NS_18	6.6.3.3.11	28	10, 15, 20	≥1	≤4
NS_19	6.6.3.3.12	44	10, 15, 20	Table 6.2.4-14	
...					
NS_32	–	–	–	–	–

For example, a UE receiving NS_03 from the network when operating in bands 2, 4, 10, 23, 25, 35, or 36 with a 15 MHz channel bandwidth and > 8 RBs allocated is allowed to reduce its maximum power by up to an additional 1 dB above the normally allowed MPR in order to meet the additional spurious emission requirements identified in 36.101[1] Subclause 6.6.2.2.1. Table 2.1-8 identifies the additional spurious emissions and spectrum emission mask (SEM) requirements for which the network might signal the UE.

Table 2.1-8. Additional spectrum emission requirements for NS_03 (36.101 [1] Table 6.6.2.2.1-1)

Δf_{OOB} (MHz)	Spectrum emission limit (dBm)/channel bandwidth						Measurement bandwidth
	1.4 MHz	3.0 MHz	5 MHz	10 MHz	15 MHz	20 MHz	
±0–1	−10	−13	−15	−18	−20	−21	30 kHz
±1–2.5	−13	−13	−13	−3	−13	−13	1 MHz
±2.5–2.8	−25	−13	−13	−13	−13	−13	1 MHz
± 2.8–5		−13	−13	−13	−13	−13	1 MHz
±5–6		−25	−13	−25	−25	−25	1 MHz
±6–10			−25	−13	−13	−13	1 MHz
±10–15				−25	−25	−25	1 MHz
±15–20					−25	−25	1 MHz
±20–25						−25	1 MHz

Table 2.1-8 shows another consequence of LTE's channel bandwidth flexibility: the additional SEM requirements, which are a function not just of the frequency offset as in UMTS but also of the channel bandwidth. Table 2.1-9 is an example of one of the most complex A-MPR definitions, in this case for NS_12. The allowed power reduction is a function of channel bandwidth and the position and size of the RB allocation within the channel bandwidth.

Table 2.1-9. A-MPR for "NS_12" (36.101 [1] Table 6.2.4-6)

Channel BW	Parameters	Region A		Region B
1.4	RB$_{start}$	0		1–2
	L$_{CRB}$ [RBs]	≤ 3	≥ 4	≥ 4
	A-MPR [dB]	≤ 3	≤ 6	≤ 3
3	RB$_{start}$	0–3		4–5
	L$_{CRB}$ [RBs]	4–9	1–3 and 10–15	≥ 9
	A-MPR [dB]	≤ 4	≤ 3	≤ 3
5	RB$_{start}$	0–6		7–9
	L$_{CRB}$ [RBs]	≤ 8	≥ 9	≥ 15
	A-MPR [dB]	≤ 5	≥ 3	≤ 3

It should be evident at this point how complex the rules are which govern the maximum power that the UE can use under different conditions and the requirements, both static and dynamic, that have to be met. Checking for correct behavior under all possible conditions will be a substantial verification exercise.

Accuracy for Maximum Output Power

Given the complexity of the maximum power specifications, several new terms have been defined. The number of terms and their definitions have evolved from Release 8 to their current form in Release 11.

P_{CMAX} is defined in 36.101 [1] as the "configured maximum UE output power." This is the nominal power the UE chooses to set as its maximum power based on all the requirements and applicable relaxations. The UE is allowed to set P_{CMAX} between a lower limit P_{CMAX_L} and an upper limit P_{CMAX_H} such that

$$P_{CMAX\_L} \leq P_{CMAX} \leq P_{CMAX\_H}.$$

The definitions of P_{CMAX_L} and P_{CMAX_H} are affected by a number of parameters:

$$P_{CMAX\_H} = \text{MIN} \{P_{EMAX}, P_{PowerClass}\}$$

where

P_{EMAX} is defined in 36.331 [5] and is the maximum allowed power defined by higher layers,

$P_{PowerClass}$ is the nominal power defined for the UE (23 dBm for Power Class 3),

and

$$P_{CMAX\_L} = \text{MIN} \{P_{EMAX} - \Delta TC, P_{PowerClass} - \text{MAX} (MPR + A\text{-}MPR, P\text{-}MPR) - \Delta T_C\}$$

where $\Delta T_C = 1.5$ dB when an allowance applies for transmission bandwidths within 4 MHz of the band edge (see Note 2 in 36.101 [1] Table 6.2.2-1) and $\Delta T_C = 0$ dB when the band-edge allowance is not applied.

P-MPR is an additional allowance that can be applied to meet applicable electromagnetic energy absorption requirements and to address unwanted emissions and self-desense requirements in case of simultaneous transmissions on multiple radio access technologies (RATs) for scenarios not within scope of 3GPP RAN specifications. P-MPR may also be used in conjunction with proximity detection to meet electromagnetic compatibility (EMC) requirements. For cable-conducted testing P-MPR is set to 0 dB and is not considered further here. P-MPR was introduced in the P_{CMAX} equation so that the UE can report to the BS the available maximum output transmit power, which may be less than otherwise expected due to EMC reasons. Lower available power might impact uplink performance and needs to be considered by the BS for scheduling decisions.

Having defined the range P_{CMAX_L} to P_{CMAX_H} within which the UE must set P_{CMAX}, the specifications now establish the requirements for how accurate the actual maximum output power needs to be. When the UE is configured for its chosen P_{CMAX}, which is a nominal or target power, the power that is actually transmitted is defined in 36.101 [1] as P_{UMAX}. This is the "measured configured maximum UE output power"—that is to say, the power that is actually transmitted at the antenna connector (see Section 6.4.3.1) as would be measured assuming no measurement uncertainty.

The limits on P_{UMAX} are defined by extending the range P_{CMAX_L} to P_{CMAX_H} by a tolerance which varies as a function of P_{CMAX}. The tolerance is denoted as $T(P_{CMAX})$ and is given in Table 2.1-10.

Table 2.1-10. P$_{CMAX}$ tolerance (36.101 [1] Table 6.2.5-1)

PCMAX (dBm)	Tolerance T(PCMAX) (dB)
21 ≤ P$_{CMAX}$ ≤ 23	2.0
20 ≤ P$_{CMAX}$ < 21	2.5
19 ≤ P$_{CMAX}$ < 20	3.5
18 ≤ P$_{CMAX}$ < 19	4.0
13 ≤ P$_{CMAX}$ < 18	5.0
8 ≤ P$_{CMAX}$ < 13	6.0
−40 ≤ P$_{CMAX}$ < 8	7.0

The tolerance is evaluated for P$_{CMAX\_L}$ and P$_{CMAX\_H}$ independently. From these tolerances, the limits on P$_{UMAX}$ are defined as the following:

$$P_{CMAX\_L} - T(P_{CMAX\_L}) \leq P_{UMAX} \leq P_{CMAX\_H} + T(P_{CMAX\_H}).$$

The considerable complexity in defining P$_{UMAX}$ does not include the additional complexity that will be applied when the test system uncertainties are taken into account.

2.1.4.2 Base Station Transmit Power

There are several parameters used to describe the BS output power:

- Pout—the mean power of one carrier
- Pmax—the maximum total output power for the sum of all carriers
- Pmax,c—the maximum output power per carrier
- Rated total output power—the manufacturer-declared available output power for the sum of all carriers
- PRAT—the manufacturer-declared available output power per carrier.

Unlike the UE case, there are no requirements for BS maximum output power, either for all carriers or per carrier. The only requirements relate to the accuracy with which the manufacturer declares PRAT. Different PRAT limits can be applied to different BS configurations. Three different BS classes are defined based on their PRAT as given in Table 2.1-11. (Note that the term home BS in this case is equivalent to home eNB or HeNB.)

Table 2.1-11. Base station rated output power (based on 36.104 [2] Table 6.2-1)

BS class	PRAT (dBm per carrier)	Number of carriers	Tolerance	
			Normal	Extreme
Wide area BS	No upper limit	N/A	N/A	N/A
Local area BS	≤ 24	1		
	≤ 21	2		
	≤ 18	4		
	≤ 15	8	±2 dB	±2.5 dB
Home BS	≤ 20	1		
	≤ 17	2		
	≤ 14	4		
	≤ 11	8		

Regional requirements can sometimes override the 3GPP specifications; for instance, Band 34 in Japan is limited to 60 W for the 20 MHz channel bandwidth, whereas no upper limit is defined for other regions.

There are restrictions on the home BS Pout for certain deployment scenarios. For protection of an adjacent UTRA (UMTS) deployment, Pout is limited to between 8 dBm and 20 dBm as a function of the received level of the adjacent UTRA common pilot indicator channel (CPICH) and Ioh, which is defined as the total power present at the home BS antenna connector on the home BS downlink operating channel excluding the power generated by the home BS on that channel. For protection of an adjacent LTE deployment there are similar Pout restrictions, but these are based on Ioh and the received level of the cell reference signals (CRS) from the adjacent LTE channel. In order to meet the requirements on Pout the home BS ideally needs to have the ability to measure the adjacent channel signal levels although the requirement to control Pout does not mandate how the downlink power measurement is achieved. Adjacent channel measurement is not a usual capability of a BS since frequency planning is used to control adjacent channel interference.

Restrictions on the home BS apply when the adjacent channel that needs to be protected belongs to a different operator. If the adjacent channel belongs to the same operator, then the issue of interference mitigation is left to that single operator to work out.

The most stringent interference requirement for the home BS applies in the co-channel case in which an operator chooses to deploy the home BS on the same channel as the macro network using a closed subscriber group (CSG). In a CSG, macro users are not allowed to use the home BS. The Pout restrictions for this scenario are a complex function of CRS, Ioh, and Iob. This last parameter is the uplink received interference power, including thermal noise, present at the home BS antenna connector on the operating channel.

2.1.5 Output Power Dynamics

This section covers UE and BS output power dynamics.

2.1.5.1 UE Output Power Dynamics and Power Control

The UE output power dynamics cover the following areas:

- Minimum output power
- Off power
- Power time mask
- Output power control (accuracy).

2.1.5.1.1 Minimum Output Power

The UE minimum output power is defined as −40 dBm for all channel bandwidths. This is the lowest power at which the UE is required to control the power level and meet all the transmit signal quality requirements. In UMTS, the transmit signal quality requirements apply only from maximum power down to −20 dBm for QPSK and −30 dBm for 16QAM, which is well above the −50 dBm UMTS minimum power requirement. The LTE requirement that signal quality not be degraded over the full operating range puts more demands on the fidelity of the digital-to-analog convertors of the transmitter than was the case with UMTS.

2.1.5.1.2 Off Power

When commanded to switch off its transmitter, the UE output power must be less than −50 dBm. This applies for all channel bandwidths.

2.1.5.1.3 Power Time Mask

The on/off requirements for slot-based transmissions are similar to UMTS. Figure 2.1-2 shows the profile for the general on/off time mask.

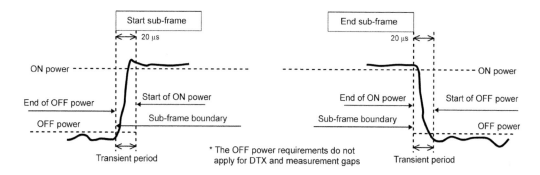

Figure 2.1-2. General ON/OFF Time Mask (36.101 [1] Figure 6.3.4.1-1)

The general requirement is used any time the signal turns on or off. The measurement duration is at least one subframe excluding any transient period. Note that the transient period is not symmetrical with the subframe boundary; the on power ramp transient period starts after the subframe boundary but the off power ramp starts after the end of the subframe. There is no ideal position for the transient period in an FDD system in which no gaps are defined, and the solution specified is a compromise. The choice made for FDD is to minimize any interference to adjacent subframes during the ramp up but allow the ramp down to be delayed until after the end of the subframe.

There is a similar mask for the physical random access channel (PRACH) shown in Figure 2.1-3, but the on period is one PRACH symbol and the on ramp is shifted to before the symbol starts, making it symmetrical with the off ramp.

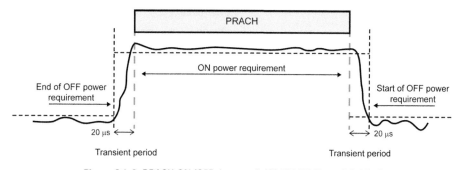

Figure 2.1-3. PRACH ON/OFF time mask (36.101 [1] Figure 6.3.4.2-1)

The time mask for the sounding reference signal (SRS) is similar although a special case for TDD is shown in Figure 2.1-4 for the dual SRS transmission in the uplink pilot timeslot (UpPTS).

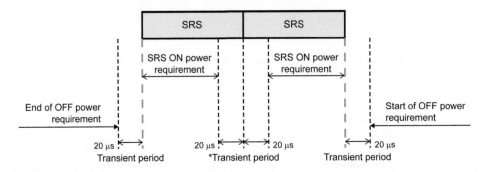

Figure 2.1-4. Dual SRS time mask for the case of UpPTS transmissions (36.101 [1] Figure 6.3.4.2.2-2)

The shift in the start of the on power requirement from the general requirement is particularly important for the SRS since the SRS symbol can be transmitted in isolation for approximately 70 μs and is used by the BS to estimate the uplink channel conditions. If this symbol were not stable for its nominal duration, the BS might get an incorrect estimate of the uplink channel. A symmetrical mask was not chosen for the general on/off case since there are times when it is important to protect the symbol just prior to a power change.

Another example of SRS protection is shown in Figure 2.1-5. This shows the time mask for FDD SRS blanking, which occurs when the UE is required to blank its output during the SRS symbol period.

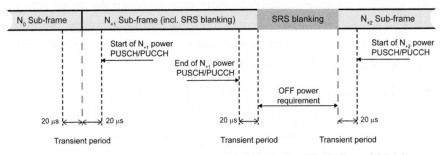

Figure 2.1-5. SRS time mask when there is FDD SRS blanking (36.101 Figure 6.3.4.4-4)

Apart from the traditional on and off transitions described thus far, other transitions from one power state to another are sometimes necessary. These include changes to the transmit power and transitions into and out of the SRS. In addition, a change to the allocated frequency can trigger a power transient due to baseband compensation for known "unflatness" in the transmission path.

A common example of a frequency-induced power transient occurs when the UE is transmitting the physical uplink control channel (PUCCH). The PUCCH generally transmits one timeslot at the lower end of the channel followed by another timeslot at the upper end. See the measurement example in Section 6.4.6.7. The PUCCH frequency hopping is shown graphically in Figure 3.2-13. The requirements for spectrum flatness in 36.101 [1] Subclause 6.5.2.4 specify that at the band edge (and for wide channels in narrow bands most of the channels are at the band edge), the UE is allowed to have a variation in power across the channel of +3 to −5 dB. This variation could be a slope of some 8 dB. In extreme conditions the allowance rises to 12 dB. When the UE transmits the PUCCH or narrow allocations of the PUSCH, it may be necessary to compensate for known flatness issues. This then creates the possibility of a power transient at baseband and RF even though the nominal power remains constant.

The requirements for maintaining PUSCH/PUCCH power accuracy apply to the second and subsequent subframes after the start of a contiguous block of subframes. This requirement also applies to non-contiguous transmissions provided the gap is less than or equal to 20 ms (two frames). There are also requirements for the PRACH that apply to the second and subsequent PRACH preambles.

2.1.5.1.4 Output Power Control

The requirements on UE output power accuracy are not particularly onerous and are shown in Table 2.1-12

Table 2.1-12. Relative power tolerance for transmission (normal conditions)
(36.101 [1] Table 6.3.5.2.1-1)

Power step ΔP (up or down) [dB]	All combinations of PUSCH/PUCCH transitions [dB]	All combinations of PUSCH/PUCCH and SRS transitions [dB]	PRACH [dB]
$\Delta P < 2$	±2.5 (Note 3)	±3.0	±2.5
$2 \le \Delta P < 3$	±3.0	±4.0	±3.0
$3 \le \Delta P < 4$	±3.5	±5.0	±3.5
$4 \le \Delta P \le 10$	±4.0	±6.0	±4.0
$10 \le \Delta P < 15$	±5.0	±8.0	±5.0
$15 \le \Delta P$	±6.0	±9.0	±6.0

Note 1: For extreme conditions an additional ± 2.0 dB relaxation is allowed.

Note 2: For operating bands under Note 2 in Table 6.2.2-1, the relative power tolerance is relaxed by increasing the upper limit by 1.5 dB if the transmission bandwidth of the reference subframes is confined within F_{UL_low} and F_{UL_low} + 4 MHz or F_{UL_high} − 4 MHz and F_{UL_high} and the target subframe is not confined within any one of these frequency ranges; if the transmission bandwidth of the target subframe is confined within F_{UL_low} and F_{UL_low} + 4 MHz or F_{UL_high} − 4 MHz and F_{UL_high} and the reference subframe is not confined within any one of these frequency ranges, then the tolerance is relaxed by reducing the lower limit by 1.5 dB.

Note 3: For PUSCH to PUSCH transitions with the allocated resource blocks fixed in frequency and no transmission gaps other than those generated by downlink subframes, DwPTS fields or Guard Periods for TDD: for a power step $\Delta P \le 1$ dB, the relative power tolerance for transmission is ±1.0 dB

From Table 2.1-12 it can be seen that even for no change to the nominal power, the relative power can vary by up to ±2.5 dB. This makes allowance for the case in which transmissions are contiguous in time but not in frequency. For changes to the configured power the allowance increases up to a maximum of ±6 dB for steps between 15 dB and 20 dB.

There is a growing trend within UE design to reduce cost through the use of multi-stage power amplifiers. These reduce the dynamic range that has to be covered in one section but also introduce the possibility of power and phase transients at the power level where the switching between gain stages takes place. This is a known issue being studied in the specifications, particularly for UL-MIMO, and it is likely that there will be requirements developed to allow for gain-stage switching with appropriate limits on hysteresis to avoid unnecessary transients.

2.1.5.2 Base Station Output Power Dynamics

The BS output power dynamics cover the following areas:

- Resource element power control dynamic range
- Total power dynamic range
- Transmitter off power and transient period.

2.1.5.2.1 Resource Element Power Control Dynamic Range

This requirement, which is shown in Table 2.1-13, specifies the range over which the BS is required to control the output power of a single resource element (RE) relative to the average RE power.

Table 2.1-13. E-UTRA BS RE power control dynamic range (36.104 [2] Table 6.3.1.1-1)

Modulation scheme used on the RE	RE power control dynamic range (dB)	
	(down)	(up)
QPSK (PDCCH)	−6	+4
QPSK (PDSCH)	−6	+3
16QAM (PDSCH)	−3	+3
64QAM (PDSCH)	0	0

Note 1: The output power per carrier shall always be less or equal to the maximum output power of the base station.

This is a necessary capability required to implement power boosting or de-boosting of particular parts of the downlink signal.

2.1.5.2.2 Total Power Dynamic Range

This requirement, shown in Table 2.1-14, is the minimum required total power dynamic range between a fully allocated signal at maximum power and a signal with only one RB allocated. The required dynamic range is 10 log (N_{RB}^{DL}). See Table 3.2-7 for the number of RBs per channel.

Table 2.1-14. E-UTRA BS total power dynamic range (36.104 [2] Table 6.3.2.1-1)

E-UTRA channel bandwidth (MHz)	Total power dynamic range (dB)
1.4	7.7
3	11.7
5	13.9
10	16.9
15	18.7
20	20

2.1.5.2.3 Transmitter Off Power and Transient Period

The off power is defined as a maximum power spectral density measured over a period of 70 μs in a square filter equal to the transmission bandwidth configuration. The off power is required to be less than −85 dBm/MHz.

There are no equivalents to the UE power time mask; however, for TDD operation the concept of transmitter transient period is defined. The transient period occurs twice during a TDD frame, first at the off-to-on transition from the uplink subframe to the downlink subframe and second at the on-to-off transition from the downlink subframe to the guard period or uplink pilot timeslot. In both cases the transient period is defined to be 17 μs.

2.1.6 Transmit Signal Quality

The transmit signal quality requirements specify the in-channel characteristics of the wanted signal. These are distinct from the out-of-channel requirements, which specify limits on unwanted emissions outside the wanted channel. This section covers transmit signal quality for the UE and BS.

2.1.6.1 UE Transmit Signal Quality

Subclause 6.5 of 36.101 [1] defines the in-channel signal quality requirements. These are split into five categories: frequency error, error vector magnitude (EVM), carrier leakage (IQ component), in-channel emissions, and EVM equalizer spectrum flatness. These measurements are fully defined in 36.101 [1] Annex F. Figure 2.1-6 shows the block diagram of the measurement points.

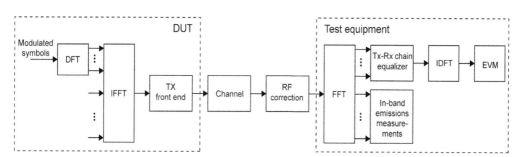

Figure 2.1-6. EVM measurement points (36.101 [1] Figure F.1-1)

The frequency error component is a residual result from measuring EVM and identical in concept to UMTS, so will not be discussed further here.

2.1.6.1.1 Error Vector Magnitude Definition

The EVM definition for LTE is similar in concept to UMTS but there are two new elements specific to SC-FDMA that need to be explained. The first relates to the presence of the TX-RX chain equalizer block from Figure 2.1-2 and the second relates to the time window over which EVM is defined.

The EVM requirements are shown in Table 2.1-15.

Table 2.1-15. Minimum requirements for error vector magnitude (36.101 [1] Table 6.5.2.1.1-1)

Parameter	Unit	Average EVM level[1]	Reference signal EVM level[2]
QPSK or BPSK	%	17.5	17.5
16QAM	%	12.5	12.5

Note 1: RMS average over 10 subframes (not necessarily contiguous) excluding transient periods
Note 2: RMS average over 60 subframes (not necessarily contiguous) excluding transient periods

The requirements apply for UE output power ≥ -40 dBm. The measurement period is one timeslot for PUSCH and PUCCH and one preamble sequence for the PRACH. If an SRS symbol is inserted into the frame, the measurement interval is shortened accordingly. The measurement period is also shortened when the mean power between slots, the modulation depth, or the allocation between slots is expected to change. For the PUSCH, the reduced measurement period is the on period defined for the power time mask less a further 5 μs. The exclusion period is applied after the inverse discrete Fourier transform (IDFT) shown in Figure 2.1-6. For the PUCCH, the measurement interval is reduced by one symbol adjacent to the boundary where the power change is expected to occur.

The stated intention in early Release 8 to provide requirements for 64QAM has been dropped and no uplink 64QAM requirements have been developed.

EVM Equalizer Definition

In UMTS, the EVM measurement was defined through a root raised cosine (RRC) filter matched to the filter defined for UE transmissions. This same filter was assumed in the UMTS BS and was required in order to optimize the received signal quality. In LTE no such transmit filter is defined, which opens up a significant new challenge in determining how to specify transmitter performance. In real-life operation the BS will attempt to determine the amplitude and phase characteristics of the transmitted signal as seen through the imperfect radio channel. It is essential for accurate demodulation that this equalization process take place, but the LTE specifications for the BS do not define the method or a reference receiver. This has partly to do with the complexity of the problem, which is a function of noise and dynamic channel conditions. As a result the equalization process is considered proprietary to each implementer and is therefore undefined within the standards.

The lack of a standard equalizer through which the uplink signal quality is measured presents a problem for the EVM definition. It is known that metrology-grade test equipment working in non-real-time can perform iterative corrections on the signal that are not possible for the BS to mimic in a live network. For this reason it is necessary to define a reference equalizer that can constrain the amount of correction but still be somewhat representative of what might be achieved in real operation. At the very least this equalizer definition will provide a stable reference against which alternative receiver designs can be compared.

One of the challenges in defining an equalizer for the uplink is that the signal contains noise. In a test environment this noise is primarily generated as a result of any crest factor reduction techniques used in the UE such as baseband clipping. In real operation the uplink will be further degraded by interference.

Noise can always be averaged but short uplink signal transmissions do not make this easy. With intra-subframe hopping enabled the UE can be transmitting one RB (180 kHz for 0.5 ms) at one end of the channel and the next RB could be 20 MHz away. Although it is possible to average such signals, the errors are not correlated so the end result may not improve. For this reason the EVM definition is based on the smallest possible transmission of one RB.

The only part of this signal that is known is the RS symbol, which in a normal cyclic prefix (CP) is the fourth of seven symbols transmitted within each active timeslot allocation (see Figure 3.2-13). The RS represents a known amplitude and phase on each subcarrier. The subcarriers are spaced at 15 kHz intervals across the transmission bandwidth. For one RB this represents only 12 data points in frequency lasting around 70 μs and is considered insufficient to provide a stable reference for the equalizer. To allow for more averaging, the EVM definition makes use of the six data symbols in each timeslot to provide a more stable time-averaged reference. This makes EVM measurements vulnerable to data decode errors since unlike the RS pattern, the data is not known in advance. However, provided the noise is below a critical threshold, the addition of the data symbols to the averaging improves the measurement accuracy.

EVM Window Length

The other difference between UMTS EVM and LTE EVM lies in the timing of the measurement. For successful decoding of the CDMA signals used in UMTS the decoder has to be precisely aligned to within a few ns of the signal timing, otherwise the perceived EVM rises sharply and decode errors occur. For CDMA there is only one point in time at which the signal looks its best. In OFDMA and SC-FDMA systems the situation is very different. The symbols are much longer and have an additional extension, the CP, which adds redundancy in the time domain to mitigate multipath distortion. It is sufficient to say at this point that without multipath, the signal can be successfully decoded over a range of timing equal to the length of the CP.

When multipath is present, the error-free decoding window reduces in size such that even if the receiver observes multiple delayed signals up to the length of the CP, there will always be an error-free symbol position for decode. However, if the signal (rather than the channel) contains any time-domain distortion, this distortion eats into the effective length of the CP.

When the transmission system is constructed, there is always a trade-off between in-channel signal quality (EVM) and out-of-channel signal quality (spectral regrowth, etc.). When the channel and duplex filters are designed to meet the out-of-channel performance, the quality of the in-channel signal can degrade. This degradation can include time-domain effects that resemble multipath inter-symbol interference (ISI) distortion on the signal. Such a signal measured through a perfect channel would no longer be error-free over the full range of the CP. To limit the amount of time-domain distortion in the signal, the EVM is measured at two points on either side of the ideal timing.

Table 2.1-16 shows the EVM window length as a function of channel bandwidth.

Table 2.1-16. EVM window length for normal CP (36.101 [1] Table F.5.3-1)

Channel bandwidth MHz	Cyclic prefix length N_{cp} for symbol 0	Cyclic prefix length[1] N_{cp} for symbols 1 to 6	Nominal FFT size	Cyclic prefix for symbols 1 to 6 in FFT samples	EVM window length W	Ratio of W to CP for symbols 1 to 6[2]
1.4			128	9	5	55.6
3			256	18	12	66.7
5	160	144	512	36	32	88.9
10			1024	72	66	91.7
15			1536	108	102	94.4
20			2048	144	136	94.4

Note 1: The unit is number of samples, sampling rate of 30.72 MHz is assumed.
Note 2: These percentages are informative and apply to symbols 1 through 6. Symbol 0 has a longer CP and therefore a lower percentage.

It can be seen from Table 2.1-16 that the narrower channel bandwidths are allowed to use up much more of the useful CP length than the wider channels. This reflects the challenge of designing suitable filters for the narrower bands that meet both the in-channel and out-of-channel requirements. Example measurements of EVM versus time on distorted signals are given in Section 6.4.6.

2.1.6.1.2 Carrier Leakage (IQ Component) Definition

Signal distortions in the IQ plane such as IQ offset lead to local oscillator (LO) leakage, also known as carrier leakage or carrier feed-through. This distortion shows up in the frequency domain as energy at the center of the channel, although if the transmission is allocated across the center of the channel, the error energy is spread by the SC-FDMA processing across the allocation and is not visible. Table 2.1-17 shows the requirements for relative carrier leakage power (RCLP).

Table 2.1-17. Minimum requirements for relative carrier leakage power (36.101 [1] Table 6.5.2.2.1-1)

Output power	Relative limit (dBc)
> 0 dBm	−25
−30 dBm ≤ output power ≤ 0 dBm	−20
−40 dBm ≤ output power < −30 dBm	−10

Note how the requirement is relaxed at low signal power. This is because the mechanisms that create RCLP tend to be linked with residual errors in the IQ modulator, and these errors become more apparent at low signal powers. The RCLP is removed from the signal prior to applying the equalizer for the calculation of EVM.

2.1.6.1.3 In-Band Emissions

The EVM definition is a measure of the quality of the allocated part of the signal. In UMTS the allocated part by definition was the entire signal, since all transmissions occupied the entire channel bandwidth. With SC-FDMA this is no longer the case since normally only a portion of the channel bandwidth is allocated to the UE. In such circumstances the unallocated part of the channel is available for use by other UEs. Therefore it is necessary to specify limits on the amount of power the UEs may transmit into the unallocated RBs. This specification of power at frequencies other than the wanted frequencies is similar in concept to the spectrum emission mask (SEM) and adjacent channel leakage ratio (ACLR) requirements for adjacent channel protection except that now the unallocated resources need to be protected within the channel bandwidth.

Table 2.1-18 defines limits for three different types of in-band emissions: general, IQ image, and carrier leakage (formerly called the DC component). The limit that applies for any specific unallocated RB is the highest value calculated for all three types.

Table 2.1-18. Minimum requirements for in-band emissions (36.101[1] Table 6.5.2.3.1-1)

Parameter description	Unit	Limit		Applicable frequencies
General	dB	Max $\{-25 - 10 \log_{10} (N_{RB}/L_{CRBs})$, $20 \cdot \log_{10} EVM - 3 - 5 \cdot (|\Delta_{RB}| - 1)/L_{CRBs}$, -57 dBm/180 kHz $- P_{RB}\}$		Any non-allocated RB measured relative to average allocated RB power
IQ Image	dB	-25		Any non-allocated image RB measured relative to average allocated RB power
Carrier leakage	dBc	-25	Output power > 0 dBm	Any non-allocated RB containing or adjacent to the carrier feedthrough measured relative to total allocated RB power
		-20	-30 dBm ≤ output power ≤ 0 dBm	
		-10	-40 dBm ≤ output power < -30 dBm	

General In-Band Emissions

The formula that defines the requirements for the general in-band emissions in Table 2.1-18 is particularly complicated and deserves further explanation. The general requirement can be considered as the in-band noise component for unallocated RBs that are not covered by the more specific allowances for direct current (DC) at the channel center and for the image RB. In specifying a general noise component, it is tempting to set a fixed noise floor for each unallocated RB. This would work well for narrow allocations in which the overall noise generated in the channel would be the sum of all the noise from many UEs. However, when the transmission from one UE gets wider, a fixed noise limit becomes more challenging to achieve and less important to the system performance. If the UE occupies almost the entire channel, there are fewer UEs elsewhere in the cell generating noise, and so the general limit is relaxed for the unallocated RBs when the number of allocated RBs is higher. Due to the shape of the SC-FDMA spectrum, the unallocated RBs closest to the allocated RBs have a higher noise limit. The general requirement also takes into account the modulation depth being evaluated. The use of higher-order modulation implies better channel conditions, which in turn imply a lower path loss to the BS. The significance of any noise generated is therefore larger and so there is a component in the requirement that lowers the allowed noise when the EVM limit is lower.

The formula for the general requirement takes into account all these factors starting with the size of the allocation L_{CRBs} relative to the maximum number of RB in the channel N_{RB}. When the allocation reaches the maximum $L_{CRBs} = N_{RB}$, the second term reaches an upper limit of -25 dB, representing a relaxation for wide allocations. The third term starts with a level defined by the EVM limit for the modulation depth being analyzed, which is then reduced as a function of the separation Δ_{RB} between the last allocated RB and the non-allocated RB being measured. The final term sets an absolute noise floor of -57 dBm/180 kHz, which becomes relevant when P_{RB}, the average power per allocated RB, is less than -27 dBm.

IQ Image In-Band Emissions

One of the classic OFDM distortion mechanisms, which also applies to SC-FDMA, is the presence of an image of the allocated RB reflected around the center of the channel. Quadrature error and IQ gain imbalance are usually the cause. The mathematics of this mechanism are described in "Effects of physical layer impairments on OFDM systems" [6]. Measurement examples are discussed in Section 6.4.6.9. The limit for image power is set at −25 dB.

Carrier Leakage In-Band Emissions

The distortion mechanisms that create the carrier leakage component are the same as those that create RCLP. The difference, however, is that RCLP is a requirement of the wanted signal whereas the carrier leakage component of the in-band emissions is specified in terms of its impact on other UE. For this reason the way that the carrier leakage component is measured differs from that of RCLP. In the case of RCLP, the signal is first passed through an equalizer and then the carrier leakage component is estimated prior to its removal. The carrier leakage component is measured as it is seen by the victim UE, without equalization.

The RB over which the carrier leakage component needs to be evaluated depends on the number of RBs supported by the channel bandwidth (see Table 3.2-7). The subcarriers used for the SC-FDMA uplink are offset from the channel center frequency by half a subcarrier spacing (7.5 kHz). For channel bandwidths of 3 MHz, 5 MHz, and 15 MHz there are an odd number of RBs. In this case the carrier leakage is contained within the central RB, and only this RB needs to be measured for the carrier leakage component.

For the 1.4 MHz, 10 MHz, and 20 MHz bandwidths there are an even number of RBs, which means that the carrier leakage falls in between the RBs on either side of the center frequency. The carrier leakage impacts both RBs so both need to be measured for the carrier leakage component. The situation for the downlink is different since it has one subcarrier reserved at the channel center frequency for carrier leakage and therefore is orthogonal to the RBs allocated on either side.

2.1.6.1.4 EVM Equalizer Spectrum Flatness

The fact that EVM is measured through an equalizer makes it desirable to define an unequalized spectrum flatness limit. Without such a limit, large variations in power across the channel could be removed by the EVM measurement and thus go unnoticed. In practice, the equalizer in the BS will be limited in performance due to noise in the channel and so an additional flatness requirement helps constrain the signal quality when the equalizer is unable to correct for large errors.

The EVM equalizer spectrum flatness measurement is a residual result of the calculation of EVM equalizer coefficients. Table 2.1-19 defines the requirements, which are given as the maximum allowed peak-to-peak (p–p) ripple in the EVM equalizer coefficients.

Table 2.1-19. Minimum requirements for EVM equalizer spectrum flatness (normal conditions) (36.101 [1] Table 6.5.2.4.1-1)

Frequency range	Maximum ripple (dB)
If FUL_Meas − FUL_Low ≥ 3 MHz and if FUL_High − FUL_Meas ≥ 3 MHz (Range 1)	4 (p-p)
If FUL_Meas − FUL_Low < 3 MHz or if FUL_High − FUL_Meas < 3 MHz (Range 2)	8 (p-p)

Note: FUL_Meas refers to the subcarrier frequency for which the equalizer coefficient is evaluated .
Note: FUL_Low and FUL_High refer to each E-UTRA frequency band specified in 36.101 [1] Table 5.5-1.

The requirements are split into two ranges. For RB more than 3 MHz from the band edge the requirement is 4 dB but near the band edge the requirements are relaxed considerably to allow for the additional effect of the duplex filter, which has to attenuate the out-of-channel emissions falling into the adjacent band. It can be seen that at the band edge there could be up to 8 dB of unflatness across the channel. Since LTE supports up to 20 MHz channel bandwidths and since some bands are not much larger than this, it can be seen that the band-edge condition can apply to a large proportion of the channels. A graphical explanation of the use of ranges is shown in Figure 2.1-7. The relaxed extreme temperature condition (ETC) limits are in brackets.

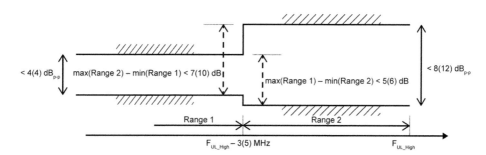

Figure 2.1-7. Limits for EVM equalizer spectral flatness with the maximum allowed variation of the coefficients indicated (the ETC minimum requirement within brackets) (36.101 [1] Figure 6.5.2.4.1-1)

It is also important to consider the absolute phase correction applied by the equalizer, but currently no requirements are defined for this.

2.1.6.2 Base Station Transmit Signal Quality

2.1.6.2.1 Error Vector Magnitude

The requirements for the downlink in-channel signal quality are less complex than for the uplink and are covered by EVM. Although not always fully allocated, the downlink signal bandwidth is nearly always at the full channel bandwidth due to the presence of the reference signal subcarriers, which are always transmitted. The consistency of the signal means that its quality can be defined over a full 10 ms frame thus avoiding the noise concerns when the 0.5 ms uplink bursts are evaluated.

The requirements for EVM are defined for the PDSCH only as 17.5% (QPSK), 12.5% (16QAM), and 8% (64QAM). Figure 2.1-8 shows the reference point for the EVM definition.

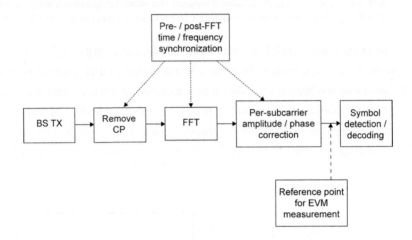

Figure 2.1-8. Reference point for EVM measurement (36.104 [2] Figure E.1-1)

The downlink definition is similar to the uplink definition in that it is necessary to specify EVM through an equalizer, described in Figure 2.1-8 as a per-subcarrier amplitude and phase correction. As with the uplink, the equalizer coefficients are calculated from the distortions seen at the receiver of the ideal reference signals transmitted as part of the signal. However, due to the way the downlink reference signals are mapped, the downlink equalizer definition is quite different from that used for the uplink. The pattern of the reference signal for a two-antenna system is shown later in Figure 2.4-3.

A fully unconstrained equalizer would take the amplitude and phase of every reference symbol and then interpolate amplitude and phase correction values for the intermediate subcarriers to correct the entire signal. Although it is possible to construct such an equalizer in the conformance test system, this unconstrained equalizer would provide optimistic results compared to what might be provided by a realistic UE equalizer. Therefore a simpler constrained equalizer has been defined using moving average smoothing in the frequency domain. The RS on the reference subcarriers from one 10 ms frame are averaged in time into a linear array, which results in one amplitude and phase correction value for every third subcarrier. A moving average is then calculated by sliding a window of length 19 across the array according to Figure 2.1-9.

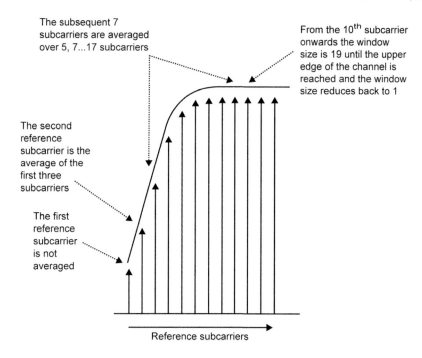

The subsequent 7 subcarriers are averaged over 5, 7...17 subcarriers

From the 10th subcarrier onwards the window size is 19 until the upper edge of the channel is reached and the window size reduces back to 1

The second reference subcarrier is the average of the first three subcarriers

The first reference subcarrier is not averaged

Reference subcarriers

Figure 2.1-9. Reference subcarrier smoothing in the frequency domain (36.104 [2] Figure E.6-1)

At the channel edges the width of the moving average gradually decreases until at the outermost reference subcarriers there is no averaging. The effect this has on the equalizer performance is that in the middle of the signal where it is expected to be flat there is the least amount of correction whereas at the channel edge where filter roll-off is expected, the equalizer can do the most correction. The only downside of shortening the moving average window is that the equalization at the outermost subcarriers is more susceptible to noise. There is an EVM window length requirement for the downlink similar to that described in Section 2.1.6.1.1 for the uplink.

2.1.6.2.2 Time Alignment Error

For TX diversity or MIMO transmission there is a requirement that the maximum timing difference between transmitter branches be less than 65 ns. The potential exists in any multi-antenna transmitter to control the direction of the signal by use of phase shifting. For this to work the transmitter branches have to be time- and phase-aligned. Requirements for this more advanced form of spatial control are being studied for Release 12. See Section 8.5.7.1.

2.1.6.2.3 Downlink RS Power Accuracy

The power of the reference signals transmitted on the downlink is required for path loss calculation by the UE. For this reason the transmitted power of an RS has to be within 2.1 dB of the signaled power of that RS on the physical downlink shared channel (PDSCH).

2.1.7 Unwanted Emissions

Sections 6.6 and 6.7 of 36.101 [1] and 36.104 [2] cover output RF spectrum emissions for the UE and BS, respectively. These sections specify the following requirements for the unwanted aspects of the transmitted signal:

- Occupied bandwidth
- Operating band unwanted emissions
- ACLR
- Spurious emissions
- Transmit intermodulation.

2.1.7.1 Occupied Bandwidth

This is a largely historical requirement from the days of FM radio systems when it was necessary to specify the bandwidth occupied by 99% of the signal energy to prevent over-modulation. In modern digital systems, if ACLR requirements are met, then usually it is not possible for the system to fail the occupied bandwidth requirements, making the latter redundant. However, some regulatory authorities still expect an occupied bandwidth requirement for all systems and LTE provides this using the definitions in ITU-R Recommendation SM.328 [7]. The requirement states that more than 99% of the energy of a fully allocated LTE signal has to fall within the assigned channel bandwidth. In reality, due to the use of OFDMA and SC-FDMA for the downlink and uplink, the actual occupied bandwidth will be closer to the narrower transmission bandwidth configuration than the channel bandwidth itself (see Table 3.2-7).

2.1.7.2 Operating Band Unwanted Emissions

2.1.7.2.1 UE Spectrum Emission Mask

Spectrum emission mask provides a generic way of specifying unwanted emissions close to the carrier resulting from imperfections in the signal generation process. Emissions occurring far from the carrier are specified under the spurious emissions category. The requirements are shown in Table 2.1-20. The frequency frequency offset Δf_{OOB} is the offset below and above the wanted channel.

Table 2.1-20. General E-UTRA spectrum emission mask (36.101 [1] Table 6.6.2.1.1-1)

Δf_{OOB} (MHz)	Spectrum emission limit (dBm)/channel bandwidth						Measurement bandwidth
	1.4 MHz	3.0 MHz	5 MHz	10 MHz	15 MHz	20 MHz	
± 0–1	−10	−13	−15	−18	−20	−21	30 kHz
+ 1–2.5	−10	−10	−10	−10	−10	−10	1 MHz
± 2.5–2.8	−25	−10	−10	−10	−10	−10	1 MHz
± 2.8–5		−10	−10	−10	−10	−10	1 MHz
± 5–6		−25	−13	−13	−13	−13	1 MHz
± 6–10			−25	−13	−13	−13	1 MHz
± 10–15				−25	−13	−13	1 MHz
± 15–20					−25	−13	1 MHz
± 20–25						−25	1 MHz

2.1.7.2.2 Base Station Operating Band Unwanted Emissions

The operating band unwanted emissions requirements for the BS are concerned with emissions up to 10 MHz outside of the operating band and are conceptually similar to the SEM requirements for the UE. However, due to coexistence issues the BS specifications are far more complicated to describe since they are a function of BS class, channel bandwidth, operating band, and geographic region. The requirements extend to 27 tables in 36.104 [2] Section 6.6.3 and are split into Category A requirements covering all regions and more stringent Category B requirements applicable for Europe.

2.1.7.3 Adjacent Channel Leakage Ratio

2.1.7.3.1 UE Adjacent Channel Leakage Ratio

The UE ACLR emission requirements provide an alternative way of measuring the same impairments covered by SEM but specified in a way that makes the effect of emissions on adjacent channels easier to understand. The SEM measurement bandwidth is narrower than that of ACLR, which makes SEM a harder requirement to meet. Adjacent channel leakage power ratio is defined as the ratio of the filtered mean power centered on the assigned channel frequency to the filtered mean power centered on an adjacent channel frequency. As Table 2.1-21 shows, the ACLR requirements for protection of adjacent LTE carriers (E-UTRA$_{ACLR1}$) are defined for all LTE channel bandwidths.

Table 2.1-21. General requirements for E-UTRA$_{ACLR}$ (36.101 [1] Table 6.6.2.3.1-1)

	Channel bandwidth/E-UTRA$_{ACLR1}$/Measurement bandwidth					
	1.4 MHz	3.0 MHz	5 MHz	10 MHz	15 MHz	20 MHz
E-UTRA$_{ACLR1}$	30 dB	30 dB	30 dB	30 dB	30 dB	30 dB
E-UTRA channel measurement bandwidth	1.08 MHz	2.7 MHz	4.5 MHz	9.0 MHz	13.5 MHz	18 MHz
Adjacent channel center frequency offset [MHz]	+1.4/−1.4	+3.0/−3.0	+5/−5	+10/−10	+15/−15	+20/−20

Then a further set of ACLR requirements (UTRA$_{ACLR1}$ and UTRA$_{ACLR1}$) is defined using UTRA channel bandwidths for the protection of adjacent and alternate UTRA channels. Since there are two UTRA channel bandwidths (5 MHz and 1.28 MHz), and since both adjacent and alternate channels are specified, the result is quite complex as shown in Table 2.1-22. Note that the ratio is now between powers measured in different bandwidths.

Table 2.1-22. Requirements for UTRA$_{ACLR1/2}$ (36.101 [1] Table 6.6.2.3.2-1)

	Channel bandwidth/UTRA$_{ACLR1/2}$/Measurement bandwidth					
	1.4 MHz	3.0 MHz	5 MHz	10 MHz	15 MHz	20 MHz
UTRA$_{ACLR1}$	33 dB	33 dB	33 dB	33 dB	33 dB	33 dB
Adjacent channel center frequency offset [MHz]	0.7+BW$_{UTRA}$/2 / −0.7−BW$_{UTRA}$/2	1.5+BW$_{UTRA}$/2 / −1.5−BW$_{UTRA}$/2	+2.5+BW$_{UTRA}$/2 / −2.5−BW$_{UTRA}$/2	+5+BW$_{UTRA}$/2 / −5−BW$_{UTRA}$/2	+7.5+BW$_{UTRA}$/2 / −7.5−BW$_{UTRA}$/2	+10+BW$_{UTRA}$/2 / −10−BW$_{UTRA}$/2
UTRA$_{ACLR2}$			36 dB	36 dB	36 dB	36 dB
Adjacent channel center frequency offset [MHz]			+2.5+3*BW$_{UTRA}$/2 / −2.5−3*BW$_{UTRA}$/2	+5+3*BW$_{UTRA}$/2 / −5−3*BW$_{UTRA}$/2	+7.5+3*BW$_{UTRA}$/2 / −7.5−3*BW$_{UTRA}$/2	+10+3*BW$_{UTRA}$/2 / −10−3*BW$_{UTRA}$/2
E-UTRA channel measurement bandwidth	1.08 MHz	2.7 MHz	4.5 MHz	9.0 MHz	13.5 MHz	18 MHz
UTRA 5 MHz channel measurement bandwidth (Note 1)	3.84 MHz	3.84 MHz	3.84 MHz	3.84 MHz	3.84 MHz	3.84 MHz
UTRA 1.6 MHz channel measurement bandwidth (Note 2)	1.28 MHz	1.28 MHz	1.28 MHz	1.28 MHz	1.28 MHz	1.28 MHz

Note 1: Applicable for E-UTRA FDD coexistence with UTRA FDD in paired spectrum.
Note 2: Applicable for E-UTRA TDD coexistence with UTRA TDD in unpaired spectrum.

2.1.7.3.2 Base Station Adjacent Channel Leakage Ratio

The approach taken to define BS ACLR requirements is similar to that used for the UE requirements. For all adjacent and alternate LTE and UTRA channels the ACLR requirement is 45 dB. This is considerably tighter than for the UE due partly to the higher powers involved and also their being fewer power, cost, and size restrictions on the BS design.

2.1.7.4 Spurious Emissions

2.1.7.4.1 UE Spurious Emissions

The UE spurious emission requirements specify emissions that are outside the bandwidth covered by the SEM and ACLR out-of-band emission requirements. The boundary between the out-of-band domain and the spurious domain is shown in Table 2.1-23 for the six LTE channel bandwidths.

Table 2.1-23. Boundary between E-UTRA Δf_{OOB} and spurious emission domain (36.101 [1] Table 6.6.3.1-1)

Channel bandwidth	1.4 MHz	3.0 MHz	5 MHz	10 MHz	15 MHz	20 MHz
Δf_{OOB} (MHz)	2.8	6	10	15	20	25

The spurious emission requirements are developed in accordance with the ITU-R Recommendation SM.329-10 [8], and more specifically, define exceptions to deal with coexistence issues between LTE operating bands. The basic requirements shown in Table 2.1-24 cover emissions from 9 kHz up to 12.75 GHz with the exception of the frequencies close to the carrier, which are covered by the out-of-band domain. The upper limit of 12.75 GHz is also extended to cover the fifth harmonic for Bands 22, 42, and 43.

Table 2.1-24. Spurious emissions limits (36.101 [1] Table 6.6.3.1-2)

Frequency range	Maximum level	Measurement bandwidth	Note
9 kHz ≤ f < 150 kHz	−36 dBm	1 kHz	
150 kHz ≤ f < 30 MHz	−36 dBm	10 kHz	
30 MHz ≤ f < 1000 MHz	−36 dBm	100 kHz	
1 GHz ≤ f < 12.75 GHz	−30 dBm	1 MHz	
12.75 GHz ≤ f < 5th harmonic of the upper frequency edge of the UL operating band in GHz	−30 dBm	1 MHz	1

Note 1: Applies for Band 22, Band 42, and Band 43

Stating the general case for UE spurious emissions is quite straightforward. However, as with many of the radio requirements, the exceptions to the general rule for band coexistence are more involved. Table 2.1-25 is an excerpt from 36.101[1] Table 6.6.3.2-1 showing the entries for Bands 1 and 2.

Table 2.1-25. Requirements (excerpt from 36.101 [1] Table 6.6.3.2-1)

E-UTRA band	Protected band	Spurious emission Frequency range (MHz)	Maximum level (dBm)	MBW (MHz)	Note
1	E-UTRA Band 1, 7, 8, 11, 20, 21, 22, 26, 38, 40, 42, 43	F_{DL_low}–F_{DL_high}	−50	1	
	E-UTRA Band 3, 9, 34	FDL_low–FDL_high	−50	1	15
	E-UTRA Band 33	FDL_low–FDL_high	−50	1	3
	E-UTRA Band 39	FDL_low–FDL_high	−50	1	3
	Frequency range	1884.5–1915.7	−41	0.3	6, 8, 15
2	E-UTRA Band 4, 5, 10, 12, 13, 14, 17, 22, 23, 24, 26, 41, 42	F_{DL_low}–F_{DL_high}	−50	1	
	E-UTRA Band 2, 25	F_{DL_low}–F_{DL_high}	−50	1	15
	E-UTRA Band 43	F_{DL_low}–F_{DL_high}	−50	1	2

Note 2: As exceptions, measurements with a level up to the applicable requirements defined in 35.101 [1] Table 6.6.3.1-2 are permitted for each assigned E-UTRA carrier used in the measurement due to 2nd, 3rd, or 4th harmonic spurious emissions. An exception is allowed if there is at least one individual RB within the transmission bandwidth (see 36.101 [1] Figure 5.6-1) for which the 2nd, 3rd or 4th harmonic totally or partially overlaps the measurement bandwidth (MBW).

Note 3: To meet these requirements some restriction will be needed for either the operating band or protected band.

Note 6: Applicable when NS_05 in Section 6.6.3.3.1 is signaled by the network.

Note 8: Applicable when coexistence with PHS system operating in 1884.5–1915.7MHz.

Note 15: These requirements also apply for the frequency ranges that are less than Δf_{OOB} (MHz) in 36.101 [1] Table 6.6.3.1-1 and Table 6.6.3.1A-1 from the edge of the channel bandwidth.

There are additional special requirements to deal with certain network signaling cases (see Table 2.1-7) as well as exceptions for carrier aggregation and UL-MIMO conditions.

2.1.7.4.2 Base Station Spurious Emissions

The BS spurious emissions apply from 9 kHz to 12.75 GHz with the exception of the 10 MHz below and above the operating band. The 12.75 GHz upper limit is extended for the fifth harmonic of Bands 22, 42, and 43. As with the out-of-band emissions, the requirements are split between general category A requirements for the whole world (9 to 13 dBm) and tighter category B requirements for Europe (−30 dBm to −36 dBm). In addition to the general requirements there exist numerous more stringent requirements for the protection of other receivers. The first and most stringent of these is to protect the receiver of any BS sharing the same antenna as the transmitter. This restricts any spurious emissions in the receive band to as low as −96 dBm for the wide area BS. Many other requirements are defined for coexistence with systems operating in different frequency bands. An example of these requirements is shown in Table 2.1-26. Further requirements exist for co-location with other BS where the emission limits are typically −96 dBm or −98 dBm.

Table 2.1-26. BS spurious emission limits for E-UTRA BS for coexistence with systems operating in other frequency bands (extract from 36.104 [2] Table 6.6.4.3.1-1)

System type for E-UTRA to coexist with	Frequency range for coexistence requirement	Maximum level	Measurement bandwidth	Note
GSM900	921–960 MHz	−57 dBm	100 kHz	This requirement does not apply to E-UTRA BS operating in Band 8.
GSM900	876–915 MHz	−61 dBm	100 kHz	For the frequency range 880–915 MHz, this requirement does not apply to E-UTRA BS operating in Band 8, since it is already covered by the requirement in Subclause 6.6.4.2.
DCS1800	1805–1880 MHz	−47 dBm	100 kHz	This requirement does not apply to E-UTRA BS operating in Band 3.
DCS1800		−61 dBm	100 kHz	This requirement does not apply to E-UTRA BS operating in Band 3, since it is already covered by the requirement in Subclause 6.6.4.2.

2.1.7.5 Transmit Intermodulation

2.1.7.5.1 UE Transmit Intermodulation

The transmit intermodulation requirement ensures that the UE transmitter does not generate unwanted intermodulation products when an interfering signal is present at the transmit antenna. This can happen when the UE power amplifier is working in a non-linear region. The interfering signal is a continuous wave (CW) sinusoid at a level 40 dB below the wanted signal and at a frequency offset depending on the wanted channel bandwidth. The requirement for the level of the intermodulation products is shown in Table 2.1-27.

Table 2.1-27. Transmit Intermodulation (36.101 [1] Table 6.7.1-1)

BW channel (UL)	5 MHz		10 MHz		15 MHz		20 MHz	
Interference signal frequency offset	5 MHz	10 MHz	10 MHz	20 MHz	15 MHz	30 MHz	20 MHz	40 MHz
Interference CW signal level	−40 dBc							
Intermodulation product	−29 dBc	−35 dBc	−29 dBc	−35 dBc	−29 dBc	−35 dBc	−29 dBc	−35 dBc
Measurement bandwidth	4.5 MHz	4.5 MHz	9.0 MHz	9.0 MHz	13.5 MHz	13.5 MHz	18 MHz	18 MHz

2.1.7.5.2 Base Station Transmit Intermodulation

There are no specific BS transmit modulation requirements beyond the emission requirements already specified in 36.104 [2] Section 6; however, these limits have to be met with a 5 MHz modulated LTE interferer at 2.5 MHz, 7.5 MHz, and 12.5 MHz offset from the carrier at a level 30 dB below the wanted signal. In order to limit test times it is usual to retest the spurious emissions in the presence of the interferer only at those frequencies where intermodulation products would be expected to fall.

2.1.8 Receiver Requirements

The receiver requirements are defined in 36.101 [1] Section 7 for the UE and 36.104 [2] Section 7 for the BS. They cover the following categories:

- Reference sensitivity
- Maximum input level (UE only)
- Adjacent channel selectivity
- Blocking
- Spurious response(UE only)
- Intermodulation
- Spurious emissions
- Receiver image (UE only).

2.1.8.1 Reference Sensitivity

2.1.8.1.1 UE Reference Sensitivity

The most basic receiver requirement is reference sensitivity, which is the level (REFSENS) at which the UE reaches at least 95% of the maximum throughput defined for the specified RMC. This level is specified as a function of operating band, channel bandwidth, and modulation depth. An excerpt for Bands 1 and 2 with QPSK modulation is shown in Table 2.1-28.

Table 2.1-28. Reference sensitivity QPSK PREFSENS (excerpt from 36.101 [1] Table 7.3.1-1)

E-UTRA Band	1.4 MHz (dBm)	3 MHz (dBm)	5 MHz (dBm)	10 MHz (dBm)	15 MHz (dBm)	20 MHz (dBm)	Duplex mode
1			−100	−97	−95.2	−94	FDD
2	−102.7	−99.7	−98	−95	−93.2	−92	FDD
3	−101.7	−98.7	−97	−94	−92.2	−91	FDD
4	−104.7	−101.7	−100	−97	−95.2	−94	FDD
5	−103.2	−100.2	−98	−95			FDD
...							
33			−100	−97	−95.2	−94	TDD
34			−100	−97	−95.2		TDD
35	−106.2	−102.2	−100	−97	−95.2	−94	TDD
36	−106.2	−102.2	−100	−97	−95.2	−94	TDD
37			−100	−97	−95.2	−94	TDD

Note 1: The transmitter shall be set to P_{UMAX} as defined in Subclause 6.2.5.
Note 2: Reference measurement channel is A.3.2 with one sided dynamic OCNG Pattern OP.1 FDD/TDD as described in Annex A.5.1.1/A.5.2.1.
Note 3: The signal power is specified per port.

It should be noted that not all combinations of bands and bandwidths have requirements.

2.1.8.1.2 Base Station Reference Sensitivity

The BS reference sensitivity requirements are specified in a manner similar to the UE with some minor terminology differences. The RMCs used are defined in 36.104 [2] Annex A and are referred to as FRCs. The specified reference sensitivity at which the BS must reach 95% of the maximum FRC throughput is referred to as $P_{REFSENS}$. For wide area BS, $P_{REFSENS}$ is typically −101.5 dBm for channel bandwidths of 5 MHz and higher. A relaxation of 8 dB exists for the local area BS and the home BS. The specified FRC may be narrower than the channel (e.g., 25 RB for 20 MHz) and the requirements apply for any position of the FRC within the channel.

2.1.8.1.3 Base Station Dynamic Range

This requirement applies for a fixed level of the FRC with a co-channel AWGN interferer at a level below the wanted signal. For the 20 MHz channel bandwidth these levels are −70.2 dBm and −76.4 dBm, respectively. With this interference the BS is required to reach 95% of the FRC maximum throughput.

2.1.8.2 UE Maximum Input Level

The maximum input level is an important requirement when the UE is operating in close proximity to a base station and a strong wanted signal is present. This requirement, shown in Table 2.1-29, tests the saturation performance of the UE receiver.

Table 2.1-29. Maximum input level (36.101 [1] Table 7.4.1-1)

Rx parameter	Units	Channel bandwidth					
		1.4 MHz	3 MHz	5 MHz	10 MHz	15 MHz	20 MHz
Power in transmission bandwidth configuration	dBm	−25					

Note 1: The transmitter shall be set to 4 dB below P_{CMAX_L} at the minimum uplink configuration specified in 36.101 [1] Table 7.3.1-2 with P_{CMAX_L} as defined in subclause 6.2.5.
Note 2: Reference measurement channel is Annex A.3.2: 64QAM, R=3/4 variant with one sided dynamic OCNG Pattern OP.1 FDD/TDD as described in Annex A.5.1.1/A.5.2.1.

2.1.8.3 Channel Selectivity

2.1.8.3.1 UE Adjacent Channel Selectivity

The adjacent channel selectivity (ACS) requirement tests the ability of the UE receiver to attenuate strong signals in the adjacent channel. Signal attenuation can occur within the operating band due to adjacent LTE carriers or due to strong interferers from other systems present just outside the operating band. Unlike most requirements, ACS is not directly measureable as it relates to an internal attribute of the receiver. For that reason, the requirement to reach 95% throughput of the RMC is specified at the extreme ends of the operating range: first at REFSENS + 14 dB (Case 1) and then with the interferer at −25 dBm (Case 2). The ACS requirements are given in Tables 2.1-30 and the Case 1 parameters are shown in Table 2.1-31.

Table 2.1-30. Adjacent channel selectivity (36.101 [1] Table 7.5.1-1)

Rx parameter	Units	Channel bandwidth					
		1.4 MHz	3 MHz	5 MHz	10 MHz	15 MHz	20 MHz
ACS	dB	33.0	33.0	33.0	33.0	30	27

Table 2.1-31. Test parameters for adjacent channel selectivity, Case 1 (36.101 [1] Table 7.5.1-2)

Rx parameter	Units	Channel bandwidth					
		1.4 MHz	3 MHz	5 MHz	10 MHz	15 MHz	20 MHz
Power in transmission BW configuration	dBm	REFSENS + 14 dB					
$P_{Interferer}$	dBm	REFSENS +45.5dB	REFSENS +45.5dB	REFSENS +45.5dB	REFSENS +45.5dB	REFSENS +42.5dB	REFSENS +39.5dB
$BW_{Interferer}$	MHz	1.4	3	5	5	5	5
$F_{Interferer}$ (offset)	MHz	1.4+0.0025 / −1.4−0.0025	3+0.0075 / −3−0.0075	5+0.0025/ −5−0.0025	7.5+0.0075 / −7.5−0.0075	10+0.0125 / −10−0.0125	12.5+0.0025 / −12.5−0.0025

Note 1: The transmitter shall be set to 4 dB below PCMAX_L at the minimum uplink configuration specified in 36.101 [1] Table 7.3.1-2 with PCMAX_L as defined in subclause 6.2.5.
Note 2: The interferer consists of the reference measurement channel specified in Annex A.3.2 with one sided dynamic OCNG Pattern OP.1 FDD/TDD as described in Annex A.5.1.1/A.5.2.1 and set-up according to Annex C.3.1.

2.1.8.3.2 Base Station In-Channel Selectivity

The in-channel selectivity (ICS) requirement tests the ability of the BS to reject a strong interferer within the wanted channel but on an unallocated RB. For the wide area BS 20 MHz channel bandwidth, the wanted signal is a 25 RB FRC at −98.5 dBm on one side of the channel center frequency (F_c) and the interferer is a time-aligned LTE signal of 25 RB at −77 dBm adjacent to the wanted signal on the other side of F_c. The BS is required to reach at least 95% of the maximum throughput of the FRC.

The requirement is important for handling deployment issues in which partial frequency reuse is involved since in such cases strong interferers might occur within the wanted channel but on an unallocated RB. For example, a UE operating near maximum power on the edge of a macrocell in close proximity to a home BS on the same channel but using a different RB might experience strong in-channel interference on its unallocated RBs.

2.1.8.3.3 Base Station Adjacent Channel Selectivity

The BS ACS requirement is similar in concept to that defined for the UE. For the wide area BS 20 MHz channel bandwidth, the wanted signal is set at $P_{REFSENS}$ +6 dB and a 5 MHZ LTE adjacent channel interferer is set at −52 dBm. The BS is expected to meet 95% of the maximum throughput of the FRC.

2.1.8.4 Blocking

2.1.8.4.1 UE Blocking

The UE blocking requirements are split into three categories:
- In-band blocking
- Out-of-band blocking
- Narrowband blocking.

In-band blocking tests the ability of the UE receiver to handle a fixed-level, strongly modulated interferer at frequency offsets up to 15 MHz above or below the wanted channel. The requirement is specified in terms of an allowed degradation in the reference sensitivity, which is 6 dB for the 10 MHz channel bandwidth in the presence of a 5 MHz wide interferer at −56 dBm offset 12.5 MHz, or −44 dBm for an interferer offset by 7.5 MHz.

Out-of-band blocking tests the ability of the UE to handle a fixed-level CW interferer at frequencies beyond 15 MHz from the wanted channel. A similar requirement is specified in terms of a relaxation in the reference sensitivity. Interferers are specified for three frequency ranges: −44 dBm interferer from 15 to 60 MHz; −30 dBm interferer from 60 to 85 ; and −15 dBm interferer for the remaining frequencies down to 1 MHz and up to 12.75 GHz. The third frequency range is very large and presents practical issues with testing. Therefore normally only specific frequencies identified as likely to be problematic based on the UE architecture are chosen to be tested. For Cases 1 to 3 there can be up to $\max (24, 6 \cdot [N_{RB}/6])$ frequencies at which the UE does not pass the test. The value of N_{RB} is determined according to the channel bandwidth (see Table 3.2-7).

The narrowband blocking is a special case of CW blocking in which the offset frequency is just outside of the assigned channel bandwidth. For example, in the 10 MHz case, the frequency offset is 5.2125 MHz, which is about 200 kHz closer than an adjacent channel signal. Again, the requirement is specified in terms of a relaxation in the reference sensitivity, which for the 10 MHz case is 13 dB for an interferer at −55 dBm.

2.1.8.4.2 Base Station Blocking

The BS blocking requirements are split into three categories:
- In-band blocking
- Out-of-band blocking
- Narrowband blocking.

The in-band blocking requirements generally apply within 10 or 20 MHz of the wanted band edge depending on the band. The interferer is a 5 MHz modulated LTE signal except for the channel bandwidths below 5 MHz. For the wide area BS, most requirements are for an interferer of −43 dBm at which the BS must meet 95% throughput of the FRC at $P_{REFSENS}$ + 6 dB. The requirements for the local area BS and home BS are more stringent at −35 dBm and −27 dBm, respectively. This is due to the increased probability of interference in smaller cell deployments. There are additional CW blocking requirements for co-location with BS in other bands which are much more stringent at +16 dBm for wide area BS dropping to typically −6 dBm for local area BS.

The out-of-band blocking requirements apply for a CW interferer at −15 dBm from 1 MHz to 12.75 GHz excluding the frequencies covered by the in-band blocking requirements.

The narrowband blocking requirements are specified as a special case of ACS in which the adjacent channel interferer is represented by a single RB. This RB is positioned 342.5 kHz from the wanted channel edge and then at eight further offsets up to 4.32 MHz.

2.1.8.5 UE Spurious Response

For those frequencies that fail the out-of-band blocking test, the UE is retested using a CW interferer at −44 dBm, the level at which the UE must meet 95% of the maximum throughput for the specified relaxation in reference sensitivity.

2.1.8.6 Intermodulation

2.1.8.6.1 UE Intermodulation

The receiver intermodulation requirements test the ability of the UE receiver to handle two interferers close to the wanted signal. One interferer is a CW signal positioned at an offset frequency based on the channel bandwidth. A second modulated interferer is placed at twice the offset frequency. This combination means that third order intermodulation products generated in the UE receiver will fall onto the wanted signal and degrade performance. The requirements, shown in Table 2.1-32, are again specified in terms of an allowed degradation in reference sensitivity.

Table 2.1-32. Wideband intermodulation (36.101 [1] Table 7.8.1.1-1)

Rx parameter	Units	Channel bandwidth					
		1.4 MHz	3 MHz	5 MHz	10 MHz	15 MHz	20 MHz
Power in transmission bandwidth configuration	dBm	REFSENS + channel bandwidth specific value below					
		12	8	6	6	7	9
$P_{Interferer\,1}$ (CW)	dBm	−46					
$P_{Interferer\,2}$ (modulated)	dBm	−46					
$BW_{Interferer\,2}$		1.4	3	5			
$F_{Interferer\,1}$ (offset)	MHz	−BW/2 −2.1 / +BW/2+ 2.1	−BW/2−4.5 / +BW/2+4.5	−BW/2−7.5 / +BW/2+7.5			
$F_{Interferer\,2}$ (offset)	MHz	$2*F_{Interferer\,1}$					

Note 1: The transmitter shall be set to 4dB below P_{CMAX_L} at the minimum uplink configuration specified in 36.101 [1] Table 7.3.1-2 with P_{CMAX_L} as defined in Subclause 6.2.5.

Note 2: Reference measurement channel is specified in Annex A.3.2 with one sided dynamic OCNG Pattern OP.1 FDD/TDD as described in Annex A.5.1.1/A.5.2.1.

Note 3: The modulated interferer consists of the reference measurement channel specified in Annex A.3.2 with one sided dynamic OCNG Pattern OP.1 FDD/TDD as described in Annex A.5.1.1/A.5.2.1 with set-up according to Annex C.3.1. The interfering modulated signal is 5 MHz E-UTRA signal as described in Annex D for channel bandwidth ≥ 5 MHz.

The ability to handle intermodulation from in-band interferers depends mainly on the linearity of the UE receiver. For interferers that are out of band, additional protection is provided by the UE's band filter.

2.1.8.6.2 Base Station Intermodulation

The BS receiver intermodulation requirements are in two parts. The first part is similar to the requirements for the UE except with slightly different interferer levels of −52 dBm for a wide area BS at $P_{REFSENS}$ +6 dB, −44 dBm for a local area BS at PREFSENS + 6 dB, and −36 dBm for a home BS at $P_{REFSENS}$ + 14 dB. The second part is a narrowband intermodulation requirement in which the modulated interferer has only one RB allocated and both the CW and modulated interferer are located much closer to the wanted channel.

2.1.8.7 Receiver Spurious Emissions

2.1.8.7.1 UE Receiver Spurious Emissions

This requirement is similar in concept to that specified for the transmitter but the limits as shown in Table 2.1-33 are significantly lower than the −36 dBm general threshold for the transmitter spurious emissions.

Table 2.1-33. General receiver spurious emission requirements (36.101 [1] Table 7.9.1-1)

Frequency band	Measurement bandwidth	Maximum level	Note
30MHz ≤ f < 1GHz	100 kHz	−57 dBm	
1GHz ≤ f ≤ 12.75 GHz	1 MHz	−47 dBm	
12.75 GHz ≤ f ≤ 5th harmonic of the upper frequency edge of the DL operating band in GHz	1 MHz	−47 dBm	1

Note 1: Applies only for Band 22, Band 42, and Band 43.

2.1.8.7.2 Base Station Receiver Spurious Emissions

The BS receiver spurious emissions requirements are identical to those for the UE at −57 dBm in 100 kHz up to 1 GHz and −47 dBm in 1 MHz up to 12.75 GHz.

2.1.8.8 UE Receiver Image

The final receiver requirement is a new one introduced as a result of carrier aggregation. It is specified for intra-band contiguous aggregation bandwidth Class C (two 20 MHz component carriers). One of the possible UE architectures for such a signal is that of a wideband receiver (40 MHz). In such a receiver it is expected that an RB received on one carrier will have an image response in the other carrier. The requirement specifies that any such image be at least 25 dB below the wanted signal power for all receiver levels up to −22 dBm. Tests for this requirement are yet to be developed and as with ACS, this is not a directly measurable requirement.

2.1.9 Performance Requirements

2.1.9.1 UE Receiver Performance Requirements

The UE performance requirements set minimum levels of performance for the UE's reception of various physical channels according to the downlink RMC configurations in 36.101 [1] A.3 and the channel propagation conditions in Annex B. Requirements cover all the essential downlink data and control channels at the supported modulation depths of QPSK, 16QAM, and 64QAM. The conditions under which performance is measured take into account the various antenna configurations of receive diversity, transmit diversity, single-user spatial multiplexing (open- and closed-loop), and multi-user spatial multiplexing. There are further options for cell-specific and user-specific reference signals. The UE performance requirements will be discussed further in Sections 6.5, 6.6, 6.7, and 7.2.

2.1.9.2 Base Station Receiver Performance Requirements

The BS performance requirements cover reception of the PUSCH, PUCCH, and PRACH. Base station receiver testing is discussed further in Sections 6.5, 6.6, 6.7, and 7.2.

2.1.10 Multi-Standard Radio

Release 9 introduced the concept of multi-standard radio (MSR). This was in recognition of the evolution of BS technology, which enabled more than one carrier from the same or another radio access technology to be operated from a single base station using a wideband transceiver. The introduction of MSR has not resulted in new radio requirements as such but it has changed the way in which the existing radio requirements for LTE and other systems such as UMTS and GSM have been interpreted for the purposes of conformance testing. As such, a new MSR conformance testing specification 37.141 was created. In Release 11 the MSR concept is extended for non-contiguous cases in which the different RATs are not located in the same band. The subject of MSR is covered further in Section 6.4.7.

2.1.11 Carrier Aggregation

One of the most significant air interface enhancements to the LTE specifications is the introduction of carrier aggregation in Release 10. The principles of CA are outlined in 36.807 [3]. Carrier aggregation is based on the carriers defined for Release 8, which are termed component carriers (CC). The purpose of CA is to increase the available bandwidth to and from the UE by aggregating two to five CCs to create instantaneous bandwidth of up to 100 MHz maximum. The channel bandwidths of the aggregated carriers do not have to be the same; e.g., a 10 MHz carrier can be aggregated with a 20 MHz carrier. The aggregation can be contiguous—i.e., the carriers occupy adjacent channels within one band (intra-band)—or they can be non-contiguous within one band or between bands (inter-band).

Table 2.1-34 defines six CA bandwidth classes. Class A represents the Release 8 single carrier case. Class B is intended for dual-carrier scenarios in which the combined bandwidth of the carriers does not exceed the 20 MHz of a Release 8 single carrier. This is expected to be a common deployment scenario for operators owning small quantities of spectrum in different bands. The remaining four classes cover the cases in which the addition of carriers increases the aggregated transmission bandwidth configuration beyond the 20 MHz single carrier limit of Release 8.

Table 2.1-34. CA bandwidth classes and corresponding nominal guard bands 3 (36.101 [1] Table 5.6A-1)

CA bandwidth class	Aggregated transmission bandwidth configuration	Maximum number of CC	Nominal guard band, BW_{GB}
A	$N_{RB,agg} \leq 100$	1	$0.05BW_{Channel(1)}$
B	$N_{RB,agg} \leq 100$	2	FFS
C	$100 < N_{RB,agg} \leq 200$	2	$0.05\ max(BW_{Channel(1)},\ BW_{Channel(2)})$
D	$200 < N_{RB,agg} \leq [300]$	FFS	FFS
E	$[300] < N_{RB,agg} \leq [400]$	FFS	FFS
F	$[400] < N_{RB,agg} \leq [500]$	FFS	FFS

Note 1: $BW_{Channel(1)}$ and $BW_{Channel(2)}$ are channel bandwidths of two E-UTRA component carriers according to 35.101 [1] Table 5.6-1.

There are an enormous number of possible CA scenarios but only a small number will lead to the specification of performance requirements. Through Release 11, the specification of minimum performance requirements for the air interface is limited to dual-carrier CA. Each individual CA scenario needs to be studied in order to identify the combination of requirements necessary to ensure commercially viable deployment. Most of the CA tradeoffs are made on the UE side since the UE has limited power and space to implement a multi-carrier transceiver.

Table 2.1-35 lists the intra-band CA scenarios for which requirements are being developed in Release 11. Technical Report 36.823 acts as a skeleton report for the other TRs, which cover contiguous scenarios for Bands 7, 38, and 41 and non-contiguous scenarios for Bands 3 and 25. The non-contiguous scenarios are more complicated in terms of the impact on device architecture and requirements.

Table 2.1-35. Intra-band CA scenarios for Release 11

Work item name	Technical Report number
LTE Carrier Aggregation Enhancements (skeleton TR for intra-band)	36.823
LTE Advanced Carrier Aggregation in Band 7	36.831
LTE Advanced Carrier Aggregation in Band 38	36.830
LTE Advanced Carrier Aggregation in Band 41	36.827
Intra-band Non-contiguous Carrier Aggregation in Band 25	36.841

The inter-band CA scenarios in Release 11 are being studied in 36.850 and are shown in Table 2.1-36. Inter-band CA is considerably more complicated than intra-band CA and so for the purposes of characterizing the different combinations, five inter-band CA classes have been identified.

- Class A1, low-high band combination without harmonic relation between bands
- Class A2, low-high band combination with harmonic relation between bands
- Class A3, low-low or high-high band combination without intermodulation problem (low order IM)
- Class A4, low-low or high-high band combination with intermodulation problem (low order IM)
- Class A5, combination except for A1 to A4 (similar to mid band combinations).

The classes A2 and A4 require special study and may require alternative UE architectures.

Table 2.1-36. Inter-band CA scenarios for Release 11

First band	Second band	Inter-band Class
1	18	Class A1
1	19	Class A1
2	17	Class A1
3	5	Class A1
3	5 (two uplink carriers)	Class A1
3	20	Class A1
4	5	Class A1
4	13	Class A1
7	20	Class A1
3	8	Class A2
4	12	Class A2
4	17	Class A2
1	7	Class A3
3	7	Class A3
4	7	Class A3
5	12	Class A3
5	17	Class A3
8	20	Class A4
1	21	Class A5
11	18	Class A5

A consequence of carrier aggregation on the RF requirements is that many single-band requirements have to be modified to take account of what is practical to implement. There are also some new definitions and measurements required. For the base station, CA can be seen as a special case of multi-standard radio (see Section 2.1.10), and for that purpose additions have been made to 36.104 [2] for the support of non-contiguous intra-band CA which include the following:

- Introduction of definition of sub-block bandwidth for intra-band non-contiguous spectrum
- Clarification on requirements for contiguous and non-contiguous spectrum
- Introduction of time-alignment error requirement for intra-band non-contiguous operation
- Clarification of occupied bandwidth and ACLR requirements for non-contiguous spectrum
- Introduction of cumulative ACLR (CACLR) requirement for intra-band non-contiguous operation
- Clarification of operating band unwanted emissions and transmitter intermodulation requirements for non-contiguous spectrum
- Clarification of ACS, narrowband blocking, blocking, and receiver intermodulation requirements for non-contiguous spectrum.

For the UE, the introduction of CA has implications on most of the transmitter and receiver requirements in 36.101 [1] Sections 6 and 7, including maximum output power and output power dynamics, transmit signal quality, SEM, ACLR, spurious emissions, reference sensitivity, and many of the other receiver requirements.

In general for the transmitter the existing requirements still apply per carrier although there are some exceptions. For example, the in-band emission requirements for transmit signal quality are specified for the intra-band contiguous CA case to take account of the different ways in which the UE is designed. The UE can implement intra-band CA either by aggregating two separate transmitters or by using a single wideband transmitter. The interaction between the carriers and the resulting spurious products are different in each case. The in-band emission requirements have been written with this in mind and are specified for both carriers active but only one carrier allocated. There are also differences in the number of exceptions for the IQ image and carrier leakage requirements.

Additionally there are special cases in which network signaling requirements interact with CA. An example for CA class 1C contiguous allocation is given in Table 2.1-37.

Table 2.1-37. Contiguous allocation A-MPR for CA_NS_01 (36.101 [1] Table 6.2.4A.1-1)

CA_1C	RB$_{Start}$	L$_{CRB}$ [RBs]	RB$_{start}$ + L$_{CRB}$ [RBs]	A-MPR for QPSK and 16QAM[dB]
	0–30 and 170–199	> 0	N/A	[≤ 10]
100 RB/100 RB	31–105	> 80	N/A	[≤ 5]
	105–169	N/A	> 170	[≤ 3]
	0–13 and 137–149	> 0	N/A	[≤ 10]
75 RB/75 RB	13–79	> 55	N/A	[≤ 6]
	80–136	N/A	> 137	[≤ 2]

Note 1: RB$_{start}$ indicates the lowest RB index of transmitted resource blocks.
Note 2: L$_{CRB}$ is the length of a contiguous resource block allocation.
Note 3: For intra-subframe frequency hopping which intersects regions, notes 1 and 2 apply on a per slot basis.
Note 4: For intra-subframe frequency hopping which intersects regions, the larger A-MPR value may be applied for both slots in the subframe.

Some of the proposed relaxations are quite substantial (up to 10 dB) indicating the considerable strain that certain combinations of CA put on the UE design. Operating the UE under such conditions is therefore limited to small cell deployments where maximum power handling is not critical.

There are also implications from CA for many of the BS radio requirements including the new concept of CACLR, which defines the ACLR requirements as the addition of emissions from multi-carrier signals on either side of a gap between the carriers.

2.1.12 Uplink MIMO

Uplink MIMO is introduced in Release 10 and is a significant development for the UE since it is the first time that multiple transmitters are specified. Uplink MIMO has implications for UE design and gives rise to some changes in the requirements. For example, the maximum output power, output power dynamics, and power control requirements are affected.

The only new requirement is for time alignment between transmitter branches. The requirement is provisionally set at 130 ns. MIMO is discussed further in Sections 2.4, 6.7, and 6.9.

2.2 Orthogonal Frequency Division Multiplexing

Orthogonal frequency division multiplexing (OFDM) is the modulation scheme chosen for the LTE downlink. It is a digital multi-carrier scheme that uses a large number of closely spaced subcarriers to carry data and control information. Each individual subcarrier is modulated at a low symbol rate with a conventional modulation format such as quadrature amplitude modulation (QAM). The combination of the many low-rate subcarriers provides overall data rates similar to conventional single-carrier modulation schemes using the same bandwidth. Today, OFDM is widely used in applications from digital television and audio broadcasting to wireless networking and wired broadband Internet access.

2.2.1 History of OFDM

OFDM was proposed as a mathematical possibility as far back as 1957 with "Kineplex," a multi-carrier high frequency (HF) modem designed by Mosier and Clabaugh, although the first patented application was not until 1966 when Chang of Bell Labs filed US patent 3488445. The first practical implementation of an OFDM system came in 1985 when Telebit introduced the "Trailblazer" range of modems that reached speeds of 9600 bps. This highlighted one of the key advantages of OFDM: its ability to perform well through a low quality channel—in this case telephone lines—thereby outperforming existing solutions. From this early beginning, OFDM has become the technology that now delivers up to 10 Mbps over digital subscriber lines (DSL). It is also used in systems that communicate over domestic power lines.

The 1980s and early 1990s saw a number of experimental broadcast systems, with companies including Thomson-CSF and TDF in France and BBC Research in the UK. The first international standard to specify OFDM was digital audio broadcast (DAB) in 1995, the outcome of the European Eureka147 project, and this was followed two years later by the digital video broadcast-terrestrial (DVB-T) standard. Both DAB and DVB-T are now in widespread use. In addition to the use of OFDM in unidirectional broadcast technologies, parallel work throughout the 1990s led in 1999 to the first OFDM-based wireless LAN (WLAN) standard, IEEE 802.11a. This was followed in succession by 802.11g, 802.11n (adding MIMO), and 802.16d (fixed WiMAX™), although the most widely deployed WLAN standard is still 802.11b, which uses direct sequence spread spectrum.

The use of OFDM for cellular systems was first briefly considered back in the late 1980s as a candidate technology for GSM but was quickly dropped due to lack of cost-effective computing power. A decade later, OFDM was seriously considered as one of the candidates for 3GPP's UMTS but was ruled out in favor of wideband code division multiple access (W-CDMA). Again the decision was influenced by the cost of computing power and the associated power consumption in the terminals.

However, with today's availability of small, low-cost, low-power chipsets, OFDM has become the technology of choice for the next generation of cellular wireless. The first cellular system to adopt OFDM was 802.16e (Mobile WiMAX™). It was followed soon after by 802.20, the basis for 3GPP2's Ultra-Mobile Broadband (UMB), and most recently by 3GPP for the long-term evolution of UMTS. It now seems apparent that the evolution of these newest so-called 3.9G systems toward 4G will not result in any change to the underlying air interface, so OFDM will likely be the technology of choice for cellular wireless systems well into the future. The new OFDM cellular systems all focus on delivering high-speed data services and have similar goals in terms of improving spectral efficiency, with the widest bandwidth systems providing the highest single-user data rates.

2.2.1 OFDM Basic Signal Construction

The basic OFDM signal comprises a large number of closely spaced continuous wave (CW) tones in the frequency domain. The most basic form of modulation applied to the subcarriers is square wave phase modulation, which produces a frequency spectrum represented by a sinc or $\frac{sin(x)}{x}$ function that has been convolved around the subcarrier frequency. A truncated sinc function is shown in Figure 2.2-1.

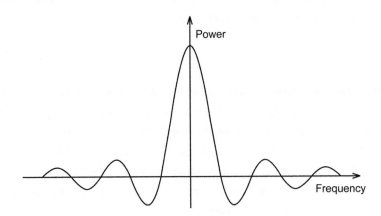

Figure 2.2-1. Spectrum of a single modulated OFDM subcarrier (truncated)

The rate of change of the phase modulation will determine the position of the zero crossings in frequency. The trick that makes OFDM a practical transmission system is to link the subcarrier modulation rate to the subcarrier spacing such that the nulls in the spectrum of one subcarrier line up with the peaks of the adjacent subcarriers.

For standard LTE each modulating symbol lasts 66.7 μs. By setting the subcarrier spacing to be 15 kHz, which is the reciprocal of the symbol rate, the peaks and nulls line up perfectly such that at any subcarrier frequency, the subcarriers are orthogonal; i.e., there is no interference between them. This can be seen in Figure 2.2-2.

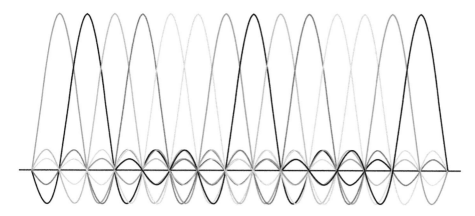

Figure 2.2-2. Spectrum of multiple OFDM subcarriers of constant amplitude

In Figure 2.2-2 each subcarrier has the same magnitude, which is the case when any of the LTE-supported constant amplitude modulation formats are used: Zadoff-Chu sequences, binary phase-shift keying (BPSK), and quadrature phase-shift keying (QPSK). It is also possible for the subcarriers to vary in amplitude since LTE also supports 16 quadrature amplitude modulation (16QAM) and 64 quadrature amplitude modulation (64QAM).

Compared to the 3.84 Msps of UMTS, the 15 ksps subcarrier symbol rate of LTE is very low, but in the same 5 MHz channel bandwidth, LTE can simultaneously transmit 300 subcarriers to provide an aggregate 4.5 Msps rate. Thus on first inspection, CDMA and OFDM have similar capacity for carrying data.

2.2.2 Guard Intervals and Immunity From Delay Spread

In 1971 Weinstein and Ebert proposed the introduction of a guard interval between each symbol to reduce the inter-symbol interference (ISI) caused by delay spread in the transmission channel. To illustrate the principle of ISI, consider the simple five-tap delay profile in Figure 2.2-3. This shows the amplitude and phase response of delayed copies of the transmitted signal, which arrive at the receiver having taken different paths through the transmission channel. By definition, the first detected path is assigned a relative amplitude of 1 and phase of 0 degrees. The X-axis is given in units of symbol length, so for this example the difference between the earliest and latest components — known as the delay spread — is 15% of the symbol length.

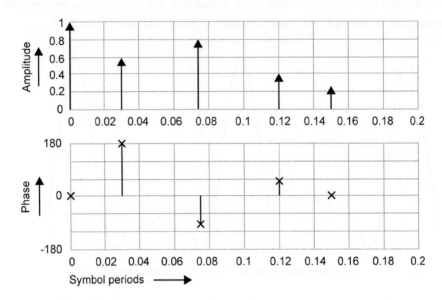

Figure 2.2-3. Amplitude and phase response for a five-tap channel delay profile

Figure 2.2-4 shows the effect of passing one subcarrier of an OFDM signal through this channel and how the delay spread creates ISI at the symbol boundaries. The ideal received signal is shown in the top trace of Figure 2.2-4. This signal represents two adjacent BPSK symbols with a 180 degree phase shift between them. The number of baseband cycles per symbol is shown as five, indicating that this is the fifth subcarrier on either side of the channel center frequency. The next four traces show the amplitude, phase, and timing of the delayed copies of the ideal signal at the receiver. The bottom trace represents the composite received signal, being the sum of the five components. In order to show the extent of the delay spread, the dotted line in the bottom trace is a copy of the ideal signal. From this it is evident that during the period from 1.05 to 1.2, the received signal is severely distorted and of negative value to the demodulation process. The received signal is undistorted only after the last component has arrived and before the first component leaves. For this delay profile the undistorted symbol is reduced to 85% of its transmitted length.

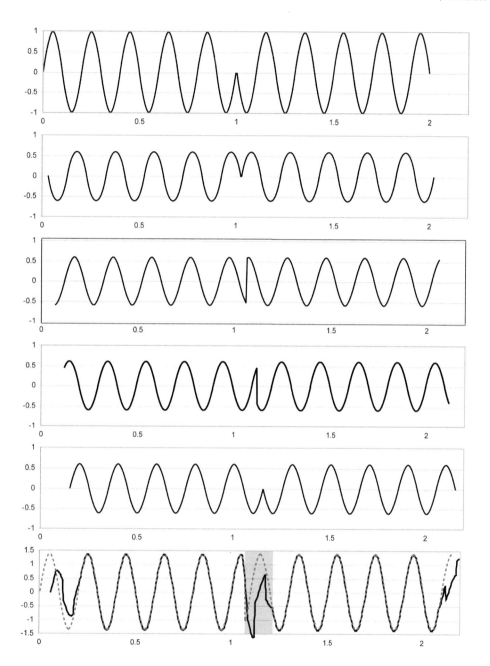

Figure 2.2-4. Inter-symbol interference caused by delay spread

In order to optimize demodulation performance and reject inter-carrier interference (ICI), the symbol must be sampled for exactly its nominal length. At baseband, each successive subcarrier has one more cycle during the OFDM symbol period. When error energy from an adjacent subcarrier is added to the wanted subcarrier, the only time period for which the signals are orthogonal (multiply and integrate to zero) is at the reciprocal of the subcarrier spacing.

The CP adds redundancy through repetition of the signal rather than by adding any new information. When the CP is added, it guarantees that the symbol will be undistorted for at least its nominal symbol length in the presence of multipath up to the length of the CP. By correctly aligning with the signal timing, the receiver is then able to sample the signal for exactly one nominal symbol period. This allows the receiver to avoid the frequency domain ICI while at the same time avoiding all the time domain ISI due to multipath.

For the example channel in Figure 2.2-3, the CP would have to be at least 15% of the symbol length. The choice of CP in cellular systems depends on the propagation conditions and cell size. Typical figures are in the region of 5 μs, which represents 1.5 km of path delay difference. Note that this delay spread is the difference in path length and not the absolute path length. For LTE, the standard CP is set to 4.69 μs creating an extended symbol of some 71.35 μs. An obvious consequence of adding redundancy to the symbol is a loss in capacity due to a lower symbol rate, in this case a reduction of about 7 %. Thus there is a tradeoff between the amount of protection from delay spread and the consequent loss of capacity.

It is interesting to contrast the way OFDM and CDMA deal with multipath distortion. The symbol length in CDMA systems is the reciprocal of the chip rate. All lower-rate data is spread (multiplied) up to this fixed system chip rate by use of spreading codes. For UMTS, which uses the 3.84 Mcps W-CDMA air interface, this spreading results in a symbol length of 260 ns and a 3 dB bandwidth prior to filtering of 3.84 MHz. A 5 μs delay spread on this system would create serious ISI extending to around 20 CDMA symbols. Since the delay spread would be nearly 20 times the symbol length, it is impractical to consider extending the symbol with a CP in the way described for OFDM. This is why CDMA systems have to rely on rake receivers and frequency-domain equalizers to untangle the ISI. In CDMA systems, the wider the channel bandwidth, the higher the chip rate and the worse the ISI becomes. This is why it is impractical to design CDMA systems with channel bandwidths much wider than the 5 MHz of today's UMTS.

This situation is in sharp contrast to OFDM, whose symbol length is set not by the reciprocal of the channel bandwidth but rather by the subcarrier spacing. This makes OFDM systems highly scalable in the frequency domain with no impact on the symbol length. The 15 kHz subcarrier spacing in LTE provides a symbol length that is 256 times longer than is used for W-CDMA, and the sheer length of the OFDM symbol makes it feasible to extend the symbol by 4.69 μs with a loss of only 7% in system capacity.

It is reasonable to ask why OFDM does not continue to make the symbols even longer by using ever-narrower subcarrier spacing. However, there are practical considerations that limit how close the subcarriers can become. The first factor is phase noise, which causes the energy of each subcarrier to leak into the adjacent subcarriers causing ICI. The second factor is the consequence of frequency errors between the transmitter and receiver, which can shift energy sideways in the frequency domain also causing ICI.

Simple frequency errors due to Doppler shift or other single-sided errors can be corrected using normal closed-loop frequency tracking methods. Even then, the closer the subcarrier spacing, the better these tracking loops have to perform.

There is, however, a double-sided frequency error that can occur if the UE simultaneously receives downlink signals from behind and in front as a result of reflections. If the UE is moving towards or away from the reflector, one signal could be reduced in frequency while the other is increased. This creates a symmetrical image in the frequency domain that cannot be removed using a simple frequency tracking loop. This kind of ICI is possible to remove using digital techniques but the process can require huge amounts of computing power if the ICI becomes significant. This and the earlier reasons are why practical subcarrier spacing in OFDM systems do not get much lower than the 7.5 kHz defined for LTE's optional multimedia broadcast over single frequency network (MBSFN) service.

2.2.3 Example of OFDM Signal Generation

Figure 2.2-5 illustrates the principles of OFDM signal generation using a simple four subcarrier example.

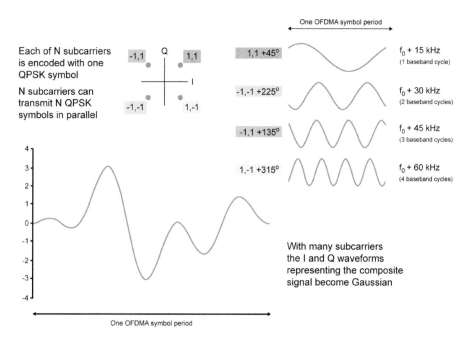

Figure 2.2-5. Example of OFDM signal generation using four subcarriers

Since four subcarriers are used in this example, it will be possible to transmit four data symbols in parallel. The four possible data symbols are represented by phases of the subcarrier, which in the IQ plane are represented as (1, 1), (−1, 1), (−1, −1), and (1, −1). One of these symbols will be mapped to each subcarrier for the duration of an OFDM symbol. The four possible subcarrier phases are shown.

Each subcarrier is defined in the frequency domain by a vector, which represents the amplitude and phase of the data symbol that has been mapped to it. After symbol mapping, the subcarriers are each converted to time domain waveforms using an inverse fast Fourier transform (IFFT). At this point the CP is inserted for each waveform and then the waveforms are vector-summed to produce the composite waveform for transmission.

The lower trace in Figure 2.2-5 shows the composite waveform (without the CP, for simplicity) resulting from the parallel transmission of the four symbols shown. Due to the summation of the four waveforms, the peak of the signal now reaches nearly four times the voltage of one subcarrier. With increasing numbers of subcarriers the I and Q waveforms that represent the composite signal become Gaussian, which creates a chi-square power distribution.

This simple example illustrates one of the key disadvantages of OFDM, which is that the peak to average power ratio (PAPR) of the composite signal can have peaks exceeding 12 dB above the average signal power, which presents significant challenges for transmission by an eNB operating at high power. There are many steps that can be taken to reduce the signal peaks and to extend the useful operating range of the power amplifier. These are discussed further in Sections 6.2.9 and 6.4.1.4.

2.2.4 Comparing CDMA and OFDM

The attributes of CDMA technology upon which UMTS is based and the corresponding attributes of OFDM are summarized in Table 2.2-1.

Table 2.2-1. Comparison of CDMA and OFDM

Attribute	CDMA	OFDM
Transmission bandwidth	Full system bandwidth	Variable up to full system bandwidth
Frequency-selective scheduling	Not possible	A key advantage of OFDM although it requires accurate real-time feedback of channel conditions from receiver to transmitter
Symbol period	Very short; inverse of the system bandwidth	Very long; defined by subcarrier spacing and independent of system bandwidth
Equalization	Difficult above 5 MHz	Easy for any bandwidth due to signal representation in the frequency domain
Resistance to multipath	Difficult above 5 MHz	Completely free of multipath distortion up to the CP length
Suitability for MIMO	Requires significant computing power due to signal being defined in the time domain	Ideal for MIMO due to signal representation in the frequency domain and possibility of narrowband allocation to follow real-time variations in the channel
Sensitivity to frequency domain distortion and interference	Averaged across the channel by the spreading process	Vulnerable to narrowband distortion and interference
Separation of users	Scrambling and orthogonal spreading codes	Frequency and time although scrambling and spreading can be added as well

The early commercial uses of OFDM have been primarily for broadcast and wired applications in which interference has not been a major factor. OFDM is used in some of the WLAN protocols, which are typically found in hotspot deployments. The first cellular communications deployment of OFDM was based on the 802.16e (Mobile WiMAX) standard and has now been followed by broad deployment of LTE. Cellular systems are different from broadcast and hotspot systems in that they have to operate seamlessly across a wide area, including the area around the cell boundaries where signal levels are at their lowest and inter-cell interference is at its highest.

It is expected that at the cell edge, OFDM will be more difficult to operate in the downlink than CDMA. CDMA relies on scrambling codes to provide protection from inter-cell interference at the cell edge, whereas OFDM has no such intrinsic feature. In addition, the interference profile at the cell edge for CDMA is relatively stable across frequency and can be modeled and removed using interference cancelling receivers. The situation for OFDM is more complex because the presence or absence of interference is a function of narrowband scheduling in the adjacent cell, and the resulting noise profile is likely to be far less stable and predictable. One solution to this is to use some form of frequency planning at the cell edge. Figure 2.2-6 gives one example of how this might be done. The white central area of the cell is where the entire channel bandwidth would get used and the colored areas show a frequency reuse pattern with a repetition factor of four.

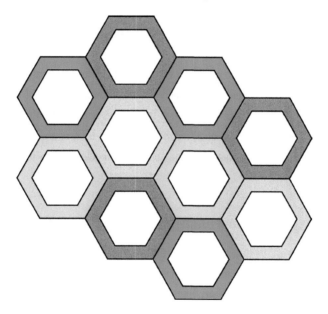

Figure 2.2-6. Example of cell-edge frequency planning to mitigate inter-cell interference

This approach can significantly reduce cell-edge interference, although it remains a challenge to know the location of the UE for scheduling purposes. Moreover, such a reuse scheme has a direct impact on cell edge capacity, in this case reducing it by a factor of four. Operating OFDM efficiently at the cell edge is likely to require significant network optimization after initial deployment. Uplink OFDM is simpler than CDMA since UE transmissions are orthogonal in time and frequency.

2.2.5 Orthogonal Frequency Division Multiple Access

Until now the discussion has been about OFDM, but LTE uses a variant of OFDM for the downlink called orthogonal frequency division multiple access (OFDMA). Figure 2.2-7 compares OFDM and OFDMA.

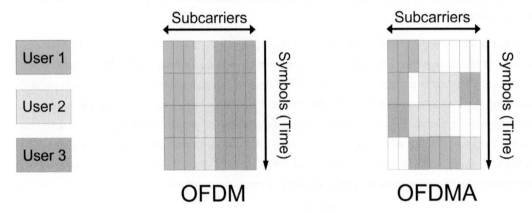

Figure 2.2-7. OFDM and OFDMA subcarrier allocation

With standard OFDM, the subcarrier allocations are fixed for each user and performance can suffer from narrowband fading and interference. OFDMA incorporates elements of time division multiple access (TDMA) so that the subcarriers can be allocated dynamically among the different users of the channel. The result is a more robust system with increased capacity. The capacity comes from the trunking efficiency gained by multiplexing low rate users onto a wider channel to provide dynamic capacity when needed, and the robustness comes from the ability to schedule users by frequency to avoid narrowband interference and multipath fading.

2.3 Single-Carrier Frequency Division Multiple Access

The high peak-to-average power ratio (PAPR) associated with OFDM led 3GPP to look for an alternative modulation scheme for the LTE uplink. SC-FDMA was chosen since it combines the low PAPR techniques of single-carrier transmission systems such as GSM and CDMA with the multipath resistance and flexible frequency allocation of OFDMA. The reduction in PAPR of SC-FDMA over OFDMA is best described using the cubic metric (CM) described in Section 2.1.4.1. Table 2.3-1 shows the power amplifier (PA) back-off that is required based on the CM for different SC-FDMA and OFDMA waveforms.

Table 2.3-1. Comparing SC-FDMA and OFDMA with the cubic metric

Modulation depth	Cubic metric	
	SC-FDMA	OFDMA
QPSK	1.2	4
16QAM	2.2	4
64QAM	2.4	4

A mathematical description of an SC-FDMA symbol in the time domain is given in 36.211 [9] Subclause 5.6. A brief description is as follows: data symbols in the time domain are converted to the frequency domain using a discrete Fourier transform (DFT); once in the frequency domain they are mapped to the desired location in the overall channel bandwidth before being converted back to the time domain using an inverse FFT (IFFT). Finally, the CP is inserted. SC-FDMA is sometimes called discrete Fourier transform spread OFDM (DFT-S-OFDM) because of this process, although this terminology is becoming less common.

2.3.1 OFDMA and SC-FDMA Compared

A graphical comparison of OFDMA and SC-FDMA as shown in Figure 2.3-1 is helpful in understanding the differences between these two modulation schemes. As will be described in Section 3.2, real uplink SC-FDMA signals are allocated in units of 12 adjacent subcarriers known as resource blocks (RBs). However, for clarity, this example uses only four (M) subcarriers over two symbol periods with the payload data represented by QPSK modulation.

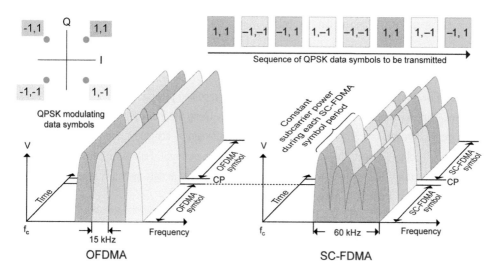

Figure 2.3-1. Comparison of OFDMA and SC-FDMA transmitting a series of QPSK symbols

On the left side of Figure 2.3-1, M adjacent 15 kHz subcarriers—already positioned at the desired place in the channel bandwidth—are each modulated for the OFDMA symbol period of 66.7 μs by one QPSK data symbol. In this four subcarrier example, four symbols are taken in parallel. These are QPSK data symbols so only the phase of each subcarrier is modulated and the subcarrier power remains constant between symbols. After one OFDMA symbol period has elapsed, the CP is inserted and the next four symbols are transmitted in parallel. For visual clarity, the CP is shown as a gap; however, it is actually filled with a copy of the end of the next symbol, which means that the transmission power is continuous but has a phase discontinuity at the symbol boundary. To create the transmitted signal, an IFFT is performed on each subcarrier to create M time-domain signals. These in turn are vector-summed to create the final time-domain waveform used for transmission.

In contrast, SC-FDMA signal generation begins with a special precoding process but then continues in a manner similar to OFDMA. However, before getting into the details of the generation process, it is helpful to describe the end result as shown on the right side of Figure 2.3-1. The most obvious difference between the two schemes is that OFDMA transmits the four QPSK data symbols in parallel, one per subcarrier, while SC-FDMA transmits the four QPSK data symbols in series at four times the rate, with each data symbol occupying a wider M x 15 kHz bandwidth.

Visually, the OFDMA signal is clearly multi-carrier with one data symbol per subcarrier, but the SC-FDMA signal appears to be more like a single-carrier (hence the "SC" in the SC-FDMA name) with each data symbol being represented by one wide signal. Note that OFDMA and SC-FDMA symbol lengths are the same at 66.7 µs; however, the SC-FDMA symbol contains M "sub-symbols" that represent the modulating data. It is the parallel transmission of multiple symbols that creates the undesirable high PAPR of OFDMA. By transmitting the M data symbols in series at M times the rate, the SC-FDMA occupied bandwidth is the same as multi-carrier OFDMA but, crucially, the PAPR is the same as that used for the original data symbols. Adding together many narrowband QPSK waveforms in OFDMA will always create higher peaks than would be seen in the wider bandwidth, single carrier QPSK waveform of SC-FDMA. As the number of subcarriers M increases, the PAPR of OFDMA with random modulating data approaches Gaussian noise statistics but, regardless of the value of M, the SC-FDMA PAPR remains the same as that used for the original data symbols.

2.3.2 SC-FDMA Signal Generation

As noted earlier, SC-FDMA signal generation begins with a special precoding process. Figure 2.3-2 shows the first steps, which create a time-domain waveform of the QPSK data sub-symbols. Using the four color-coded QPSK data symbols from Figure 2.3-1, the process creates one SC-FDMA symbol in the time domain by computing the trajectory traced by moving from one QPSK data symbol to the next. This is done at M times the rate of the SC-FDMA symbol such that one SC-FDMA symbol contains M consecutive QPSK data symbols. Time-domain filtering of the data symbol transitions occurs in any real implementation, although it is not discussed here.

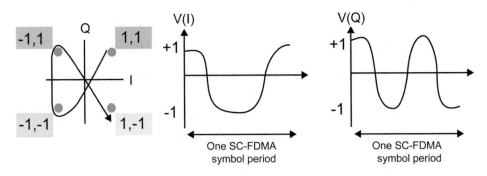

Figure 2.3-2. Creating the time-domain waveform of an SC-FDMA symbol

Once an IQ representation of one SC-FDMA symbol has been created in the time domain, the next step is to represent that symbol in the frequency domain using a DFT. This is shown in Figure 2.3-3. The DFT sampling frequency is chosen such that the time-domain waveform of one SC-FDMA symbol is fully represented by M DFT bins spaced 15 kHz apart, with each bin representing one subcarrier in which amplitude and phase are held constant for 66.7 µs.

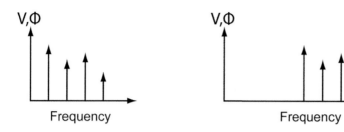

Figure 2.3-3. Baseband and frequency-shifted DFT representations of an SC-FDMA symbol

A one-to-one correlation always exists between the number of data symbols to be transmitted during one SC-FDMA symbol period and the number of DFT bins created. This in turn becomes the number of occupied subcarriers. When an increasing number of data symbols are transmitted during one SC-FDMA period, the time domain waveform changes faster, generating a higher bandwidth and hence requiring more DFT bins to fully represent the signal in the frequency domain. Note that in Figure 2.3-3 there is no longer a direct relationship between the amplitude and phase of the individual DFT bins and the original QPSK data symbols. This differs from the OFDMA example in which data symbols directly modulate the subcarriers.

The next step of the signal-generation process is to shift the baseband DFT representation of the time-domain SC-FDMA symbol to the desired part of the overall channel bandwidth. Because the signal is now represented as a DFT, frequency-shifting is a simple process achieved by copying the M bins into a larger DFT space of N bins. This larger space equals the size of the system channel bandwidth, of which there are six to choose from in LTE, spanning 1.4 to 20 MHz. The SC-FDMA signal, which is almost always narrower than the channel bandwidth, can be positioned anywhere in the channel bandwidth, thus executing the frequency division multiple access (FDMA) essential for efficiently sharing the uplink between multiple users.

To complete SC-FDMA signal generation, the process follows the same steps as for OFDMA. Performing an IDFT converts the frequency-shifted signal to the time domain and inserting the CP provides the fundamental robustness of OFDMA against multipath. The relationship between SC-FDMA and OFDMA is illustrated in Figure 2.3-4.

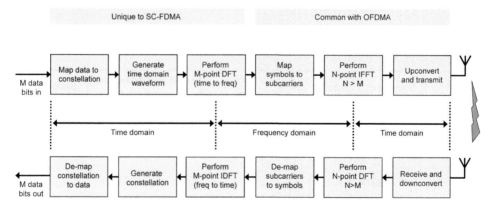

Figure 2.3-4. Simplified model of SC-FDMA and OFDMA signal generation

2.3.3 SC-FDMA Resistance to Multipath

At this point, it is reasonable to ask how SC-FDMA can be resistant to multipath when the data symbols are still short. In OFDMA, the modulating data symbols are constant over the 66.7 µs OFDMA symbol period, but an SC-FDMA symbol is not constant over time since it contains M sub-symbols of much shorter duration. The multipath resistance of the OFDMA demodulation process seems to rely on the long data symbols that map directly onto the subcarriers. Fortunately, it is the constant nature of each subcarrier, not the data symbols, that provides the resistance to delay spread. As shown in Figure 2.3-1 and Figure 2.3-3, the DFT of the time-varying SC-FDMA symbol generated a set of DFT bins constant in time during the SC-FDMA symbol period, even though the modulating data symbols varied over the same period. It is inherent to the DFT process that the time-varying SC-FDMA symbol—made of M serial data symbols—be represented in the frequency domain by M time-invariant subcarriers. Thus, even SC-FDMA with its short data symbols benefits from multipath protection.

It may seem counterintuitive that M time-invariant DFT bins can fully represent a time-varying signal. However, the DFT principle is simply illustrated by considering the sum of two fixed sine waves at different frequencies. The result is a non-sinusoidal time-varying signal, fully represented by two fixed sine waves.

2.3.4 Analysis of SC-FDMA Signals

Table 2.3-2 summarizes the differences between the OFDMA and SC-FDMA modulation schemes. When OFDMA is analyzed one subcarrier at a time, it resembles the original data symbols. At full bandwidth, however, the signal looks like Gaussian noise in terms of its PAPR statistics and the constellation. The opposite is true for SC-FDMA. In this case, the relationship to the original data symbols is evident when the entire signal bandwidth is analyzed. The constellation (and hence low PAPR) of the original data symbols can be observed rotating at M times the SC-FDMA symbol rate (ignoring the 7% rate reduction that is due to adding the CP). When analyzed at the 15 kHz subcarrier spacing, the SC-FDMA PAPR and constellation are meaningless because they are M times narrower than the information bandwidth of the data symbols.

Table 2.3-2 Analysis of OFDMA and SC-FDMA at different bandwidths

Modulation format	OFDMA		SC-FDMA	
Analysis bandwidth	15 kHz	Signal bandwidth (M * 15 kHz)	15 kHz	Signal bandwidth (M * 15 kHz)
Peak-to-average power ratio	Same as data symbol	High PAPR (Gaussian)	Lower than data symbol (not meaningful)	Same as data symbol
Observable IQ constellation	Same as data symbol at 1/66.7 µs rate	Not meaningful (Gaussian)	Not meaningful (Gaussian)	Same as data symbol at M/66.7 µs rate

2.3.5 Clustered SC-FDMA

Although SC-FDMA provides a significant advantage over OFDMA in terms of power back-off (nearly 3 dB for QPSK), it does so by sacrificing OFDMA's ability to schedule the transmission on any RB. For a fully allocated uplink this has no significance. For a partially allocated uplink, however, single cluster SC-FDMA in which all the RBs have to be contiguous requires that the eNB scheduler select the optimal starting RB for the transmission knowing that all the RBs then have to be allocated contiguously. In an OFDM system, the scheduler would be able to allocate the RBs according to the frequency selectivity of the channel. For instance, if the channel were seen to have a null in the middle, it would be advantageous to split the uplink transmission into two parts, with both scheduled for the more attractive part of the channel.

In order to close the gap between the spectral efficiency of OFDMA and the PAPR advantage of SC-FDMA, the concept of clustered SC-FDMA was introduced in Release 10 as part of the submission for the ITU's IMT-Advanced program. The principle behind clustered SC-FDMA is to allow some flexibility in uplink scheduling but not the arbitrarily flexible scheduling that is possible with OFDMA and which creates the highest PAPR. From a block diagram perspective, the only change from Figure 2.3-4 is in the process of mapping and de-mapping the symbols to the subcarriers. With SC-FDMA, the symbols are all mapped to adjacent subcarriers whereas with clustered SC-FDMA, the symbols are mapped in two or more non-adjacent groups. Although any number of groups could be used, it was decided for Release 10 to restrict the number of clusters to two. Table 2.3-3 shows the impact this has on the PAPR as calculated by the CM.

Table 2.3-3. Comparing SC-FDMA and two cluster SC-FDMA

Modulation depth	Cubic metric	
	SC-FDMA	Two cluster SC-FDMA
QPSK	1.2	2
16QAM	2.2	2.6
64QAM	2.4	2.75

It can be seen from Table 2.3-3 that two cluster SC-FDMA adds just over 1 dB to the PAPR of single cluster SC-FDMA. If the number of clusters gets beyond six, the PAPR starts to look like OFDMA. The uplink spectral efficiency of clustered SC-FDMA is similar to that of OFDMA, which is around 15% better than SC-FDMA depending on the assumptions. The cost in terms of implementation is that slightly more PA back-off is required and there are further issues in dealing with in-channel intermodulation products caused by the presence of two discrete carriers within the channel. These aspects are discussed further in Section 2.1.

2.4 Multi-Antenna Operation and MIMO

This section describes the multi-antenna mechanisms adopted by LTE to increase coverage and physical layer capacity. It focuses largely on the air interface as many of the operational details of the system are left to the designers of the eNB.

Adding antennas to a radio system gives the possibility of fundamental performance improvements because the radiated signals will take different physical paths and undergo variations in polarization. There are three main multi-antenna application categories. The first makes direct use of path diversity in which one radiated path may be subject to fading loss and another may not. Diversity can be introduced at the transmitter, the receiver, or both simultaneously. The second application is to apply beamsteering by controlling the phase relationship of the electrical signals radiated at the antennas to direct transmitted energy toward the physical location of the receiver. A third application is to put to use the path diversity—introduced by separating the antennas in space or by polarization—to enable the use of spatial multiplexing. Spatial multiplexing allows for the simultaneous transmission of more than one stream of data in the same time and frequency.

The term multiple input multiple output (MIMO) is typically used to describe spatial multiplexing and it is worth some explanation here to avoid potential confusion. The I and the O in MIMO refer to the channel through which the radio signal propagates and not to inputs or outputs of the devices at either end of the link. The simultaneous use of transmit diversity with receive diversity on a radio link—i.e., multiple input single output (MISO) plus single input multiple output (SIMO)—is sometimes referred to as MIMO, but this is a degenerate case that does not exploit spatial multiplexing since only one data stream is transmitted. The more typical use of the term MIMO involves the simultaneous transmission of more than one data stream into the channel in the same frequency and time, which creates the potential for increased throughput and spectral efficiency. In this book the term MIMO will be used to mean spatial multiplexing unless otherwise stated.

The terms beamsteering and beamforming are also worth some explanation as they are often used in the context of MIMO systems. Beamsteering, as its name implies, is associated with the physical steering of the composite radio signal by use of coherent phase-shifting of a signal sent to multiple antennas in some form of spatial array. Beamsteering has been variously used in other radio systems, and because the principle is rooted in the underlying physics of the design of the antenna arrays, beamsteering is often considered a proprietary implementation choice rather than being explicitly defined within a radio standard. Beamforming on the other hand is a more complex technique that is used to enhance single layer and spatial multiplexing performance. In the simplest form of spatial multiplexing, the individual layers are directly mapped to individual antennas. This is sometimes referred to as open-loop or non-precoded spatial multiplexing. With the addition of beamforming, the transmitted signals are further modified in amplitude and phase and additionally cross-coupled between the antennas in order to adapt to the channel propagation conditions as seen by the receiver. In this sense it is not the physical direction of the signal that is being controlled as with beamsteering, but an attempt to match the Eigen modes of the channel so that the transmitted layers appear decorrelated at the receiver antennas. In summary, beamsteering can be considered a subset of the more general beamforming case. In this book the term beamsteering will be used when precode weightings are limited to phase-shifting only, and beamforming will be used when precode weightings can include both magnitude and phase.

The following overview of multi-antenna techniques gives more details of how spatial multiplexing works. The distinction between diversity and spatial multiplexing is also explained. Next comes a description of the signals and the hardware configurations for multi-antenna downlink operation, including the diversity techniques used at the eNB and UE and in single-user MIMO (SU-MIMO). The section continues with the uplink and explains how multi-user MIMO (MU-MIMO) operates. These sections include basic information about the expected physical layer performance of the links. Terms such as antenna port, layer, precoding and codeword are introduced. Having described the system operation and how MIMO in particular works, the section concludes with a description of the main features of open- and closed-loop operation and how diversity, beamforming, and MIMO can be combined.

2.4.1 Overview of Multi-Antenna Techniques

As shown in Figure 2.4-1, there are four ways to make use of the radio channel. For simplicity only those using one or two antennas are shown although LTE now supports up to eight layers in the downlink and four in the uplink. The mode of operation changes once there is more than one antenna, and in theory, any number could be used.

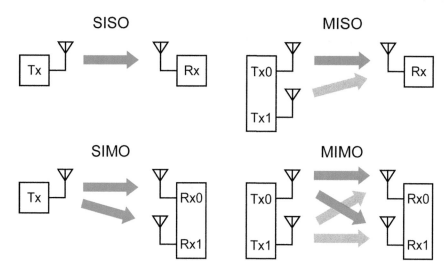

Figure 2.4-1. Radio channel access modes

The terms that describe the access modes, such as MISO and MIMO, use the labels input and output to refer to the channel, not the transmitters or receivers. The overall channel includes the transmission medium (the air), the transmit and receive antennas, and the cabling and analog circuits connected to the antennas. Thus antennas are vital components in the link. With multi-antenna operation, the physical relationship between the antennas becomes a new variable to deal with, affecting the relationship between the paths that the signals take. The relationship between the paths is referred to as correlation. The significance of the inclusion of analog circuits will become apparent when some of the LTE design challenges and measurements are covered in Chapter 6.

Employing a single antenna at the transmitter and a single antenna at the receiver, single input single output (SISO) is the most basic radio channel access mode. It is the default configuration referred to elsewhere in this book, and it gives the baseline for assessing the performance improvements possible when more antennas are used.

SIMO describes receive diversity, a method that isn't generally dependent on the technology being used. SIMO is suited to low signal-to-noise ratio (SNR) conditions; for example, due to cell edge operation or fading. For a specified modulation and coding scheme (MCS) there is no improvement in peak data rates over SISO, but SIMO is more robust in difficult radio conditions since SIMO operates at a lower SNR than SISO. Since receive diversity is a baseline UE requirement for LTE, SIMO becomes the baseline radio access mode for the downlink. Diversity reception is not mandated for the eNB but is almost always employed for performance reasons.

The colored arrows in the MISO and MIMO cases indicate the use of different user data for each transmitter. MISO is a transmit diversity technique that only requires a single receive antenna. It has been used for some time in cellular systems with Alamouti space time block coding (STBC), where it can offer significant gains in signal robustness under fading channel conditions but, like SIMO, does not increase peak data rates for a given MCS.

The use of STBC involves the duplication of data onto multiple antennas. The signals for additional antennas are distinguished by a combination of reversing the time allocation and applying a complex conjugation to part of the signal. In LTE, MISO is implemented using space frequency block coding (SFBC). This uses the Alamouti principle but copies data onto different frequencies (resource elements) instead of using blocks of time.

More than one receive antenna can be used with MISO, but it is important to note that the simultaneous use of transmit and receive diversity, MISO plus SIMO, although qualifying as a MIMO antenna implementation, does not imply that spatial multiplexing is being employed. There may be two transmitters and two receivers involved but still only one stream of data.

When the channel propagation conditions are favorable, spatial multiplexing can increase spectral efficiency since multiple data streams, each with unique data, can be transmitted simultaneously in the same time and frequency. This potential for improved spectral efficiency is statistical and opportunistic since it is only possible under instantaneous decorrelated channel conditions with sufficient SNR. It is also possible to send a single data stream across a spatially multiplexed channel to achieve another form of diversity transmission (see Section 2.4.4.6).

2.4.2 Spatial Multiplexing (MIMO) Operation

The basics of spatial multiplexing operation can be understood by using a static, four port network to represent the channel, as shown in Figure 2.4-2. In this figure, two signals containing different user data are simultaneously transmitted and received. At this point, there is no need to specify the intended destination of the data. It could be intended for one user or several users.

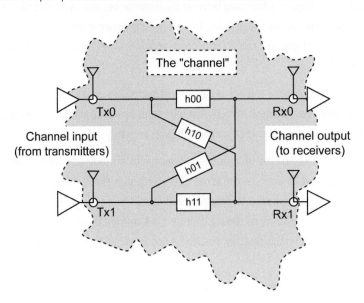

Figure 2.4-2. Basic 2 x 2 spatial multiplexing channel configuration

In the ideal case, to use the same frequency and time simultaneously, isolated connections would be established from transmitter 0 to receiver 0 and transmitter 1 to receiver 1. In practice, although antenna polarization may partially achieve this isolation, there will be coupling between the signals as soon as they are transmitted. The challenge, therefore, is to reverse the coupling in the radio channel after the signals have been received. As with other radio systems such as IEEE 802.11 and 802.16, LTE uses a "non-blind" technique. Predefined orthogonal training signals are transmitted from each antenna. The receiver knows which training signal was used for each antenna and therefore can calculate the channel amplitude and phase responses, h00, h10 and h11, h01. Note that a convention of the channel matrix definition is to specify the receiver first; i.e., hR,T. In this way the receiver can calculate the transformation that the signals from each antenna have undergone. Since the unknown data is sent at or around the same time as the known training signals, the receiver can assume that the part of the signal containing the unknown user data from each antenna has undergone the same transformation as the known part of the signal. In essence, spatial multiplexing is using a "trick": known training signals are mixed with the randomly varying data in such a way that the unknown data can be recovered.

Conceptually, the simplest way to recover the unknown data is to multiply the received signals by the inverse of the channel matrix. In practice this zero-forcing technique is vulnerable to noise, and more sophisticated techniques can be used that involve the minimizing of errors during the recovery process.

A key point to note about MIMO is that there must be at least as many receiving antennas as there are transmitted data streams. However, this number of streams should not be confused with the number of transmitting antennas, which may be higher than the number of streams if transmit diversity is mixed with spatial multiplexing. The minimum number of receivers is determined by what is mathematically required for the calculation of the channel matrix H. If there are fewer receivers than transmitted data streams, it is mathematically impossible to resolve the channel matrix meaning the additional transmit data streams contribute interference to the other streams rather than additional information.

This simplified description of 2 x 2 spatial multiplexing operation intentionally does not consider the source of the data. There is a lot of flexibility in how the two data streams can be used. In LTE, the source data for each stream can have different modulation and coding and does not need be associated with a single user.

It is necessary to consider how to design the training signals to suit the characteristics of the radio channel. In IEEE 802.11n, training signals take the form of a preamble. For more rapidly changing channels, and to suit the frame structure of the signal, LTE interleaves the known signal, called the reference signal (RS), throughout the frame in both frequency and time. The RS definition is different for the downlink and uplink.

Figure 2.4-3 shows how the individual symbols of the reference signal are allocated to subcarriers for a two-antenna downlink signal. Note how the RS symbols are orthogonal on each antenna port in both frequency and time. To see the full range of downlink RS allocations for single, dual, and quad antenna configurations, see 36.211 [9] Subclause 6.10.

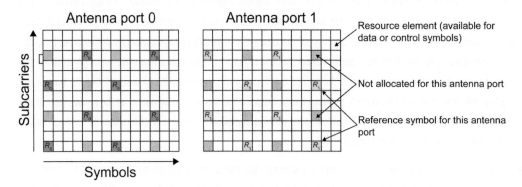

Figure 2.4-3. Orthogonal structure of downlink reference symbols for dual antenna
(adapted from 36.101 [1] Figure 6.10.1.2-1)

In Release 8, the reference signals are either cell-specific or (optionally) UE-specific. Cell-specific RS (CRS) are common to all the UE in the cell. The UE-specific reference signals, also known as demodulation reference signals (DMRS), are precoded in the same way as the associated PDSCH, to suit the prevailing channel condition. Because the user data is precoded in the same way as the UE-specific RS, the UE can directly decode the data using the UE-specific RS without the eNB having to communicate what precoding may have been applied. Release 9 extends the use of UE-specific RS to dual layer transmissions, and Release 10 extends this concept further to support up to eight layer transmissions. The advantages of UE-specific RS in terms of optimizing the downlink signal by enabling non-codebook-based beamforming are offset by the need to allocate more downlink resources to carry the UE-specific RS.

The RS allocation for the LTE uplink is very different from the downlink allocation. For data transmission the RS occupies all subcarriers for one symbol of each timeslot, excluding those situations in which control channel and shared channel are transmitted simultaneously. This is explained in more detail in Section 3.2.8. As indicated in Figure 2.4-4, uplink SU-MIMO shares many processing steps with downlink MIMO. However, there are some clear differences such as the transform precoder block, which is unique to the uplink, and the way in which the antennas are identified. In the downlink, orthogonal time and frequency allocations for the RS are used to identify each antenna, whereas uplink SU MIMO relies on different Zadoff-Chu cyclic shift phase sequences contained within the sounding reference signal (SRS) orthogonal transmissions from each physical antenna. Similarly, different Zadoff-Chu cyclic shift phase sequences plus additional orthogonal coding are applied to the DMRS transmissions associated with each spatial layer. This use of different codes in the same frequency and time is similar to the approach used for 802.11n.

As with any radio signal, signal recovery depends on the signal to noise ratio. The Shannon-Hartley capacity theorem predicts the error-free capacity C of a radio channel as

$$C = B \left[\log_2 (1 + SNR) \right]$$

where

C = channel capacity in bits per second,

B = occupied bandwidth in Hz, and

SNR = The linear signal-to-noise ratio.

The performance of a spatial multiplexing system introduces additional simultaneous paths plus a further dependency, which is the cross-coupling of interfering signals between the different paths from each transmitting antenna to each receiving antenna through the radio channel. The long-form version of the channel capacity theorem can be written as

$$C = B \left[\log_2 (1+(\sigma/N) \, \rho_1{}^2) + \log_2 (1+(\sigma/N) \, \rho_2{}^2) \right]$$

where

σ/N = signal to noise ratio and ρ = a singular value of the channel matrix, H.

It is useful to highlight the potential asymmetry in performance between the streams in a spatially multiplexed link. In the ideal but impractical case of no cross-coupling, the values of ρ_i will be 1, 1 indicating a doubling of channel capacity. However, in the case of total in-phase coupling, the values of ρ_i will be 2, 0 indicating that the capacity has dropped back to that of a SISO channel. Note that in either case, for a fair comparison, the equivalent SISO transmitter power is shared between each transmitted stream.

The potential increase in instantaneous system capacity can be derived from the ratio of singular values of H, also known as the condition number. The condition number can also be used to indicate the increase in SNR needed to recover the spatially multiplexed signal, relative to the SISO case.

From the above the following can be concluded:
- For the 2 x 2 case the increase in channel capacity will not exceed twice the SISO case, and achieving this may require a substantial improvement in SNR at the receivers if the values of ρ_i are much less than 1.
- If the matrix coefficients are known by the transmitters, the asymmetry in stream performance can allow a higher order modulation format on the stronger stream, or the outgoing signals can be modified (precoded) to equalize the performance between the streams. Precoding requires real-time feedback from receiver to transmitter, so this is also known as closed-loop spatial multiplexing. For effective precoding, the relative signal phase between transmitters does not have to be known but it must be stable over the time interval of the feedback process.

The long form capacity equation shows the situation for a snapshot in time of the channel. In practice, the highly variable nature of the channel and the impact of the antenna configurations need to be included. More details on this and the use of channel correlation factors are provided in Section 6.8.

The use of MIMO in non-OFDM systems such as CDMA is possible, as evidenced by its use in UMTS from Release 7 for HSDPA, although the processing to recover the same quality of channel information is more difficult. OFDM is particularly well-suited to MIMO operation because the channel is defined by a single vector coefficient for each subcarrier, which makes the required digital processing in the frequency domain much more straightforward than in systems such as CDMA that are defined in the time domain.

2.4.3 LTE Terminology for Multiple Antennas

The terms codeword, layer, and precoding have been adopted specifically for LTE to refer to signals and their processing. Figure 2.4-4 shows the processing steps to which they refer. The terms are used in the following ways:

- Codeword: A codeword represents a channel-encoded and rate-matched user data transport block, protected by a HARQ process before it is formatted for transmission. One or two codewords, CW0 and CW1, can be used depending on the prevailing channel conditions and transmission mode (TM). In the most common case of downlink SU-MIMO, two codewords are sent to a single UE, but in the case of downlink MU-MIMO, each codeword is intended for one UE only.

- Layer: The term layer is synonymous with stream. The number of layers is denoted by the symbol ν (pronounced nu). The number of layers is always less than or equal to the number of transmit antennas. For Release 8 downlink spatial multiplexing, at least two layers must be supported with fallback to single layer when channel conditions do not favor multiple layers. Up to four layers are supported in Release 8, which is increased to eight layers for Release 10. For Release 8 uplink spatial multiplexing only one layer per UE is allowed, but this can enable two-layer MU-MIMO requiring two UEs. Release 10 uplink spatial multiplexing introduces SU-MIMO for two or four layers with fallback to single layer as required.

- Precoding: Precoding modifies the layer signals before transmission. This may be done for diversity, beamforming, or spatial multiplexing. As noted earlier, the MIMO channel conditions may favor one layer (data stream) over another. If the eNB is given information about the channel—e.g., information sent back from the UE—it can add complex cross-coupling to counteract the imbalance in the channel.

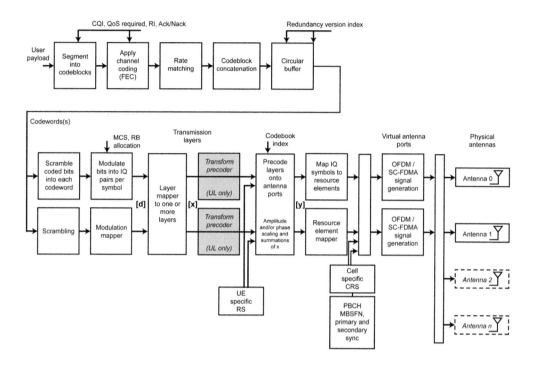

Figure 2.4-4. Signal processing for transmit diversity and spatial multiplexing (MIMO)
(Adapted from 36.211 [6] Figure 6.3-1)

The symbols d, x, and y are used in the specifications to denote signals before and after layer mapping and after precoding, respectively. Transform precoding applies only to the uplink. Reference signal insertion shown applies to the downlink.

2.4.4 LTE Downlink Transmission Modes

LTE specifies the way the downlink radio channel will be used by defining a set of transmission modes in 36.213 [10] Section 7.1. There are nine different modes, all of which have different benefits in different use cases depending on the radio environment. These modes and their use cases are summarized in Table 2.4-1.

Table 2.4-1. LTE Downlink transmission modes and use cases

Transmission mode	Description	Primary benefit	Typical use cases				
			SINR at UE Rx	Multipath/ scattering	Correlation (channel + TX and Rx antennas)	UE speed	Cell size
1	Basic SIMO	Basic single transmit antenna operation	Low–med	Low	High	High	Med–large
2	Transmit diversity	Improved signal robustness in low power/SINR conditions	Low	Low (cell-edge, rural)	High (cell-edge, rural)	High	Large
3	Open-loop SU-MIMO	Potential for increased throughput in good conditions	High	High (urban)	Low (urban)	High	Small
4	Closed-loop SU-MIMO	Potential for increased throughput in good conditions	High	High (urban)	Low (urban)	Low	Small
5	MU-MIMO	Improved cell spectral efficiency	High	Med–high	Low–med	Low	Small–med
6	Closed-loop Rank 1 beamsteering	Improved signal robustness	Low	Low (cell-edge, rural)	High (cell-edge, rural)	Low	Large
7	UE-specific RS beamforming	Improved signal robustness with non-codebook precoding	Low	Low (cell-edge, rural)	High (cell-edge, rural)	High	Large
8	Dual layer UE-specific RS beamforming	As 7 with potential for increased throughput or increased cell capacity	Med–high	Med–high	Low–med	Low–med	Small–med
9	Up to eight layer UE-specific RS beamforming	As 8 with potential for up to increased throughput or increased cell capacity	High	High	Low	Low (static)	Small

The Release 8 specifications describe the first seven transmission modes, Release 9 adds mode 8, and Release 10 adds mode 9. Transmission modes give two choices for the way data is manipulated before it is delivered to the physical antennas. When using more than one transmit antenna, one of these choices always allows a fallback to single rank transmit diversity (MISO).

An antenna "port" does not always mean the same thing as a physical antenna. The term antenna port refers to the use of a particular set of reference signals, which are multiplexed onto physical antennas to suit the transmission mode. An example is port 5, used in TM7.

The same coded signal is adjusted in phase and fed to multiple physical antennas to create a beamformed signal. This concentrates the transmitted power towards a specific direction or location, while the UE needs only to estimate the channel from one group of reference signals.

The way in which the different transmission modes are applied to the various downlink signals depends on what use the transmission mode makes of transmit diversity, spatial multiplexing, or cyclic delay diversity (CDD). These three techniques are applied to the physical signal or physical channel, according to Table 2.4-2. The signal and channel definitions are discussed fully in Section 3.2. Some details, such as the reference signal not being subject to any addition processes, give important benefits for measurements and will be discussed further in Section 6.8.

Table 2.4-2. Summary of diversity and spatial multiplexing techniques applied to LTE downlink signals

Physical signal or physical channel	Transmit diversity (SFBC)	Spatial multiplexing	CDD
Cell specific reference signal	No	No	No
UE-specific reference signal	No	Yes	No
Primary synchronization signal	No	No	No
Secondary synchronization signal	No	No	No
Physical broadcast channel	Yes	No	No
Physical downlink control channel	Yes	No	No
Physical hybrid ARQ indicator channel	Yes*	No	No
Physical control format indicator channel	Yes	No	No
Physical multicast channel	No**	No	No
Physical downlink shared channel	Yes	Yes	Yes

*Precoding type depends on PHICH group number.
**The PMCH does not use TX diversity in the sense of SFBC from one location but does use a form of Tx diversity due to multi-point transmission from adjacent cell sites.

The broadest range of transmission schemes apply to the PDSCH transmissions used during user data transfer to the UE. The master information block (MIB) carried on the downlink indicates the number of cell-specific RS transmit antennas being used by the eNB, and the UE is semi-statically configured according to the transmission mode chosen by eNB. The transmission mode is communicated to the UE in the radio resource control (RRC) connection setup message as part of the downlink antenna configuration information. Associated with the use of each transmission mode are two downlink control information (DCI) formats. These allow the eNB to quickly adapt the downlink transmission per subframe within the configured transmission mode to ensure that the UE can optimally interpret and recover downlink signals correctly without having to switch between modes. Table 2.4-3 provides a more detailed summary of the features of the transmission modes configured by the cell radio network temporary identity (C-RNTI) parameters. DCI format 1A is typically used as the fallback scheme to single layer transmit diversity when the channel is not able to support the more advanced capability of the transmission mode.

Table 2.4-3. LTE Downlink transmission modes (adapted from 36.213 [10] Table 7.1-5)

Transmission mode	DCI format	Transmission scheme of PDSCH corresponding to PDCCH	Channel use
Mode 1	DCI format 1A	Single-antenna port, port 0	SIMO
	DCI format 1	Single-antenna port, port 0	SIMO
Mode 2	DCI format 1A	Transmit diversity	MISO
	DCI format 1	Transmit diversity	MISO
Mode 3	DCI format 1A	Transmit diversity	MISO
	DCI format 2A	Large delay CDD or transmit diversity	MIMO
Mode 4	DCI format 1A	Transmit diversity	MISO
	DCI format 2	Closed-loop spatial multiplexing or transmit diversity	MIMO
Mode 5	DCI format 1A	Transmit diversity	MISO
	DCI format 1D	Multi-user MIMO	MU-MIMO
Mode 6	DCI format 1A	Transmit diversity	MISO
	DCI format 1B	Closed-loop spatial multiplexing using a single transmission layer	MISO
Mode 7	DCI format 1A	If the number of PBCH antenna ports is one, single-antenna port, port 0, is used; otherwise transmit diversity	SIMO or MISO
	DCI format 1	Single-antenna port, port 5 (UE-specific RS)	MISO
Mode 8	DCI format 1A	If the number of PBCH antenna ports is one, single-antenna port, port 0, is used; otherwise transmit diversity	SIMO or MISO
	DCI format 2B	Dual layer transmission, port 7 and 8, or single-antenna port, port 7 or 8	MU-MIMO or SU-MIMO
Mode 9	DCI format 1A	Non-MBSFN subframe: If the number of PBCH antenna ports is one, single-antenna port, port 0, is used; otherwise transmit diversity. MBSFN subframe: single-antenna port, port 7	SIMO or MISO
	DCI format 2C	Up to 8 layer transmission, ports 7–14. Up to 8 layer transmission, ports 7–14	MU-MIMO or SU-MIMO

The following sections describe the properties of the different modes in more detail.

2.4.4.1 Transmission Mode 1: UE Diversity Reception

UE diversity reception refers to the SIMO mode and is mandatory for the UE. It is typically implemented using maximum ratio combining (MRC).

In a cellular environment, the signal from a single receive antenna will suffer level fluctuations due to various types of fading. A two antenna example is shown in Figure 2.4-5. With the broadband nature of the wider LTE channel bandwidths; there may also be a noticeable frequency dependency on the signal level. By combining the signal received from both antennas, the UE can recover a more robust signal. Receive diversity provides up to 3 dB of gain in low SNR conditions. Note that the use of a lower performance, secondary antenna may be of value for diversity reception but is likely to cause problems in a MIMO receiver, as MIMO requires matched receivers for best performance.

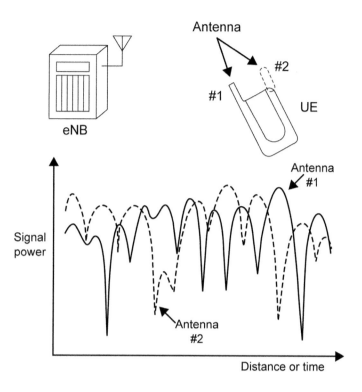

Figure 2.4-5 Example of diversity reception through loosely correlated paths

2.4.4.2 Transmission Mode 2: SFBC Diversity Transmission

Transmission mode 2 uses a transmit diversity technique, SFBC, at the eNB. This contrasts with the space time block coding method used in 802.16, which takes pairs of OFDM symbols and transmits them in reverse time order on the antennas. In LTE, a single codeword is mapped onto two or four layers, which directly relates to the number of transmitters available. Data is interleaved onto different subcarriers on each antenna, according to the expressions in Table 2.4-4.

In the table, the letter d denotes the input modulation symbol (codeword) and x denotes the modulation symbol mapped onto the subcarrier of a layer. Thus even numbered modulation symbols are mapped to even layers, and odd symbols to odd layers.

Table 2.4-4. Codeword to layer mapping for transmit diversity (Ref 36.211 [6] Table 6.3.3.3-1)

Number of layers	Number of codewords	Codeword-to-layer mapping $i = 0,1..., M_{\text{symb}}^{\text{layer}} - 1$	
2	1	$x^{(0)}(i) = d^{(0)}(2i)$ $x^{(1)}(i) = d^{(0)}(2i+1)$	$M_{\text{symb}}^{\text{layer}} = M_{\text{symb}}^{(0)}/2$
4	1	$x^{(0)}(i) = d^{(0)}(4i)$ $x^{(1)}(i) = d^{(0)}(4i+1)$ $x^{(2)}(i) = d^{(0)}(4i+2)$ $x^{(3)}(i) = d^{(0)}(4i+3)$	$M_{\text{symb}}^{\text{layer}} = M_{\text{symb}}^{(0)}/4$

This open-loop diversity technique is identical in concept to that used for UMTS. However, for reasons of simplicity the more complex closed-loop transmit diversity techniques from UMTS have not been defined for LTE, partly because LTE has closed-loop methods defined for MIMO, which are considered more important. Two or four transmitter diversity is supported. Figure 2.4-6 shows the processing steps for the four transmitter case.

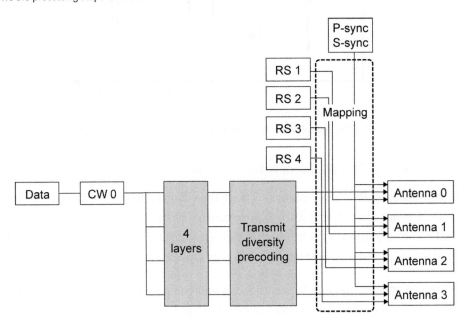

Figure 2.4-6. Configuration used for four transmitter diversity

Transmit diversity precoding is automatically applied for all control channels regardless of the physical downlink shared channel precoding mode that is used. A modified MISO scheme is used for the multimedia broadcast over single frequency network (MBSFN) scheme in which a network of many eNBs transmits a common signal to improve cell edge performance for broadcast services.

2.4.4.3 Transmission Mode 3: MIMO With Cyclic Delay Diversity

Transmission mode 3 employs a technique known as cyclic delay, which introduces a time delay between multi-antenna signals. This technique is used in several other radio systems, including 802.11n and 802.16. Cyclic delay is used to reduce the effects of possible unwanted signal cancellation that can occur if the same signal is transmitted from multiple antennas when the channel is relatively flat in the frequency domain. The addition of a delay—typically on the order of a few microseconds— to one of the transmit paths introduces a frequency dependent phase shift as shown in Figure 2.4-7. When the signals from two transmitters combine at the receiver, peaks and nulls are the result, depending on the exact phase of the paths for a given frequency (subcarrier). Operationally, there are tradeoffs in choosing the length of delay, with no single value suiting all situations and channel bandwidths.

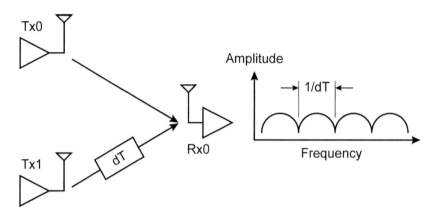

Figure 2.4-7. Impact of adding time delay to one path

Transmission mode 3 uses a type of cyclic delay known as large delay CDD. It is larger than the cyclic delay used by 802.11n or 802.16, and it is used in a different way. (The concept of small delay CDD was removed from the LTE specifications in March 2008.) The intent of large CDD is to evenly distribute the probability of symbol decoding errors across each of the transmitted spatial multiplexing layers. This open-loop transmission technique is used when channel conditions are changing faster than the UE can report through the UL control precoding matrix indicator (PMI) feedback process. This technique ensures optimal decoding error correction performance observed at the UE receiver device as a result of the channel encoding protection applied by the eNB. The reference signal subcarriers do not have CDD applied, which allows the UE to report the actual channel conditions to the scheduler in the eNB, which in turn uses that information to determine when best to switch between open-loop (TM3) and closed-loop (TM4) transmission modes for that specific UE. As shown in Table 2.4-5, the delay is expressed as a phase shift for adjacent subcarriers.

Table 2.4-5. Cyclic phase shifts for two, three, and four eNB antennas

Number of transmitters	Phase shift per subcarrier (overall delay)
2	$180°\ (t_{symbol}/2)$
3	$120°\ (t_{symbol}/3)$
4	$90°\ (t_{symbol}/4)$

Expressed in terms of time, the two antenna case CDD is 33.33 μs, which is half the symbol length. In order to accommodate such long delays without affecting the delay spread of the channel, the CDD is applied prior to the addition of the CP. When the UE reports that the channel is capable of supporting more than one spatial layer, but with dynamically changing channel matrix conditions, the eNB precodes the downlink using CDD and transmits one layer per transmit antenna. TM3 is typically used when the rate of change of the channel is too high to use PMI feedback to control the precoding according to the instantaneous channel conditions. On average, the frequency selectivity created in the channel by using large CDD provides sufficient diversity at the receiver for spatial multiplexing to provide data throughput gains in high SINR conditions. If the radio conditions do not look favorable, TM3 falls back to rank 1 SFBC transmit diversity.

2.4.4.4 Transmission Mode 4: Single-User MIMO

Figure 2.4-8 shows how both codewords are used for a single user to provide downlink SU-MIMO, TM4. It is also possible for the codewords to be allocated to different users to create downlink MU-MIMO, which is described in the next section.

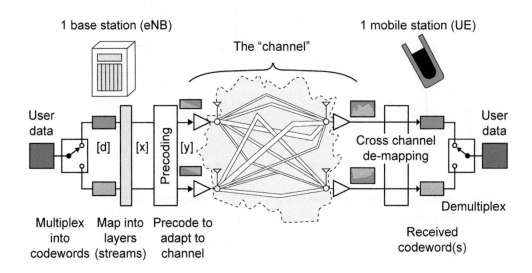

Figure 2.4-8. SU-MIMO in the downlink with two antennas; codebook 0 shown

Depending on the channel information available at the eNB, the modulation and the precoding of the layers may be different to equalize the performance.

The precoding choices are defined in the codebook, which is essentially a lookup table. The codebook is used to quantize the available options and thus limit the amount of information fed back from the receiver to the transmitter. Some of the precoding choices are straightforward; for example, codebook index (CI) 0 for TM3 is a direct mapping of codewords to layers and CI 1 applies what is referred to in 802.11 as spatial expansion. Table 2.4-6 shows the codebook choices for one and two layers. Precoding with one layer is limited to beamsteering using a 0 degree, ±90 degree, or 180 degree phase shift. The two-layer case uses the more advanced technique of beamforming with cross-coupling between the layers.

In operation, the UE reports the rank and PMI to the eNB scheduler with the rank and codebook index most closely matching the channel. The system can be configured for multiple codebook values, one for each resource block group. To use this information while it is still valid, the eNB scheduler has to respond rapidly, within milliseconds, depending on the rate of change of the channel. If the UE is instructed to provide channel information more regularly, the information will be more accurate but the proportion of resources used for signaling will increase and place higher demands on the eNB.

Table 2.4-6. Codebook for transmission on antenna ports 0,1 (36.211 [6] Table 6.3.4.2.3-1)

Codebook index	Number of layers v	
	1	2
0	$\frac{1}{\sqrt{2}}\begin{bmatrix}1\\1\end{bmatrix}$	$\frac{1}{\sqrt{2}}\begin{bmatrix}1 & 0\\0 & 1\end{bmatrix}$ *
1	$\frac{1}{\sqrt{2}}\begin{bmatrix}1\\-1\end{bmatrix}$	$\frac{1}{2}\begin{bmatrix}1 & 1\\1 & -1\end{bmatrix}$
2	$\frac{1}{\sqrt{2}}\begin{bmatrix}1\\j\end{bmatrix}$	$\frac{1}{2}\begin{bmatrix}1 & 1\\j & -j\end{bmatrix}$
3	$\frac{1}{\sqrt{2}}\begin{bmatrix}1\\-j\end{bmatrix}$	-

*Codebook 0 for two layers is used only by TM3 which uses large CDD

2.4.4.5 Transmission Mode 5: Multi-User MIMO

The primary difference between SU-MIMO and MU-MIMO is in the distribution of the codewords between UEs. For SU-MIMO the codewords are all intended for a single UE and have the effect of multiplying the potential peak data rates by the number of codewords, which may be 2, 4, or 8. The eNB transmitter and the UE receiver must have at least as many transmit antennas as codewords in order that the codewords can be decoded. For MU-MIMO, there can be as many codewords as eNB transmit antennas but the destination of the codewords is toward different UE with each UE attempting to decode only the codeword intended for it. This means the peak data rate per UE is not increased over basic SISO but the capacity gain of the cell has the potential to multiply by the number of codewords. An additional benefit is that unlike SU-MIMO in which the number of UE receive antennas must be at least as great as the number of codewords, MU-MIMO can operate with single receiver UEs.

The challenge for MU-MIMO is how to precode the downlink codewords in order that they are received by each UE without interference from the other codewords. To do this requires accurate channel state information at the eNB transmitter (CSIT). In SU-MIMO this CSIT is provided by the individual UE but for MU-MIMO no such feedback is possible. However, the physical separation between the UEs selected for MU-MIMO means that the channels to each UE are likely to be decorrelated already due to the physical separation. This makes the need for precise beamforming at the eNB less important. The challenge for the eNB is to pick pairs (or larger groups) of UEs that are decorrelated. The CSI feedback provided by a UE configured for MU-MIMO is the same as that defined for SU-MIMO with rank = 1; i.e., the UE selects the desired CQI and PMI based on a single codeword. The eNB can then select pairs of UEs that report orthogonal PMI values. The only indication the UE gets that another UE is simultaneously scheduled is the presence of a bit in the DCI format 1D, which indicates that the downlink power has been offset by 3 dB relative to the RS to account for the sharing of power. Otherwise MU-MIMO in Release 8 is transparent to the UE.

TM5 can be described as single layer beamforming with one layer per beam. In Release 9, dual layer beamforming using UE-specific RS is introduced in TM8. TM8 extends MU-MIMO to up to four UEs and has the added advantage that it is totally transparent since the UE-specific nature of the RS mean that no power PDSCH offset needs to be signaled. TM8 is described more fully in Section 2.4.4.8.

2.4.4.6 Transmission Mode 6: Closed-Loop Rank 1 Spatial Multiplexing

Transmission mode 6 is a beamforming case of transmit diversity implemented by transmitting the same signal into two antennas with phase changes applied during precoding. The PMI feedback from the UE is used to select between two different transmit phases, which creates a simple binary beamsteering effect. The end result is a special case of beamforming in which the per antenna beam weightings are limited to a fixed set of codebook-based relative phase shifts only, with constant magnitudes. The fallback mode for TM6 used when the channel conditions are changing too fast for PMI to operate is the traditional SFBC transmit diversity scheme.

2.4.4.7 Transmission Mode 7: UE-specific RS for Single Layer (Antenna Port 5)

The PDSCH coding that has been described thus far for the precoded transmission modes TM3 to TM6 has been, for reasons of limiting control signaling bandwidth, quantized by the use of codebooks. This is efficient but limits the potential of what might be achieved by utilizing arbitrary coding schemes capable of taking full advantage of the radio propagation conditions to maximize received signal quality. Another limiting factor of the TM3 to TM6 modes is that although the UE benefits from PDSCH coding, the CRS remain uncoded since they are provided as a reference for all users in the cell. This use of CRS is efficient in terms of required downlink resources but ultimately limits the potential gains from beamforming.

Transmission mode 7 introduces the concept of UE-specific RS. These are also known as demodulation RS or DMRS since their primary purpose is to aid the process of demodulation in the UE, whereas CRS are additionally used by the UE for channel state reporting. UE-specific RS are covered in Section 3.2.12.1. Since the UE-specific RS are transmitted only for the benefit of one UE, they can be precoded to optimize the signal quality at the receiver. The optimal precoding is the same as that used for the PDSCH, which by its nature is always UE-specific. By using the same precoding for both the RS and the PDSCH, the UE can directly decode the PDSCH using the UE-specific RS as the reference without needing prior knowledge of the precoding used. This freedom from signaling the precoding means that although the uplink CSI is still quantized into codebooks to conserve signaling resources, the downlink is freed from the restrictions of codebooks and can implement arbitrary precoding with complete transparency toward the UE. This additional degree of freedom optimizes the downlink coding and releases downlink signaling bandwidth, but it has the downside of requiring additional downlink resources to transmit the UE-specific RS. However, for cells without large numbers of UE, the use of UE-specific RS provides an overall benefit.

The subject of beamforming is covered in more depth in Section 6.9 with TM7 and TM8 covered in some detail in Section 6.9.3.

2.4.4.8 Transmission Mode 8: UE-Specific RS for Dual Layers (Antenna Ports 7 and 8)

In Release 9 the UE-specific RS concept was extended to support two spatial layers using TM8. These are assigned to virtual antenna ports 7 and 8. As described in Section 2.4.4.5, the dual layer capability of TM8 also extends the potential for downlink MU-MIMO from two to four UE. It is also possible to mix and match combinations—for example, one antenna beam might support two different UEs in MU-MIMO mode while the other beam supports a single UE with two layers. Section 6.9.3 provides a much more detailed description of TM8 with measurement examples.

2.4.4.9 Transmission Mode 9: UE-Specific RS for Eight Layers (Antenna Ports 9 to 14)

Prior to the introduction of TM9 the LTE downlink was limited to four layers. A number of enhancements were introduced in Release-10 to accommodate spatial multiplexing up to eight layers. This included TM9 and a number of elaborations to the RS structure including the introduction of channel state information RS (CSI-RS) covered in Section 3.2.12.4. As with TM8, the use of TM9 is flexible in that it can support different combinations of SU-MIMO and MU-MIMO up to a maximum of eight layers. This is important since although the implementation of eight layers at the eNB is within reach, the development of UE with eight receive antennas is much more problematic, unless the UE is not constrained is size. By allowing the available layers to be used for MU-MIMO, the potential within the downlink coding to support eight layers can be taken advantage of by UE that do not themselves have to support all eight layers at once.

It is significant that the support of downlink eight layer transmission has been defined only for the UE-specific RS and not for the CRS. The reason is that the overhead to support eight layer CRS continuously on the downlink was considered too high. Instead, the UE-specific RS required for eight layer transmission are scheduled only when required. This approach provides the necessary environment to support a dedicated link; however, to report the eight-layer channel conditions, the UE must be provided with CSI-RS for the purpose of CSI reporting when connected and in the idle state. Unlike the CRS which are continuously transmitted, these CSI-RS are scheduled only as required so as not to consume too many downlink resources.

Backward compatibility with earlier LTE releases is an essential principle of the continued development of LTE and for that reason the two-layer design of the Release-10 UE-specific RS is the same as that of Release 9. However, for the support of higher layers, it was considered impractical to continue to extend the frequency and time division orthogonal approach and so for the four and eight layer UE-specific RS, code domain orthogonality was introduced.

2.4.5 Multiple Antenna Operation in the Uplink

There are three types of multiple antenna operation defined for the uplink:
- Receive diversity at the eNB
- SU-MIMO for single UE
- MU-MIMO for multiple UEs.

Receive diversity at the eNB is nothing new and will not be discussed further.

2.4.5.1 Single-User MIMO in the Uplink

Uplink SU-MIMO is defined in Release 10. To implement SU-MIMO the UE requires two or four transmitters. For a handheld device, this is a significant development challenge in terms of cost, size, and battery consumption, particularly in light of the additional challenge of carrier aggregation. Also, the increased data rates in the uplink that might be possible from SU-MIMO are not as important as they are in the downlink due to asymmetrical traffic distribution. Furthermore, if the system is deployed to be uplink-performance-limited, it may be impractical to increase the transmit power from the UE sufficiently to achieve the SINR needed at the eNB receivers, although techniques such as envelope tracking may make higher powers more practical.

2.4.5.2 Multi-User MIMO in the Uplink

Although a UE typically has a single transmitter in its baseline configuration, it nevertheless is still capable of supporting MU-MIMO in a way similar to that described in Section 2.4.4.5 for the downlink. As suggested by Figure 2.4-2 and the description of MIMO, spatial multiplexing—unlike the receive function—does not require the transmitters to be in the same physical device or location. Thus uplink MIMO can be implemented using two transmitters belonging to two different UEs. This creates the potential for an increase in uplink capacity, although an individual user will see no increase in data rates. See Figure 2.4-9.

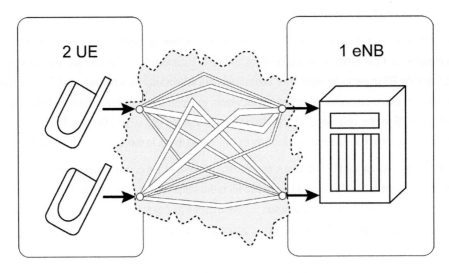

Figure 2.4-9. Multi-user MIMO in the uplink

The fact that the transmitters are physically separate has two consequences. First, there is no possibility of precoding since the source data cannot be shared between the two UEs to create the necessary cross-coupling of the data streams. This reduces the potential gains that co-located transmitters may have had. Second, the separation of the transmitters increases the probability that the radio channels seen by the eNB will be uncorrelated. Indeed, when the eNB has to select two UEs for pairing with MU-MIMO, the primary criterion will be the presence of de-correlated channels. Any potential gains lost through lack of precoding will be more than compensated for by the gains likely from better channel de-correlation. MU-MIMO therefore is a valuable technique for improving uplink capacity.

OFDM signal recovery is tolerant of small timing and frequency errors. Normal uplink operation will result in each UE adjusting its frequency quite precisely to that of the eNB. The eNB will also instruct the UE to adjust its timing and power so that all signals arrive at the eNB receiver at approximately the same level and time. With the antennas located in different devices, the transmit paths are assumed to be uncorrelated. These conditions give the eNB scheduler the opportunity to control two UEs to transmit data simultaneously using the same subcarriers.

Multi-user MIMO involves the simultaneous transmission of codewords via layers from different UEs at the same time and frequency. The use of normal radio management techniques will ensure adequate frequency, timing, and power alignment of the signals received at the eNB. Aligning the received power from the UEs at the eNB will be the most difficult thing to control if the potential capacity gains are to be realized.

As stated earlier, precoding cannot be used for MU-MIMO because the transmitters do not have access to each other's signals. Even if they did, precoding still would not work because it involves matching the phase of the transmitted signals to that of the channel, and the phase between the two UEs is uncontrolled. However, the eNB will support receive diversity with two or four antennas, and the latter case will help improve performance with MU-MIMO.

2.4.6 Cooperative MIMO

Cooperative MIMO is sometimes referred to as network MIMO. It extends the MIMO concept by coordinating different eNBs to provide beamforming and spatial multiplexing capabilities between adjacent cells with the primary goal of improving the spectral efficiency of cell-edge users. Within LTE, cooperative MIMO has been under consideration for some time and is currently a Release 11 study item under the name coordinated multi-point transmission (CoMP). The primary challenge for cooperative MIMO is the need to share vast quantities of baseband data between the transmitting entities. Within the confines of a single device, whether a UE or eNB, this sharing can be accomplished on-chip or between modules. In the cooperative MIMO case, however, the distances between transmitting elements may be hundreds of meters or even several kilometers. The provision of sufficient backhaul transmission bandwidth with the necessary latency of perhaps 1 ms is a considerable technical and commercial challenge. CoMP is discussed more fully in Section 8.4.4.

2.4.7 Combining Multi-Antenna Techniques

With a matched channel and good SNR, spatial multiplexing offers incremental benefits while making use of the additional hardware already required for diversity techniques. The performance loss from path correlation can be mitigated by adding diversity to spatial multiplexing. Figure 2.4-10 describes the mapping of two codewords using serial-to-parallel (S/P) converters to three or four layers, with 16 indices being available in the codebook. This configuration enables the limited use of beamforming.

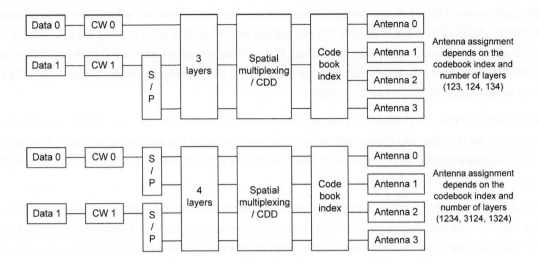

Figure 2.4-10. Processing steps to map two codewords to three or four antennas

2.4.8 Conclusion

The multiple antenna techniques used in LTE include transmit and receive diversity, beamforming, and spatial multiplexing. Diversity techniques increase the robustness of the signal path but do not increase the data rates. Beamforming techniques improve the SINR observed at the receiver device as a result of the multi-antenna beamforming gain. Spatial multiplexing leverages the addition of transmit and receive antennas to increase the fundamental channel capacity. Suitable channel conditions are needed to make this practicable, and LTE supports the combination of transmit diversity or beamforming along with spatial multiplexing to improve the likely performance.

LTE uses multi antenna techniques dynamically, placing considerable demands on the eNB and UE to report the correct channel state information and react to it appropriately. These topics are considered in further detail in Sections 6.7, 6.8, and 6.9.

2.5 References

[1]   3GPP TS 36.101 V11.2.0 (2012-09) UE Radio Transmission and Reception

[2]   3GPP TS 36.104 V11.2.0 (2012-09) Base Station Radio Transmission and Reception

[3]   3GPP TR 36.807 V10.0.0 (2012-07) User Equipment (UE) Radio Transmission and Reception

[4]   ETSI TR 102 735 V7.0.1 (2007-08) Band-specific Requirements for UMTS

[5]   3GPP TS 36.331 V11.1.0 (2012-09) Radio Resource Control (RRC) Protocol Specification

[6]   Cutler, Bob, "Effects of physical layer impairments on OFDM systems," RF Design, pp. 36-44, May 2002. Available from http://rfdesign.com/images/archive/0502Cutler36.pdf.

[7]   ITU-R Recommendation SM.329-10, "Unwanted emissions in the spurious domain"

[8]   ITU-R recommendation SM.328: "Spectra and bandwidth of emissions"

[9]   3GPP TS 36.211 V11.0.0 (2012-09) Physical Channels and Modulation

[10] 3GPP TS 36.213 V11.0.0 (2012-09) Physical layer procedures

Links to all reference documents can be found at www.agilent.com/find/ltebook.

Chapter 3

Physical Layer

3.1 Introduction to the Physical Layer

This chapter describes the LTE (E-UTRA) physical layer design. A general description of the LTE physical layer can be found in 36.201 [1] with the detailed design found in the 36.2XX technical specifications as follows:

    36.211 Physical channels and modulation [2]

    36.212 Multiplexing and channel coding [3]

    36.213 Physical layer procedures [4]

    36.214 Physical layer measurements [5].

Due to the complexity and scope of the physical layer no book of this length can begin to serve as a comprehensive reference. The goal here is to provide an introduction to the subject that will facilitate further study of the specifications themselves.

The physical layer covers the downlink transmission from the evolved node B (eNB) base transceiver station to the user equipment (UE), and the uplink transmission from the UE to eNB. As discussed in the previous chapter, 3GPP has selected a new OFDMA modulation scheme for the physical layer. This is one of the key differences between LTE and existing 3G systems such as UMTS and cdma2000, which are based on CDMA.

3.2 Physical Channels and Modulation

3.2.1 General Description of the Radio Interface

The physical layer supports two multiple access schemes: OFDMA on the downlink and SC-FDMA on the uplink. In addition, both paired and unpaired spectra are supported by using frequency division duplexing (FDD) and time division duplexing (TDD), respectively.

The LTE air interface needs to be described in both the time and frequency domains. The frame structure defines the frame, slot, and symbol in the time domain. Two types of frame structures are defined: type 1 for FDD and type 2 for TDD. Although the downlink and uplink utilize different multiple access schemes, they share a common frame structure.

In order to simplify the system, LTE supports only packet-switched communication carried by shared channels, so LTE does not have any dedicated channels.

Two types of physical layer channels are defined: physical channels, which carry information originating from higher layers, and physical signals, which are generated in the physical layer and are used for system synchronization, cell identification, and radio channel estimation.

LTE and the Evolution to 4G Wireless: Design and Measurement Challenges, Second Edition, Edited by Moray Rumney.
Copyright Agilent Technologies, Inc. 2013. Published by John Wiley & Sons, Ltd.

3.2.2 Frame Structure

The frame structure defines frame, subframe, slot, and symbol in the time domain. The time length is expressed in time units of $T_s = 1/(15000 \times 2048) = 32.55$ ns unless otherwise stated.

3.2.2.1 FDD Frame Structure

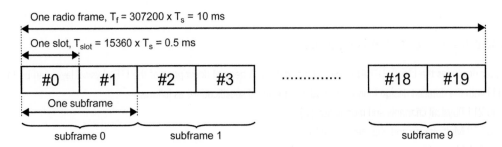

Figure 3.2-1. Frame structure type 1 (FDD mode) (36.211 [2] Figure 4.1-1)

Frame structure type 1 is defined for FDD mode. Each radio frame is 10 ms long and consists of 10 subframes. Each subframe contains two slots. In FDD, both uplink and downlink have the same frame structure but use different spectra.

3.2.2.2 TDD Frame Structure

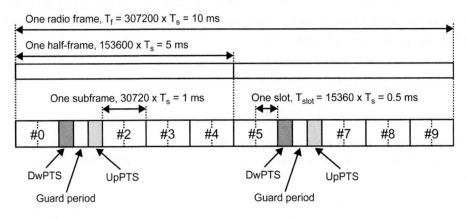

Figure 3.2-2. Example of frame structure type 2 (TDD mode) (Based on 36.211 [2] Figure 4.1-1)

Frame structure type 2 is defined for TDD mode. There are seven configurations defined for frame structure type 2 as shown in Table 3.2-1. Each radio frame is 10 ms long and consists of two half frames. Each half frame contains five subframes. Subframe #1 and sometimes subframe #6 consist of three special fields: downlink pilot timeslot (DwPTS), guard period (GP), and uplink pilot tmeslot (UpPTS). The length of DwPTS, GP, and UpPTS is given by Table 3.2-2. The total length of DwPTS, GP, and UpPTS is equal to $30720^*T_s = 1ms$.

The other eight subframes in the radio frame hold two slots each. "D" denotes a subframe reserved for downlink transmissions, "U" denotes a subframe reserved for uplink transmissions, and "S" denotes a special subframe with the three fields DwPTS, GP, and UpPTS.

Table 3.2-1. Uplink-downlink configurations (36.211 Table 4.2-2)

Uplink-downlink configuration	Downlink-to-uplink switch-point periodicity	Subframe number									
		0	1	2	3	4	5	6	7	8	9
0	5 ms	D	S	U	U	U	D	S	U	U	U
1	5 ms	D	S	U	U	D	D	S	U	U	D
2	5 ms	D	S	U	D	D	D	S	U	D	D
3	10 ms	D	S	U	U	U	D	D	D	D	D
4	10 ms	D	S	U	U	D	D	D	D	D	D
5	10 ms	D	S	U	D	D	D	D	D	D	D
6	5 ms	D	S	U	U	U	D	S	U	U	D

The flexible assignment for downlink or uplink slot direction in a frame enables asymmetric data rates. Depending on the switch-point periodicity of 5 ms or 10 ms, there can be one or two changes of direction within the frame, providing considerable deployment flexibility. It is necessary, however, to coordinate the frame structure configuration between adjacent cells to avoid simultaneous transmit and receive on the same frequency and time.

Table 3.2-2. Configuration of special subframe (lengths of DwPTS/GP/UpPTS) (Based on 36.211 [2] Table 4.2-1)

Special subframe configuration	Normal cyclic prefix			Extended cyclic prefix		
	DwPTS	GP	UpPTS	DwPTS	GP	UpPTS
0	$6592*T_s$	$21936*T_s$		$7680*T_s$	$20480*T_s$	
1	$19760*T_s$	$8768*T_s$		$20480*T_s$	$7680*T_s$	$2560*T_s$
2	$21952*T_s$	$6576*T_s$	$2192*T_s$	$23040*T_s$	$5120*T_s$	
3	$24144*T_s$	$4384*T_s$		$25600*T_s$	$2560*T_s$	
4	$26336*T_s$	$2192*T_s$		$7680*T_s$	$17920*T_s$	
5	$6592*T_s$	$19744*T_s$		$20480*T_s$	$5120*T_s$	$5120*T_s$
6	$19760*T_s$	$6576*T_s$		$23040*T_s$	$2560*T_s$	
7	$21952*T_s$	$4384*T_s$	$4384*T_s$	$12800*T_s$	$12800*T_s$	
8	$24144*T_s$	$2192*T_s$		-	-	-
9	$13168*T_s$	$13168*T_s$		-	-	-

3.2.3 Slot Structure

3.2.3.1 OFDM Symbol and Cyclic Prefix

One of the key advantages in OFDM systems (including SC-FDMA in this context) is the ability to protect against multipath delay spread. As discussed in Section 2.2.2, the long OFDM symbols allow the introduction of a guard period between each symbol to eliminate inter-symbol interference due to multipath delay spread. If the guard period is longer than the delay spread in the radio channel, and if each OFDM symbol is cyclically extended into the guard period (by copying the end of the symbol to the start to create the cyclic prefix), then the inter-symbol interference can be completely eliminated.

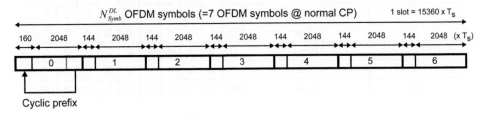

Figure 3.2-3. OFDM symbol structure for normal cyclic prefix case (downlink)

Figure 3.2-3 shows the seven symbols in a slot for the normal cyclic prefix case. The length of the cyclic prefix is shown for the uplink in Table 3.2-3 and the downlink in Table 3.2-4. In the latter table, Δf represents the 15 kHz or 7.5 kHz subcarrier spacing.

Table 3.2-3. SC-FDMA cyclic prefix length (uplink) (36.211 [2] Table 5.6-1)

Configuration	Cyclic prefix length $N_{CP, l}$
Normal cyclic prefix	160 for $l = 0$ 144 for $l = 1, 2, ..., 6$
Extended cyclic prefix	512 for $l = 0, 1, ..., 5$

Table 3.2-4. OFDM cyclic prefix length (downlink) (36.211 [2] Table 6.12-1)

Configuration		Cyclic prefix length $N_{CP, l}$
Normal cyclic prefix	Δf = 15 kHz	160 for $l = 0$ 144 for 1, 2, ..., 6
Extended cyclic prefix	Δf = 15 kHz	512 for 0, 1, ..., 5
	Δf = 7.5 kHz	1024 for 0, 1, 2

The normal cyclic prefix of 144 x T_s (4.69 μs) protects against multipath delay spread of up to 1.4 km. Note that the delay spread represents the variation in path delay in the cell and not the cell size, which is likely to be larger. The longest cyclic prefix provides protection for delay spreads up to 10 km.

3.2.3.2 Resource Element and Resource Block

A resource element is the smallest unit in the physical layer and occupies one OFDM or SC-FDMA symbol in the time domain and one subcarrier in the frequency domain. This is shown in Figure 3.2-4.

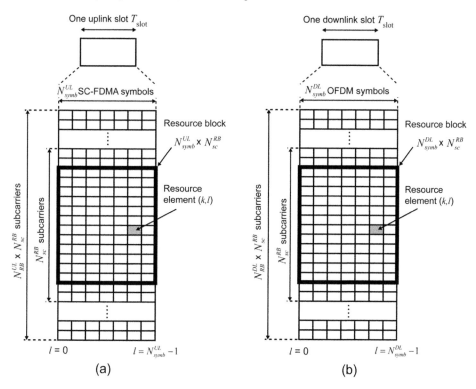

Figure 3.2-4. Resource grid for uplink (a) and downlink (b) (36.211 [2] Figures 5.2.1-1 and 6.2.2-1)

A resource block (RB) is the smallest unit that can be scheduled for transmission. An RB physically occupies 0.5 ms (= 1 slot) in the time domain and 180 kHz in the frequency domain. The number of subcarriers per resource block N_{sc}^{RB} and the number of symbols per resource block N_{Symb}^{UL} and N_{Symb}^{DL} vary as a function of the cyclic prefix length and subcarrier spacing, as shown in Tables 3.2-5 and 3.2-6.

Table 3.2-5. Resource block parameters for the uplink

Configuration	N_{sc}^{RB}	N_{Symb}^{UL}
Normal cyclic prefix	12	7
Extended cyclic prefix	12	6

Table 3.2-6. Resource block parameters for the downlink

Configuration		N_{sc}^{RB}	N_{Symb}^{DL}
Normal cyclic prefix	$\Delta f = 15$ kHz	12	7
Extended cyclic prefix	$\Delta f = 15$ kHz		6
	$\Delta f = 7.5$ kHz	24	3

The obvious difference between the uplink and downlink is that the downlink transmission supports 7.5 kHz subcarrier spacing, which is used for multimedia broadcast over single frequency network (MBSFN). The 7.5 kHz subcarrier spacing means that the symbols are twice as long, which allows the use of a longer CP to combat the higher delay spread seen in the larger MBSFN cells.

The uplink resource grid consists of $N_{RB}^{UL} \times N_{sc}^{RB}$ subcarriers in the frequency domain and N_{Symb}^{UL} SC-FDMA symbols in the time domain, where N_{RB}^{UL} denotes the uplink transmission bandwidth, expressed in multiples of N_{sc}^{RB}. The unit N_{sc}^{RB} defines the number of subcarriers per 180 kHz RB, which for the uplink is always 12 due to the 15 kHz subcarrier spacing. The unit N_{Symb}^{UL} denotes the number of SC-FDMA symbols in an uplink slot, which varies as a function of the CP length. See Table 3.2-5.

Similarly, the downlink consists of $N_{RB}^{DL} \times N_{sc}^{RB}$ subcarriers in the frequency domain and N_{Symb}^{DL} OFDM symbols in the time domain, where N_{RB}^{DL} denotes the downlink transmission bandwidth, expressed in multiples of N_{sc}^{RB}. For the downlink, the unit N_{sc}^{RB} is either 12 or 24 depending on the subcarrier spacing of 15 kHz or 7.5 kHz. The unit N_{Symb}^{DL} denotes the number of OFDM symbols in a downlink slot, which varies as a function of the CP length and subcarrier spacing as defined in Table 3.2-6.

3.2.4 Configurable Channel Bandwidth

Unlike CDMA, OFDM easily enables flexible transmission bandwidths. In CDMA systems, the transmission bandwidth is fixed and determined by the inverse of the chip rate. In OFDM systems, the subcarrier spacing is determined by the inverse of the FFT integration time. However, the number of subcarriers and hence the transmission bandwidth can be determined independently. This gives LTE the flexibility to have six different transmission bandwidth configurations from 1.4 MHz to 20 MHz, presenting more options for system deployment. The terminology to describe the flexibility of the LTE air interface is given in Figure 3.2-5.

The channel bandwidth defined in MHz represents the nominal occupied channel, which in effect becomes the channel spacing. The transmission bandwidth configuration defined in units of RB represents the maximum number of RB that can be transmitted for any channel bandwidth. The transmission bandwidth also defined in RB represents the number of RB allocated to any specific transmission and can vary from one up to the maximum RB allowed for that channel bandwidth.

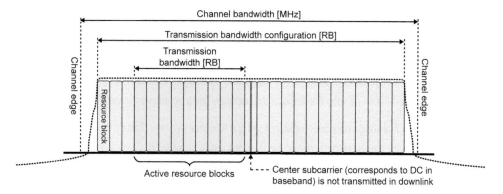

Figure 3.2-5. Definition of channel bandwidth and transmission bandwidth configuration for one E-UTRA carrier (36.101 [6] Figure 5.6-1)

The different channel bandwidths and associated transmission bandwidth configurations are given in Table 3.2-7.

Table 3.2-7. Transmission bandwidth configuration (based on 36.101 [6] Table 5.6-1)

Channel bandwidth (MHz)	1.4	3	5	10	15	20
Transmission bandwidth configuration (MHz)	1.08	2.7	4.5	9	13.5	18
Transmission bandwidth configuration (N_{RB}^{UL} or N_{RB}^{DL}) (RB)	6	15	25	50	75	100

3.2.5 Downlink Physical Signals and Channels

Table 3.2-8 defines the physical signals and physical channels for downlink transmission. These will be explained in the following sections.

Table 3.2-8. List of downlink physical signals and physical channels

Physical signals	Physical channels
Primary synchronization signal	Physical downlink shared channel (PDSCH)
Secondary synchronization signal	Physical broadcast channel (PBCH)
Cell-specific reference signal (CRS)	Physical downlink control channel (PDCCH)
MBSFN reference signal	Physical multicast channel (PMCH)
UE-specific reference signal	Physical control format indicator channel (PCFICH)
Positioning reference signal (PRS)	Physical hybrid automatic request (ARQ) indicator channel (PHICH)
Channel state information (CSI) reference signal (CSI-RS)	

3.2.5.1 Primary Synchronization Signal

Both primary and secondary synchronization signals are designed to be detected by all types of UE. They are transmitted twice per 10 ms radio frame. The synchronization signals always occupy the central 62 subcarriers of the channel, which makes the cell search procedure the same regardless of the channel bandwidth. Although 72 subcarriers (6 RB) are available, only 62 subcarriers are used so that the UE can perform the cell search procedure using an efficient length 64 FFT.

The primary synchronization signal subcarriers are modulated using a frequency-domain Zadoff-Chu sequence. Each subcarrier has the same power level with its phase determined by the root index number in a sequence generator as defined in 36.211 [2] Subclause 6.11.1. Three different cell identities are used for the primary synchronization signal. The root index number corresponds to the cell identity $N_{ID}^{(2)}$. Figure 3.2-6 shows an example of a constellation plot of the primary synchronization signal.

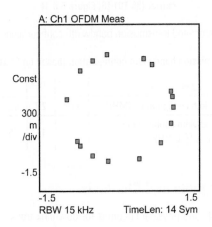

Figure 3.2-6. I/Q constellation sample of a primary synchronization signal

3.2.5.2 Secondary Synchronization Signal

The secondary synchronization signal is used to identify cell-identity groups. The number and position of subcarriers are the same as for the primary synchronization signal: that is, the central 62 subcarriers. The sequence generation function utilizes an interleaved concatenation of two length-31 binary sequences as defined in 36.211 [2] Subclause 6.11.2. The secondary synchronization signal gives a cell-identity group number from 168 possible cell identities $N_{ID}^{(1)}$.

In the cell search procedure, the primary synchronization signal is used first. The UE determines the timing and center frequency by detecting the primary synchronization signal. There are 504 unique cell identities, N_{ID}^{cell}, grouped into 168 unique cell-identity groups, $N_{ID}^{(1)}$, with each group containing three unique identities $N_{ID}^{(2)}$. The primary synchronization signal gives the identity information, which is one of the three unique identities $N_{ID}^{(2)}$. The procedure is further described in Section 3.5.1. Figure 3.2-7 shows an example of a constellation plot of the secondary synchronization signal.

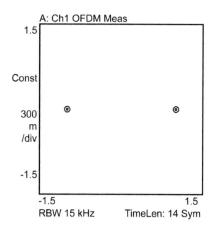

Figure 3.2-7. I/Q constellation of a secondary synchronization signal

This concept is similar to the Scrambling Code Group and cell search procedure in UMTS (FDD). In UMTS, the secondary synchronization code gives the Scrambling Code Group, and then the primary scrambling code (one of 512 scrambling codes) can be determined.

3.2.5.3 Reference Signals

There are five types of downlink reference signals. In Release 8 there are the cell-specific reference signal (CRS), the MBSFN reference signal, and the UE-specific reference signal (sometimes known as the demodulation or DMRS). In Release 9 the positioning reference signal (PRS) was added and in Release 10, the channel state information reference signal (CSI-RS) was added. This section describes the general cell-specific reference signal. The other reference signals are described in Section 3.2.12.

The CRS plays a very important role, as it enables the UE to mitigate amplitude, phase, and timing errors in the received signal that can be attributed to in-band flatness error introduced by the radio channel and impairments from the eNB transmitter. The reference signals hold known amplitude and phase, and they are uniformly allocated every six subcarriers in the frequency domain and every two symbols per slot in the time domain. From these references the UE can calculate corrections and thus minimize the probability of demodulation errors. The CRS is present in every frame. It is used by the UE to demodulate other downlink signals and also to report the downlink channel state information on the uplink.

For the normal cyclic prefix, the sequence generation is a product of a two-dimensional orthogonal sequence and a two-dimensional pseudo-random sequence. The orthogonal sequence holds three different sequences that correspond to the cell identity given by the primary synchronization signal. At the same time, the pseudorandom sequence holds 168 different sequences that correspond to the cell-identity group given by the secondary synchronization signal. In the cell search procedure, after detecting the primary and secondary synchronization signals, a UE can calculate which reference signal is being used in the cell.

3.2.5.4 Physical Broadcast Channel

The physical broadcast channel (PBCH) is the physical channel that carries the broadcast channel (BCH) transport channel. The BCH carries various cell-specific contents and is used for all types of UE. As with the synchronization signals, the PBCH is transmitted in the center of the channel but it occupies 6 RB (72 subcarriers), which is the whole of the narrowest channel bandwidth. The PBCH supports only the QPSK modulation scheme. The PBCH is located in slot #1 at OFDM symbols #0, #1, #2, and #3.

3.2.5.5 Physical Downlink Shared Channel

The physical downlink shared channel (PDSCH) is the physical channel that carries the traffic data. As its name suggests, this channel is shared in the time domain between multiple users. The PDSCH supports QPSK, 16QAM, and 64QAM modulation schemes and carries the downlink shared channel (DL-SCH) or paging channel (PCH) transport channels.

Various multi-antenna techniques can be applied for the PDSCH. The multi-antenna operations and MIMO are described in more detail in Sections 2.4, 6.8, and 6.9.

3.2.5.6 Physical Downlink Control Channel

The physical downlink control channel (PDCCH) is the physical channel that carries the channel allocation and control information. It consists of one or more consecutive control channel elements (CCEs), where a control channel element corresponds to nine resource element groups. The number of OFDM symbols allocated for the PDCCH is given by the control format indicator (CFI) carried on the PCFICH. The CFI can take the values 1, 2, and 3. For transmission bandwidth configurations greater than 10 RB (1.8 MHz), the number of PDCCH symbols per subframe is the CFI value. For transmission bandwidth configurations less than or equal to 10 RB, the number of PDCCH symbols per subframe is the CFI value +1. The PDCCH supports only the QPSK modulation scheme. Multiple PDCCHs can be transmitted in a subframe.

3.2.5.7 Physical Multicast Channel

The physical multicast channel (PMCH) is the physical channel that carries the multicast channel (MCH) transport channel. The PMCH is similar to the PDSCH except that it carries information to multiple users for point to multi-point broadcast services. It uses QPSK, 16QAM, or 64QAM modulation.

3.2.5.8 Physical Control Format Indicator Channel

The physical control format indicator channel (PCFICH) is the physical channel that carries the number of OFDM symbols used for transmission of PDCCHs in a subframe. PCFICH is located at OFDM symbol #0 of every subframe, and the assignment to the subcarriers is determined by cell ID information.

3.2.5.9 Physical Hybrid ARQ Indicator Channel

The physical hybrid automatic repeat request (ARQ) indicator channel (PHICH) is the physical channel that carries the hybrid ARQ indicator (HI). The HI contains the acknowledgement/negative acknowledgement (ACK/NACK) feedback to the UE for the uplink blocks received by the eNB.

3.2.6 Example FDD Downlink Mapping to Resource Elements

The primary and secondary synchronization signals, reference signals, PDSCH, PBCH, and PDCCH are almost always present in a downlink radio frame. There is a priority rule for allocation (physical mapping) as follows. Signals (reference signal, primary and secondary synchronization signals) take precedence over the PBCH. The PDCCH takes precedence over the PDSCH. The PBCH and PDCCH are never allocated to the same resource elements, thus they are not in conflict.

Figures 3.2-8 and 3.2-9 show an LTE FDD mapping example. The primary synchronization signal is mapped to the last symbol of slot #0 and slot #10 in the central 62 subcarriers. The secondary synchronization signal is allocated in the symbol just before the primary synchronization signal. The reference signals are located at symbol #0 and symbol #4 of every slot. The reference signal takes precedence over any other allocation. The PBCH is mapped to the first four symbols in slot #1 in the central 6 RB. The PDCCH can be allocated to the first three symbols (four symbols when the number of RB is equal or less than 10) of every subframe as shown in Figure 3.2-8. The remaining unallocated areas can be used for the PDSCH. Note how the five subcarriers on either side of the primary and secondary synchronization signals remain unallocated .

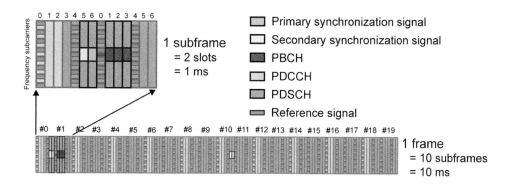

Figure 3.2-8. Example of downlink mapping (normal cyclic prefix)

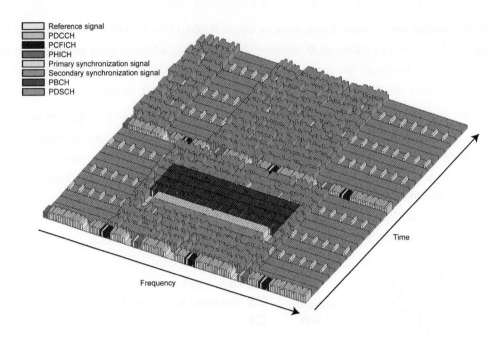

Figure 3.2-9. Example of downlink mapping showing frequency (subcarriers) vs. time

3.2.7 OFDM (Downlink) Baseband Signal Generation

Figure 3.2-10 shows the major stages in OFDM signal generation. The first stage is to scramble the incoming bit stream before modulation mapping. In scrambling, pseudo-random sequence generation is used; in this case a length-31 Gold sequence. At the second stage, mapping of these bits to symbols is applied. For example, the PDSCH can use QPSK, 16QAM, or 64QAM, which require 2 bits, 4 bits, or 6 bits per symbol, respectively. At the third stage, layer mapping is applied to support various antenna configurations. At the next stage, precoding is applied to adjust the phase and amplitude for each antenna and adjust the total power of multiple antennas. The phase rotation schemes differ by application; that is, by cyclic delay diversity, transmit diversity, and so on. At the last two stages, resource mapping is applied and the OFDM signal is generated for each antenna. In receiving, the signal demodulation follows the inverse process.

Up to four antennas can be configured in Releases 8 and 9, and up to eight antennas in Release 10 (LTE-Advanced).

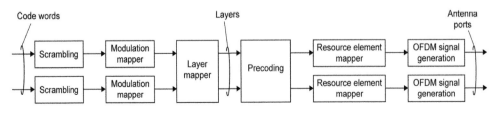

Figure 3.2-10. Downlink OFDM signal generation (36.211 Figure 6.3-1)

3.2.8 Uplink Physical Signals and Channels

Table 3.2-9 shows the signals and channels defined for the uplink.

Table 3.2-9. List of uplink physical signals and physical channels

Physical signals	Physical channels
Demodulation reference signal (DMRS) for PUSCH/PUCCH	Physical uplink shared channel (PUSCH)
Sounding reference signal (SRS)	Physical uplink control channel (PUCCH)
	Physical random access channel (PRACH)

3.2.8.1 Demodulation Reference Signal

The DMRS is used for synchronization and uplink channel estimation. There are two types of DMRS; one for the PUSCH and one for the PUCCH. The DMRS for the PUSCH are assigned to SC-FDMA symbol #3 (normal CP case) and SC-FDMA symbol #2 (extended CP case) in a PUSCH slot. The DMRS for the PUCCH is assigned according to the PUCCH format and cyclic prefix mode. For example, when the PUCCH format is set to 1/1a/1b and normal cyclic prefix is selected, the DMRS for the PUCCH is assigned to the SC-FDMA symbols #2, #3, and #4 in a PUCCH slot.

3.2.8.2 Sounding Reference Signal

The eNB can request transmission of the SRS, which allows the eNB to estimate the uplink channel characteristics for arbitrary channel bandwidths. This estimate cannot be done using the PUCCH demodulation reference signal that is fixed to the bandwidth of the associated PUSCH/PUCCH. The SRS length, frequency-domain starting position, and other parameters are determined by the SRS configuration.

3.2.8.3 Physical Uplink Shared Channel (PUSCH)

The physical uplink shared channel (PUSCH) is the physical channel that carries the traffic data. The PUSCH supports QPSK, 16QAM, and 64QAM modulation, although there are no performance requirements for 64QAM. The PUSCH carries the uplink shared channel (UL-SCH) transport channel and the uplink control information (UCI). Mapping of the PUSCH is shown in Figure 3.2-11.

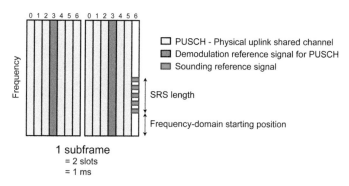

Figure 3.2-11. Mapping of PUSCH, demodulation reference signal for the PUSCH, and sounding reference signal

3.2.8.4 Physical Uplink Control Channel (PUCCH)

The physical uplink control channel (PUCCH) is the physical channel that carries the uplink control information such as scheduling requests, hybrid ARQ acknowledgement/negative acknowledgement (HARQ ACK/NACK), and channel quality indicator (CQI). The PUCCH is transmitted exclusively with the PUSCH from the same UE. The PUCCH supports BPSK and QPSK modulation schemes. A mapping for PUCCH format 1a/1b is shown in Figure 3.2-12. Other PUCCH formats exist that use the inner resource block (RB).

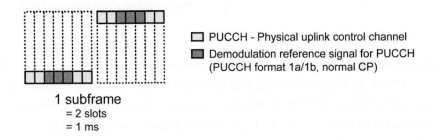

1 subframe
= 2 slots
= 1 ms

□ PUCCH - Physical uplink control channel

■ Demodulation reference signal for PUCCH
(PUCCH format 1a/1b, normal CP)

Figure 3.2-12. Example of PUCCH mapping and demodulation reference signal for PUCCH

3.2.9 Example Uplink Mapping to Resource Elements

Figure 3.2-13 shows an example of an uplink mapping. The spectrum for the uplink is shared by multiple UE in the time domain and frequency domain. A resource is usually allocated for a UE as a unit of RB. In some cases, the same RB in time is allocated to multiple UE, which are identified using orthogonal spreading codes as in CDMA. The constellations for PUCCH and the DMRS for PUCCH and PUSCH may be rotated based on parameters given by higher layers; for example, cyclic shift and sequence index. The constellations shown in Figure 3.2-13 are without rotation. The PUCCH modulation is more fully described in Section 6.4.6.7.

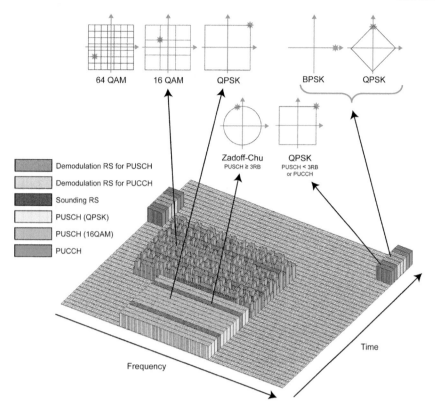

Figure 3.2-13. Example of uplink mapping showing frequency (subcarriers) vs. time

3.2.10 Physical Random Access Channel (PRACH)

The physical random access channel (PRACH) is the physical channel that initiates communication with the eNB. The PRACH allows the eNB to calculate the time delay to the UE and identify it prior to establishing a packet connection. To establish a connection with the eNB the UE initiates the random access procedure by sending a random access preamble using the PRACH. The physical layer random access preamble consists of a cyclic prefix and a sequence part, as shown in Table 3.2-10. This preamble is orthogonal to other uplink user data to allow the eNB to differentiate each UE. The subcarrier spacing for the PRACH is 1.25 kHz for formats 0 to 3, and 7.5 kHz for format 4. See example in Figure 3.2-14. Format 4 is used for the frame structure type 2 (TDD) only. The random access procedure is described more fully in Section 3.5.3.

Table 3.2-10. Random access preamble parameters (36.211 [2] Table 5.7.1-1)

Preamble format	Cyclic prefix (T_{CP})	Sequence part (T_{SEQ})
0	$3168*T_s$	$24576*T_s$
1	$21024*T_s$	$24576*T_s$
2	$6240*T_s$	$2*24576*T_s$
3	$21024*T_s$	$2*24576*T_s$
4 (Frame structure type 2 only)	$448*T_s$	$4096*T_s$

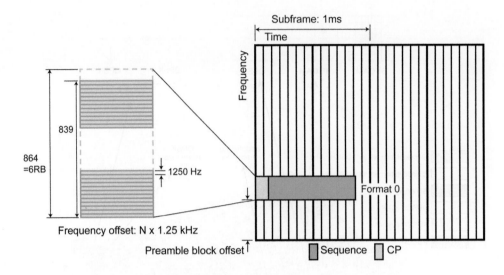

Figure 3.2-14. Example physical random access channel (PRACH), format 0

3.2.11 Example TDD Mapping to Resource Elements

Figures 3.2-15 and 3.2-16 show examples of 5 ms and 10 ms TDD switch point periodicity. The primary synchronization signal is mapped to the third symbol of slot #2 and slot #12 in the central 62 subcarriers. The secondary synchronization signal is allocated in the last symbol of slot #1 and slot #11. The reference signals are located at symbol #0 and symbol #4 of every slot. The reference signal takes precedence over any other allocation. The PBCH is mapped to the first four symbols in slot #1 in the central 6 RB. The PDCCH can be allocated to the first three symbols of every subframe as shown here. The remaining unallocated areas can be used for the PDSCH.

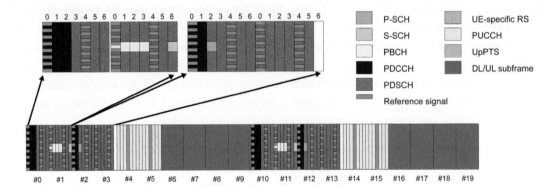

Figure 3.2-15. Example of LTE TDD 5 ms switch periodicity mapping

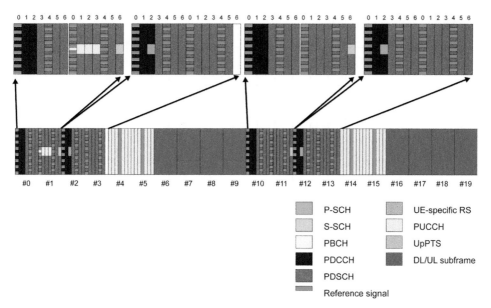

	P-SCH		UE-specific RS
	S-SCH		PUCCH
	PBCH		UpPTS
	PDCCH		DL/UL subframe
	PDSCH		
	Reference signal		

Figure 3.2-16. Example of LTE TDD 10 ms switch periodicity mapping

3.2.12 Additional Reference Signals

In addition to the basic CRS described in Section 3.2.5.3, LTE supports several other RS used for different purposes. These are the UE-specific RS, the MBSFN RS, the PRS, and the CSI-RS.

3.2.12.1 UE-Specific Reference Signal

The purpose of the UE-specific RS is to optimize the RS for a specific UE by precoding the UE-specific RS resource elements with the same gain and phase shifts used for the PDSCH dedicated to that UE. In this way the UE no longer needs to be told how the PDSCH has been precoded and can directly decode the PDSCH using the similarly precoded UE-specific RS. This is clearly a benefit to the specific UE but it also means that the REs allocated to the UE-specific RS are of no use to other UEs in the serving cell. The use of UE-specific RS should improve the link budget for a specific UE but at the cost of consuming a small portion of the downlink capacity. Figure 3.2.17 shows the mapping of the UE-specific RS for single antenna transmission (virtual antenna port 5) in Release 8. This allocation is orthogonal in frequency and time to the CRS shown in Figure 2.4-3. Since the UE is not told by the eNB what precoding has been applied to the UE-specific RS, these RS can not be used by the UE to report channel state information.

The UE-specific RS is used in FDD mode when the uplink and downlink are on different frequencies, which makes it difficult for the eNB to estimate the downlink channel conditions from the uplink signal. In TDD mode this is not an issue since the eNB can directly calculate the precoding required for the downlink from the signal received on the uplink.

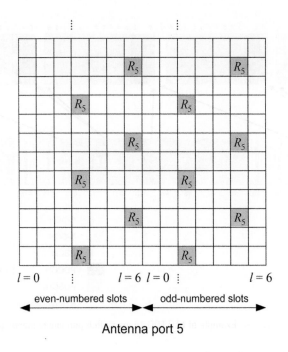

Antenna port 5

Figure 3.2-17. Mapping of UE-specific reference signals, antenna port 5 (normal cyclic prefix) (from 36.211 [2] Figure 6.10.3.2-1)

In Release 9 the UE-specific RS was extended for two antenna operation (ports 7 and 8). The eNB can apply eigenvectors on each physical antenna to achieve the maximum signal-to-noise ratio (SNR) at the UE receiver. This scheme is also known as dual layer beamforming and in Release 10 was further extended to eight antennas (ports 7 through 14). It is described further in Sections 2.4 and 6.9.

3.2.12.2 MBMS Single Frequency Network Reference Signal

The MBMS single frequency network (MBSFN) reference signal is transmitted with the PMCH. In MBSFN, multiple time-synchronized eNBs transmit the same information with the same time-frequency resources. The signal quality and coverage can be improved by combining multiple signals in the UE. This scheme is similar to what is done for multipath combining.

When the UE is located near an eNB, the arrival time difference among multiple signals is relatively large. Extended CP is used to resolve any large time difference. MBSFN reference signal mapping is shown in Figure 3.2-18.

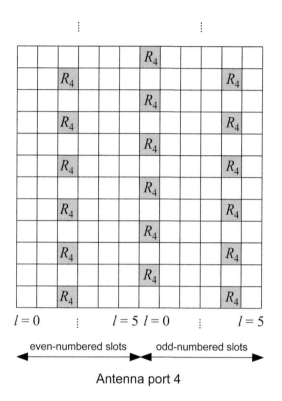

Antenna port 4

Figure 3.2-18. Mapping of MBSFN reference signals (extended cyclic prefix, $\Delta f = 15$ kHz)
(from 36.211 [2] Figure 6.10.4.2-1)

3.2.12.3 Positioning Reference Signal

The positioning reference signal was added in Release 9 to support the observed time difference of arrival (OTDOA) positioning scheme. The UE measures the time differences between the signals arriving from at least three eNBs and determines position by the intersection of arcs. Positioning reference signals are allocated on virtual antenna port 6 as shown in Figure 3.2-19.

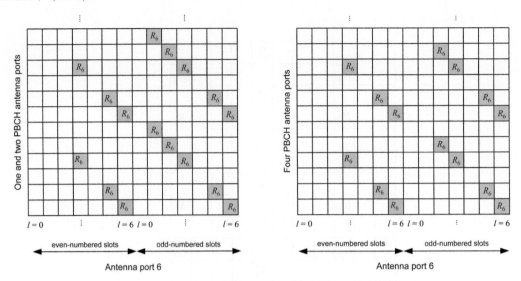

Figure 3.2-19. Mapping of positioning reference signals (normal cyclic prefix)
(from 36.211 [2] Figure 6.10.4.2-1)

3.2.12.4 CSI Reference Signal

Support for up to eight layers was introduced in Release 10. In principle this would require the extension of the CRS from four layers to eight layers, consuming significant downlink resources in every subframe, regardless of whether any UE were actually allocated an eight-layer downlink transmission. As an alternative to a direct extension of the CRS, Release 10 introduced the CSI-RS. This reference signal performs the same basic function as the CRS; that is, it provides a known amplitude and phase reference to the UE. However, the CSI-RS has two distinct differences from the CRS. First, the CSI-RS can be scheduled as required rather than being present in every frame. Second, the CSI-RS is used only for reporting of channel state information by the UE on the uplink and (unlike the CRS) is not used for demodulation.

Figure 3.2-20 shows the CSI-RS mapping for eight antennas on virtual antenna ports 15 through 22. The CSI-RS can be allocated with a periodicity of 5, 10, 20, 40 and 80 subframes. This range helps optimize the trade-offs between the accuracy of channel sounding with the overhead of transmitting the CSI-RS. Further MIMO discussion is available in Sections 2.4, 6.8, and 6.9.

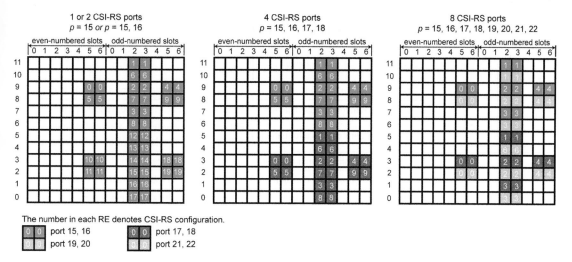

The number in each RE denotes CSI-RS configuration.

port 15, 16   port 17, 18
port 19, 20   port 21, 22

Figure 3.2-20. Mapping of CSI reference signals (CSI configuration 0, normal cyclic prefix)

3.3 Multiplexing and Channel Coding

This section describes the air interface multiplexing and channel coding according to 36.212 [3]. It covers transport channel (TrCH) and control channel information. Channel coding is a combination of error detection, segmentation, error correcting, rate matching, concatenation, and interleaving. The main focus here is on FDD mode; note that the FDD and TDD modes of LTE differ less than do the FDD and TDD modes of UMTS.

3.3.1 Mapping to Physical Channels

The physical layer offers data transport services to the higher layers through transport channels and control information channels. Table 3.3-1 shows the mapping of these to the corresponding physical channels.

Table 3.3-1. Mapping of uplink/downlink to corresponding physical channels
(36.212 [3] Tables 4.1-1, 4.1-2, 4.2-1, 4.2-2)

Radio link	Channel type	Channel	Channel name	Physical channel	Channel name
Uplink	TrCH	UL-SCH	Uplink shared channel	PUSCH	Physical uplink shared channel
		RACH	Random access channel	PRACH	Physical random access channel
	Control information	UCI	Uplink control information	PUCCH, PUSCH	Physical uplink control channel, physical uplink shared channel
Downlink	TrCH	DL-SCH	Downlink shared channel	PDSCH	Physical downlink shared channel
		BCH	Broadcast channel	PBCH	Physical broadcast channel
		PCH	Paging channel	PDSCH	Physical downlink Shared channel
		MCH	Multicast channel	PMCH	Physical multicast channel
	Control information	CFI	Control format indicator	PCFICH	Physical control format indicator channel
		HI	HARQ indicator	PHICH	Physical HARQ indicator channel
		DCI	Downlink control information	PDCCH	Physical downlink control channel

3.3.2 Channel Coding, Multiplexing, and Interleaving

3.3.2.1 CRC Calculation

Error detection is provided on transport blocks through a Cyclic Redundancy Check (CRC). The entire transport block is used to calculate the CRC parity bits which are appended to the transport block. The parity bits are generated by one of the following cyclic generator polynomials. The CRC24A generator is shown in Figure 3.3-1.

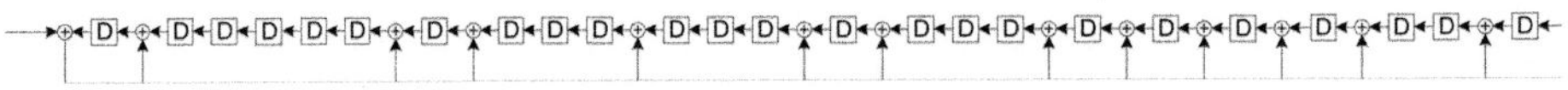

Figure 3.3-1. CRC24A generator

The mathematical equations for CRC24A and some of the other CRC codes are as follows:

CRC24A: $g_{CRC24A}(D) = [D^{24} + D^{23} + D^{18} + D^{17} + D^{14} + D^{11} + D^{10} + D^7 + D^6 + D^5 + D^4 + D^3 + D + 1]$ for a CRC length $= 24$

CRC24B: $g_{CRC24B}(D) = [D^{24} + D^{23} + D^6 + D^5 + D + 1]$ for a CRC length $= 24$

CRC16: $g_{CRC16}(D) = [D^{16} + D^{12} + D^5 + 1]$ for a CRC length $= 16$

CRC8: $g_{CRC8}(D) = [D^8 + D^7 + D^4 + D^3 + D + 1]$ for a CRC length $= 8$

Errors detected on the transport channels and control information are reported to higher layers. Table 3.3-2 shows the usage of CRC calculation and CRC scrambling for transport channels and control information.

Table 3.3-2. Usage of CRC calculation and CRC scrambling

Radio link	Channel	CRC calculation scheme	CRC scrambling
Uplink	UL-SCH	Transport Block CRC: CRC24A Code Block CRC: CRC24B	Not applicable
	UCI	CRC8	Not applicable
Downlink	DL-SCH	Transport Block CRC: CRC24A Code Block CRC: CRC24B	Not applicable
	BCH	CRC16	The CRC bits are scrambled by the PBCH CRC mask, which is according to the eNB transmit antenna configuration.
	PCH	Transport Block CRC: CRC24A Code Block CRC: CRC24B	Not applicable
	MCH	Transport Block CRC: CRC24A Code Block CRC: CRC24B	Not applicable
	DCI	CRC16	The CRC bits are scrambled by the antenna selection mask and Radio Network Temporary Identity (RNTI).

3.3.2.2 Code Block Segmentation and Code Block CRC Attachment

If the input bit sequence of code block segmentation is larger than the maximum code block size of 6144 bits, segmentation of the input bit sequence is performed and an additional CRC sequence equal to 24 bits (CRC24B) is attached to each of the code blocks. If the input bit sequence of code block segmentation is less than 40 bits, filling bits are added. The code block segmentation and transport block/code block CRC attachment process is shown in Figure 3.3-2.

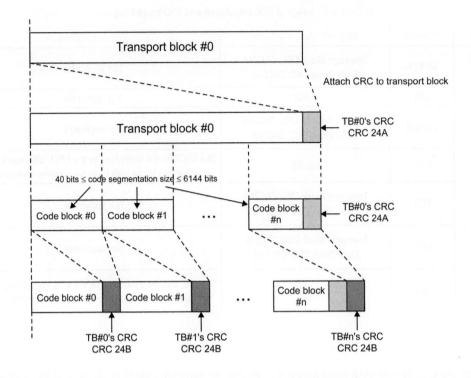

Figure 3.3-2. Code block segmentation and Transport Block (TB)/Code Block (CB) CRC

3.3.2.3 Channel Coding

Table 3.3-3. Usage of channel coding scheme and coding rate for transport channel and control information (36.212 [3] Tables 5.1.3-1 and 5.1.3-2)

Channel type	Channel	Coding scheme	Coding rate
TrCH	UL-SCH	Turbo coding	1/3
	DL-SCH		
	PCH		
	MCH		
	BCH	Tail biting convolutional coding	1/3
Control information	DCI	Tail biting convolutional coding	1/3
	CFI	Block code	1/16
	HI	Repetition code	1/3
	UCI	Block code	Variable
		Tail biting convolutional coding	1/3

Channel coding gives forward error correction (FEC) to the transport channel and control information. There are two types of channel coding: tail biting convolutional coding and turbo coding. Table 3.3-3 shows which channel coding schemes and coding rates are used for the transport channel and control information. Coding rate is defined as the ratio of the input bits to the output bits, and it represents the loss in capacity caused by adding redundancy and hence robustness to the signal.

A tail biting convolutional code with constraint length 7 and coding rate 1/3 is used for control information coding as shown in Figure 3.3-3.

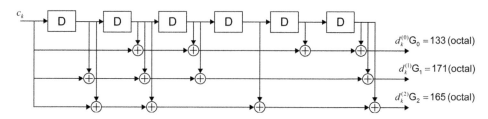

Figure 3.3-3. Rate 1/3 tail biting convolutional encoder (36.212 [3] Figure 5.1.3-1)

Turbo coding is used mainly for transport channel coding. The scheme of the turbo encoder is a parallel concatenated convolutional code (PCCC) with two 8-state constituent encoders and one turbo code internal interleaver as shown in Figure 3.3-4.

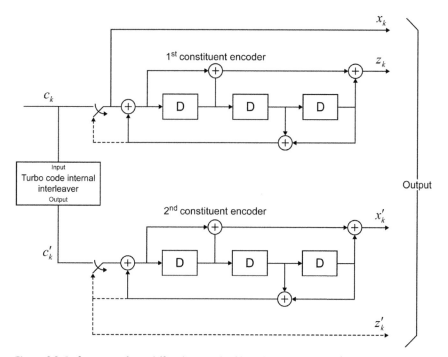

Figure 3.3-4. Structure of rate 1/3 turbo encoder (dotted lines apply for trellis termination only) (36.212 [3] Figure 5.1.3-2)

3.3.2.4 Rate Matching for Turbo-Coded Transport Channels

Rate matching provides data length control between the transport channel and the physical channel and hybrid ARQ (HARQ) using a virtual circular buffer as shown in Figure 3.3-5.

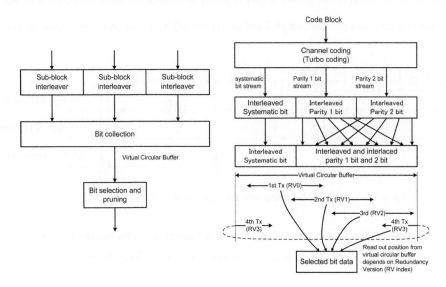

Figure 3.3-5. Rate matching for turbo-coded transport channels

3.3.2.5 Rate Matching for Convolutional-Coded Transport Channels and Control Information

The coding process for rate matching of convolutional-coded transport channels and control information is shown in Figure 3.3-6.

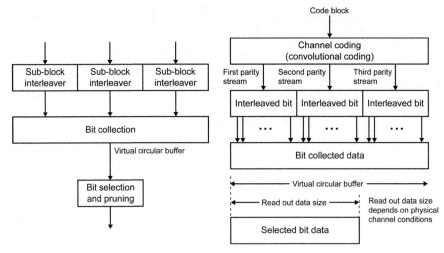

Figure 3.3-6. Rate matching for convolutional-coded transport channels and control channels

3.3.2.6 Code Block Concatenation

The code blocks from the rate matching outputs are the inputs to the code block concatenation process. The code block concatenation sequentially concatenates these different code blocks as shown in Figure 3.3-7.

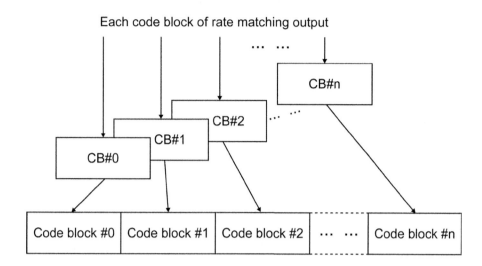

Figure 3.3-7. Code block concatenation

3.3.3 Uplink Shared Channel

Figure 3.3-8 shows the uplink transport channel and control information and the uplink physical channel processing for all of the uplink. The uplink shared channel (UL-SCH) is a transport block with dynamic transport format size. The coded UL-SCH transport block is mapped to the physical uplink shared channel every 1 ms transmission time interval (TTI).

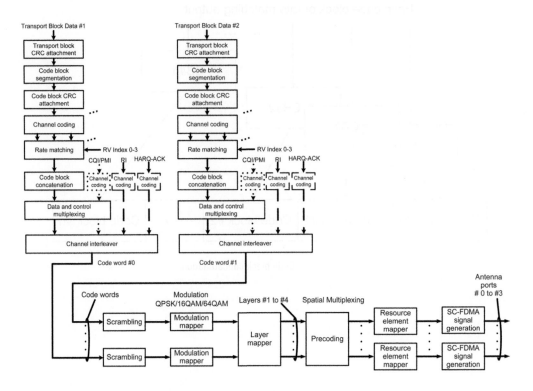

Figure 3.3-8. **Overview of uplink transport channel and control information and uplink physical channel processing (based on 36.211 [2] Figure 5.3-1 and 36.212 [3] Figure 5.2.2-1)**

The UL-SCH transport channel coding, shown in Figure 3.3-9, comprises the transport block CRC attachment, code block segmentation, code block CRC attachment, turbo coding, rate matching, and code block concatenation. Parameters carried on the UL-SCH include the channel quality indicator (CQI), precoding matrix indicator (PMI), rank indication (RI), and redundancy version (RV). These parameters are discussed in Section 3.4. The HARQ process controls which RV index is to be used for the transmission. The channel feedback parameters CQI and PMI have two channel coding cases, one for a payload size less than or equal to 11 bits and the other for a payload size greater than 11 bits. In the first case the channel coder uses simple block coding. The second case involves CRC attachment, convolutional coding, and rate matching. The RI is also coded and multiplexed onto the UL-SCH.

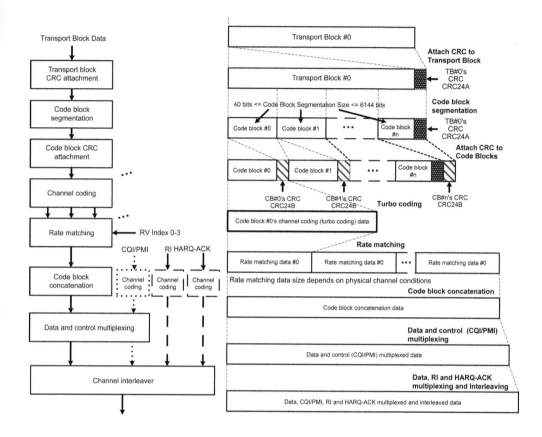

Figure 3.3-9. Transport channel processing for UL-SCH

3.3.3.1 Uplink Control Information Mapping

Uplink control information (UCI) can be mapped to the PUCCH and PUSCH on the physical layer. For PUCCH the UCI coding process depends on the type of control information being scheduled as shown in Table 3.3-4. In the case of spatial multiplexing, CQI includes the necessary feedback information of PMI and RI. The HARQ response from the UE for the downlink data transmission integrity contains one ACK/NACK bit per HARQ process. The number of UCI bits "A" to encode varies and they are processed using a (20, A) code, which is a linear combination of the 13 basis sequences defined in 36.212 [3] Table 5.2.3.3-1.

Uplink control information can be carried on any scheduled PUSCH, even if no transport data is available to be carried.

Table 3.3-4. Relation between UCI and PUCCH formats (based on 36.211 [2] Table 5.4-1)

Control information	PUCCH format	Modulation scheme	Number of bits per subframe, M_{bit}
Scheduling request	1	N/A	N/A
HARQ-ACK (1 bit)	1a	BPSK	1
HARQ-ACK (2bits)	1b	QPSK	2
CQI	2	QPSK	20
CQI + HARQ-ACK (1bit)	2a	QPSK + BPSK	21
CQI + HARQ-ACK (2bits)	2b	QPSK + QPSK	22
HARQ-ACK (up to 10 bits for FDD), HARQ-ACK (up to 20 bits for TDD)	3	QPSK	48

3.3.4 Downlink Transport Channel and Control Information Overview

Figure 3.3-10 shows the downlink transport channel coding and the physical channel processing for all of the downlink. In the case of spatial multiplexing, two code words can be used whereupon scrambling and modulation can be independent for each code word.

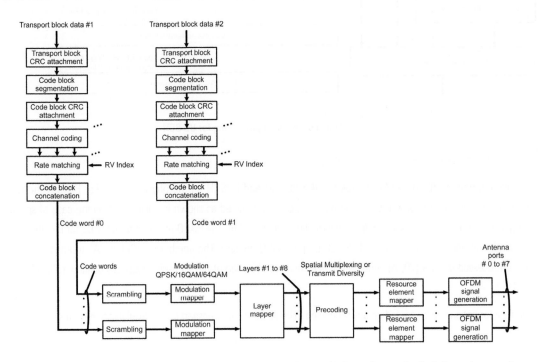

Figure 3.3-10. Overview of downlink transport channel and control information to physical channel processing
(Based on 36.211 [2] Figure 6.3-1)

3.3.5 Downlink Transport Channel Coding

3.3.5.1 Broadcast Channel

The broadcast channel (BCH) is a predefined transport format of fixed size and carries the master information block (MIB) broadcast to the entire coverage area of the cell. The MIB includes a limited number of the most essential and most frequently transmitted parameters: downlink transmission bandwidth, PHICH duration, PHICH resource, and system frame number (SFN). The BCH is mapped to the physical broadcast channel (PBCH) in the physical layer. The coded BCH transport block is mapped to four subframes (four OFDM symbols of slot #1 within subframe #0) within a 40 ms BCH TTI. Each subframe is self-decodable and can be blind-detected. The BCH transport channel coding comprises the CRC attachment, convolutional coding, and rate matching. No HARQ or channel interleaving is applied. The BCH-specific coding is PBCH CRC mask scrambling (CRC + PBCH CRC mask) Mod 2 as shown in Figure 3.3-11 and Table 3.3-5.

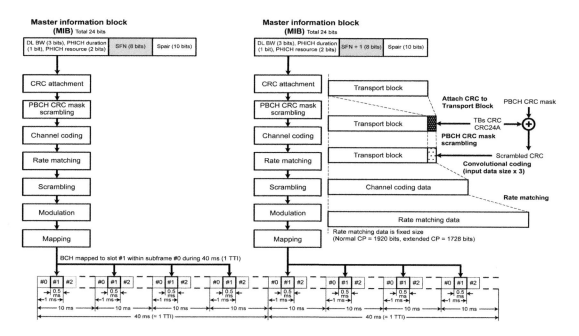

Figure 3.3-11. Transport channel processing and mapping for BCH

Table 3.3-5. CRC mask for PBCH (36.212 [3] Table 5.3.1.1-1)

Number of transmit antenna ports at eNB	PBCH CRC mask $< x_{ant,0}, x_{ant,1}, ..., x_{ant,15} >$
1	<0, 0, 0, 0, 0, 0, 0, 0, 0, 0, 0, 0, 0, 0, 0, 0>
2	<1, 1, 1, 1, 1, 1, 1, 1, 1, 1, 1, 1, 1, 1, 1, 1>
4	<0, 1, 0, 1, 0, 1, 0, 1, 0, 1, 0, 1, 0, 1, 0, 1>

3.3.5.2 Downlink Shared Channel

The downlink shared channel (DL-SCH) has up to two transport blocks of dynamic transport format size. The coded DL-SCH transport block is mapped to the PDSCH every 1 ms TTI. The DL-SCH transport channel coding, shown in Figure 3.3-12, comprises the transport block CRC attachment, code block segmentation, code block CRC attachment, turbo coding, rate matching, and code block concatenation. The DL-SCH is used for a variety of cases. One of these is transmission of system information, which is divided into the MIB and a number of system information blocks (SIBs). The system information is shown in Figure 3.3.13.

3.3.5.3 Paging Channel

The paging channel (PCH) carries information for broadcast in the entire coverage area of the cell. The coded PCH transport block is mapped to the PDSCH every 1 ms TTI. The PCH transport channel coding, shown in Figure 3.3-12, comprises the transport block CRC attachment, code block segmentation, code block CRC attachment, turbo coding, rate matching, and code block concatenation.

3.3.5.4 Multicast Channel

The multicast channel (MCH) carries information for broadcast in the entire coverage area of the cell. The MCH transport block is mapped to the PMCH every 1 ms TTI. The PMCH can only be transmitted in the MBSFN region of an MBSFN subframe with full resource blocks of the system bandwidth and the PMCH using extended cyclic prefix. The MCH transport channel coding, shown in Figure 3.3-12, comprises the transport block CRC attachment, code block segmentation, code block CRC attachment, turbo coding, rate matching, and code block concatenation. The redundancy version 0 is used for the PMCH.

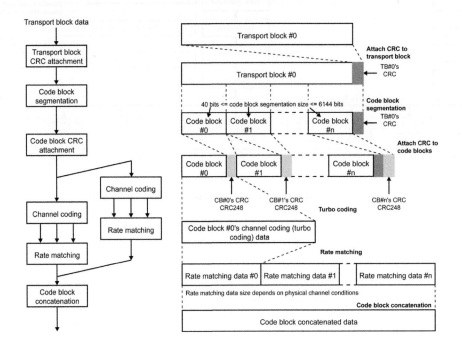

Figure 3.3-12. Transport channel processing for DL-SCH, PCH, and MCH

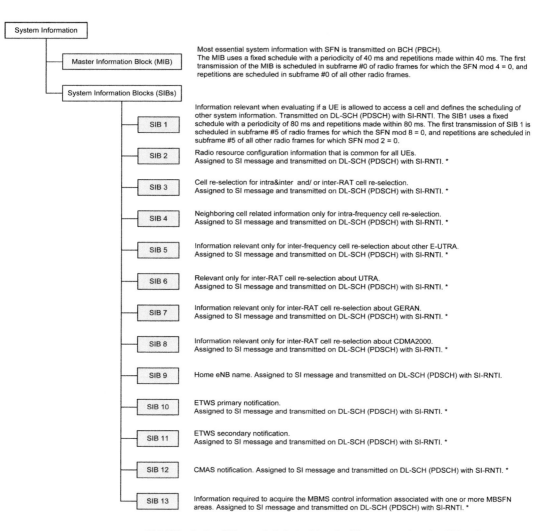

Figure 3.3-13. System Information

3.3.6 Downlink Control Information Coding

The downlink control information (DCI) is mapped to the PDCCH in the physical layer. The DCI carries information regarding the following:

- Transport format information: modulation scheme, coding scheme, redundancy version, new data indicator, cyclic shift for demodulation RS, UL index, CQI request downlink assignment index, HARQ process number, code word information

- Resource allocation information: RB assignment, hopping resource allocation, localized/distributed virtual resource block (VRB) assignment flag, HARQ information)

- Transmit power control (TPC) command

Each DCI carries its own radio network temporary identity (RNTI) to identify the target user. The DCI can have several formats depending on the transmission mode as shown in Table 3.3-6. DCI coding, shown in Figure 3.3-14, comprises the CRC attachment, convolutional encoding, and rate matching. The coded data is multiplexed with other data onto the PDCCH. DCI-specific coding includes CRC scrambling (CRC + RNTI + antenna selection mask) module 2 shown in Figure 3.3-14.

Table 3.3-6. DCI transmission mode and format (based on 36.213 [4] Table 7.1-1)

Transmission mode	Condition of transmission mode	Reference DCI format
1	Single-antenna port; port 0	1, 1A
2	Transmit diversity	1, 1A
3	Open-loop spatial multiplexing	2A
4	Closed-loop spatial multiplexing	2
5	Multi-user MIMO	1D
6	Closed-loop rank=1 precoding	1B
7	Single-antenna port; port 5	1, 1A
8	Dual layer transmission; port 7 or 8 or single-antenna port; port 7 or 8	1A, 2B
9	Multi-layer transmission (up to 8); ports 7–14	1A, 2C

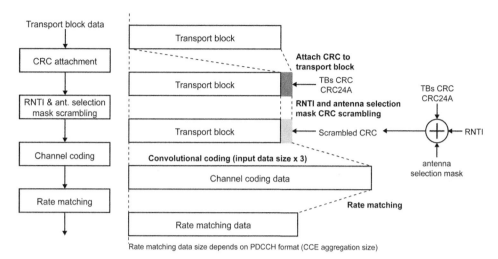

Figure 3.3-14. Control information processing for DCI

DCI formats are defined and transmit these contents as follows.

DCI Format 0: Used for scheduling of PUSCH in one UL cell. Carrier Indicator, Format 0/1A Flag, Frequency Hopping Flag, RB Assignment and Hopping Resource Allocation, MCS Index, NDI, TPC Command for scheduled PUSCH, Cyclic Shift for DMRS and OCC Index, UL Index (TDD only), DAI (TDD only), CQI Request, SRS Request, Multi-cluster Flag.

DCI Format 1: Used for scheduling of one PDSCH codeword in one cell. Carrier Indicator, Resource Allocation Header, Resource Block Assignment, MCS Index, HARQ Process Number, NDI, RV Index, TPC Command for PUCCH, DAI (TDD only).

DCI Format 1A: Used for compact scheduling of one PDSCH codeword in one cell. Carrier Indicator, Flag for Format 0/1A, Localized/Distributed VRB Assignment Flag, Resource Block Assignment, Preamble Index, PRACH Mask Index. Used for random access procedure initiated by a PDCCH order. Flag for Format 0/1A, Localized/Distributed VRB Assignment Flag, Resource Block Assignment, MCS Index, HARQ Process Number, NDI, RV Index, TPC Command for PUCCH (RA-RNTI/P-RNTI/SI-RNTI; N_1A_PRB), DAI (TDD only).

DCI Format 1B: Used for compact scheduling of one PDSCH codeword in one cell with precoding information. Carrier Indicator, Localized/Distributed VRB Assignment Flag, Resource Block Assignment, MCS Index, HARQ Process Number, NDI, RV Index, TPC Command for PUCCH, DAI (TDD only), Transmitteed PMI (TPMI) Information for Precoding, PMI Confirmation for Precoding.

DCI Format 1C: Used for very compact scheduling of one PDSCH codeword. Gap Indicator, Resource Block Assignment, MCS. Used for notifying MCCH change. Information for MCCH Change Notification.

DCI Format 1D: Used for compact scheduling of one PDSCH codeword in one cell with precoding and power offset information. Carrier Indicator, Localized/Distributed VRB Assignment Flag, Resource Block Assignment, MCS Index, HARQ Process Number, NDI, RV Index, TPC Command for PUCCH, Downlink Assignment Index (TDD only), TPMI Information for Precoding, Downlink Power Offset.

DCI Format 2: Carrier Indicator, Resource Allocation Header, Resource Block Assignment, TPC Command for PUCCH, Downlink Assignment Index (TDD only), HARQ Process Number, Transport Block to Codeword Swap Flag, MCS Index/NDI/RV Index for Transport Block 1, MCS Index/NDI/RV Index for Transport Block 2, Precoding Information.

DCI Format 2A: Carrier Indicator, Resource Allocation Header, Resource Block Assignment, TPC Command for PUCCH, Downlink Assignment Index (TDD only), HARQ Process Number, Transport Block to Codeword Swap Flag, MCS Index/NDI/RV Index for Transport Block 1, MCS Index/NDI/RV Index for Transport Block 2, Precoding Information.

DCI Format 2B: Carrier Indicator, Resource Allocation Header, Resource Block Assignment, TPC Command for PUCCH, Downlink Assignment Index (TDD only), HARQ Process Number, Scrambling Identity, Transport Block to Codeword Swap Flag, MCS Index/NDI/RV Index for Transport Block 1, MCS Index/NDI/RV Index for Transport Block 2

DCI Format 2C: Carrier Indicator, Resource Allocation Header, Resource Block Assignment, TPC Command for PUCCH, Downlink Assignment Index (TDD only), HARQ Process Number, Antenna Port/Scrambling Identity/Number of Layers, Transport Block to Codeword Swap Flag, MCS Index/NDI/RV Index for Transport Block 1, MCS Index/NDI/RV Index for Transport Block 2.

DCI Format 3: Used for the transmission of TPC commands for PUCCH and PUSCH with 2-bit power adjustments. 2-bits TPC Command for PUCCH/PUSCH (TPC #1, #2,...).

DCI Format 3A: Used for the transmission of TPC commands for PUCCH and PUSCH with single bit power adjustments. 1-bit TPC Command for PUCCH/PUSCH (TPC #1, #2,...).

DCI Format 4: Used for the scheduling of PUSCH in one UL cell with multi-antenna port transmission mode. Carrier Indicator, Resource Block Assignment and Hopping Resource Allocation, TPC Command for scheduled PUSCH, Cyclic Shift for DMRS and OCC Index, UL Index (TDD only), DAI (TDD only), CQI Request, SRS Request, Multi-cluster Flag, MCS/NDI for Transport Blcok1, MCS/NDI for Transport Block2, Precoding Information and Number of Layer.

In the physical downlink control channel (PDCCH), each separate control channel has its own set of x-RNTI. A shared channel is used for a variety of cases. This Is why the RNTI is used to identify information dedicated a target UE. RNTI types, values, usage, associated transport channels, and associated logical channels are listed in Table 3.3-7.

Table 3.3-7. RNTI values and usages (based on 36.321 [7] Table 7.1-1 and 7.1-2)

RNTI	Value (hexa-decimal)	Usage	Transport channel	Logical channel
P-RNTI	FFFE	Paging and system information change notification	PCH	PCCH
SI-RNTI	FFFF	Broadcast of system information	DL-SCH	BCCH
M-RNTI	FFFD	MCCH Information change notification	N/A	N/A
RA-RNTI	0001-003C	Random access response	DL-SCH	N/A
Temporary C-RNTI	0001-003C 003D-FFF3	Contention resolution(when no valid C-RNTI is available)	DL-SCH	CCCH
Temporary C-RNTI	0001-FFF3	Msg3 transmission	UL-SCH	CCCH, DCCH, DTCH
C-RNTI	0001-FFF3	Dynamically scheduled unicast transmission	UL-SCH	DCCH, DTCH
C-RNTI	0001-FFF3	Dynamically scheduled unicast transmission	DL-SCH	CCCH, DCCH, DTCH
C-RNTI	0001-FFF3	Triggering of PDCCH ordered random access	N/A	N/A
Semi-persistent scheduling C-RNTI	0001-FFF3	Semi-persistently scheduled unicast transmission (activation, reactivation and retransmission)	DL-SCH, UL-SCH	DCCH, DTCH
Semi-persistent scheduling C-RNTI	0001-FFF3	Semi-persistently scheduled unicast transmission (deactivation)	N/A	N/A
TPC-PUCCH-RNTI	0001-FFF3	Physical layer uplink power control	N/A	N/A
TPC-PUSCH-RNTI	0001-FFF3	Physical layer uplink power control	N/A	N/A

3.3.6.1 Control Format Indicator

The control format indicator (CFI) is mapped to the physical control format indicator channel (PCFICH) in the physical layer. The CFI defines how many OFDM symbols are used for the PDCCHs in a subframe. The CFI channel coding, shown in Figure 3.3-15, is a block code with coding rate 1/16. The strong coding indicates the importance of this channel since any decode errors will result in failure to read the PDCCH correctly. The CFI can take the values 1, 2, and 3. For transmission bandwidth configurations greater than 10 RB (1.8 MHz), the number of PDCCH symbols per subframe is the CFI value. For transmission bandwidth configurations less than or equal to 10 RB, the number of PDCCH symbols per subframe is the CFI value plus 1.

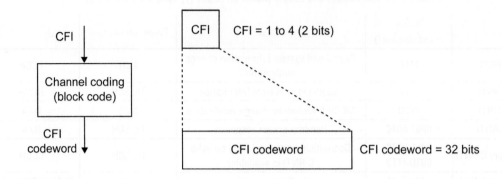

Figure 3.3-15. CFI coding

3.3.6.2 HARQ Indicator

The HARQ indicator (HI) carries hybrid ARQ ACK/NACKs, which take value HI = 1 for a positive acknowledgement (ACK) and HI = 0 for a negative acknowledgement (NACK). The HI is mapped to the PHICH in the physical layer using simple repetition coding to produce a coding rate of 1/3 as shown in Figure 3.3.16. The coded data is further multiplexed onto the PHICH.

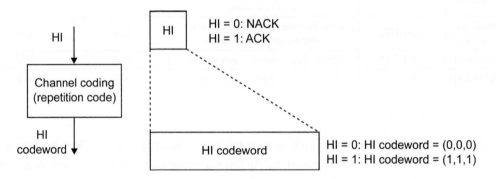

Figure 3.3-16. HI coding and codeword

3.4 Introduction to Physical Layer Signaling

Throughput and latency are two important measures of performance for digital communication systems. High throughput is needed if someone wants to download large files, and superb throughput is possible by shipping a box of DVDs overnight. But the delay (latency) in this case may not be acceptable. Low latency is essential for some applications, such as web browsing and peer-to-peer games, because they require a quick response to a user's requests.

Many of the underlying network protocols, such as the transmission control protocol (TCP), work by passing short messages back and forth for every request at the application level. A simple request by the user may result in hundreds of TCP messages being sent and received. Therefore the time needed to send each of these messages can be multiplied many times.

LTE employs a number of mechanisms in the physical layer to improve both the latency and the overall throughput of the system. This section describes the hybrid automatic repeat requests (HARQ) and adaptive modulation and coding (AMC) procedures as well as the parameters used by the system to report channel state information.

3.4.1 ARQ and HARQ

Automatic repeat request (ARQ) is a layer 2 protocol that has been used for years in the telecommunications industry to ensure that data is sent reliably from one node to another. It uses error detecting (ED) codes such as CRC and a sliding window to identify when an error has occurred in a transmission. If an error occurs, the destination requests a retransmission from the source. The ARQ protocol suffers under poor channel conditions due to excessive retransmissions.

3.4.1.1 Type-I HARQ

Prior to HSDPA and HSUPA, UMTS Rel-99 used Type-I HARQ to overcome ARQ's high retransmission rates by adding forward error correction (FEC) in the form of convolutional or turbo codes. Although the FEC significantly improved the probability of successful transmission in poor signal conditions, the increased redundancy of the FEC code unnecessarily and significantly decreased the amount of user data that was transmitted under strong signal conditions for a given bandwidth.

3.4.1.2 Type-II HARQ

Type-II HARQ is used in HSDPA, HSUPA, HSPA+, and LTE to get around this performance limitation. Like Type-I HARQ, Type-II HARQ uses FEC and ED, but it does so in an iterative manner. On the first transmission of a packet, a subset of the coded bits is transmitted with sufficient information for the receiver to decode the original information bits of the packet and the CRC, but probably with only a small amount of redundancy. This results in high efficiency under good channel conditions in which minimal protection is needed. If the packet is not decoded correctly, a retransmission is triggered. However, rather than resend the same data, the HARQ process selects a different set of coded bits representing the original information bits and the destination node adds this new information to what it received during the first transmission. This process is known as incremental redundancy. With each retransmission the effective code rate decreases until the destination has enough information to decode the packet correctly.

Type-II HARQ uses a mother code that can be punctured to achieve the desired code rate. For LTE this mother code is a rate 1/3 turbo code. This code contains systematic bits, which means that the data and CRC input bits are also present in the output. The first transmission of the packet sends most of these systematic bits plus some redundancy bits. Subsequent retransmissions send fewer of the systematic bits and more of the redundancy bits. The different transmitted versions of the packet that contain different mixes of redundancy and systematic bits are called redundancy versions (RVs). LTE uses four RVs. These are repeatedly sequenced through until the packet is either received correctly or a maximum number of retransmissions have been sent, at which time HARQ declares a failure and leaves it up to ARQ running in the radio link control (RLC) layer to try again with a new packet. More details are included later in Section 3.4.2, Adaptive Modulation and Coding.

3.4.1.3 HARQ Processes

HARQ uses a stop-and-wait protocol. It transmits a packet, waits until it receives an acknowledgement (ACK) or negative acknowledgement (NACK) from the destination, and then sends the next packet. This is shown in Figure 3.4-1. As a result of processing and transmission delays, this cycle takes several subframes. Since it takes only one subframe to transmit the data, quite a bit of bandwidth is underutilized. To fully utilize this bandwidth, HARQ makes use of multiple processes offset in time from each other. Each process transmits a packet to the destination receiver, and before the next transmission time allocation for that process arrives, the HARQ process will have received the ACK or NACK from the destination transmitter and created the next packet for (re)transmission.

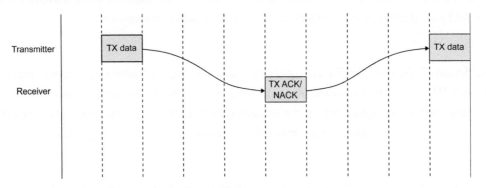

Figure 3.4-1. A single HARQ process

It should be noted that LTE TDD supports a configurable number of HARQ processes with varying timing requirements.

For the FDD uplink the number of HARQ processes is fixed at eight per UE. Thus there are always eight subframes between transmission opportunities for a specific uplink HARQ process. The FDD downlink also supports up to eight HARQ processes per UE, but each process can transmit on any subframe. This technique is more flexible but requires that the HARQ process number be transmitted with the downlink resource assignment. This extra step is not necessary in the uplink since the process number is automatically defined by the position in the frame.

One HARQ entity exists in the eNB for each UE, managing all uplink HARQ processes associated with that UE. Another entity, or the same entity, manages all downlink HARQ processes. The specifications are somewhat ambiguous in this regard, allowing a single HARQ entity to handle both uplink and downlink HARQ processes. However, the specifications clearly do not allow multiple HARQ entities for one UE in one direction.

3.4.1.4 Synchronous vs. Asynchronous HARQ

The downlink uses asynchronous Type-II HARQ transmission. The receiver does not know ahead of time what's being transmitted, so the HARQ process identifier and the RV must be sent along with the data. See Figure 3.4-2. The RV specifies which combination of systematic and redundancy bits are being sent to the UE. This is done through the PDSCH resource allocation messages sent on a PDCCH simultaneous to the corresponding PDSCH transmission. The advantage of such a scheme is that the scheduling algorithm has considerable freedom in deciding which UEs and HARQ processes are scheduled during any one subframe.

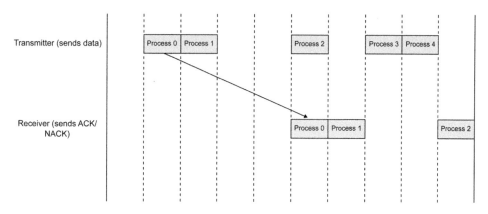

Figure 3.4-2. Downlink HARQ processes associated with a particular UE transmit when data is available and scheduled

In contrast the uplink uses synchronous HARQ transmission. Transmission is synchronous in the sense that once a HARQ process starts, any required retransmissions occur every eighth subframe, as shown in Figure 3.4-3. By using a fixed interval, the uplink does not have to specify a HARQ identifier in the PDCCH allocation message. However, the scheduling algorithm in the uplink is not quite as flexible as that in the downlink as a consequence of this fixed interval.

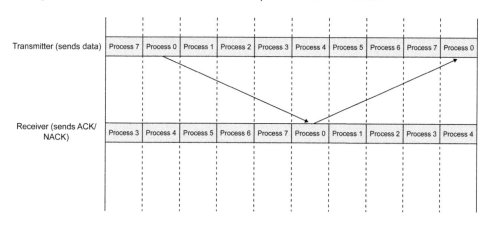

Figure 3.4-3. Retransmissions of an uplink HARQ process that occur every eighth subframe

3.4.2 Adaptive Modulation and Coding

Adaptive modulation and coding (AMC) attempts to match the transmissions from a HARQ process to the channel conditions. Under good channel conditions a higher order modulation format such as 64QAM is used with less redundancy in the initial transmission, allowing a larger transport block to be carried in the allocated channel resources. Under poorer channel conditions a lower order modulation format such as QPSK and possibly more redundancy bits are sent initially to improve the probability of reception. This means, however, that a smaller transport block must be sent. A very low packet error rate implies that the modulation format is too low or that too much redundancy has been used, both of which reduce the transport block size and resulting throughput. At the other extreme, if the packet error rate is high, meaning there are frequent retransmissions of the same data, then either the modulation depth is too high or too little redundancy is being used. Although the packets are large, the error rate is large as well and the corresponding throughput is low. The goal of AMC is to maximize overall throughput, and this occurs at a packet error rate that is slightly greater than zero. A figure of 10% is often the target error rate.

AMC can work only if the eNB is informed of the channel quality seen by the UE. This is accomplished through channel quality indicator (CQI) information sent from the UE in the uplink control information (UCI) scheduled on the PUCCH or PUSCH. CQI reports can be sent periodically or aperiodically according to the configuration by the radio resource control (RRC) higher layer. CQI reporting is defined in 36.213 [4] Subclause 7.2. The topic is discussed more fully in Section 3.4.6.

Given the wide bandwidths supported by LTE and the fact that a particular UE is usually allocated to only a portion of the subcarriers, it is advantageous to allocate the UE to those subcarriers that the UE can best receive. Such frequency-dependent scheduling can only be accomplished if the eNB scheduling algorithm is informed of the channel quality corresponding to different portions of the downlink as seen by the UE. LTE accomplishes this through subband CQI reports in which the UE sends CQI information for independent subbands within the downlink channel. An LTE UE can be configured by the eNB to send either subband or wideband CQI reports. Wideband reports cover the entire channel and reflect the average channel conditions. Wideband reports are used when the channel conditions are changing too fast for the subband reports to be acted upon. The reported CQI, either subband or wideband, is simply an index of the CQI table specified in 36.213, shown in Table 3.4-1.

Table 3.4-1. Four-bit CQI Table (36.213 [4] Table 7.2.3-1)

CQI index	Modulation	Coding rate x 1024	Efficiency
0	out of range		
1	QPSK	78	0.1523
2	QPSK	120	0.2344
3	QPSK	193	0.3770
4	QPSK	308	0.6016
5	QPSK	449	0.8770
6	QPSK	602	1.1758
7	16QAM	378	1.4766
8	16QAM	490	1.9141
9	16QAM	616	2.4063
10	64QAM	466	2.7305
11	64QAM	567	3.3223
12	64QAM	666	3.9023
13	64QAM	772	4.5234
14	64QAM	873	5.1152
15	64QAM	948	5.5547

Each row of the table consists of a specific modulation and code rate. For each CQI reporting period the UE is required to return the highest CQI index that would have resulted in an error probability of less than 10% for a single PDSCH transport block transmitted using the reported modulation and code rate.

It is not mandated that the eNB use the modulation and code rate associated with the received CQI for the target 10% error probability. The eNB's scheduling algorithm uses the CQI as a guide in selecting from the transport block size table (defined in 36.213 [4] Table 7.1.7.2.1-1), which specifies the permissible combinations of transport block size, resource allocation, and modulation scheme that are available for transmission. For example, when the rate of change of the CQI reports exceeds the ability of the HARQ process to keep up, the eNB might use some form of statistical averaging or peak detection of the CQI reports to determine the optimal parameters for transmission. Thus to maintain the high quality of service (QoS) required by a service such as voice over internet protocol (VoIP), the eNB might sacrifice some capacity by selecting a lower error rate target to minimize retransmissions.

Retransmissions of a particular HARQ process use the same modulation and coding scheme as the initial transmission. Each subsequent retransmission simply reduces the effective code rate through incremental redundancy. In the event that the channel characteristics change considerably after the initial transmission, the medium access control (MAC) higher layer is free to abort that transmission and start a new one using a more appropriate modulation and coding scheme. This is done by toggling the new data indication (NDI) bit in the PDSCH or PUSCH allocation message sent by the PDCCH.

LTE uses a clever algorithm to implement incremental redundancy and adaptive coding, as illustrated in Figure 3.4-4.

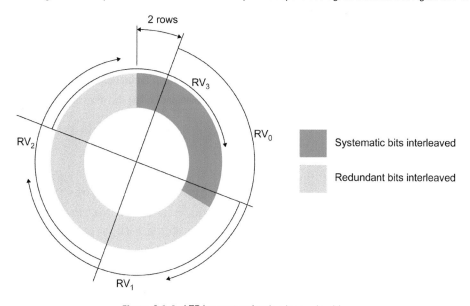

Figure 3.4-4. LTE incremental redundancy algorithm

All the redundant bits from the turbo encoder are included in the circular buffer for the uplink. The number of redundant bits included in the circular buffer for the downlink is configured by the eNB.

The systematic bits from the turbo encoder are interleaved and placed into a circular buffer called the soft buffer. The redundancy bits are then interleaved and placed after the systematic bits. All the redundancy bits are included in the soft buffer used for uplink transmissions, but the number of redundancy bits included for downlink transmissions is defined by upper layers. Bits are copied from the buffer starting at a position that is dependent on the RV. The starting position for RVn is approximately n/4 of the way around the circular buffer, plus a fixed offset of two interleaved rows. The number of bits pulled from the circular buffer for each RV is dependent on the target code rate. For poor channel conditions the code rate approaches 0.1, in which case the entire soft buffer is transmitted multiple times for each RV. In excellent channel conditions the code rate approaches 0.92, which means that the number of bits transmitted in each RV is slightly more than the number of bits in the transport block.

3.4.3 ARQ Assist

An ARQ receiver has to rely on a skipped sequence number to identify when a packet has been corrupted in transit. This technique assumes that a subsequent packet will be received that identifies the gap in the sequence numbers. Once this gap has been identified, a retransmission can be requested immediately by the receiver.

In practice, ARQ on its own cannot fully rely on skipped sequence numbers. Suppose that the last packet transmitted was corrupted. In that case a subsequent packet will not be received to identify the gap in sequence numbers. The receiver will simply wait idly for the next packet and the system will stall. To avoid this problem ARQ is assisted by a timer that starts when a packet is transmitted. If the sender does not receive either an ACK or a NACK response from the receiver before the timer expires, a retransmission is attempted. In most cellular standards the ARQ timer is configurable but typically is set for around 100 ms, which is a long time by LTE standards.

Recall that a HARQ process will repeatedly send a packet approximately every 8 ms until it is acknowledged by the receiver. After a configurable number of attempts HARQ will give up and inform ARQ that the transmission attempt has failed. ARQ can immediately attempt a retransmission, possibly with different modulation and coding. No timer is needed and the delay is minimized.

3.4.4 Technical Challenges

The LTE specifications impose constraints on the UE and eNB regarding the amount of time they have to complete the HARQ process. The receiver has three subframes in which to decode the packet, check the CRC, and encode the ACK/NACK. See Figure 3.4-5. Assuming that the transmitter sent the data in subframe n, the ACK/NACK must be sent back to the transmitter in subframe n+4.

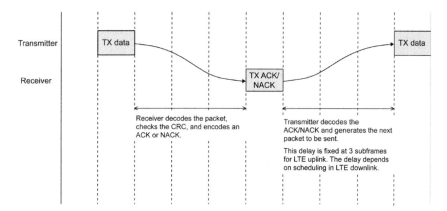

Figure 3.4-5. A single HARQ process

The transmitter now also has three subframes in which to decode the ACK/NACK returned from the receiver, construct the next transport block based on the ACK/NACK (this is a job for RLC and MAC), and finally encode the transport block if it hopes to transmit this HARQ process in subframe n+8. This short turnaround time is required in the uplink, but can be delayed in the downlink through scheduling. A lower cost eNB such as a Home eNB (femtocell) might use this technique to reduce the processing demands it imposes on itself. Although the specifications allow the eNB to retransmit a particular HARQ process earlier than subframe n+8, such an early retransmission would impose significantly more technical challenges on the eNB processing elements.

Given that each subframe is utilized by a different HARQ process, some assumptions can be made regarding the execution times allowed for each of the processing steps listed above. If only one processing unit for each step is multiplexed in time between all HARQ processes, the computations associated with each step cannot exceed 1 ms. Should any one of the steps exceed 1 ms, the processing unit won't keep up with the continuous flow of HARQ information in consecutive subframes.

3.4.5 Summary of HARQ and AMC

Type-II HARQ and AMC work together to provide a very adaptive transport mechanism in LTE. Adaptive modulation and coding tunes the initial HARQ transmission to use a coding rate that results in approximately the ideal packet error ratio from a throughput perspective. Type-II HARQ then uses incremental redundancy to add redundancy bits for each successive retransmission, thereby reducing the effective code rate until the packet can be decoded correctly. The result, although not perfect, is a means of optimizing the overall throughput over wide ranges of dynamically changing channel conditions while minimizing latency.

3.4.6 Uplink Control Signaling

3.4.6.1 Introduction to Uplink Control Signaling

Reporting the channel conditions experienced by the receiver represents the baseline for many adaptation algorithms such as frequency selective scheduling, MIMO precoding, adaptive modulation and coding (AMC), and rank adaptation.

The LTE specifications define four indicators that reflect the channel state information (CSI) as seen by the recipient of the communication (that is, by the UE): the channel quality indicator (CQI), the rank indication (RI), the precoding matrix indicator (PMI), and the precoding type indicator (PTI). The reporting of these indicators is done either aperiodically using the PUSCH or periodically using the PUCCH or PUSCH.

For calculation of these indicators, the UE will generally perform measurements on the cell-specific reference signals (CRS), although the CSI reference signals (CSI-RS) are introduced in Release 10 and can be used in transmission mode 9 (up to 8 layer transmission, ports 7–14). The CSI-RS allows the eNB more flexibility in configuring higher layer operation. See Sections 2.4.4.9 and 3.2.12.4 for more details.

For MIMO operation the selection of closed-loop mode or open-loop mode depends on the rate of change in the channel conditions. For low speed UEs, closed-loop MIMO with precoding is beneficial as is frequency-selective RB scheduling. Indeed, the likelihood of variable MIMO conditions across the channel means closed-loop MIMO works best in conjunction with frequency selective scheduling. For moderate to high speed UEs, the preferred mode of operation is space frequency block coding (SFBC) transmit diversity or spatial multiplexing with a fixed precoding matrix (open-loop MIMO).

3.4.6.2 Channel Quality Indicator

The CQI plays a central role in LTE scheduling and link adaptation. The CQI is computed at the UE receiver for each codeword on either the full transmission bandwidth configuration (wideband CQI) or on groups of resource blocks known as subbands. The CQI reflects the level of noise and interference experienced by the receiver on a particular portion of the channel. It can be thought of as a measure of signal-to-interference plus noise ratio (SINR) but in fact is coded in terms of the modulation and coding scheme (MCS) required for a particular error-rate probability. Moreover, some compression is used when reporting subband CQI using differential encoding.

Depending on the reporting mode (all subbands, wideband, or preferred subbands) the CQI is used by the eNB as an input to the process for scheduling traffic. When the eNB receives the indices of those subbands in which the receiver experiences the highest CQI, it can allocate those subbands to efficiently schedule user transmissions. This technique is called frequency-selective scheduling.

3.4.6.3 Rank Indication

The RI is a value computed by the UE representing the preferred number of layers to be used in the next downlink transmissions to the UE. The value reflects the actual usable rank of the channel; in other words, the maximum number of spatially parallel transmissions supported by the channel but limited to the UE capabilities. The maximum number of layers is eight, but some UE categories support only one, two, or four layers. See 36.306 [8] Table 4.1-1.

Typically, the RI is computed as follows: for each possible rank, the subbands are ordered by decreasing level of throughput. The M best subbands are selected and the sum throughput is derived. The rank maximizing the sum throughput is chosen and fed back to the transmitter. Computing the rank on the best subbands is justified by the fact that the UE is likely to be scheduled on these subbands.

3.4.6.4 Precoding Matrix Indicator

One of the techniques used in LTE to improve single user spatial multiplexing (SU-MIMO) performance is precoding. Precoding involves adapting the transmitted signal to the current CSI. In the FDD mode, downlink channel conditions can be very different from uplink channel conditions due to the duplex frequency spacing. Thus a method for reporting the CSI to the transmitter is needed. The PUSCH and PUCCH are used to carry this information. In theory, a full CSI report for each resource block at the maximum update rate achieves the best performance, but the accompanying overhead in the uplink channel is unacceptable. Compression mechanisms have therefore been designed to decrease the amount of feedback. These introduce important loop parameters such as granularity, feedback rate, and codebook size. There are many methods that can be used for compression but the one chosen for LTE is linear codebook-based precoding. In codebook-based techniques, the preferred precoding matrix is chosen from a finite set of predefined matrices (the codebook) that are based on some selection criteria.

The size of the codebook defined for LTE varies with the number of antenna ports and rank. For transmissions using two antenna ports, a codebook of four unitary matrices is defined, while in the four antenna port case the codebook size is equal to 16. The largest codebook size, in which eight antenna ports are used, is 109. When the number of layers is smaller than number of antenna ports (i.e., when transmit diversity is being used), subsets of the full rank codebook are defined..

The codebook for four antenna transmission is based on Householder reflection matrices. The codebook matrices W_n are derived from unit vectors u_n by the following equation:

$$W_n = I - \frac{2}{\|u_n\|^2} u_n u_n^H$$

The codebook and unit vectors definitions can be found in 36.211 [2] Subclause 6.3.4.2.3.

The PMI is defined as the index to the preferred precoding matrix within the codebook matrices. The PMI is 4 bits wide for four antenna ports or 2 bits wide for two antenna ports. For eight antenna ports, there are two separate PMI values reported, each 4 bits wide, the combination of which allows a single precoding matrix to be selected. The PMI selection process is not specified and is left as an implementation choice. It will usually be based on metrics such as SINR maximization and sum throughput maximization.

To further reduce the amount of feedback, the PMI is compressed in both the time and frequency domains. Since adjacent resource blocks are likely to share the same PMI, they are grouped in subbands and one PMI is computed per subband. A subband is defined as a localized frequency resource unit for which a separate CQI is reported. In practice, a subband is a group of k adjacent resource blocks. The early LTE specifications considered distributed RB allocation as a means of achieving frequency diversity but dropped this in favor of localized allocation in conjunction with frequency-selective scheduling.

Defining the optimal size k of a subband in terms of performance and overhead is called the granularity tradeoff. The optimal granularity is highly dependent on the coherence bandwidth of the channel. For frequency-selective channels in which conditions vary significantly across the channel, subband size is preferably small while for flat-fading (or correlated-fading) channels, a wideband value (one value for all subbands) is enough. To reflect this, various reporting modes have been defined.

The fastest periodic PMI reporting rate is once every 2 ms, although aperiodic reports can be requested every 1 ms subframe. Longer periodicity values are possible up to a 160 ms interval. The reporting rate is semi-statically configured by higher layers.

For transmission mode 9, two PMIs are reported. The PTI can be used to indicate which PMI is currently being reported.

3.4.6.5 Methods of Reporting

The CQI, PMI, PTI, and RI are reported in control-indication fields on either the PUCCH (periodically) or the PUSCH (aperiodically). The time and frequency resources for reporting the indicators are managed and semi-statically configured by the eNB. Whether to use PUCCH or PUSCH depends on the existence of a UL-SCH; that is, whether the user is already scheduled on the PUSCH for that particular subframe. The periodic reporting on PUCCH is used only on those subframes in which the UE has no PUSCH allocation and in the case of carrier aggregation is permitted only on the primary cell. If the user has been allocated a PUSCH, the CQI, PMI, PTI, and RI are sent over that channel even if there is no additional user data to transmit. The reporting can be roughly categorized in three modes:

- Wideband
- UE-selected
- eNB-selected (PUSCH only).

In wideband mode, the CQI report is valid for the entire transmission bandwidth configuration (the entire channel). This is referred to as reporting over the entire set of subbands (S). One value of CQI and one or multiple values of PMI are reported.

In UE-selected mode, the reports are derived on the so-called best-M subbands. The UE selects the M subbands (usually covering 25% of the bandwidth) with the highest CQI. A single CQI plus PMI is computed reflecting the channel conditions as if using these preferred bands only. Moreover, wideband values are also reported.

In eNB-selected mode, CQI and PMI are reported per subband. To help reduce the uplink overhead, differential encoding with respect to the wideband value is used.

The subband size as well as the M number of preferred bands are signaled by higher layers and vary with the channel bandwidth.

3.4.6.5.1 Aperiodic Reporting

Aperiodic reporting is done exclusively on the PUSCH and is triggered by an indication sent in the scheduling grant. When the UE has been assigned more than one serving cell, the indication may trigger a report on the channel conditions of more than one cell, although there will only be at most one indication sent in a subframe. The modes available when reporting on the PUSCH are shown in Table 3.4-2.

Table 3.4-2. CQI and PMI feedback types for PUSCH reporting modes (36.213 [4] Table 7.2.1-1)

		PMI feedback type		
		No PMI	Single PMI	Multiple PMI
PUSCH CQI feedback type	Wideband (wideband CQI)			Mode 1-2
	UE-selected (subband CQI)	Mode 2-0		Mode 2-2
	Higher-layer-configured (subband CQI)	Mode 3-0	Mode 3-1	

PUSCH Modes 2-0 and 3-0

These modes are used in single-stream and open-loop systems (SISO, transmit diversity, and open-loop spatial multiplexing) or in the transmission mode 8 (dual-layer transmission, ports 7 and 8) or transmission mode 9 if the UE is configured without PMI/RI reporting. The PMI is not reported and a wideband CQI plus a subband CQI (mode 3-0, differential encoding) or a best-M subband CQI (mode 2-0) are reported for the first codeword.

PUSCH Modes 1-2, 2-2, and 3-1

These modes are used in combination with closed-loop spatial multiplexing, closed-loop rank-one precoding, or transmission mode 8 or 9 if the UE is configured with PMI/RI reporting. One wideband CQI value is computed for all three modes. In addition, a best-M subband CQI is provided for the mode 2-2 case, and a per-subband CQI is provided for the mode 3-1 case. Each CQI value is reported per codeword and the per-subband CQI values differentially encoded relative to the reported wideband CQI value. The PMI is reported once for the full system bandwidth (mode 3-1), once for each subband (mode 1-2, mode 2-2) and once assuming transmission on the best-M subbands (mode 2-2).

The RI is reported in all modes (except for closed-loop rank-one precoding) and is computed over the full system bandwidth. In UE-selected modes, the preferred M subband locations are also reported by the UE using a uniquely defined combinatorial index (label).

3.4.6.5.1 Periodic Reporting

Periodic reporting may be sent on the PUCCH, or it may be sent on the PUSCH if it has been scheduled in the same subframe in which the reporting is due. Since the PUCCH supports only very low data rates, some of the aperiodic reporting modes are unavailable or modified accordingly. In the case of multiple subband reports, CQI values are given for particular parts of the bandwidth (called bandwidth parts or BPs). Each BP spans one or several consecutive subbands in the frequency domain. An example of the concept is given in Figure 3.4-6 for a 10 MHz bandwidth (50 RB). There are 6 RB per subband, nine subbands, and three BPs.

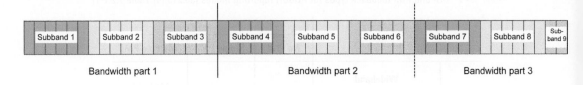

Figure 3.4-6. Illustration of the concept of subbands and bandwidth parts

The periodicity, the reporting offsets, and the reporting modes are parameters configured by higher layers. The RI reports are configured so that their periodicity is a multiple of the CQI and PMI reporting periods. Strict priority rules are applied to avoid collision between the reports.

The modes available when reporting on the PUCCH are shown in Table 3.4-3.

Table 3.4-3. CQI and PMI feedback types for PUCCH reporting modes (36.213 [4] Table 7.2.2-1)

		PMI feedback type	
		No PMI	Single PMI
PUCCH CQI feedback type	**Wideband** (wideband CQI)	Mode 1-0	Mode 1-1
PUCCH CQI feedback type	**UE-selected** (subband CQI)	Mode 2-0	Mode 2-1

PUCCH Modes 1-0 and 2-0

These modes are used in single stream and open-loop systems. The following indicators are reported:

- RI (open-loop spatial multiplexing only) calculated on the full system bandwidth
- Wideband CQI (conditioned by RI, computed on the full system bandwidth)
- Additional UE-selected subband CQI wideband report (mode 2-0 only). Selection of preferred subbands is done within a bandwidth part. One CQI value is computed assuming transmission in the preferred subbands only. Reports for each BP are given consecutively on successive reporting instances. To indicate the location of the preferred subbands, a bit label is also reported.

Modes 1-1 and 2-1

These modes are used in combination with closed-loop spatial multiplexing, closed-loop rank-one precoding, multi-user MIMO (MU-MIMO), transmission mode 8, or transmission mode 9. The following indicators are reported:

- RI calculated on the full system bandwidth
- PMI calculated on the full system bandwidth, conditioned by RI; for transmission mode 9, a second PMI may also be reported
- Wideband CQI (conditioned by RI, computed on the full system bandwidth assuming that single matrix precoding was applied)
- 3-bit wideband spatial differential CQI (if RI > 1)
- Additional UE selected subband CQI wideband report (mode 2-1 only) as in mode 2-0 above
- Additional 3 bit spatial differential CQI (mode 2-1 only), calculated by taking the difference between the CQI value for codeword 1 and the CQI value for codeword 2 (assuming use of latest matrix precoding and full system bandwidth).

Scheduling

Each mode defines a number of reports, combining several report types in a report instance. The parameters for the scheduling of the different reports are semi-statically configured by higher layers as follows:

J = number of bandwidth parts

K = number of reports of subband CQI needed per BP

N_P = reporting period between two CQI reports

$H = J*K + 1$ = reporting instances required for a complete CQI/PMI report
    ($J*K$ reports of subband CQIs + 1 report of wideband CQI/PMI)

N_P*H = reporting period of complete CQI/PMI (wideband + subbands)

M_{RI} (multiple of P) = reporting period of RI

$N_{OFFSET,RI}$ = offset between RI report and wideband CQI/PMI report

Figure 3.4-7 depicts a Mode 2-1 report with the following parameters:

$N_P = 2, J = 3, K = 1, H = 4, M_{RI} = 8, N_{OFFSET,RI} = 1.$

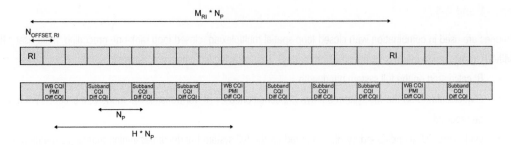

Figure 3.4-7. Reporting scheduling example for Mode 2-1 and 10 MHz BW (RI>1)

3.5 Physical Layer Procedures

This section describes two important physical layer procedures:

- Cell search and synchronization using the primary and secondary synchronization signals in conjunction with the BCH
- The random access procedure using the PRACH.

3.5.1 Synchronization and Cell Search

Cell search is the procedure by which a UE acquires time and frequency synchronization with a cell and detects that cell's physical layer Cell ID. LTE cell search supports a scalable overall transmission bandwidth corresponding to six or more resource blocks. LTE cell search is based on the downlink physical signals: the primary and secondary synchronization signals, the reference signals, and the physical broadcast channel (PBCH). Figure 3.5-1 shows the downlink mapping of these signals.

As shown in Figure 3.5-1 and described more fully in Section 3.2.5, the primary synchronization signal and secondary synchronization signal are transmitted over the center 62 subcarriers (930 kHz) in the first and sixth subframe of each 10 ms frame. The reference signals are transmitted on every sixth subcarrier across the entire channel. Neighbor cell search is also based on the same downlink signals as the initial cell search. The presence of these signals at the center of the channel means that one set of procedures will work for all six supported LTE channel bandwidths of 1.4 MHz, 3 MHz, 5 MHz, 10 MHz, 15 MHz, and 20 MHz.

From Release 10 onwards, the possibility exists for carrier aggregation (CA) wherein the component carriers (CC) defined in Release 8 can be aggregated either contiguously or non-contiguously to create a wider total allocated bandwidth. In order to maintain backwards compatibility with Release 8, the use of CA does not introduce new bandwidths beyond the original six defined for the Release 8 CCs.

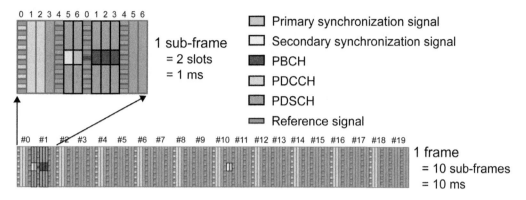

Figure 3.5-1. Downlink mapping showing primary and secondary synchronization signals

3.5.1.1 Cell Identification

There are 504 unique physical layer cell identities in 168 unique physical layer cell identity groups, each group containing three unique identities. The grouping is such that each physical layer cell identity is part of one and only one physical layer cell identity group. A physical layer cell identity is thus uniquely defined by a number in the range of 0 to 167, representing the physical layer cell identity group, and a number in the range of 0 to 2, representing the physical layer identity within the physical layer cell identity group. Refer to 36.211 [2] Subclause 6.11.

3.5.1.2 Cell Search Procedure

The initial cell search begins when the UE is switched on causing the universal subscriber identity module (U-SIM) to issue a cell search request procedure. At first the UE does not have information of nearby cells, so it starts to look for the strongest cells in the band of interest. When the UE finds the strongest subcarriers carrying the synchronization signals and the PBCH, the frequency synchronization procedure begins. Figure 3.5-2 outlines the process.

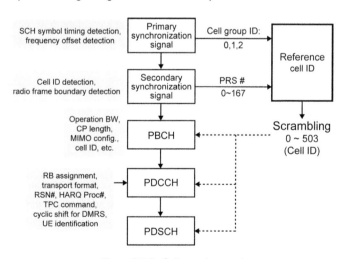

Figure 3.5-2. Cell search procedure

143

The UE first looks for the primary synchronization signal from which it will be able to find the exact carrier frequency and the timing of slot 0 or 10. At this point the UE cannot tell which half of the frame structure it has found since the primary synchronization signal is repeated twice per frame. By trial and error the UE will determine the cyclic prefix (CP) configuration. The next step is to decode the primary synchronization signal to determine to which of three identities the cell belongs. The primary synchronization signal is encoded as a Zadoff-Chu sequence of phase shifts. Each of the 62 subcarriers has the same power level with its phase determined by the root index number in a Zadoff-Chu sequence generator as defined in 36.211 [2] Subclause 6.11.1. This produces the characteristic circular constellation pattern seen earlier in Figure 3.2-6. The Zadoff-Chu sequence is chosen because of its excellent autocorrelation properties and the low cross correlation with the other sequences. This makes it easy for the UE to decode to which of the three identities the cell belongs.

Next the UE decodes the secondary synchronization signal, which is transmitted one OFDM symbol before the primary synchronization signal. The secondary synchronization signal is encoded as an interleaved concatenation of two length-31 binary sequences as defined in 36.211 [2] Subclause 6.11.2. This creates a BPSK pattern as seen earlier in Figure 3.2-7. The first 31 subcarriers use one sequence and the second 31 carriers use the other sequence. By cyclic-shifting each sequence, it is easy to create the necessary 168 unique identity groups. Once the secondary synchronization signal is decoded, the UE will have removed the uncertainty in the frame timing and be able to calculate the cell ID from the 504 possibilities.

With the cell ID and frame timing known the UE can read the PBCH. The PBCH carries only the most basic information about the cell (bandwidth, frame number, etc.). To decode information such as the public land mobile network (PLMN) identity, the UE must further decode the system information (SI) messages, which are carried on the broadcast control channel (BCCH) logical channel sent over the downlink shared channel (DL-SCH) transport channel. Once a valid PLMN identity is known, the UE can start to register with the cell.

The point in the cell search procedure at which the UE makes use of the reference signals may vary. The reference signals do not carry any unique information but provide a known phase and amplitude reference essential for reliable decoding of the downlink in difficult channel conditions. The RS cannot be used until after the cell ID is determined, but it is possible to read the PBCH without the aid of the RS. However, when the UE starts to decode other parts of the downlink that are not in the center of the channel—for example, the PDCCH—it is essential for the RS phase and amplitude reference to be used to ensure reliable decoding.

With the introduction of CA in Release 10, the cell with which the UE registers is known as the primary cell (PCell), and if an additional carrier is aggregated, it is known as the secondary cell (SCell).

3.5.2 Random Access and Paging

Refer to 36.213 [3] Subclause 6 and 36.300 [9] Subclause 10.1.5. The physical layer random access preamble (PRACH) occupies six contiguous RB and can be allocated anywhere across the channel and in any of the 10 subframes as shown in Figure 3.5-3.

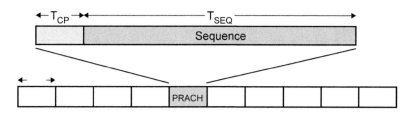

Figure 3.5-3. PRACH structure

The PRACH subframe consists of a cyclic prefix of length T_{CP} and a sequence part of length T_{SEQ}. The parameter values, listed in Table 3.5-1, depend on the preamble format, which is controlled by the higher layers.

Table 3.5-1. Random access preamble parameters

Preamble format	Cyclic prefix (T_{CP})	Sequence part (T_{SEQ})
0	$3168*T_s$	$24576*T_s$
1	$21024*T_s$	$24576*T_s$
2	$6240*T_s$	$2*24576*T_s$
3	$21024*T_s$	$2*24576*T_s$
4 (frame structure type 2 only)	$448*T_s$	$4096*T_s$

For PRACH preamble format 0, T_{SEQ} represents a long Zadoff-Chu sequence of $24756*T_S$ or 800 µs. The PRACH is created in the frequency domain. When converted to the time domain, a cyclic shift can be applied and a CP is added. The UE transmits the PRACH at zero timing advance (TA) but it is received at the eNB sometime later. To prevent interference between PRACH from two UE scheduled in adjacent subframes but possibly at very different positions within the cell, it is necessary to insert a guard period in the form of the CP such that the eNB can decode the PRACH regardless of how late it arrives, up to the length of the CP. For PRACH format 0, T_{CP} is 3168 T_S or 103 µs. This represents a round-trip distance of 30 km, which indicates a maximum cell radius of 15 km.

The transmission of a random access preamble, if triggered by the MAC layer, is restricted to certain time and frequency resources. These resources are enumerated in increasing order of the subframe number within the radio frame and the physical resource blocks in the frequency domain such that index 0 corresponds to the lowest numbered physical resource block and subframe within the radio frame.

For preamble formats 0 to 3, there is at most one random access resource per subframe for FDD. Table 3.5-2 lists the subframes in which random access preamble transmission is allowed for a given PRACH configuration. The start of the random access preamble is aligned with the start of the corresponding uplink subframe at the UE assuming a timing advance of zero. For configurations 0, 1, 2, and 15, the UE may for handover purposes assume an absolute value of the relative time difference between the radio frame in the current cell and the target cell of less than $153600*T_S$.

Table 3.5-2. PRACH timing for different PRACH configuration

PRACH configuration	System frame number	Subframe number
0	Even	1
1	Even	4
2	Even	7
3	Any	1
4	Any	4
5	Any	7
6	Any	1, 6
7	Any	2 ,7
8	Any	3, 8
9	Any	1, 4, 7
10	Any	2, 5, 8
11	Any	3, 6, 9
12	Any	0, 2, 4, 6, 8
13	Any	1, 3, 5, 7, 9
14	Any	0, 1, 2, 3, 4, 5, 6, 7, 8, 9
15	Even	9

For TDD, the PRACH configuration index specifies whether the system frame number that should be used is even or odd, but does not directly specify a subframe number. Instead it specifies an offset into the first or second half of the frame. This fits better with the fact that not all uplink subframes are available in all of the uplink-downlink configurations. For TDD, the PRACH configuration index may also specify a frequency offset.

3.5.2.1 Application of PRACH in LTE

The Zadoff-Chu sequences of the PRACH preamble are excellent for auto-correlation. In the eNB, the cross-correlation of the received signal with respect to the ideal preamble is performed in order to determine the transmission delay. This allows the UE TA to be set in such a way that future transmissions from different UEs will arrive aligned to the uplink frame structure at the eNB.

A preamble root sequence is a non-repeating waveform with a basic length of 800 μs, which fits into one subframe. It can be used to identify the timing of a UE for distances up to 120 km (240 km round trip) from the base station. If the eNB receives two peaks within the time range of one cell, it will consider the condition to be a collision and not reply to either UE.

In case of large cells, the eNB allocates many root sequences for the PRACH to differentiate between the UEs. Each additional root sequence makes the PRACH detection process for the eNB harder. The situation becomes easier when the cells are smaller because then a smaller number of PRACH root sequences can be cyclically shifted in time relative to each other to differentiate between the UEs. When a UE is transmitting a cyclically shifted sequence, the eNB will have a late peak but will know that this peak was actually generated by a UE close by.

3.5.2.2 Detection Effort of the PRACH Preambles

There is a tradeoff between the cell size and the detection effort of the 64 PRACH preambles. Table 3.5-3 shows the relation between effort, cell radius, the number of root sequences, and the number of cyclic shifts per root sequence.

Table 3.5-3. PRACH detection effort vs. cell radius

Cell radius (Km)	Number of PRACH preambles and detection effort in %	No. of cyclic shifts per root preamble
0–1.875 km	1—100%	64
1.875–3.75 km	2—200%	32
3.75–7.5 km	4—400%	16
7.5–15 km	8—00%	8
15–30 km	16—1600%	4
30–60 km	32—3200%	2
> 60 km	64—6400%	1

When the cell radius is smaller than 1.875 km, only one single root sequence is needed within 64 possible cyclic shifts to achieve the lowest PRACH detection effort.

3.5.2.3 Random Access Procedure

At the beginning of the random access procedure, the UE is informed of the parameters and the allowed location in frequency and time of the PRACH. The procedure is similar to UMTS whereby the UE randomly chooses between the root sequences and their cyclically shifted versions and transmits periodically according to a power ramping-procedure until the UE receives a response from the eNB. This is shown in Figure 3.5-4.

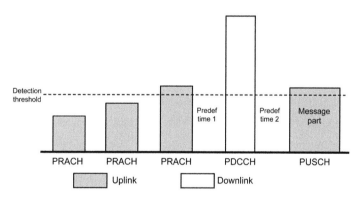

Figure 3.5-4 Random access procedure

The procedure in LTE varies slightly from UMTS, however, because there is no acquisition indictor channel (AICH) and so the eNB responds on the PDCCH to provide the sequence number (SN), the TA, and the allocated resources on the PUSCH. The UE responds on the PUSCH with the random access message. Consequently, the random access channel (RACH) messages are scheduled similar to the way normal uplink packets are scheduled.

3.6 Physical Layer Measurements and Radio Resource Management

The UE and the eNB are required to make physical layer measurements of the downlink and uplink radio characteristics, respectively. The measurement definitions are specified in 36.214 [5]. The measurements are reported to the higher layers and are used for a variety of purposes including intra- and inter-frequency handover, inter-radio access technology (inter-RAT) handover, timing measurements, and other purposes in support of radio resource management (RRM).

Although the physical layer measurements are defined in 36.214 [5], the measurement conditions and accuracy requirements are provided in Subclauses 9 and 10 of the RRM specification 36.133 [10]. The broader subject of RRM will be discussed in Section 3.6.3.

3.6.1 UE Physical Layer Measurements

The UE physical layer measurements are all either measures of absolute power or power ratios. They are defined for operation within LTE-only systems. In addition, to enable interworking of LTE with other radio access technologies, LTE UE must have the ability to measure equivalent parameters from the other systems LTE is defined to work with. These are UMTS FDD, UMTS TDD, GSM, and cdma2000 based systems.

3.6.1.1 Reference Signal Receive Power

Reference signal receive power (RSRP) is the most basic of the UE physical layer measurements and is the linear average (in watts) of the downlink reference signals (RS) across the channel bandwidth. Since the RS exist only for one symbol at a time, the measurement is made only on those resource elements (RE) that contain cell-specific RS. It is not mandated for the UE to measure every RS symbol on the relevant subcarriers. Instead, accuracy requirements have to be met. There are requirements for both absolute and relative RSRP. The absolute requirements range from ±6 dB to ±11 dB depending on the noise level and environmental conditions. Measuring the difference in RSRP between two cells on the same frequency (intra-frequency measurement) is a more accurate operation for which the requirements vary from ±2 dB to ±3 dB. The requirements widen again to ±6 dB when the cells are on different frequencies (inter-frequency measurement).

Knowledge of absolute RSRP provides the UE with essential information about the strength of cells from which path loss can be calculated and used in the algorithms for determining the optimum power settings for operating the network. Reference signal receive power is used both in idle and connected states. The relative RSRP is used as a parameter in multi-cell scenarios.

3.6.1.2 Reference Signal Receive Quality

Although RSRP is an important measure, on its own it gives no indication of signal quality. Reference signal receive quality (RSRQ) provides this measure and is defined as a ratio of RSRP and E-UTRA carrier received signal strength indicator (RSSI). The formula is as follows:

$$N \times \text{RSRP}/(\text{E-UTRA carrier RSSI})$$

where N is the number of RBs of the E-UTRA carrier RSSI measurement bandwidth.

The E-UTRA carrier RSSI parameter represents the entire received power measured only on OFDM symbols containing reference symbols for antenna port 0, including the wanted power from the serving cell as well as all co-channel power and other sources of noise. Measuring RSRQ becomes particularly important near the cell edge when decisions need to be made, regardless of absolute RSRP, to perform a handover to the next cell. Reference signal receive quality is used only during connected states. Intra- and inter-frequency absolute RSRQ accuracy varies from ±2.5 to ±4 dB, which is similar to the inter-frequency relative RSRQ accuracy of ±3 dB to ±4 dB.

3.6.1.3 UTRA FDD CPICH Received Signal Code Power

Received signal code power (RSCP) comes from UMTS and is a measure of the absolute power of one code channel within the overall UTRA CDMA signal. UTRA FDD CPICH RSCP is therefore a measure of the code power of the common pilot indicator channel (CPICH) and is used for interworking between LTE and UMTS. It has the same basic function as RSRP in LTE and is used in LTE inter-RAT idle and inter-RAT connected states.

3.6.1.4 UTRA FDD Carrier Received Signal Strength Indicator

UTRA FDD received signal strength indicator (RSSI) is also inherited from UMTS. It is a measure of the total received power, including thermal noise and noise generated in the receiver, within the bandwidth defined by the receiver pulse shaping filter. It is the UTRA equivalent of the E-UTRA carrier RSSI defined as part of RSRQ.

3.6.1.5 UTRA FDD CPICH E_C/N_0

This final measurement from UMTS is the ratio of the CPICH to the power density in the channel. If receive diversity is not being used by the UE, CPICH E_C/N_0 is the same as CPICH RSCP divided by RSSI. A typical value in a UMTS cell without significant noise would be around -10 dB; indicating that the CPICH had been set 10 dB below the total power of the cell. The UTRA FDD CPICH E_C/N_0 is used in LTE inter-RAT idle and connected states.

3.6.1.6 GSM Carrier RSSI

When LTE has to interwork with GSM-based systems including GPRS and E-GPRS (EDGE), the GSM version of RSSI must be measured. GSM RSSI is measured on the broadcast control channel (BCCH). It is used in LTE inter-RAT idle and connected states.

3.6.1.7 UTRA TDD Carrier RSSI

This measurement is used for interworking with UTRA TDD systems and performs the same basic function as the other RSSI measurements. It is used in LTE inter-RAT idle and connected states.

3.6.1.8 UTRA TDD P-CCPCH RSCP

This measurement is the UTRA TDD equivalent of RSRP. It is a measure of the code power of the primary common control physical channel (P-CCPCH) and is used in LTE inter-RAT idle and connected states.

3.6.1.9 cdma2000 1xRTT Pilot Strength

This measurement is the RSRP equivalent for cdma2000-based technologies. These technologies all share the same radio transmission technology (RTT) bandwidth based on the 1.2288 Mcps chip rate that is referred to as "1x." Multicarrier versions of cdma2000 such as 3xRTT have been standardized but no multicarrier measurement is yet defined. The cdma2000 pilot is carried on Walsh code 0, typically at around -7 dB from the total downlink power.

3.6.1.10 cdma2000 High Rate Packet Data Pilot Strength

High rate packet data (HRPD) systems including 1xEV-DO Releases 0, A, and B do not use the code domain pilot signal defined for the speech-capable cdma2000. The cdma2000 HRPD pilot is defined in the time domain, existing for 9.375% of the frame. Its measurement is therefore necessary for LTE interworking with HRPD systems and is another version of LTE RSRP.

3.6.1.11 Additional UE Measurements for the Support of Positioning

In Release 9 a number of new measurements were defined for the support of positioning. (See Section 8.2.4.)

- Reference signal time difference (RSTD) is a measure of the time difference between the RS of different cells used for the observed time difference of arrival (OTDOA) positioning system.

- Global navigation satellite system (GNSS) timing of cell frames for UE positioning is a measurement between an E-UTRA cell as detected by the UE and the GNSS time for a specific GNSS identified by the UE.

- The UE GNSS code measurement is a measure of the spreading code phase of a GNSS satellite signal.

- The UE Rx–Tx time difference is a measure of the time difference between a downlink radio frame and the transmission time of the equivalent uplink frame.

3.6.2 Evolved Node B Physical Layer Measurements

There are fewer physical layer measurements for the eNB than for the UE, primarily because the base station is not mobile and does not need to measure non-LTE systems.

3.6.2.1 Downlink RS Tx Power

The first eNB measurement is different in two respects from the UE measurements described so far: first, it describes the eNB transmission itself rather than a transmission from another entity, and second, it is not so much a measurement as a report generated by the eNB reflecting the transmitted power. Even so, the report has to be accurate and take into account losses between the baseband (where power is defined) through the transmit chain to the antenna connector.

3.6.2.2 Received Interference Power

The uplink received interference power is a measure of the interference power and thermal noise within an RB that is not scheduled for transmission within the cell. The absolute accuracy has to be ± 4 dB for interference measured between -117 dBm and -96 dBm. This measure will be used to identify narrowband co-channel interference from neighbor cells on the same frequency.

3.6.2.3 Thermal Noise Power

The uplink thermal noise power measurement is a broadband version of received interference power and is measured optionally at the same time under the same conditions. The definition is $(N_0 * W)$ where N_0 is the white noise power spectral density and W is the transmission bandwidth configuration (see Figure 3.2-5).

3.6.2.4 Additional eNB Measurements for the Support of Positioning

The additions in Release 9 for the support of positioning by the eNB are as follows:

- Timing advance (TA), which comes in two forms. Type 1 is the eNB Rx–Tx timing difference added to the UE Rx–Tx timing difference. Type 2 is just the eNB Rx–Tx timing difference.

- eNB Rx–Tx timing difference, which is a measure of the time difference between an uplink radio frame and the transmission time of the equivalent downlink frame.

- E-UTRAN GNSS timing of cell frames for UE positioning, which is a measurement between the transmit timing of an E-UTRA cell and a specific reference time of a GNSS cell.

- Angle of arrival (AoA), which defines the estimated angle of a user with respect to a reference direction relative to geographic North in a counter-clockwise direction.

3.6.3 Radio Resource Management

Having introduced the underlying physical layer measurements that support RRM, it is possible to describe how they are used in the operation of the system. The subject of RRM, specified in 36.133 [10], covers a broad range of procedures that can seem rather impenetrable upon first study. A good analogy for RRM in cellular systems is to consider air transportation. Suppose the base station could be equated with an airport and the UE with an airplane. Provided we have two airports and one plane, we have the beginnings of a transportation system. Such a simple system could be operated with almost no procedures since the one plane could take off and land knowing that there would be no "interference" from other air traffic.

However, to turn such a simple example into something resembling a high-capacity transportation system requires that there be many airports and even more planes. In such a crowded environment it becomes evident that to avoid collisions, an air traffic control (ATC) system is necessary. To the extent to which the analogy is useful, RRM provides for the cellular industry the same essential traffic management functions that ATC provides for the air transportation industry.

The RRM requirements in 36.133 [10] are divided into two major parts. First are the individual performance requirements for the core functions supporting RRM. These are defined in Subclauses 4 through 10. Second, Annex A provides normative test case descriptions that will be used as the basis for the RRM conformance tests described in Section 7.2.8. These test cases combine many of the underlying core requirements into typical operating scenarios, which is preferable to testing each function individually. Subclauses 9 and 10 specify requirements for the physical layer measurements described in Section 3.6.1. The remainder of this chapter will describe the core requirements in 36.133 [10] Subclauses 4 through 8.

3.6.3.1 E-UTRAN RRC_IDLE State Mobility

Refer to 36.133 [10] Subclause 4. This section covers the two most basic of procedures carried out when the UE is in the idle state (not on a call). These procedures are cell selection, which is performed after the UE is switched on, and cell reselection, which is the procedure performed when the UE moves from one cell to another. See Figure 3.6-1.

Cell Selection

The complexity of the processes for cell selection can be seen in the idle state transition diagram in Figure 3.6-1. The most obvious parameter to specify for cell selection performance is the time taken to camp onto an appropriate cell for a given radio scenario. One of the most complex scenarios commonly occurs when the UE is switched on in a rich radio environment; for example, in a foreign airport where competition for roaming customers can be fierce. There are many ways of configuring parameters in the network that can influence the behavior of the UE when it initially chooses a cell on which to camp. It is perhaps due to the complexity of the cell selection process that in UMTS, no requirements were specified. It is likely that LTE will take the same approach. This might seem surprising but as things stand, this aspect of UE performance is left as a competitive rather than a regulated issue.

Cell Reselection

For cell reselection, the situation is quite different as LTE specifies numerous performance requirements for this process. When the UE is camped on the serving cell, it will be commanded to detect, synchronize, and monitor intra-frequency, inter-frequency, and inter-RAT cells to determine whether a more suitable cell on which to camp can be found. Sometimes the serving cell will provide a neighbor list for the intra-frequency and inter-frequency LTE cells, but at other times only the carrier frequency and bandwidth will be provided. The rules for neighbor cell reporting allow the UE to limit its measurement activity in complex situations.

The goal of the cell reselection process is the evaluation of the cell selection criterion S for each detected cell. This measure, which is defined in 36.304 [11] Subclause 5.2.3.2, is based on relative and absolute power measurements and is used to determine the most favorable cell for the UE to camp on. The cell reselection performance requirements are defined in terms of three time allowances: the time allowed to detect and evaluate S for a new cell, the time allowed to reevaluate S for an existing cell, and the maximum time allowed between measurements of the cell. One of the important parameters impacting cell reselection performance is the discontinuous reception (DRX) cycle length. This is the time between attempts by the UE to measure cells other than the serving cell. There is clearly a tradeoff between a long DRX cycle that does not interrupt normal operation in the serving cell but gives slow reselection times, and a much shorter DRX cycle that speeds up detection but interrupts normal operation. The defined DRX cycle lengths in seconds are 0.32, 0.64, 1.28, and 2.56.

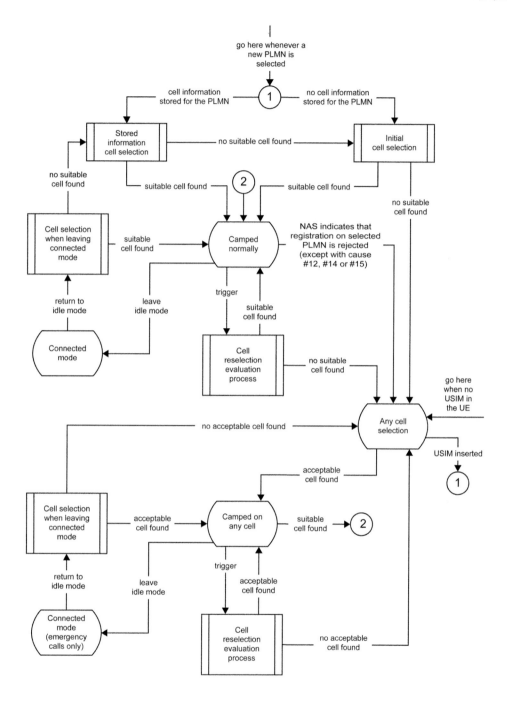

Figure 3.6-1: RRC_IDLE Cell Selection and Reselection (36.304 [8] Figure 5.2.2-1)

The cell re-selection rules are complex and are only briefly described here.

The UE is required to continuously monitor the serving cell and should it fail to fulfill the cell selection criteria, the UE has to immediately measure all the neighbor cells indicated by the serving cell, regardless of any rules currently limiting UE measurements.

The UE is required to identify detectable intra-frequency E-UTRAN cells and measure the RSRP without prior knowledge of the physical cell identity. Cells are considered detectable if they exceed certain absolute power and SNR limits. For detectable cells, the key performance requirement is the time allowed to evaluate the cell selection criterion S. The rules for inter-frequency E-UTRAN cells have additional complexity but the key performance requirement remains the time taken to evaluate S.

As might be expected the situation gets significantly more complex for inter-RAT cell reselection. Cell reselection performance requirements exist for UTRA FDD, UTRA TDD, GSM, HRPD, and cdma2000 1xRTT. The specification of further RATs is likely in the future.

3.6.3.2 E-UTRAN RRC_CONNECTED State Mobility

Refer to 36.133 [10] Subclause 5. The requirements for mobility while connected are more generally known by the term "handover." The combinations of handover for which performance requirements have been defined fall into two categories:

E-UTRAN handover
    E-UTRAN FDD to FDD
    E-UTRAN FDD to TDD
    E-UTRAN TDD to FDD
    E-UTRAN TDD to TDD.

Handover to another RAT
    E-UTRAN to UTRAN FDD
    E-UTRAN to UTRAN TDD
    E-UTRAN to GSM
    E-UTRAN to HRPD
    E-UTRAN to cdma2000 1xRTT.

For each scenario two performance parameters are defined. These are the handover delay and the interruption time. Both parameters are necessary as the first is a measure of the delay from the start of the process to its completion and needs to be kept low, while the second parameter is the shorter period of time during which communication is interrupted.

3.6.3.3 RRC Connection Mobility Control

Refer to 36.133 [10] Subclause 6. The requirements for RRC connection mobility control are for RRC reestablishment following a failure in the RRC connection, random access and connection release with redirection. The most likely causes of an RRC connection failure are if the radio link drops below an acceptable quality or if a handover fails. The requirements are written based on the time allowed to reestablish the RRC connection. The procedure is specified in 36.331 [12] Subclause 5.3.7.

The reestablishment delay is determined by four parameters: the number of frequencies being monitored, the time to search each frequency, the time to read the system information from each cell, and the delay in the RACH procedure. For simple cases in which the target cell is known by the UE and has been recently measured, the delay may be as short as 160 ms. More difficult situations that require searching for a suitable cell on which to reestablish the link could be in the order of one second per frequency searched.

The requirements for random access relate to correct behavior when a random access response and other messages are received from the eNB.

The timing requirements for connection release with redirection are for cases in which the UE is required to move to a UTRAN FDD, UTRAN TDD, or GERAN cell.

3.6.3.4 Timing and Signalling Characteristics

Refer to 36.133 [10] Subclause 7.

UE Transmit Timing

A critical performance requirement in any wireless system is the ability of the UE to maintain timing synchronization with the base station. The unit for measuring timing is T_s, where $T_s = 1/(15000*2048)$ seconds. The timing reference point for the UE is the first detected path from the eNB. The nominal transmit timing of the UE is specified in advance of this reference time as $N_{TA}*T_s$, where N_{TA} is the timing advance parameter.

Requirements exist for the initial timing accuracy, the maximum step in any one timing adjustment, and finally the maximum and minimum timing adjustment rates. These requirements are necessary in order that the worst case timing error between the eNB and UE is bounded. Timing errors can be caused by large changes in multipath delay (as with shadow fading) or by a handover to a cell with different timing.

The initial timing accuracy requirement is $\pm12*T_s$ and should this be exceeded, the UE is required to adjust its timing to get to within the allowed range. During the adjustment process the maximum allowed step size is $\pm2*T_s$ and the rate of change has to be between $2*T_s$ and $7*T_s$ seconds per second.

UE Timer Accuracy

Many of the RRM processes require that the UE start and stop various timers. For timers of less than four seconds the accuracy is fixed at 0.1 seconds and for longer timers the UE is given a greater allowance of 0.25%. These are not critical figures but are specified in order to give guidance to the UE designer about the precision required for timer implementation.

Timing Advance

The timing advance process is specified in 36.321 [12] Subclause 5.2. When the UE receives a new timing advance command in frame number n it is required to implement the new timing in frame $n+6$ to an accuracy of $\pm 4*T_s$.

Cell Phase Synchronization Accuracy (TDD)

In TDD systems, this requirement controls the frame start timing for any two cells that share the same frequency and overlap coverage areas. It is necessary to control the timing between such cells to avoid the transmission from one cell occurring at the same time as reception by the other. The requirement for wide area base stations with cell radius ≤ 3 km is ≤ 3 μs. This requirement is relaxed to ≤ 10 μs for cell radius above 3 km. For home base stations, the timing requirement is ≤ 3 μs for a cell radius of ≤ 500 m. For larger cells, the requirement is ≤ 1.33 μs plus $T_{propagation}$, which is the propagation delay between the home base station and the cell selected as the network listening synchronization source.

Synchronization Requirements for E-UTRAN to cdma2000 1xRTT and HRPD Handover

For a successful handover to cdma2000 1xRTT and HRPD it is necessary for the UE to know the CDMA system timing reference. This is achieved by the eNB providing the timing via a system information message. Once the UE knows the system timing, it can report the timing of the target system's pilot signals. The basic requirement is for the eNB to be within ±10 μs of the CDMA system time. The eNB should be synchronized to GPS time and maintain ±10 μs accuracy for a period of up to 8 hours in case GPS synchronization is lost. The eNB also has to ensure that the message transmitting the CDMA system time is transmitted within 10 μs of the expected time.

Radio Link Monitoring

The UE is required to monitor the quality of the downlink for the purposes of determining if the radio link is good enough to continue transmission. This is done through the parameters Q_{out} and Q_{in}. The threshold for Q_{out} is defined as the level at which the downlink radio link cannot be reliably received. There is no direct measure, but the assumption is that Q_{out} corresponds to an approximate 10% block error ratio of a hypothetical PDCCH transmission, taking into account a number of network settings and radio conditions. Q_{in} is defined as having a much higher probability of reception than Q_{out}. The Q_{in} threshold is nominally a 2% block error ratio of the hypothetical PDCCH for a defined set of network settings and radio conditions. The requirements for the UE to monitor the radio link quality are specified in terms of how long the UE takes to switch off when the quality drops below Q_{out} and how long it takes for the UE to switch back on when the quality rises above Q_{in}.

3.6.3.5 UE Measurements Procedures in RRC_CONNECTED State

Refer to 36.133 [10] Subclause 8. The requirements for connected state mobility have been introduced in Section 3.6.3.2 but the discussion was limited to the physical process of performing the handover. However, in cellular systems it is not the handover itself that is difficult; it is knowing when and where to make the handover. An analogy is changing lanes while driving on a multi-lane road. The action of turning the steering wheel is easy. The difficult bit, especially in heavy traffic, is knowing when and where to make the change. To make good handover decisions requires knowledge of the environment and this applies equally to cellular radio and motoring.

By measuring and reporting the radio environment when in a connected state, the UE provides the system with the raw material needed to make the correct handover decisions. Many parameters can be measured, and the rules for how and when to gather and report these parameters are complex.

The requirements, which are split according to RAT, are the following: E-UTRA intra-frequency, E-UTRA inter-frequency, inter-RAT UTRA FDD, UTRA TDD, and GSM. The parameters to measure are defined in 36.214 [5] and are introduced here in Section 3.6.1. The required measurement accuracy is defined in 36.133 [10] Subclause 9.

With the exception of intra-frequency measurements, it is not possible for the UE to gather information on different frequencies or RATs without implementing a transmission gap. During this period the UE is able to retune its receiver (DRX) to monitor other frequencies. The options for configuring the UE can become quite complex, especially when the radio environment includes multiple bands and RATs. Tradeoffs have to be made between the desire for full knowledge of the radio environment, which requires frequent gaps, and the desire for less interruption and fewer measurements, which leads to slower and less optimized handover decisions.

3.7 Summary

It should be evident from this chapter that LTE employs some very advanced tools in the physical layer to optimize throughput and latency. The challenge with these tools will be to optimize their use in the dynamic channel conditions that are the norm in cellular wireless. Many of the projected gains for LTE over existing systems rely on the potential advantages of having a wider channel available and taking advantage of transitory variations in channel conditions including spatial multiplexing. These opportunities will become advantages only when the closed-loop algorithms to control these mechanisms are optimized. It is worth pointing out that there is no intention within the scope of the 3GPP specifications to provide such algorithms. They are intentionally left to the implementation and so will remain a competitive differentiator in the market.

3.8 References

[1]   3GPP TS 36.201 V11.0.0 (2012-09) LTE Physical Layer–General description
[2]   3GPP TS 36.211 V11.0.0 (2012-09) Physical Channels and Modulation
[3]   3GPP TS 36.212 V11.0.0 (2012-09) Multiplexing and Channel Coding
[4]   3GPP TS 36.213 V11.0.0 (2012-09) Physical Layer Procedures
[5]   3GPP TS 36.214 V11.0.0 (2012-09) Physical Layer Measurements
[6]   3GPP TS 36.101 V11.2.0 (2012-09) UE Radio Transmission and Reception
[7]   3GPP TS 36.321 V11.0.0 (2012-09) Medium Access Control (MAC) Protocol Specifications
[8]   3GPP TS 36.306 V11.1.0 (2012-09) UE Radio Access Capabilities
[9]   3GPP TS 36.300 V11.3.0 (2012-09) E-UTRA, E-UTRAN, Overall Description; Stage 2
[10] 3GPP TS 36.133 V11.2.0 (2012-09) Requirements for support of Radio Resource Management
[11] 3GPP TS 36.304 V11.1.0 (2012-09) UE Procedures in Idle Mode
[12] 3GPP TS 36.331 V11.1.0 (2012-09) Radio Resource Control (RRC) Protocol Specification

Links to all reference documents can be found at www.agilent.com/find/ltebook.

Chapter 4

Upper Layer Signaling

4.1 Access Stratum

The access stratum (AS) contains the functionality associated with access to the radio network and the control of active connections between a UE and the radio network. The AS consists of a user plane and a control plane. The user plane is mainly concerned with carrying user data—e.g., internet protocol (IP) packets—through the access stratum. The control plane is concerned with controlling the connection between the UE and the network.

An overall description of the evolved universal terrestrial radio access (E-UTRA) and the evolved universal terrestrial radio access network (E-UTRAN) can be found in 36.300 [1]. In contrast to earlier 3GPP protocols, LTE has located the entire access stratum inside one network entity: the base transceiver station or evolved node B (eNB). The aim of this is to simplify the architecture and speed up the control signaling, leading to improved overall performance of LTE. The majority of the physical layer (L1) is explained in Chapter 3; however, there is an important L1/L2 control protocol called the downlink control information (DCI) that resides in the physical layer and is covered in detail in Section 4.1.3. The remainder of the AS covers four protocols: the medium access control (MAC) in 36.321 [2]; the radio link control (RLC) in 36.322 [3]; the packet data convergence protocol (PDCP) in 36.323 [4]; and the radio resource control (RRC) in 36.331 [5].

4.1.1 User Plane

Figure 4.1-1 shows the protocol stack for the user plane, which is divided into three sub-layers: the PDCP, the RLC, and the MAC. These protocols carry data from one side of the radio network to the other. The PDCP and RLC protocols are similar in the eNB and the UE, performing such functionality as header compression, ciphering, and running acknowledged-mode protocols. The MAC protocols, however, behave differently in the eNB and the UE. This is mainly because the scheduler runs in the eNB, determining from subframe to subframe which UEs in the cell should be allowed access to the network.

LTE and the Evolution to 4G Wireless: Design and Measurement Challenges, Second Edition, Edited by Moray Rumney.
Copyright Agilent Technologies, Inc. 2013. Published by John Wiley & Sons, Ltd.

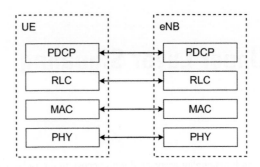

Figure 4.1-1. User plane protocol stack

4.1.2 Control Plane

Figure 4.1-2 shows the protocol stack for the control plane. The PDCP, RLC, and MAC protocols behave exactly as they do in the user plane, although in the control plane their function is to carry control messaging from the RRC, which may contain non-access stratum (NAS) messaging rather than user data. In the control plane, however, the RRC is unique, performing such functions as broadcast messaging and connection control. The NAS is described in Section 4.2.

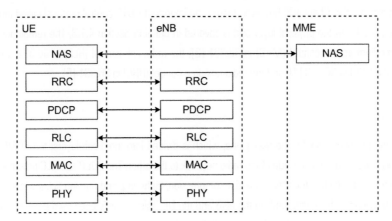

Figure 4.1-2. Control plane protocol stack

4.1.3 Physical Layer Downlink Control Information

The physical layer was described in Chapter 3. There are protocol messages sent from the eNB to the UE that are carried directly on the physical downlink control channel (PDCCH) without passing through the MAC. These messages are key to the establishment of LTE connections and are known as downlink control information (DCI) messages. The presence of DCI was introduced in Chapter 3 but is now more fully described.

4.1.3.1 Downlink Control Information Functions

A DCI message performs one of six functions:
- Providing downlink scheduling information
- Providing uplink scheduling information
- Providing uplink power control information
- Notifying the user that there has been a change to the MCCH
- Requesting that the UE transmit aperiodic CQI reports
- Requesting the UE to transmit a PRACH, a procedure known as a PDCCH order.

The eNB normally transmits many DCI messages per subframe, each using a different PDCCH. Each message is intended to be received by one or many UEs. A UE does not know which PDCCH channels have been used on a particular subframe by the eNB and, of those used, whether the PDCCH contains a DCI message intended for that UE. To receive DCI messages a UE must perform a large number of blind decodes every subframe. The DCI messages intended for that UE will be decoded successfully, whereas those not intended for that UE will fail the cyclic redundancy code (CRC) check.

Release 10 introduces cross-carrier scheduling. A DCI message transmitted on one serving cell may provide uplink or downlink scheduling information for another. This relationship between cells is established when a secondary cell (SCell) is added to the configuration assigned to a UE. Uplink or downlink scheduling information for the primary cell (PCell) is always provided on the PCell.

4.1.3.2 Identification of a DCI Message

The intended recipient or recipients of a DCI message are distinguished by use of a different radio network temporary identity (RNTI), which is an identifier used by the UE MAC. This identifier is encoded into the CRC of the message. A UE will be able to successfully decode only those DCI messages that contain an RNTI the UE is expecting to receive.

A UE may expect several RNTI values at any given time, and some RNTI values will be expected by all the UEs in a cell. The DCI messages containing downlink scheduling for system information (SI) and paging (P) messages are transmitted using special RNTI values known as the SI-RNTI and P-RNTI. Uplink scheduling messages, transmitted in response to a physical random access channel (PRACH) preamble sent by a UE are identified using an RNTI known as the random access RNTI (RA-RNTI), which is derived from the time and frequency resource used to transmit the PRACH. The MBMS RNTI (M-RNTI) tells the UEs that are listening for messages on the multicast control channel (MCCH).

An established downlink shared channel (DL-SCH) or uplink shared channel (UL-SCH) is identified with an RNTI unique to a particular UE (the cell RNTI, or C-RNTI). No other UE in the cell looks for DCI messages coded with that unique RNTI value.

4.1.3.3 Downlink Scheduling Information

One reason for sending DCI messages is to transmit downlink scheduling information to the UE. The UE needs this information to successfully decode the data that is being transmitted to it on the physical downlink shared channel (PDSCH) of the same subframe.

Before the UE can find and decode the PDSCH data, it must know the following:

- Which resource blocks are carrying data allocated for this UE. These resource blocks may or may not be contiguous in frequency.
- Which modulation and transport block size to use on these resource blocks.
- Which redundancy version to use during the rate-matching process.
- Whether or not to use spatial multiplexing.
- If spatial multiplexing is used, what pre-coding matrix to apply to the data.
- For UEs of Release 10 and later, which serving cell the PDSCH is being transmitted on. This may not be the same cell on which the DCI was transmitted.

After successfully decoding the PDSCH data, the UE then needs to know:

- The downlink hybrid automatic repeat request (HARQ) process for which this data is intended.
- Whether this data is new or a retransmission of previously transmitted data.
- Whether to ACK or NACK the transmission(s). For TDD, due to the asymmetric nature of some of the uplink-downlink configurations, several PDSCH transmissions may need to be acknowledged at once.

4.1.3.4 Uplink Scheduling Information

The DCI messages providing uplink scheduling information are addressed to only one UE at a time. These messages tell the UE how it should transmit information on the physical uplink shared channel (PUSCH). In contrast to the DCI messages that provide downlink scheduling information and which refer to a PDSCH in the same subframe, the uplink scheduling information messages refer to the PUSCH subframe that comes a number of subframes after the subframe containing the scheduling message. For FDD, this is always four subframes after. For TDD, this number varies depending on the uplink-downlink configuration.

Prior to transmitting data on the PUSCH, the UE must know the following:

- Which resource blocks to use. The resource blocks are either a single cluster of resource blocks that are contiguous in frequency, or, from Release 10 onwards, two separate clusters of resource blocks contiguous in frequency.
- Which modulation and transport block size to use.
- Which redundancy version to use.
- Whether or not to use PUSCH hopping.
- Whether to send new data or retransmit old data.
- Whether or not to alter the power that is used to transmit the PUSCH.
- For UEs of Release 10 and later, whether to use uplink spatial multiplexing or not.
- If spatial multiplexing is used, what precoding matrix to apply to the data.

The UE may also be requested to send an aperiodic channel state information (CSI) report or, for UEs of Release 10 or later, an SRS transmission.

4.1.3.5 Power Control Information

The downlink and uplink scheduling information messages described above include power control parameters such that the UE will know what power level to use on the PUSCH or PUCCH. There are cases, however—such as semi-persistent scheduling or non-adaptive retransmissions—in which the UE may be sending data for a number of transmission time intervals (TTIs) without receiving downlink or uplink scheduling information messages. In such cases, the eNB may still want to alter the power levels used by the UE. Therefore, a simple type of DCI message is available that has the sole purpose of adjusting the power level. Two such DCI messages are defined, allowing the power level to be adjusted with different levels of granularity.

4.1.4 Medium Access Control (MAC)

The medium access control (MAC) layer is a protocol layer that runs in both the UE and the eNB. It has different behaviors when running in each, generally giving commands in the eNB and responding to them in the UE. As the name suggests, MAC arbitrates and controls access to the shared transmission medium.

The main functions of the MAC layer are:

- Providing data transfer services to the RLC via logical channels.
- Multiplexing data from one or more logical channels into transport blocks. These are delivered to the physical layer on transport channels.
- Error correction through HARQ.
- Deciding which UEs will be allowed to send or receive data on the shared physical resource (eNB MAC only).
- Transport format selection—that is, choosing the modulation and coding rate that will be used to send data to the UE (eNB MAC only).

4.1.4.1 Logical Channels and Transport Channels

Logical channels are the service access points (SAPs) provided by the MAC to the RLC. Logical channels are distinguished by the type of information transferred. Transport channels are SAPs used by the MAC and provided by the physical layer to achieve data transfer. Transport channels are distinguished by how the data is transferred. The MAC provides a transport format (TF) that specifies how the transport channel is mapped to the physical layer. Table 4.1-1 and Figure 4.1-3 show the mapping between the logical channels and the transport channels for the uplink.

Table 4.1-1. Uplink logical channel to transport channel mapping

Logical channel	Transport channel	Purpose of logical channel
Common control channel (CCCH)	UL-SCH	Carries RRC signaling before the UE has been identified; e.g., for connection setup
Dedicated control channel (DCCH)	UL-SCH	Carries signaling from the RRC
Dedicated traffic channel (DTCH)	UL-SCH	Carries user data

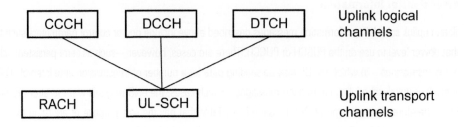

Figure 4.1-3. Uplink logical channel mapping onto transport channels

Table 4.1-2 and Figure 4.1-4 show the mapping between the logical channels and the transport channels for the downlink.

Table 4.1-2. Downlink logical channel to transport channel mapping

Logical channel	Transport channel	Purpose of logical channel
Broadcast control channel (BCCH)	Broadcast channel (BCH)	Carries master information block
BCCH	Downlink shared channel (DL-SCH)	Broadcast of system information messages
Paging control channel (PCCH)	Paging channel (PCH)	Carries paging messages
CCCH	DL-SCH	Carries RRC signaling before the UE has been identified; e.g., for connection setup
DCCH	DL-SCH	Carries signaling from the RRC
DTCH	DL-SCH	Carries user data
Multicast control channel (MCCH)	Multicast channel (MCH)	Carries MBMS-related control messaging
Multicast traffic channel (MTCH)	MCH	Carries MBMS data

4.1.4.2 Random Access, Scheduling Request, Timing Alignment, and Contention Resolution Support

The random access procedure enables the UE to establish initial contact with the network, which is usually the first thing a UE does after acquiring system information. Other scenarios in which the procedure is used are after failure of the radio link to reacquire a connection with the network, and during the handover procedure.

Random access can be contention-based or non-contention-based. For the contention-based procedure, the physical random access channel (PRACH) preamble is chosen by the UE from a set of preambles whose configuration is broadcast in the SI messages. Since there is a possibility of two UEs choosing the same preamble at the same time, there are a few subsequent steps to allow the network to uniquely identify each UE. In contrast, for non-contention based random access, the eNB MAC assigns a dedicated preamble to each UE, which allows the UE to be uniquely identified from the start of the procedure.

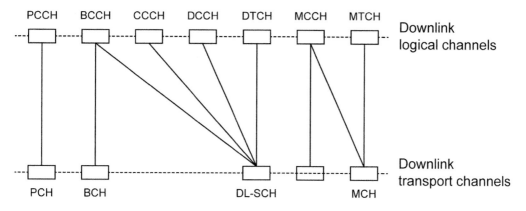

Figure 4.1-4 Downlink logical channel mapping onto transport channels

The UE decodes the random access channel (RACH) transport channel configuration options from the SI messages. This informs the UE of the allowed random access configurations in the cell. The random access procedure is initiated by the UE's MAC transmitting a PRACH preamble on the RACH. The eNB MAC can choose to respond to the UE's MAC PRACH preamble with an uplink grant including an estimate of the UE's timing alignment, or with a back-off value to the UE not to perform another random access for the signaled back-off duration. The uplink grant transmission is done on the DL-SCH transport channel addressed with a random access radio network temporary identity (RA-RNTI).

If the eNB MAC provides an uplink grant, the UE schedules a UL-SCH transmission and starts a timing alignment timer. The eNB MAC periodically sends the timing advance MAC control element to the UE, whereupon the UE applies the timing advance and restarts its timing alignment timer. The UE MAC can also signal a scheduling request (SR) to signal to the eNB MAC that the UE needs more uplink resources. The SR can be used only if the UE's timing alignment timer is running. After the UE's timing alignment timer expires, the UE has to use the random access procedure before scheduling further uplink transmissions.

The MAC also participates in the contention resolution procedure. The UE MAC starts the contention resolution timer after performing a scheduled transmission in response to a random access uplink grant. If, during the period that the contention resolution timer is running, the UE MAC detects its cell radio network temporary identity (C-RNTI) on the PDCCH or detects a match of the UE's contention resolution identity in a DL-SCH transmission addressed to the UE's temporary C-RNTI, then the UE's contention has been successfully resolved.

The random access procedure is supported only on a PCell.

4.1.4.3 Paging and Discontinuous Reception Support

Paging is used by the network to locate a UE in RRC_IDLE state within a tracking area. The MAC provides support for transmitting paging information (PI) over the PCCH logical channel. The PCCH transmission is done over the PCH transport channel in transparent mode (that is, the MAC does not insert or remove any headers) and is signaled using P-RNTI on the PDCCH on a PCell.

The MAC also supports discontinuous reception (DRX) in order to conserve the UE's battery power. With DRX the UE does not have to monitor PDCCH transmissions on every subframe. The eNB MAC takes account of the UE's DRX configuration before scheduling any transmissions or receptions for the UE.

4.1.4.4 DL and UL Data Transfer, Multiplexing and De-multiplexing Support

Perhaps the most important function of the MAC is to enable downlink and uplink data transfer. Data transmissions in the downlink support the CCCH, DCCH, and DTCH logical channels that are carried on the DL-SCH transport channel and the MCCH and MTCH logical channels that are carried on the MCH. Downlink assignments for these transmissions are signaled over the PDCCH using one of several methods: C-RNTI, semi persistent scheduling (SPS) C-RNTI, temporary C-RNTI, M-RNTI, or RA-RNTI. Data transmissions in the uplink support the CCCH, DCCH, and DTCH logical channels and are made over the UL-SCH transport channel. Uplink grants for these network-scheduled transmissions are signaled over the PDCCH using the C-RNTI, SPS C-RNTI, or temporary C-RNTI.

The MAC protocol data units (PDUs) are variable-size, byte-aligned data units consisting of a MAC header, zero or more MAC service data units (SDUs), zero or more MAC control elements (MCE), and optional padding. See Figure 4.1-5. The MAC header is a combination of MAC PDU sub-headers in which each sub-header identifies a MAC SDU, an MCE, or padding. One or two MAC PDUs (or transport blocks) can be transmitted in a subframe depending on the physical layer transport format.

The transmitting MAC multiplexes SDUs from multiple logical channels plus any MCEs and padding (if required) to form a PDU. The PDU is then de-multiplexed at the receiving MAC. The SDUs are delivered to the corresponding logical channels and the MCEs are acted upon by the receiving MAC.

MAC control elements are control messages sent to the peer MAC. Information such as timing alignment, contention resolution, power headroom reporting, and DRX commands are sent using these messages.

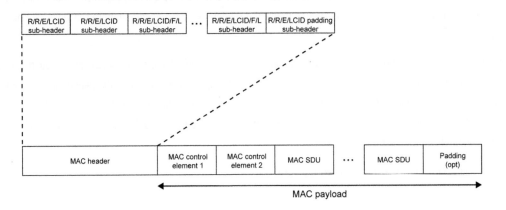

Figure 4.1-5 Example of MAC PDU consisting of a MAC header, MAC control elements, MAC SDUs, and padding (36.321 [2] Figure 6.1.2-3)

4.1.4.5 Logical Channel Prioritization and Quality of Service Support

The MAC supports quality of service (QoS) by prioritizing data transmissions of logical channels based on their configured priority and prioritized bit rate (PBR). Logical channel prioritization is specified for the UE MAC. In practice, the eNB MAC also performs logical channel prioritization. Logical channels are served in order of their priority such that on average their PBR targets are met. Any remaining resources are distributed in order of priority. In other words, a logical channel will get any remaining resources if all logical channels with higher priority have no data to send. These guidelines are enforced over several subframes on average, in order to avoid unnecessary RLC segmentation and padding.

4.1.4.6 Scheduling and Radio Resource Allocation

The scheduling function rests in the network scheduler, and scheduling decisions are signaled to the UE MAC. The eNB scheduler performs resource block (RB) allocation to individual transport blocks in a subframe and modulation and coding rate selection for transport blocks. The TF defines the physical layer coding for a transport block. The number of resource blocks and an index modulation and coding scheme (IMCS) parameter that identifies the modulation and transport block size are signaled to the UE MAC.

Various means are defined to signal scheduling decisions (uplink grants and downlink assignments) to the UE MAC. The most flexible method is dynamic scheduling in which a downlink assignment is sent for each DL-SCH transmission and an uplink grant is sent by the eNB MAC for each UL-SCH transmission. Although this scheme provides full flexibility on each subframe, it can generate excessive control information.

A more efficient method is semi-persistent scheduling in which a downlink assignment and uplink grant apply to N transmissions or receptions instead of applying to just one transmission or reception. The value N is configured and signaled by the RRC. The result reduces control information but also reduces flexibility. Another option is TTI bundling in the uplink in which a number of subframes are combined and only one HARQ feedback is sent for the entire bundle. This is especially useful for limiting the required power for transmission; for example, at the cell edge.

Factors affecting network scheduling decisions include the number of UEs in a cell, the bandwidth of the cell, the amount of buffered data available to the transmitting MAC, the capability of the UE to support advanced modulation schemes, the number of layers supported by the UE for spatial multiplexing, and the prevailing radio channel conditions as reported by the UE or as measured by the network.

The eNB MAC is also responsible for scheduling reception of the MCH. Although RRC-level messaging will have specified the subframes in which the MCH may be transmitted, the eNB MAC is responsible for specifying which subframes will contain MTCH and which will contain MCCH.

From Release 10 onwards, the eNB may also activate and deactivate SCells that have been assigned to the UE using RRC-level messaging. SCells are initially deactivated when they are first assigned and will be activated and deactivated depending on the volume of data that needs to be transferred to the UE or the radio conditions (which may change for each cell over time)—a cell-by-cell version of DTX/DRX.

4.1.4.7 Measurements to Support Scheduling and Radio Resource Allocation

To assist the network scheduler, the UE MAC performs and reports channel measurements. The channel measurements are described in detail in Section 3.4. Buffer status reports inform the network of the state of transmission buffers in the UE; i.e., the amount of data buffered. Power headroom reports inform the network of the difference between the UE's current transmit power and its maximum output power.

The channel quality indicator (CQI) report uses measurements of the downlink radio conditions to report to the scheduler a combination of modulation and coding rate that would have resulted in a 10% block error ratio had this modulation and coding been used during the period covered by the CQI report. The CQI can be provided for the entire channel bandwidth (wideband CQI) or for a specified part of the bandwidth (subband CQI). The subbands for which CQI is reported can be network-selected or UE-selected. Subband CQI helps the network scheduler in performing frequency-selective scheduling.

4.1.5 Radio Link Controller (RLC)

The radio link controller (RLC) acts as an interface and buffer between the higher layers of the protocol stack (usually PDCP in the user plane) and the MAC layer, which has almost no buffering capability and acts more as a router than anything else. The RLC's main functions are the following:

- Passing SDUs received from the higher layers to the RLC's peer via the MAC. Data units are passed as PDUs.
- Receiving PDUs from the MAC and extracting SDU data from them to be passed to higher layers.
- Re-ordering of received PDUs.
- Acknowledged mode operation (depending on configuration).

SDUs are simple blocks of binary data arranged in octet (or character) arrays. They are the basic units managed by the RLC.

The RLC maintains a count of the amount of data it has stored for transmission. This is called its buffer occupancy (BO). The BO is incremented when SDUs are received from the higher layers and decremented when PDUs are sent to the MAC for transmission. The MAC can read the BO values of all RLCs. When there is an opportunity to schedule data, the MAC can decide which RLC to request data from and how large a PDU it should request.

There are optimal sizes for PDUs that give the best data transfer performance. These sizes depend on the configuration of the physical layer and the prevailing radio conditions at transmission time, so they can vary in size considerably. SDUs can be almost any size, so one particular task of the RLC is to either concatenate or break up SDUs into PDUs. The peer must be able to reassemble the original SDUs, so information on how the SDUs are stored in a PDU is included in the PDU as part of its header.

4.1.5.1 Operating Modes

The RLC has three modes of operation: transparent mode (TM), unacknowledged mode (UM), and acknowledged mode (AM). Each offers different levels of reliability with an obvious impact on latency.

Transparent Mode

In transparent mode the RLC neither concatenates nor breaks up SDUs. The PDUs must be the same size as the SDUs and may contain only a single SDU.

Unacknowledged Mode for PDU transmission

Depending on the relative sizes of the SDUs and PDUs, the SDUs can be concatenated so that several fit into one PDU or their contents can be split over several PDUs. Each PDU has a fixed header that contains either a 5 bit or 10 bit sequence number (SN), an extension (E) indicator bit, and a 2 bit framing indicator (FI) field. Optionally, a series of length indicators (LIs) and E bits can be included. The first octet following the header bits is the first data octet in the PDU.

The SN is incremented by 1 for every PDU. The SN allows the peer to reorder PDUs should they be received out of sequence. The FI field indicates whether the first or last octets of the SDU data in the PDU are also the start or end of an SDU. The LI indicates the end of an SDU within the PDU. The E bit indicates whether an LI/E bit pair follows the E bit. Using the LI/E bit pairs allows data from multiple SDUs to be carried by a single PDU.

Unacknowledged Mode for PDU Reception

Under normal circumstances the RLC receives PDUs in SN order. The SDU data is extracted and used to reconstruct the original SDUs. Complete SDUs are forwarded to the upper layers.

However, in unacknowledged mode there is no feedback from the peer RLC, meaning lost PDUs will not be retransmitted by the RLC. If a PDU in unacknowledged mode is lost, the SDU data it carried is also lost. Further, if the previous PDU contained the start of an incomplete SDU, then part of that SDU is also lost, so the rest of the SDU must also be discarded. Similarly, if the next received PDU does not start with data from a new SDU, then part of that SDU has also been lost and the remainder must also be discarded.

Fortunately, the MAC has HARQ capability, which can retransmit lost PDUs very quickly (though only a limited number of times). The RLC must cope with receiving valid but out-of-sequence PDUs. For this reason, the RLC must be able to distinguish between a lost PDU and a delayed PDU. This is done using the reordering timer.

When an out-of-sequence PDU is received, the reordering timer is started. If the lost PDUs are received before the timer expires, then the timer is stopped and the SDU data can be processed. If the timer expires, the PDUs are considered lost and discarded as above.

Acknowledged Mode

In acknowledged mode the reordering timer is used to cope with HARQ retransmissions but, in addition, the RLC peers acknowledge the data PDUs they receive using another type of PDU called a control PDU. Control PDUs allow the loss of a PDU to be indicated to the transmitter so that the lost PDUs can be retransmitted. Further, because the radio conditions may deteriorate, forcing the MAC to use smaller PDUs, the lost PDUs may have to be broken up into smaller segments (re-segmented) before retransmission. This can happen a number of times if radio conditions continue to deteriorate. These re-segmented PDUs can also be acknowledged using control PDUs.

4.1.5.2 PDU Formats

Acknowledged Mode PDU

The header of an AM PDU is more complex than that of a UM PDU. The AM PDU header includes a data or control (D/C) indicator bit, a re-segmentation flag (RF) bit, and a polling (P) indicator bit. The SN field is 12 bits. SDU boundaries are again indicated by FI, LI, and E bit fields. The D/C flag indicates whether the PDU is a data PDU or a control PDU. The RF flag indicates whether the data PDU format is simple or segmented. The P flag indicates that the transmitter wants to receive a status message from its peer when it receives the AM PDU.

AM PDU Segment

When the RF flag is set, the header has two extra fields, a segment offset (SO) field and a last segment field (LSF). The SO is the octet number in the original PDU at which the AM PDU segment's first octet begins. The LSF indicates that the data octets in this segment run from the SO to the end of the original PDU.

AM Control PDU

The header of a control PDU is completely different than that of a data PDU. The AM control PDU includes the following mandatory fields: a 3 bit control PDU type (CPT) field, a 12 bit acknowledged SN (ACK_SN) field, an extension 1 (E1) bit, and an extension 2 (E2) bit. It also includes three optional fields: a 12 bit negative acknowledgment SN (NACK_SN) field, a 15 bit segment offset start (SOstart) field, and a 15 bit segment offset stop (SOstop) field.

The CPT indicates the type of control PDU, the ACK_SN indicates the earliest correctly received (in sequence) PDU, and the E1 indicates that a NACK_SN follows. E2 indicates the presence of an SOstart/SOstop pair associated with the NACK_SN. NACK_SN indicates the SN of a lost PDU or PDU segment. SOstart indicates the offset of the first octet of data in the lost segmented data PDU. SOstop indicates the offset of the last octet of data in the lost segmented data PDU.

PDU Transmission

When the MAC requests a PDU for transmission, a priority order is used to decide what type of PDU to send. If a control PDU is available, it is constructed and sent. Otherwise a PDU or PDU segment previously reported as lost is retransmitted. A retransmitted PDU may be segmented or re-segmented before being retransmitted. If no control PDUs are transmitted and no retransmissions are required, then a new PDU is constructed from the available SDU data. The BO is incremented when a control PDU is ready to send or when retransmissions of data PDUs or PDU segments are required; in this way the MAC knows that this RLC has more data to send. The size of any transmitted but unacknowledged data PDUs or PDU segments is decremented from the BO.

Every data PDU or data PDU segment is stored by the transmitting entity until it has been acknowledged by the peer RLC.

PDU Reception

When a data PDU is received from the MAC, it is stored until all preceding PDUs (those out of sequence) have also been received. Then the SDU data is extracted from them and forwarded, in sequence, to the higher layers. When a data PDU segment is received, the octets of data within it are used to construct a complete PDU, which is then treated as above.

When a control PDU is received, all of the unacknowledged PDUs up to ACK_SN are discarded. Any PDUs identified by NACK_SNs or any PDU segments identified by NACK_SNs and SOstart/SOstop pairs are marked for retransmission and their sizes are added to the BO information supplied to the MAC.

4.1.6 Packet Data Convergence Protocol (PDCP)

The packet data convergence protocol (PDCP) layer acts as a portal between the various higher layers of the protocol stack (RRC, RTP, UDP, TCP, etc.) and the RLC layer.

TThe main functions of the PDCP are the following:

- Passing SDUs received from the PDCP's higher layers to its peer via the RLC. The data units are transmitted as PDUs.
- Receiving PDUs from the RLC and extracting SDU data from them to be passed to higher layers.
- Header compression/decompression.
- Ciphering/deciphering.

The use of integrity and ciphering in the PDCP layer is a significant departure from the PDCP in UMTS. The use of robust header compression version 2 (RoHCv2) is an enhancement to the use of RoHCv1 when user data is compressed.

4.1.6.1 Operating Planes

PDCP can operate on either of two planes, the control plane or the user plane. The control plane is used for RRC messages. The user plane is used for all other data.

Control Plane

A PDCP connected to the control plane carries RRC messages and is connected to an AM RLC. The SDUs are converted into PDCP data PDUs by adding a header and a tail. The header is a single octet with a 7 bit SN. The SN is incremented for every new PDU and wraps from 127 back to zero. Although the RRC messages are very important, the PDCP can afford to use a small SN because the AM RLC will recover lost PDUs and present them to the receiving PDCP in their original transmitted sequence.

RRC messages have integrity protection. A 4 octet message authentication code for data integrity (MAC-I) is generated by an algorithm using the data in the SDU and the SN as input. Another input to the integrity algorithm is a set of key values that is generated from the secret key in the USIM. This MAC-I is appended to the SDU data in the PDU.

The entire message (SDU data plus MAC-I but not the header) is then ciphered. A cipher stream that is the same length as the SDU data plus MAC-I is produced using a process similar to that used to create the MAC-I. The message bits are multiplexed with the cipher stream using an Exclusive OR (XOR) function.

In the receiver, another cipher stream is generated that is XOR'd with the message to decipher it. Then the MAC-I value is calculated from the SDU data and compared with the one in the message. If they match, the SDU data is forwarded to the higher layers.

User Plane

The user plane carries data packets and can be connected to either a UM or AM RLC. If connected to a UM RLC, it can use either a 7 or 12 bit SN. If connected to an AM RLC, it always uses a 12 bit SN. User plane SDUs are compressed using RoHCv2. The original RoHCv1 is a subset of RoHCv2. After compression, user plane data PDUs are also ciphered in the same way as control plane PDUs. User plane data PDUs are not integrity-protected. In the receiver, the PDUs are deciphered and decompressed and the SDU data is forwarded to the higher layers.

4.1.6.2 PDCP Protocol Data Units

Every PDCP PDU has a D/C indicator bit to determine whether it is a data PDU or a control PDU.

Data PDUs

Every PDCP data PDU has a header prepended that is either 1 or 2 octets in length. If the PDCP is connected to the control plane, then the header will be 1 octet in length and will contain a 7 bit SN. If the PDCP is connected to the user plane, then the header may be either 1 or 2 octets. In the latter case, the SN will be 12 bits long.

Control PDUs

Control PDUs should not be confused with the data PDUs generated by a PDCP connected to the control plane. Control PDUs do not have an SN, are neither ciphered nor integrity-protected, and are not RoHC-compressed.

There are two kinds of control PDUs: RoHC feedback PDUs and PDCP status PDUs. RoHCv2 feedback PDUs are also called interspersed RoHC feedback packets. They are used when a bidirectional RoHC connection is available and configured. The feedback modifies the behavior of the RoHC compressor, allowing better compression techniques to be used. The PDCP status PDUs are generated when a handover occurs. They carry received PDU status information between PDCP peers.

4.1.7 Radio Resource Control (RRC)

The radio resource control (RRC) layer is a layer 3 (L3) protocol in the radio interface and is located at the top of the access stratum (AS) of the air interface. The RRC provides access through which higher layer signaling entities can gain services in the form of signaling transfer from the AS.

RRC performs many functions that are required for the reliable and efficient operation of the radio resource. The RRC performs a management role coordinating the functions of the other AS layers. The RRC's main areas of functionality include:

- Broadcasting system information
- Coordinating the functions of the other AS layers
- Establishing, reestablishing, maintaining, and releasing the RRC connection between the UE and E-UTRAN
- Setting, altering or releasing the radio bearer in the user plane.
- Handling issues relating to link quality
- Configuring the UE to make measurements of other cells
- Instructing the UE to hand over to other cells
- From Release 10 onwards, adding or deleting SCells to the configuration.

4.1.7.1 System Information Broadcast

One of the main functions of the RRC on the network side is to broadcast system information (SI) about the cell or the network to all UEs attached to the cell. This system information includes the collation of information elements and the scheduling of transmitted system information blocks (SIBs). In addition to the SIBs there is a master information block (MIB).The MIB includes a limited number of the most essential and frequently transmitted parameters, and it is the first piece of information read by the UE when connecting to the cell.

System information blocks other than SIB Type 1 are carried in SI messages. Mapping of SIBs to SI messages is configured flexibly by scheduling information included in SIB Type 1 with the following restrictions: each SIB is contained in a single SI message; only SIBs with the same scheduling requirement (periodicity) can be mapped to the same SI message; and SIB Type 2 is always mapped to the SI message that corresponds to the first entry in the list of SI messages in the scheduling information. Multiple SI messages may be transmitted with the same periodicity.

Unlike previous releases of UMTS in which the MIB and SIBs can be transmitted on the BCH, in LTE only the MIB is transmitted on the BCH and the rest of the SIBs are transmitted on the DL-SCH.

4.1.7.2 Broadcast Procedure

The MIB uses a fixed schedule with a periodicity of 40 ms and repetitions made within 40 ms. The first transmission of the MIB is scheduled in subframe #0 of the radio frames for which the SFN mod 4 = 0, and repetitions are scheduled in subframe #0 of all other radio frames.

The SIB Type 1 uses a fixed schedule with a periodicity of 80 ms and repetitions made within 80 ms. The first transmission of SIB Type 1 is scheduled in subframe #5 of the radio frames for which the SFN mod 8 = 0, and repetitions are scheduled in subframe #5 of all other radio frames for which SFN mod 2 = 0.

The SI messages are transmitted within periodically occurring time domain windows (referred to as SI-windows) using dynamic scheduling. Each SI message is associated with an SI-window and the SI-windows of different SI messages do not overlap; that is, within one SI-window only, the corresponding system information is transmitted. The length of the SI-window is common for all SI messages and is configurable. Within the SI-window, the corresponding SI message can be transmitted multiple times in any subframe other than multimedia broadcast over single frequency network (MBSFN) subframes, uplink subframes in TDD, and subframe #5 of radio frames for which SFN mod 2 = 0. The UE acquires the detailed time domain scheduling (and other information such as frequency domain scheduling and the transport format used) by decoding the SI-RNTI on the PDCCH.

A single SI-RNTI is used to address the SIB Type 1 as well as all SI messages. The SIB Type 1 configures the SI-window length and the transmission periodicity for the SI messages.

4.1.7.3 RRC States

In Release 8, there are two service states defined for the UE with respect to the operation of RRC: RRC_IDLE state and RRC_CONNECTED state.

In RRC_IDLE state, the UE has no connection in place. The UE has found the system and registered on it. The UE monitors the downlink system information and paging information and makes neighbor cell measurements for cell reselection.

In RRC_CONNECTED state, the UE has established an RRC connection. The UE monitors control channels associated with the shared data channel to determine if data is scheduled for the UE, provides channel quality and feedback information, and performs neighbor cell measurements and measurement reporting.

Unlike a UMTS UE, the LTE UE does not have sub-states in RRC_CONNECTED state.

4.1.7.4 RRC Connection Control

RRC offers three signaling radio bearers (SRB) for the control plane signaling message transfer: SRB0, SRB1, and SRB2. SRB0 is used for CCCH messages whereas SRB1 and SRB2 are used for DCCH messages. SRB1 has a higher priority than SRB2, which is set up only after security has been activated. SRB1 carries RRC messages and also NAS messages if SRB2 is not yet established, whereas SRB2 carries NAS messages only. In any case NAS messages are always encapsulated into an RRC message with or without RRC protocol control information. LTE RRC allows NAS messages to be piggybacked on RRC control messages for certain procedures. Examples are initial uplink NAS messages transmitted during RRC connection setup and during downlink bearer establishment, modification, or release. These procedures have joint success or failure criteria.

RRC connection establishment is the establishment of SRB1 between the UE and the E-UTRAN so that further NAS and user plane procedures can take place. Using the information provided in SIBs, a UE sends an RRC connection request message to the E-UTRAN. The UE indicates its higher layer identity and the connection establishment cause. Lower layers perform the RACH procedure to ensure the successful transmission of the RRC connection request. The E-UTRAN can respond either with an RRC connection setup command, which configures SRB1 and the default EPS radio access bearer (RAB), or with an RRC connection reject command asking the UE to retry the connection after some wait time or to try the connection on a different frequency or radio access technology. See Figure 4.1-6.

The UE performs contention resolution before acting on the contents of the response from the E-UTRAN. Following the successful configuration of the resources, the UE transmits the message "RRC connection setup complete" to the E-UTRAN. Since each eNB can be shared by multiple public land mobile networks (PLMNs) and mobility management entities (MMEs), the UE indicates the selected PLMN and MME so that the E-UTRAN can establish the S1 connection with the desired PLMN/MME. To speed up this process, the UE can piggyback initial NAS messages onto the "RRC connection setup complete" response. Once security has been activated, all messages on SRB1 and SRB2 are integrity-protected and ciphered.

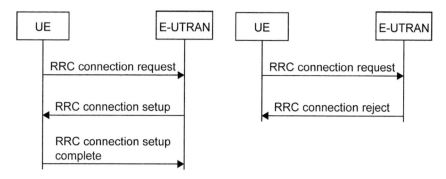

Figure 4.1-6. RRC connection setup and reject signaling (36.331 [5] Figures 5.3.3.1-1 and 5.3.3.1-2)

The RRC connection reconfiguration procedure (Figure 4.1-7) is used to establish, modify, or release signaling and user radio bearers. In LTE, RRC connection reconfiguration also involves setting up a default EPS bearer between the UE and the core network. This EPS bearer is set up on the basis of a non-guaranteed bit rate allowing the application-level signaling to take place as soon as a secure RRC connection is established.

The same RRC connection reconfiguration procedure is employed to perform handovers, NAS message transfer, and configuration of measurements. However, certain types of reconfiguration cannot be performed until the AS security has been activated. As a part of this procedure, the E-UTRAN sends the message "RRC connection reconfiguration" with the appropriate information elements. Upon a successful handover, the UE responds with the message "RRC connection reconfiguration complete."

From Release 10 onwards, the RRC connection reconfiguration procedure can be used to add or remove SCells from the configuration. These SCells will be activated or deactivated by the MAC layer, but the RRC layer will provide all the key information about the SCells, including that normally provided by SI messages. Cross-carrier scheduling may also be established, by means of which one cell provides the DCI messages that determine the uplink or downlink allocations on another cell.

To handle temporary loss of coverage during mobility or to handle failures during reconfiguration procedures, an RRC connection reestablishment procedure is introduced. This procedure is used to resume SRB1 operation and to reactivate security. However, it is not used to resume operation of any other radio bearers. The procedure can be initiated only if the AS security is active, otherwise the UE will drop all of its radio bearers and move to the RRC_IDLE state. The UE initiates the RRC connection reestablishment procedure by transmitting an RRC connection reestablishment request message to the selected cell.

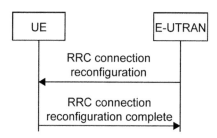

Figure 4.1-7. RRC connection reconfiguration signaling (36.331 [5] Figure 5.3.5.1-1)

If the cell can acquire a UE context based on the message sent by the UE, the E-UTRAN can reconfigure the SRB1 by sending an RRC connection reestablishment message. Alternatively the E-UTRAN can reject the reestablishment request. A MAC contention resolution similar to the one at RRC connection establishment takes place at this stage. The UE resumes SRB1 and AS security accordingly while the rest of the bearers and measurements remain suspended until an RRC connection reconfiguration takes place. Upon successful reestablishment, the UE responds with the message "RRC connection reestablishment complete." See Figure 4.1-8.

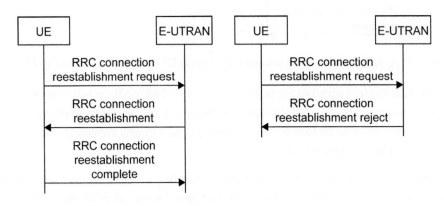

Figure 4.1-8. RRC connection reestablishment signaling (36.331 [5] Figures 5.3.7.1-1 and 5.3.7.1-2)

To tear down the RRC connection and release all radio resources associated with a UE, the E-UTRAN transmits an RRC connection release message to the UE. This message can also be used to redirect the UE to a particular RAT or cell and to update idle mode mobility information.

4.1.7.5 Inter-RAT and Intra-RAT Mobility

Idle mode mobility is controlled by the UE and is based on the SI transmitted in a cell. SIB3 carries all the information for all types of reselections: intra-frequency, inter-frequency, and inter-RAT. SIB3 parameters allow the UE to determine when to start different reselection measurements. SIB4 carries all the intra-frequency neighbor cell details eligible for reselection as well as blacklisted cells that should not be considered for reselection. In a similar manner, SIB5 carries inter-frequency neighbors and blacklisted cells, whereas one SIB per RAT is used to convey neighbor cell lists of the available RATs. A cell can be reselected only if the required criteria are satisfied. Idle mode mobility is also possible because of the redirection at RRC connection establishment or release.

Connected-mode intra-frequency, inter-frequency, and inter-RAT measurements are configured during RRC connection reconfiguration. Measurement objects, measurement gaps, and reporting requirements are also configured. Measurement reporting can be periodic or event-triggered. Different event criteria are specified to signal if the serving cell, an LTE neighbor cell, or an inter-RAT neighbor cell becomes better or worse than certain thresholds. Intra-LTE handover is achieved using the RRC connection reconfiguration procedure, which is initiated by the network based on measurements or blindly without measurements.

Inter-RAT mobility to the E-UTRAN is achieved by sending the RRC connection reconfiguration message on the source RAT. A successful handover to the E-UTRAN results in establishment of SRBs and user radio bearers and in activation of security in the E-UTRAN marked by the transmission of the RRC message "connection reconfiguration complete" by the UE. Mobility from the E-UTRAN to another RAT, whether by handover or by cell change order, is achieved by transmitting the command "mobility from E-UTRA" to the UE. If the handover is to a cdma2000 RAT, this command is preceded by a "handover from E-UTRA preparation request" message from the E-UTRAN to the UE followed by a "UL handover preparation transfer" message from the UE to the E-UTRAN. These two messages are used to acquire the non-3GPP information required for the handover from the E-UTRAN. In any case, success of handover from E-UTRAN is signaled on the target RAT. See Figures 4.1-9 and 4.1-10.

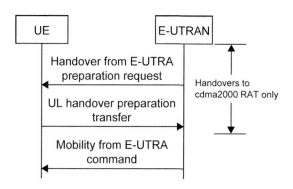

Figure 4.1-9. Handover from E-UTRAN signaling (36.331 [5] Figures 5.4.3.1-1, 5.4.4.1-1, and 5.4.5.1-1)

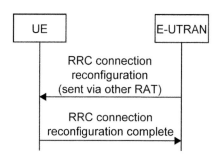

Figure 4.1-10. Handover to E-UTRAN signaling (36.331 [5] Figure 5.3.2.1-1)

4.1.7.6 Other RRC Procedures

Although it is the PDCP that actually performs ciphering and integrity check operations, it is the RRC that configures integrity protection and ciphering of the signaling plane and user plane data. The E-UTRAN RRC initiates security activation at RRC connection establishment by sending a security mode command to the UE. The UE responds with a security mode complete message upon successful activation or a security mode failure message otherwise.

The security mode command and complete messages are integrity-protected, and ciphering is applied immediately after security mode activation is complete. Only after successful security activation can SRB2 and user radio bearers be established. All further security reconfigurations such as algorithm change at handovers are handled by the RRC connection reconfiguration procedure.

In addition to messages such as "RRC connection setup complete" and "RRC connection reconfiguration," which can carry piggybacked NAS messages, the RRC provides special containers called downlink information transfer and uplink information transfer. These containers are used exclusively to transfer downlink and uplink NAS messages.

If the EPC and the E-UTRAN do not have all the information related to UE capabilities (for example, during an inter-RAT handover), the E-UTRAN RRC can request this information from the UE by transmitting a message called a UE capability enquiry. The UE sends the requested capabilities back to the network in a UE capability information message.

The paging procedure is used to notify UEs in RRC_IDLE and RRC_CONNECTED states about SI changes and initiation of mobile terminating calls. The paging procedure involves the transmission of the paging message by the E-UTRAN RRC in certain designated paging occasions. Once the UE identifies a page for itself, it reacquires SI or initiates RRC connection establishment depending on the paging cause.

The RRC also coordinates the use of MBMS in a cell. The MAC layer will configure the time domain scheduling, but the RRC layer will semi-statically configure the frequency domain scheduling and the lower layer configuration of the radio bearers that will carry MBMS data.

4.2 Non-Access Stratum

The non-access stratum (NAS) contains all the functions and protocols used directly between the UE and the core network. The main NAS specification is 24.301 [6]. The NAS protocols are transparent to the access network. In the existing GSM/UMTS core network, these protocols include call control (CC) and mobility management (MM) for the circuit-switched (CS) domain and GPRS mobility management and session management for the packet-switched (PS) domain. Various supplementary services such as short message service (SMS) are supported from either or both domains.

The core network architecture for LTE has evolved beyond the existing GSM/UMTS core network with a goal of simplifying the overall architecture. The new architecture is an all internet protocol network (AIPN) that supports both 3GPP-based (UTRAN, GERAN) and non-3GPP based (cdma2000, 802.16, etc.) radio access networks (RANs). The architecture also supports mobility between the various RANs. This evolved architecture is called the evolved packet core (EPC) network or the System Architecture Evolution (SAE), which is the original 3GPP project name.

Since the new architecture is an AIPN, the evolved core network is a PS-only network, differing significantly from all previous core networks that were built around the CS domain. The protocols defined for this new network are the evolved packet system (EPS) mobility management and EPS session management protocols. These protocols manage the mobility of the UE and the activation, modification, and deactivation of user-plane channels for transferring user data between the UE and the IP network.

he new architecture introduces a new control plane mobility management entity (MME), which implements the new rotocols. It also introduces an access gateway, which consists of two logical user plane entities, the serving gateway (S-GW) nd the packet data network gateway (PDN-GW). A detailed explanation of the new MME and an overview of the new user lane entities are given in this section. Figure 4.2-1 shows the different network elements and the interfaces connecting these ements in the new architecture.

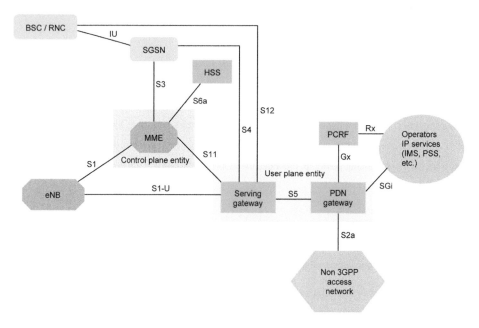

Figure 4.2-1. System architecture (based on 23.882 [7] Figure 4.2-1)

.2.1 Network Elements

here are three main network elements in the evolved packet core: the mobility management entity, the serving gateway, and e PDN gateway.

.2.1.1 Mobility Management Entity

s stated above, the MME is the control plane entity that implements the procedures used for EPS mobility management EMM) and EPS session management (ESM) in the EPS. The MME communicates with the home subscriber server (HSS) or retrieving subscription information and with the serving gateway for establishment and release of EPS bearers. The uthentication information received from the HSS is used to generate the integrity and ciphering keys, which are then used for ntegrity protection and ciphering of NAS control plane messages. The MME communicates with the eNB over the S1 interface sing the S1 application protocol. The S3 interface between the MME and the serving GPRS support node (SGSN) is used for gnaling to support mobility between 3GPP access networks. Further explanation of this entity is provided in Section 4.2.3.

4.2.1.2 Serving Gateway

The serving gateway terminates the interface towards the radio network. It handles the user plane traffic routing and the forwarding of uplink and downlink packets between the PDN gateway and the radio network. Each UE communicates with only one serving gateway and the serving gateway can communicate with different PDN gateways for different UE-specific PDN connections over the S5 interface. The serving gateway acts as the anchor for inter-eNB and inter-3GPP mobility. It communicates with the policy and charging rules function (PCRF) entity for lawful interception and charging control functions.

4.2.1.3 PDN Gateway

The PDN gateway provides PDN connectivity to the UE for different IP services provided by operators. The PDN gateway configured in the UE subscription information allocates an IP address to the UE during UE attachment to the E-UTRAN. A UE in a connected state can request connection to another PDN gateway for different IP services. The PDN gateway is responsible for uplink and downlink rate enforcement depending on the allocated maximum bit rate to the UE. It also performs the uplink and downlink service level charging and rate enforcement. The PDN gateway never changes during a session regardless of the mobility of the user. Hence it is the anchor point that manages the mobility between a 3GPP and a non-3GPP system.

4.2.2 Network Interfaces

The following interfaces have been defined for the EPC:

S1-MME: Interface for the control application protocol between the E-UTRAN and MME

S1-U: Interface for S1 user plane data for each bearer between the E-UTRAN and the serving gateway. This interface enables the serving gateway to anchor the inter-eNB handover.

S3: Interface that provides the connection between the SGSN and MME, enabling information exchange for mobility between inter-3GPP access networks

S4: Interface between the SGSN and serving gateway. It provides the user plane support for mobility support between the GPRS core and the serving gateway. It also enables the serving gateway to anchor the inter-3GPP handover.

S5: Interface that provides the user plane tunneling and tunnel management function between the serving gateway and the PDN gateway. It enables the serving gateway to connect to multiple PDN gateways for providing different IP services to the UE. It also is used for serving gateway relocation associated with UE mobility.

S6a: Interface between the MME and HSS. It is used for transfer of subscription and authentication data for authenticating and authorizing user access to the evolved packet system.

Gx: Interface that provides transfer of QoS policy and charging rules from the PCRF to the policy and charging enforcement function (PCEF) in the PDN gateway

S11: Control plane interface between the MME and serving gateway needed for EPS bearer management

SGi: Interface between the PDN gateway and the internet/intranet (equivalent to the Gi interface in GPRS).

4.2.3 Mobility Management Entity

The MME is the control plane entity that implements the EMM and ESM procedures in the EPS. The EMM procedures provide support for the mobility of the UE in the E-UTRAN, connection management services to the session management sub-layer, and control of security for the NAS protocols. These procedures enable an EPS-capable UE to move from an EMM-deregistered state to an EMM-registered state and vice versa. In EMM-deregistered state the MME has no knowledge of the location of the UE. The UE performs an attach procedure and registers its location with the MME. The UE is then in an EMM-registered state until it detaches from the network. In EMM-registered state the UE can communicate with the network for performing other procedures. For this communication to take place a signaling connection needs to be established with the registered MME. Depending on whether a signaling connection exists between the UE and MME, the UE is considered to be either in ECM-idle state (no signaling connection) or ECM-connected state (with signaling connection).

The ESM procedures are used for the handling of EPS bearer contexts. Together with the bearer control provided by the access stratum, these procedures are used for the activation, modification, and deactivation of the user plane bearers.

4.2.4 EMM Procedures

The various EMM procedures are classified as EMM common procedures, EMM specific procedures, or EMM connection management procedures.

4.2.4.1 EMM Common Procedures

EMM common procedures can be executed only when there is a signaling connection between the UE and the MME. The EMM common procedures are as follows: identification, authentication, security mode control, globally unique temporary identity (GUTI) reallocation, and EMM information.

Identification Procedure

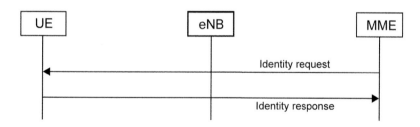

Figure 4.2-2. Identification procedure message flow (Based on 24.301 [6] Figure 5.4.4.2.1)

The identification procedure (Figure 4.2-2) is used by the network to request that a particular UE provide specific identification parameters such as the international mobile subscriber identity (IMSI) or the international mobile equipment identity (IMEI). The MME initiates this procedure by sending an identity request message to which the UE responds by sending an identity response message. These messages can be transmitted without ciphering and integrity protection.

Authentication Procedure

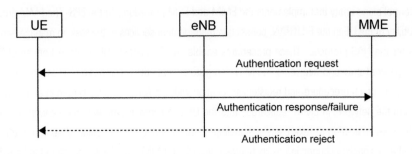

Figure 4.2-3. Authentication procedure message flow (Based on 24.301 [6] Figure 5.4.2.2.1)

The EPS authentication and key agreement (AKA) procedure is used in the E-UTRAN for mutual authentication between the UE and MME. A security context is established in both entities at the end of a successful procedure.

The MME performs the AKA procedure using the authentication vectors received from the HSS. Each authentication vector consists of four parameters: a random number (RAND), the authentication token (AUTN), the expected user response (XRES), and the intermediate access security management entity key (K_{ASME}). The MME allocates a key set identifier (KSI_{ASME}) to identify K_{ASME}. The MME prepares an authentication request message including the KSI_{ASME} and the RAND and AUTN from the selected authentication vector and transmits the message to the UE.

Upon receipt of the authentication request message, the UE verifies whether AUTN can be accepted and, if so, produces a response (RES) that is sent back to the MME in an authentication response message. From the parameter RAND, the UE generates the ciphering key (CK) and integrity key (IK). Using these keys and the serving network identity (SNid) information, which comprises mobile country code (MCC) and mobile network code (MNC), the UE generates the K_{ASME}. The K_{ASME} and the identifier KSI_{ASME} are then stored in the UE. The network can later use the KSI_{ASME} to identify the K_{ASME} stored in the UE without invoking the authentication procedure. This allows reuse of the K_{ASME} during subsequent connection setups. The UE is required to delete the K_{ASME} and reset the KSI_{ASME} when the UE is switched off or the USIM is removed.

When the MME receives the authentication response message, it compares the RES parameter with the XRES parameter of the authentication vector. If these values differ, then the MME rejects the authentication attempt by sending an authentication reject message to the UE. If the UE is not able to verify the authentication request message, then the UE transmits an authentication failure message to the MME. See Figure 4.2-3.

After a successful authentication procedure, the intermediate key K_{ASME} is used to generate three other keys: the eNB key (K_{eNB}), the NAS integrity protection key (K_{NASint}), and the NAS encryption key (K_{NASenc}). On the network side the KeNB is sent to the eNB during the initial context setup. The eNB uses K_{eNB} to generate three further keys: the uplink user data encryption key (K_{UPenc}), the RRC integrity protection key (K_{RRCint}), and the RRC encryption key (K_{RRCenc}). Similar key derivation also takes place in the UE. If a bearer is being established for emergency services, the authentication procedure does not necessarily need to succeed before a connection setup is allowed to progress.

Security Mode Control Procedure

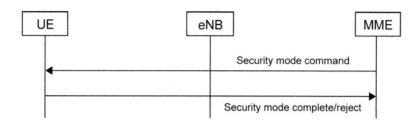

Figure 4.2-4. Security mode control procedure message flow (Based on 24.301 [6] Figure 5.4.3.2.1)

The security mode control procedure (Figure 4.2-4) is used to initialize and start the integrity and ciphering of NAS messages. This procedure is initiated by the MME when new keys are established between the UE and the MME by the EPS authentication and key agreement procedure and also when a new set of NAS algorithms is selected. The MME starts the security mode control procedure by sending the security mode command message to the UE. This message contains the replayed security capabilities of the UE, the selected NAS algorithms, and the KSI_{ASME} for identifying the K_{ASME}. The message is integrity-protected with the K_{NASint}, which is based on the K_{ASME} indicated by the KSI_{ASME} in the message.

If the UE accepts the security mode command message, it responds with the security mode complete message, which is integrity-protected using the selected NAS integrity algorithm indicated in the security mode command message and K_{NASint}. If a "non null" ciphering algorithm was indicated in the security mode command, this message will also be ciphered with the indicated ciphering algorithm and KNASenc.

If the UE does not accept the security mode command message, it responds with a security mode reject message. The UE includes an appropriate reject cause in the rejection message.

Ciphering of the NAS messages at the MME starts after receiving the NAS security mode complete message. At the UE, NAS ciphering starts before sending the NAS security mode complete message.

If a bearer is being established for emergency services, it will use a null ciphering algorithm but also a null integrity algorithm, since it will probably not have a K_{ASME} from which to derive an integrity key.

GUTI Reallocation Procedure

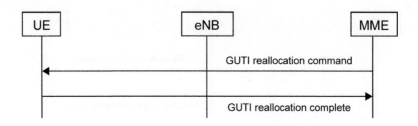

Figure 4.2-5. GUTI reallocation procedure message flow (Based on 24.301 [6] Figure 5.4.1.2.1)

The GUTI reallocation procedure is used to allocate a new GUTI and optionally to provide a new tracking area identity (TAI) list to a particular UE. This procedure can be performed implicitly along with the attach or tracking area update procedure, or independently using the GUTI reallocation command message. The MME can prepare a GUTI reallocation command message and send it to the UE only in the EMM-registered state. The UE updates the new GUTI and TAI list and responds with a GUTI reallocation complete message to the MME. See Figure 5.2-5. The GUTI reallocation procedure is usually integrity-protected and is performed in ciphered mode.

EMM Information Procedure

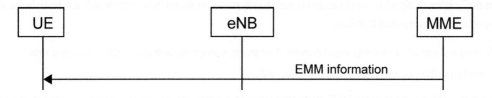

Figure 4.2-6. EMM information procedure message flow (Based on 24.301 [6] Figure 5.4.5.2.1)

The EMM information procedure enables the network to provide additional information to the UE such as network name and time zone information. This is an optional procedure and may be invoked by the network at any time during an established EMM context. See Figure 4.2-6.

4.2.4.2 EMM-Specific Procedures

The EMM-specific procedures are complete EPC procedures used for handling UE mobility in the MME. They are the attach procedure, the tracking area update procedure, and the detach procedure.

Attach Procedure

The attach procedure (Figure 4.2-7) enables a UE to register itself to an MME for receiving packet services.

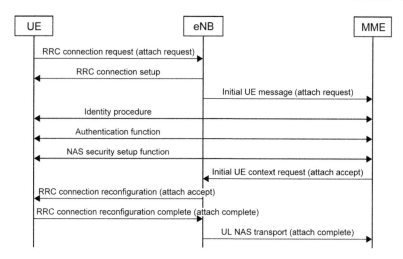

Figure 4.2-7. Attach procedure (Based on 23.401 [8] Figure 5.3.2.1-1)

After a successful attach procedure, an EMM context is established in the UE and the MME, and a default bearer is established between the UE and PDN gateway, thus enabling always-on IP connectivity to the UE. An IP address may be allocated to the UE when the default bearer is activated, or the UE can acquire an IP address after the default bearer is established using the dynamic host configuration protocol version 4 (DHCPv4) or DHCPv6.

The UE starts the attach procedure by sending an attach request message. This message may be integrity-protected if a valid NAS security context for the UE exists. The UE will include the GUTI or IMSI in this message as its identifier, which the MME uses to retrieve the UE subscription information. The UE also includes its network capability in this message so that the MME can select the appropriate security algorithms based on the UE's capabilities.

As part of the attach procedure the MME may execute the identification, authentication, or security mode control procedures. After successful handling of the attach request message, the MME will respond with an attach accept message to the UE. The MME will include a list of tracking area identities in this message, which indicates the registered area of the UE. The UE may be allocated a new GUTI as its identifier in this message.

To complete the successful attach procedure the UE will send an attach complete message to the MME. If the MME does not accept the attach request message the MME responds with an attach failure message. The MME will include an appropriate reject cause in the rejection message.

The UE includes the PDN connectivity request message within the attach request message to indicate establishment of the default bearer with the default PDN. The MME includes the activate default EPS bearer context request message to activate the default bearer in the attach accept message. The UE sends the attach complete message combined with an activated default EPS bearer context accept message.

The attach procedure may optionally also register the UE for non-EPS services—a procedure known as a combined attach. Release 9 introduces the ability to attach for emergency bearer services—which is an abbreviated attach procedure without invocation of authentication or security.

Tracking Area Update Procedure

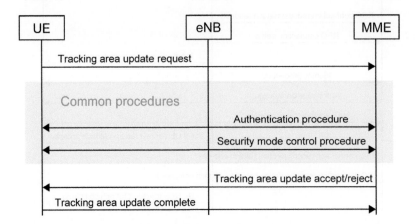

Figure 4.2-8. Tracking area update procedure message flow (Based on 24.301 [6] Figure 5.5.3.2.2.1)

The tracking area update procedure (Figure 4.2-8) is initiated by a UE in the EMM-registered state to update the UE location to the network either periodically or when the UE moves out of the registered tracking area list. This procedure may also be initiated because of an intersystem change into the E-UTRAN.

The UE starts this procedure by sending the tracking area update request message. In this message the UE includes a GUTI as the UE identifier and the last visited registered TAI. If a UE in ECM-idle mode has uplink user data pending, then the UE may request the network to reestablish the radio and S1 bearers for all active EPS bearer contexts.

If the MME accepts the tracking area update request, the MME sends a tracking area update accept message to the UE. The MME may include a new GUTI or new TAI list if either has changed. In response, the UE returns a tracking area update complete message to the MME to acknowledge the received GUTI. If the network cannot accept the tracking area update request message, the MME sends a tracking area update reject message to the UE including an appropriate rejection cause.

After the tracking area update procedure has completed, the MME normally releases the signaling connection with the UE. However, the UE can request the MME to maintain the signaling connection for following procedures.

The tracking area update procedure enables the UE and MME to synchronize the EPS bearer context status. Should there be any difference in context status due to local bearer deactivation, this difference can be corrected by sending the bearer context status information indicating the active and inactive bearers in the request and accept messages. The UE and MME can then locally deactivate the bearers marked inactive.

When the UE is in an EMM-registered state, the tracking area update procedure is always integrity-protected. If the integrity check of the tracking area update request message fails, then authentication and NAS security mode procedures are performed.

Detach Procedure

The detach procedure de-registers a UE from the registered network and moves the UE from the EMM-registered state to the EMM-deregistered state. This procedure may be initiated by the UE or by the network. The UE initiates this procedure under any of three conditions: when the UE is switched off, when the USIM card is removed, or when the EPS capability of the UE is disabled. The network initiates the detach procedure to inform the UE that it does not have access to the EPS any more.

After execution of the detach procedure, any EPS bearer contexts for the UE are deactivated locally in the MME and the UE without any peer-to-peer signaling.

UE-Initiated Detach

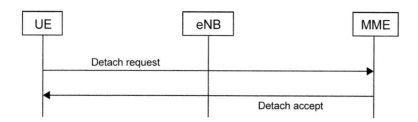

Figure 4.2-9. UE-initiated detach procedure message flow (Based on 24.301 [6] Figure 5.5.2.2.1.1)

The UE initiates the detach procedure by sending a detach request message to the MME. The UE can indicate in the detach request message whether the detach is the result of a switch-off situation. If the detach is not due to switch-off, the MME sends a detach accept message to the UE (see Figure 4.2-9); otherwise, the procedure is completed when the network receives the detach request message. The intermediate security key K_{ASME} and its identifier KSI_{ASME} are deleted in the MME and UE following the detach procedure.

MME-Initiated Detach

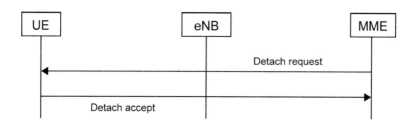

Figure 4.2-10. MME-initiated detach procedure message flow (Based on 24.301 [6] Figure 5.5.2.3.1)

TThe MME initiates the detach procedure by sending a detach request message to the UE. The UE responds with a detach accept message to the MME. See Figure 4.2-10. The MME can ask the UE to reattach after the detach procedure, in which case the UE starts the attach procedure immediately after the detach procedure.

4.2.4.3 EMM Connection Management Procedures

The EMM connection management procedures enable the UE to move from the ECM-idle state to the ECM-connected state whenever there is a need for some data or signaling transfer. The EMM connection management procedures are the paging procedure, the service request procedure, and the procedures for the transport and generic transport of NAS messages.

Paging Procedure

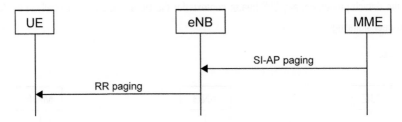

Figure 4.2-11. Paging procedure message flow

When the signaling or user data needs to be sent but there is no NAS signaling connection, the paging procedure is initiated by the network to request the establishment of an NAS signaling connection to the UE. The MME sends the S1-AP paging message to the eNB. This message includes the UE identifier and the list of TAIs to which the UE is registered. The eNB uses this information to generate a radio resource (RR) paging message in the registered tracking areas. Upon reception of a paging indication, the UE starts a service request procedure for establishment of the signaling connection and other user plane resources. See Figure 4.2-11.

Service Request Procedure

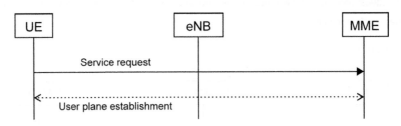

Figure 4.2-12. Service request procedure message flow

The service request procedure (Figure 4.2-12) is triggered by a UE when there is some uplink user data or signaling that needs to be transmitted or when the UE is paged by the network. The UE initiates the service request procedure by sending a service request message to the MME. If the MME accepts the service request message, the MME activates all the active EPS bearers. This initiates the establishment of user plane radio resources for the active EPS bearers in the eNB and the S1 user plane connection between the eNB and serving gateway. When the user plane radio bearers are fully established, the service request procedure is considered to be successful. The UE and network can then start transmitting data. If the MME does not accept the service request message, it sends a service reject message to the UE. This message includes an appropriate cause for rejecting the service request procedure.

Transport of NAS Messages and Generic Transport of NAS Messages Procedures

The transport of NAS messages procedure is used to carry SMS messages between the UE and the MME; the NAS messages form a wrapper around the message to be carried. The procedure can be initiated by either the UE or the MME. It can be performed only when the UE has registered both for EPS and non-EPS services—that is, has performed a Combined Attach—and is ECM-connected. Similarly, the generic transport of the NAS messages procedure allows messages from various applications—for example, a location application for sending LTE positioning protocol (LPP) messages—to be sent between the UE and the MME. The procedure can be initiated by either the UE or the MME. To perform this procedure, the UE must have registered only for EPS services and be ECM-connected.

4.2.5 ESM Procedures

The ESM procedures are used for activation, modification, and deactivation of the user plane EPS bearers, which are used for data transfer between the UE and the IP network. The ESM procedures are as follows:

- Default EPS bearer activation
- Dedicated EPS bearer activation
- EPS bearer modification
- EPS bearer deactivation
- PDN connection
- PDN disconnection.

4.2.5.1 Default EPS Bearer Activation Procedure

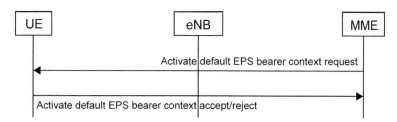

Figure 4.2-13. Default EPS bearer activation procedure message flow (Based on 24.301 [6] Figure 6.4.1.2.1)

The default EPS bearer activation procedure (Figure 4.2-13) is triggered by the MME to establish a default EPS bearer between a UE and the PDN gateway. The default bearer is used to carry all traffic which is not associated with a dedicated bearer. The default bearer is always a non-guaranteed bit rate with the resources for the IP flows not guaranteed at the eNB and with no admission control.

The MME initiates the default EPS bearer activation procedure in response to a PDN connectivity request message. This message can be sent by the UE on its own or as part of an attach request message. The MME creates an "activate default EPS bearer context request" message and sends it to the UE, either on its own or along with an attach accept message, depending on whether the PDN connectivity request message was received on its own or as part of an attach request message.

If the UE accepts the request to activate the default EPS bearer context, it responds with an "activate default EPS bearer context accept" message. If the procedure is initiated as part of the attach procedure, this message is sent as part of the attach complete message. Otherwise it is sent on its own. Failure of the attach procedure implicitly causes the default bearer activation procedure to fail. If the UE is unable to accept the bearer request, it responds with an "activate default EPS bearer context reject" message along with an appropriate rejection cause.

4.2.5.2 Dedicated EPS Bearer Activation Procedure

The dedicated EPS bearer activation procedure is used to establish a dedicated EPS bearer with specific QoS and traffic flow template (TFT) between the UE and the PDN.

Dedicated bearers are used to carry traffic for IP flows that have been identified as requiring a specific packet forwarding treatment. The dedicated bearer can be either guaranteed bit rate (GBR) or non-GBR. A GBR bearer has a guaranteed bit rate and a maximum bit rate (MBR), while more than one non-GBR bearer belonging to the same UE shares an aggregate maximum bit rate (AMBR). Non-GBR bearers can suffer packet loss under congestion, while GBR bearers are immune to such losses. The dedicated EPS bearer activation procedure is initiated by the network but may be requested by the UE by means of the UE requested bearer resource modification procedure.

Network-Initiated Dedicated EPS Bearer Activation

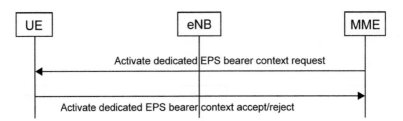

Figure 4.2-14. Network initiated EPS bearer activation procedure message flow
(Based on 24.301 [6] Figure 6.4.2.2.1)

The MME creates an "activate dedicated EPS bearer context request" message and sends it to the UE. The message includes a bearer identity, a TFT, the QoS to be allocated to the bearer and a linked EPS bearer identity for the default EPS bearer connected to the PDN. If the UE accepts the dedicated bearer request, it responds with an "activate dedicated EPS bearer context accept" message.

On failure the UE responds with an "activate dedicated EPS bearer context reject" message. This message includes the EPS bearer identity and a cause value indicating the reason for rejecting the dedicated EPS bearer context activation request. See Figure 4.2-14.

UE-Initiated Dedicated EPS Bearer Activation

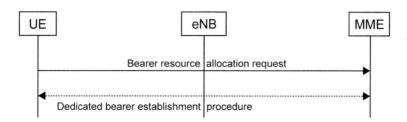

Figure 4.2-15. UE-initiated EPS bearer activation procedure message flow (Based on 24.301 [6] Figure 6.5.3.2.1)

The UE can request the establishment of a new traffic flow aggregate with a specific QoS demand and optional GBR requirement. The UE sends a "bearer resource allocation request" message to the MME. This message contains the linked bearer identity of the default bearer, which indicates the PDN to which the dedicated bearer should be established. The required QoS is also included, indicating the class of service and amount of resource that needs to be allocated.

If the MME accepts the request, it triggers the dedicated bearer activation procedure to establish the requested bearer resources. See Figure 4.2-15. If the MME does not accept the request, the MME creates a "bearer resource modification reject" message and sends it to the UE along with a cause value indicating the reason for rejecting the request.

4.2.5.3 EPS Bearer Modification Procedure

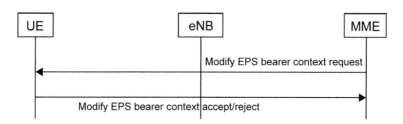

Figure 4.2-16. EPS bearer modification procedure message flow (Based on 24.301 [6] Figure 6.4.3.2.1)

The bearer modification procedure (Figure 4.2-16) is used to modify either the QoS or the TFT of an EPS bearer, or both. A UE can request that the network start a bearer modification procedure by sending a bearer resource modification request and including the bearer identity of an already established bearer.

The MME creates a "modify EPS bearer context request" message and sends it to the UE. The message contains the EPS bearer identification. Depending on whether the QoS, the TFT, or both need to be modified, the new QoS and TFT information is also included. If the modification procedure involves QoS parameters, then radio resource modification is also done between the eNB and the UE. Otherwise the PDN gateway and the UE update the modified TFT information locally.

If the UE accepts the modify request, it responds with a "modify EPS bearer context accept" message. If the UE does not accept the modify request, it sends a "modify EPS bearer context reject" message, including a cause value indicating the reason for rejecting the bearer context modification request.

4.2.5.4 EPS Bearer Deactivation Procedure

The EPS bearer deactivation procedure is used to deactivate an EPS bearer or to disconnect from a PDN by deactivating all bearers belonging to a PDN address. This procedure may be initiated on its own by the network or when requested by the UE.

Network-Initiated Bearer Deactivation

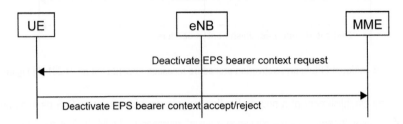

Figure 4.2-17. Network initiated bearer deactivation procedure message flow
(Based on 24.301 [6] Figure 6.4.4.2.1)

When the UE is in an ECM-connected state, the MME creates a "deactivate EPS bearer context request" message and sends it to the UE (Figure 4.2-17). This message contains the identity of the dedicated bearer to be deactivated. If all the bearers connected to a PDN need to be released, the bearer identity is the identifier of the default bearer.

After the UE receives a deactivation request, the UE deactivates the resources for the bearer or all the bearers connected to the PDN and sends a "deactivate EPS bearer context accept" message to the MME.

When the UE is in the ECM-idle state, the MME can release all the bearer contexts locally. The bearer state is synchronized between the UE and the network at the next ECM-idle to ECM-connected transition; e.g., a service request or tracking area update (TAU) procedure. When all the bearers belonging to the UE are released, the MME changes the MM state of the UE to EMM-deregistered.

UE-Initiated Bearer Deactivation

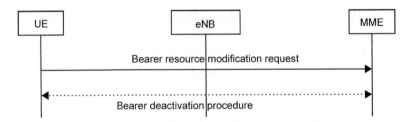

Figure 4.2-18. UE-initiated bearer deactivation procedure message flow (Based on 24.301 [6] Figure 6.5.3.2.1)

The UE can request a bearer deactivation by sending a "bearer resource modification request" message to the MME. This message contains the identity of the bearer to be released and the linked bearer identity of the default bearer connected to a PDN. On receiving this message the MME checks whether the requested bearer resource can be released, and if this is possible the MME starts the bearer deactivation procedure (Figure 4.2-18). If the MME is unable to accept the request, it sends a "bearer resource modification reject" message to the UE with a cause indicating the reason for failure.

4.2.5.5 UE-Initiated PDN Connectivity Procedure

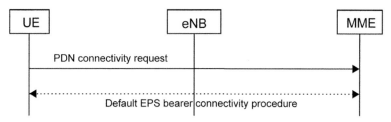

Figure 4.2-19. UE-initiated PDN connectivity procedure message flow (Based on 24.301 [6] Figure 6.5.1.2.1)

The PDN connectivity procedure (Figure 4.2-19) allows the UE to request connectivity to a PDN including allocation of a default bearer. For the first default bearer this message may be sent as part of the attach request procedure.

The UE initiates the PDN connectivity procedure by creating a PDN connectivity request message and sending it to the MME. In this message the UE includes information about the IP version capability of the IP stack in the UE. The UE may indicate whether it wants an IPv4 or IPv6 address to be allocated as part of the default bearer activation procedure or whether it wants an address assigned after the default bearer activation procedure by executing DHCPv4 or DHCPv6. The PDN connectivity request message may also include the access point name (APN) information indicating the PDN to connect with. If the APN information is not provided, then the connection is made to the default PDN. If the MME accepts the PDN connectivity request message, it starts the default EPS bearer context activation procedure. If the MME cannot accept the PDN connectivity request message, it sends a "PDN connectivity reject" message with a cause value indicating the reason for rejecting the UE-requested PDN connection.

4.2.5.6 UE-Initiated PDN Disconnection Procedure

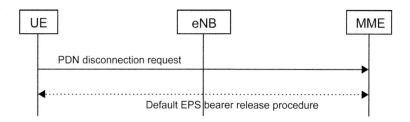

Figure 4.2-20. UE-initiated PDN disconnection procedure message flow (Based on 24.301 [6] Figure 6.5.2.2.1)

The PDN disconnection procedure (Figure 4.2-20) allows the UE to request disconnection from one PDN. All the default and dedicated bearers connected to the PDN are deleted during this procedure. The UE initiates the procedure by creating a PDN disconnection request message and sending it to the MME. This message includes a linked bearer identity, which identifies the default bearer connected to the PDN. If the MME accepts the PDN disconnection request, it initiates the bearer deactivation procedure. If the MME does not accept the PDN disconnection request message, it sends a PDN disconnection reject message to the UE with a cause indicating the reason for rejecting the request.

4.3 References

[1]   3GPP TS 36.300 V10.8.0 (2012-06) E-UTRA, E-UTRAN, Overall Description; Stage 2 (Release 10)

[2]   3GPP TS 36.321 V10.6.0 (2012-09) Medium Access Control (MAC) Protocol Specification (Release 10)

[3]   3GPP TS 36.322 V10.0.0 (2010-12) Radio Link Control (RLC) Protocol Specification (Release 10)

[4]   3GPP TS 36.323 V10.1.0 (2011-03) Packet Date Convergence Protocol (PDCP) Specification (Release 10)

[5]   3GPP TS 36.331 V10.7.0 (2012-09) Radio Resource Control (RRC) Protocol Specification (Release 10)

[6]   3GPP TS 24.301 V10.8.0 (2012-09) Non-Access-Stratum (NAS) Protocol for Evolved Packet System (EPS); Stage 3 (Release 10)

[7]   3GPP TR 23.882 V8.0.0 (2008-09) 3GPP System Architecture Evolution: Report on Technical Options and Conclusions (Release 8)

[8]   3GPP TS 23.401 V10.6.0 (2011-12) General Packet Radio Service (GPRS) enhancements for Evolved Universal Terrestrial Radio Access Network (E-UTRAN) access (Release 10)

Links to all reference documents can be found at www.agilent.com/find/ltebook.

Chapter 5

System Architecture Evolution

Development of the LTE air interface has been closely linked within 3GPP to the work on a new packet-switched system architecture initially called the System Architecture Evolution (SAE). This architecture comprises the LTE evolved UMTS radio access (E-UTRA) and evolved UMTS radio access network (E-UTRAN) as well as a new evolved packet core (EPC) network. The overall system goes by the name enhanced packet system (EPS) or LTE/SAE; these terms are often used interchangeably. The new core network is variably called the SAE/EPC, the SAE core network, or simply the EPC.

This chapter briefly introduces the EPS requirements, functions, and services. For more comprehensive information, refer to the latest versions of the standards documents listed below and at the end of this chapter.

Reference Documents

The EPS architecture supports heterogeneous access-system mobility within 3GPP and non-3GPP access systems, including fixed access systems. Architectural enhancements relating to 3GPP access systems including the E-UTRAN, legacy UTRAN, and GERAN are detailed in 23.401 [1]. Similarly, architectural enhancements relating to non-3GPP access systems including 3GPP2 cdma2000 and wireless LAN (WLAN), are detailed in 23.402 [2]. Functional descriptions are also included in 36.300 [3]. The focus of this chapter is the overall EPS architecture supporting 3GPP E-UTRAN access including E-UTRAN mobility. The remaining access systems including interworking features are referenced only as necessary for the purpose of describing the EPS architecture in a more complete manner.

5.1 Requirements for an Evolved Architecture

3GPP's overall requirements for an evolved packet system are summarized in 22.278 [4]. Different parties in the industry have provided insight into the operational and developmental challenges of current networks, and this insight has been the foundation of the ongoing work.

The overall objectives for LTE and specifically the E-UTRA and E-UTRAN are summarized in this book in Chapter 1 and detailed in Chapters 2 through 4.

LTE and the Evolution to 4G Wireless: Design and Measurement Challenges, Second Edition, Edited by Moray Rumney.
Copyright Agilent Technologies, Inc. 2013. Published by John Wiley & Sons, Ltd.

The overall objectives for the EPC, which are defined in 22.278 [4], include the following:

- Provide higher data rates, lower latency, a higher level of security, and enhanced QoS
- Support a variety of different access systems (existing and future), ensuring mobility and service continuity between these access systems
- Support access system selection based on a combination of operator policies, user preference, and access network conditions
- Realize improvements in basic system performance while maintaining the negotiated QoS across the whole system
- Provide capabilities for coexistence with legacy systems and the migration of legacy systems to the EPS.

From these objectives 3GPP established a set of requirements for the EPS, also covered in 22.278 [4] Section 5. Some of the most important items are described next.

5.1.1 User and Operational Aspects

The EPS provides the means for users to access the network with fully supported mobility across a range of access technologies. The overall system must enable full interworking not only with 3GPP systems but with non-3GPP systems as well. A minimum set of services must be supported: voice, video, messaging, and data file exchange. The EPS also has to enable efficient use of system resources, especially radio resources, through signaling and transport optimization (such as overhead, terminal power, radio resources, mobility state, and signaling load optimization). The EPS must be able to identify any device that connects via a 3GPP or 3GPP2 network. Additionally, as of 3GPP Release 11, new requirements are specified for supporting fixed mobile interworking, fixed mobile convergence, and interworking with data application providers. See 22.278 [4] Sections 5.1 and 5.2.

5.1.2 IP Support

The EPS must support increased IP traffic demand, basic IP connectivity with the UE, support of IP multicast service, and IP session control.

5.1.3 Quality of Service

The EPS has to provide quality of service (QoS) while at the same time using system resources efficiently. Quality of service has to be maintained throughout the entire EPS including the EPC, and this QoS must meet or exceed the QoS requirements specified for GSM and UMTS. Quality of service from the customer's perspective is to be considered in phases as specified in ETSI 102 250-1 [5].

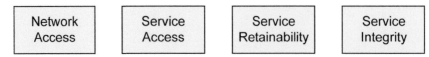

Figure 5.1-1. Phases of service use from customer's point of view (from 22.278 [4] Figure 2)

Figure 5.1-2 shows the different phases of service use from the customer's point of view. These phases are defined as follows:

- Network access: The UE display (or some other means) will inform customers that they can use the services of a particular network operator.
- Service access: If a customer wants to use a service, the network operator should provide access to the service as quickly as possible.
- Service retainability: This term describes the termination of services, which may be initiated by the customer or not.
- Service integrity: This term describes the quality of service provided to the customer during service use.

Note that the different QoS levels provided for real-time and non-real-time services will be differentiated with regard to such parameters as maximum end-to-end delay, packet size, packet drop percentage, etc. Bandwidth is not used to define a QoS level, according to 22.278 [4].

5.1.4 Support of Broadcast and Multicast Services and Emergency Calls

The EPS is required to support broadcast and multicast services that have been enhanced in certain aspects including optimized service provisioning and improved multimedia broadcast and multicast service (MBMS). This support is provided for IP multimedia subsystem (IMS) emergency calls in the packet-switched domain as defined in 22.101 [6].

5.1.5 Multiple Access and Seamless Mobility

A major requirement of the EPS is the inclusion of multiple access technologies, with mobility between heterogeneous access systems. The EPS is expected to manage handovers between all relevant technologies: 3GPP, non-3GPP mobile, and even fixed access systems.

To facilitate seamless mobility in existing networks, a function has been available in some proprietary implementations to perform "local breakout" in legacy networks. This function is now part of the requirements for the EPS. The technique can provide significant operational cost efficiencies in managing certain types of user traffic. Briefly, local breakout is a means of efficiently routing user traffic when the end points for the traffic are located within an operator-defined network region. This technique can be applied, for example, to a voice call between two users in the same cell, or to Internet access that is local to the eNB. Local breakout is fully controlled and authorized by the home public land mobile network (HPLMN).

5.1.6 Service Continuity

A key aspect of mobility is service continuity, which is the system's ability to continue providing multicast and broadcast services during a session when the access system changes, if those services are supported in the target access system.

The EPS is required to support bidirectional service continuity between cdma2000 1xRTT, cdma2000 1xEV-DO, Mobile WiMAX, and the E-UTRAN. It is also expected to manage bidirectional service continuity between Mobile WiMAX and GERAN/UTRAN packet switching. Thus the EPS should work in conjunction with almost any legacy system, enabling a truly global standard and implementation.

5.1.7 Access Network Discovery and Steering of Access

In existing W-CDMA networks, inter-radio access technology (inter-RAT) handovers are often made after the UE performs certain measurements in compressed mode. However, this mechanism reduces the performance of the ongoing transmission and consumes unnecessary resources from both the network and the handset. The EPS has a requirement to facilitate service continuation by supplying relevant access-network information to the UE in a resource-efficient and secure manner.

When the UE accesses the EPC via a non-3GPP radio access technology (RAT), the registered public land mobile network (PLMN) may request the UE to use the E-UTRA instead, for reasons of load balancing, operator policy or service mobility. If the home PLMN and the registered PLMN have different "views" of which access technology should be selected, the registered PLMN takes precedence.

5.1.8 IFOM Service Requirements

IP flow mobility (IFOM), which allows UEs to be attached simultaneously to different access network technologies, is an optional capability for multimode UEs that support 3GPP and WLAN. In this case UE attachment to 3GPP access and one WLAN access will facilitate service continuity when the UE moves from one access to the other. If the UE is under simultaneous coverage of more than one access technology, more efficient distribution of IP traffic will be possible.

The system-level description of seamless WLAN offload and IFOM between 3GPP and WLAN is covered in 23.261[7] and the description of non-seamless wireless WLAN offload in 23.402 [2].

5.1.9 Performance Requirements for the EPS

The EPS, comprising the EPC and the E-UTRA and E-UTRAN, must meet or exceed certain performance criteria as summarized below:

- Ability to support instantaneous peak packet data rates of 100 Mbps on the radio access bearer downlink to the UE and 50 Mbps on the uplink.
- Ability to provide lower user and control plane latency than existing 3GPP access networks. The maximum delay should be comparable to that of fixed broadband Internet access technologies—less than 5 ms in ideal conditions.
- Ability to support large volumes of mixed voice, data, and multimedia traffic. Enhanced load balancing and steering of roaming methods will be used to minimize cell congestion.

System complexity and mobility management signaling levels will be optimized to reduce infrastructure and operating costs. UE power consumption also will be minimized accordingly. Interrupt time during the handover of real-time and non-real-time services will be kept to a minimum, not to exceed the values defined in 25.913 [8].

5.1.10 Security and Privacy Requirements

Network security and privacy must be at least as good as in previous standards and must address security threats from the Internet as well as "traditional" threats to the telecommunication network. Security requirements include the ability to perform lawful interception.

5.1.11 Charging Requirements

The EPS will support various charging models including all those specified in 22.115 [9].

5.2 Overview of the Evolved Packet System

The previous section covered the major requirements for the EPS. The network elements and interfaces defined by 3GPP for the new system, along with legacy packet-switched (PS) and circuit-switched (CS) elements, are shown in Figure 5.2-1. Some of the major functions are discussed in this chapter, beginning with the high level functions of the overall system.

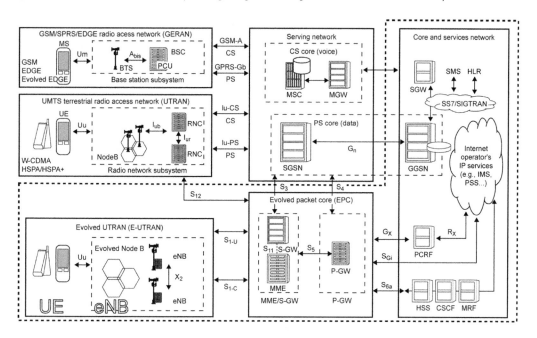

Figure 5.2-1. Overview of the EPS and legacy 3GPP RANs

5.2.1 High Level Functions of the EPS

The following logical functions are performed within the EPS and are defined in 23.401 [1].

5.2.1.1 Network Access Control

Network/access network selection functionality enables a UE to select a PLMN access network from which to gain IP connectivity. The network/access network selection procedure varies for different access technologies. For 3GPP access networks, the network selection principles are described in 23.122 [10]. The access network selection procedures are described in 36.300 [3], 43.022 [11], and 25.304 [12]. Architectural impacts stemming from support for network/access network selection procedures for non-3GPP access, and between 3GPP and non-3GPP accesses, are described in 23.402 [2].

Authentication and authorization functionality performs the identification and authentication of the service requester and the validation of the service request type to ensure that the user is authorized to use the particular network services. The authentication function is performed by the mobile management entity (MME) and HSS/HLR in association with the EPS mobility management functions.

Admission control functionality determines whether requested radio and network resources are available and, if so, reserves those resources. Radio admission control is performed, for example, at the target eNB during intra-LTE handovers. Criteria for consideration include QoS requirements and priority levels for sessions in progress and sessions to be admitted.

Policy and charging enforcement functionality includes service data flow detection, policy enforcement, and flow-based charging as defined in 23.203 [13]. This function is located at the packet data network gateway (PDN-GW) for EPS, supporting both online and offline charging functions as outlined in 32.240 [14].

Lawful interception functionality allows the network operator or service provider to make certain information available for law enforcement purposes. Lawful interception general requirements within the 3G mobile system are outlined in 33.106 [15]. The architecture and functional requirements for EPS and 3G mobile systems in general are described in 33.107 [16]. Handover interfaces for lawful interception for UMTS and EPS networks are covered in 33.108 [17].

5.2.1.2 Packet Routing and Transfer

For packet routing and transfer, IP header compression functionality optimizes the use of radio capacity by means of IP header compression techniques.

5.2.1.3 Security

Security functions guard against unauthorized EPS service usage as described in 23.401 [1] Sections 5.3.10.

Ciphering and integrity protection are performed in the MME for the non access stratum (NAS) and in the eNB for the access stratum (AS). Refer to 33.401 [18] for more details on security architecture for SAE and 33.402 [19] for security aspects of non-3GPP accesses. Network interface physical link protection is performed in network domain security (NDS) architecture as detailed in 33.210 [20].

5.2.1.4 Mobility Management

The following mobility management functions are defined in 23.401 [1]. They are used to keep track of the current location of a UE:

- UE reachability management within the ECM-IDLE state
- Tracking area list management
- Inter-eNB mobility anchor
- Inter-3GPP mobility anchor
- Idle mode signaling reduction
- Mobility restrictions
- IMS voice over PS session supported indication
- Voice domain preference and UE usage settings.

5.2.1.5 Radio Resource Management

Radio resource management (RRM) functions ensure efficient use of the available radio resources and provide mechanisms that enable the E-UTRAN to meet the radio resource requirements identified in 25.913 [8] Section 10. In particular, RRM provides the means to manage (assign, reassign, and release) radio resources, taking into account single and multi-cell aspects. Refer to 36.300 [3] for further information on the RRM in the E-UTRAN.

To support RRM in the E-UTRAN the MME provides a UE-specific RAT/frequency selection priority (RFSP) index parameter to the eNB, which maps it to a locally defined configuration in order to apply specific RRM strategies.

5.2.1.6 Network Management

The network management functions defined in 23.401 [1] provide mechanisms that support operation and maintenance (O&M) functions in the EPS. These functions are load balancing between MMEs, load rebalancing between MMEs, MME control of overload, and PDN GW control of overload.

5.2.1.7 Network Entity Selection

Packet data network gateway selection functionality allocates a PDN-GW to provide the PDN connectivity for 3GPP access. This function uses subscriber information provided in the HSS, which includes the PDN subscription contexts, hence the associated PDN GWs. Additional criteria may be used; for example, selected IP traffic offload (SIPTO) and local IP access (LIPA) per access point name (APN) configured in the SGSN or MME.

Serving gateway selection functionality allocates an available S-GW for a UE session. Selection is based mainly on network topology, by determining the S-GW that serves the UE's location. The MME must ensure that all tracking areas (TAs) in the TA list belong to the selected S-GW service area. Service area overlap between S-GWs can occur; for example, in load-balancing between S-GWs. In the case of integrated PDN-GW and S-GW nodes, the S-GW selected should preferably also be the PDN-GW for the UE. S-GW support for GPRS tunneling protocol (GTP) and proxy mobile internet protocol (PMIP) variants should also be part of the selection criteria, as subscribers of GTP-only networks can roam into PMIP networks. When SIPTO is allowed, it also is a factor in selecting the S-GW.

MME selection functionality allocates an available MME for a UE session. Selection is based mainly on network topology, determining the MME that serves the UE's location. The MME must ensure that all TAs in the TA list belong to the selected S-GW service area. Service area overlaps can occur between MMEs; for example, in load-balancing between MMEs.

SGSN selection functionality allocates an available SGSN for a UE session. Selection is based mainly on network topology, determining the SGSN that serves the UE's location. Service area overlaps are possible between SGSNs; for example, in load-balancing between SGSNs.

Policy and Charging Rules Function (PCRF) selection functionality allocates an available PCRF for a UE session.

5.2.1.8 IP Network

Directory name service (DNS) functionality resolves logical names to IP addresses for EPS nodes, including the PDN-GWs. Dynamic host configuration protocol (DHCP) functionality allows delivery of IP configuration information to the UEs. The E-UTRAN and UE support explicit congestion notification (ECN) as defined in RFC 3168 and described in 36.300 [3], 25.401 [21], and 26.114 [2].

5.2.1.9 Connection of eNB to Multiple MMEs

A single eNB may be connected to more than one MME, implying that the eNB must be able to determine which MME covers the area in which a UE is located and should receive the UE's signaling. To avoid unnecessary signaling in the core network, a UE usually continues to be served by the same MME unless relocated for various reasons such as being out of the pool area or for offloading purposes within the EPC pool. This functionality enables the eNB to select the proper MME by providing routing and related mechanisms via the globally unique temporary identity (GUTI) allocated by the serving MME.

5.2.1.10 E-UTRAN Sharing

E-UTRAN sharing allows different core network operators to connect to a shared RAN. Operators may share not only radio network elements but also radio resources. For the E-UTRAN, both a multi-operator core network (MOCN) configuration and a gateway core network (GWCN) configuration are defined in 23.251 [23].

5.2.1.11 IMS Emergency Session Support

The EPS supports emergency bearer services to support IMS emergency sessions. Emergency services are provided to normally attached UEs and, depending on local regulation, to UEs in limited service state. An overview of IMS emergency support session functions is given in 23.401 [1] Section 4.3.12 that includes expected behavior of emergency bearer support functions, architecture reference model for emergency services, mobility and access restrictions, reachability management for UE in ECM-idle state, PDN gateway selection, QoS for emergency services, load rebalancing on MMEs, PDN connection, etc. Note that in the case of any discrepancies between this overview and detailed functional and procedural descriptions found elsewhere in the specification, the latter take precedence.

5.2.1.12 Closed Subscriber Group

A closed subscriber group (CSG) is a group of subscribers for an HeNB who are permitted access one or more cells in the PLMN. CSG-related functions include subscription handling, access control, admission and rate control, charging, etc.

5.2.1.13 Location Service

The EPS supports location services. Procedures are defined in 23.271 [24].

5.2.1.14 Selected IP Traffic Offload (SIPTO)

SIPTO allows an operator to offload certain types of traffic at a network node close to a UE's point of attachment to the access network. It is achieved by selecting gateways (S-GW and P-GW) geographically close to the UE, and the function can be enabled in roaming and non-roaming scenarios. SIPTO is available for UTRAN and E-UTRAN accesses only.

5.2.1.15 Local IP Access (LIPA)

LIPA enables an IP-capable UE connected to the network via an HeNB to access other IP-enabled entities in the same residential or enterprise network without the user plane traversing the mobile operator's network other than the HeNB subsystem. This functionality uses a local gateway (L-GW) co-located with the HeNB.

5.2.1.16 Support for Machine Type Communications (MTC)

Machine-type communications are defined in Chapter 8. An overview of functions that support MTC is provided in 23.401 [1]. These functions support the MTC service requirements given in 22.368 [25], which also includes use cases and scenarios. Support functions are described for protection from MTC-related overload, optimizing periodic tracking area update (TAU) signaling, and UE configuration.

5.2.1.17 Multimedia Priority Service (MPS)

MPS subscribers have priority access to network resources in situations such as network congestion. MPS functions provide the ability to invoke, modify, maintain, and release sessions with priority and deliver prioritized media packets during congestion conditions. Service users are defined in 22.153 [26] and mechanisms for service delivery are defined in 36.331[27] and 22.011[28]. An overview in 23.401 [1] covers IMS-based MPS, priority EPS bearer services, CS fallback, network congestion controls for MPS, and load rebalancing between MMEs.

5.2.1.18 Core Network Node Resolution

This functionality allows the MME or SGSN to determine the type of signaling message sent by the UE that identifies the UE's geographical location during attach and tracking/routing area update requests.

5.2.1.19 Relaying

The relaying function allows a mobile operator to improve or extend a coverage area by wirelessly connecting a relay node (RN) to an eNB that serves the RN, called a donor eNB (DeNB). The connection uses a modified version of the E-UTRA called the Un interface, which is specified in 36.300 [3]. An overview of the relay function given in 23.401 [1] describes the relaying architecture and procedures, including RN startup and attach, and DeNB evolved radio access bearer (E-RAB) activation and modification. See also Section 8.3.3.4.

5.2.2 EPS Mobility Management and Connection Management States

EPS mobility management (EMM) states describe the relationship between the UE and the MME resulting from mobility procedures. Two main states are defined in 23.401 [1], EMM-DEREGISTERED and EMM-REGISTERED. Additional EMM main and sub-states have been specified; see 24.301 [29] for more details.

EPS connection management (ECM) states describe the signaling connectivity between the UE and the EPC. Two states are defined in 23.401 [1], ECM-IDLE and ECM-CONNECTED. These correspond to the EMM-IDLE and EMM-CONNECTED modes, respectively, described in 24.301 [29].

The EMM and ECM states are generally independent of each other; however, some state transition dependencies exist. For example, to transition from EMM-DEREGISTERED to EMM-REGISTERED, the UE has to be in the ECM-CONNECTED state.

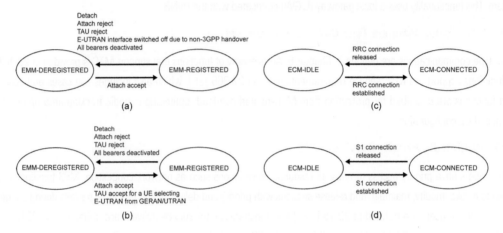

Figure 5.2-2. ECM and EMM state transition models (from 23.401 [1], Fig. 4.6.4-1 to 4.6.4-4)

Figure 5.2-2 shows the main ECM and EMM state transition models: (a) EMM state model in UE; (b) EMM state model in MME; (c) ECM state model in UE; and (d) ECM state model in MME.

Together the ECM and EMM form a multi-dimensional state model specifying varying network and UE behaviors. For example, when the UE is in the ECM-IDLE and EMM-REGISTERED states, it performs a periodic TAU; however, the TAU is not performed if the UE is in the EMM-DEREGISTERED state.

For more detail on EMM procedures see Section 4.2.4.

5.2.3 E-UTRAN

The EPS contains the E-UTRAN and EPC network elements, which are described briefly in Chapters 1 and 4 and presented here in more detail, beginning with the E-UTRAN. The overall description of the E-UTRAN is given in 36.300 [3].

5.2.3.1 E-UTRAN Architecture

The E-UTRAN consists of evolved node Bs (eNBs) that provide the E-UTRA user plane and control plane protocol terminations towards the UE. Note that the user plane protocols, described in Section 4.1 of this book, are the packet data convergence protocol (PDCP), radio link control (RLC), medium access control (MAC) and physical layer (PHY) protocols; the control plane protocol is the radio resource control (RRC) protocol. An eNB is a logical network component that serves one or more E-UTRAN cells. An eNB can support FDD mode, TDD mode, or FDD/TDD dual-mode operation.

The eNBs are interconnected by means of the X2 interface, shown in Figure 5.2-3. They are connected to the core network elements by means of the S1 interface. More specifically, they are connected to the MME by means of the S1-MME and to the serving gateway (S-GW) by means of the S1-U.

The E-UTRAN may also contain Home eNBs (HeNBs), which can be connected to the EPC either directly or via an HeNB gateway (HeNB GW), as shown in Figure 5.2-3. The HeNB GW provides additional support for a large number of HeNBs, primarily serving as an S1-MME concentrator to the HeNBs. The HeNB GW appears to the MME as an eNB, and to the HeNB as an MME. The S1 interface between the HeNB and the EPC is the same, regardless whether the HeNB is connected to the EPC via an HeNB GW or not.

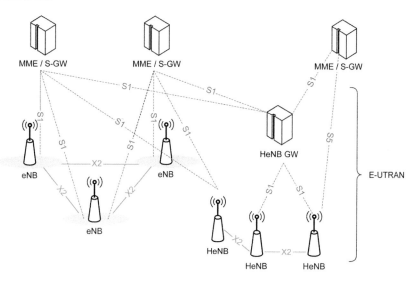

Figure 5.2-3. E-UTRAN architecture with deployed HeNB GW (from 36.300 [3] Fig. 4.6.1-2)

Note that the current version of the 3GPP specification supports direct X2-connectivity between HeNBs, independent of whether any of the involved HeNBs is connected to a HeNB GW. In Release 8 LTE, such inter-connectivity of HeNBs was not supported. Note also in Figure 5.2-3 that an HeNB operating in LIPA mode is depicted with its S5 interface.

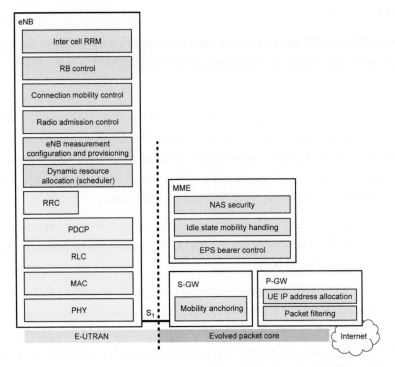

Figure 5.2-4. Functional split between E-UTRAN and EPC (from 36.300 [3] Fig. 4.1-1)

5.2.3.2 E-UTRAN Functions

Figure 5.2-4 shows the functional split in the EPS between the E-UTRAN and the EPC. This split is described in 23.401[1], 36.300 [3], and 36.401[30].

The E-UTRAN hosts the following functions listed in 36.300 [3]:

- RRM functions; specifically, the radio bearer control, radio admission control, connection mobility control, and dynamic allocation of resources to UEs in both uplink and downlink (scheduling)
- IP header compression and encryption of the user data stream
- Selection of an MME at UE attachment when no routing to an MME can be determined from the information provided by the UE
- Routing of user plane data towards the S-GW
- Scheduling and transmission of paging messages (originated from the MME)
- Scheduling and transmission of broadcast information (originated from the MME or O&M)
- Measurement and measurement reporting configuration for mobility and scheduling
- Scheduling and transmission of the public warning system (PWS) messages originating in the MME
- CSG handling
- Transport level packet marking in the uplink.

Additional functions are listed in 23.401 [1]:

- Uplink bearer level rate enforcement based on UE aggregated maximum bit rate (UE-AMBR) and maximum bit rate (MBR) via means of uplink scheduling
- Downlink bearer level rate enforcement based on UE-AMBR
- Uplink and downlink bearer level admission control
- Transport level packet marking in the uplink; e.g., setting the DiffServ code point, based on the QoS class identifier (QCI) of the associated EPS bearer
- ECN-based congestion control.

In addition to the eNB functions listed above, the DeNB provides the following for supporting relay nodes: S1/X2 proxy functionality, S11 termination, and S-GW and P-GW functionality (36.300 [3]).

5.2.4 Evolved Packet Core (EPC) Elements

The functional split in the EPS moves important control plane and gateway functions to the EPC, as shown in Figure 5.2-4. Note that the MME and the serving gateway may be implemented in a single physical node or in separate nodes; likewise, the serving gateway and packet data gateway may be implemented in one physical node or in separate nodes. The following EPC elements and interfaces are introduced in Section 4.2 of this book.

5.2.4.1 Mobility Management Entity

The mobility management entity (MME) is the control plane entity in the EPS and hosts the following functions, which are used for EPS mobility management (EMM) and EPS Session Management (ESM). These functions are detailed in 23.401 [1] and 36.300 [3].

- NAS signaling and security
- AS security control
- Inter-core-network node signaling for mobility between 3GPP access networks
- Idle mode UE reachability (including control and execution of paging retransmission)
- Tracking area list management (for UE in idle and active mode)
- Mapping from UE location to time zone and signaling UE time zone change associated with mobility
- PDN GW and serving GW selection
- MME selection for handovers with MME change
- SGSN selection for 3GPP handovers to 2G or 3G access networks
- Roaming (S6a towards home HSS)
- Authentication
- Authorization
- Bearer management functions including dedicated bearer establishment
- Warning message transfer, including selection of appropriate eNB
- Lawful Interception of signaling traffic
- Relaying support (RN attach and detach).

5.2.4.2 Serving Gateway

The serving gateway (S-GW) is the gateway that terminates the EPC interface towards the E-UTRAN. For each UE associated with the EPS, at any given point in time there will be a single S-GW hosting the following functions, which are defined in 23.401 [1] and 36.300 [3].

- Local mobility anchor point for inter-eNB handover
- Sending of one or more "end markers" to the source eNB, SGSN, or RNC immediately after switching the path during inter-eNB and inter-RAT handover
- Mobility anchoring for inter-3GPP mobility
- E-UTRAN idle mode downlink packet buffering and initiation of network triggered service request procedure
- Lawful Interception
- Packet routing and forwarding
- Transport level packet marking in the uplink and the downlink
- Accounting for inter-operator charging
- For GTP-based S5/S8, generation of accounting data per UE and bearer
- Interfacing offline charging system (OFCS) according to charging principles and through reference points specified in 32.240 [14].

5.2.4.3 Packet Data Network Gateway

The packet data network gateway (P-GW) is the gateway that terminates the SGi interface towards the PDN. If a UE accesses multiple PDNs, it may be assigned more than one P-GW. The following functions are supported as defined in 23.401 [1] and 36.300 [3].

- Per-user based packet filtering (by deep packet inspection, for example)
- Lawful interception
- UE IP address allocation
- Transport level packet marking in the uplink and downlink
- Accounting for inter-operator charging
- Uplink and downlink service level charging, gating, and rate enforcement as defined in 23.203 [13]
- Uplink and downlink rate enforcement based on APN-AMBR.

See 23.401 [1] for additional P-GW functionality.

5.2.4.4 Definitions

The following terms are important for understanding EPC functionality and processes. For a more extensive list of network architecture terms, see 23.002 [31].

MME area is the part of the network served by an MME. An MME area consists of one or several tracking areas (TAs). All cells served by an eNB are included in an MME area. There is no one-to-one relationship between an MME area and an MSC/VLR area. Multiple MMEs may share the same MME area, and MME areas may overlap each other.

MME pool area is an area where intra-domain connection of RAN nodes to multiple core network nodes is applied. Within a pool area, a UE may be served without needing to change the serving MME. An MME pool area is served by one or more MMEs (a "pool of MMEs") in parallel. MME pool areas are also a collection of complete TAs. MME pool areas may overlap each other.

S-GW service area is defined as an area within which a UE may be served without needing to change the S-GW. An S-GW service area is served by one or more S-GWs in parallel. S-GW service areas are a collection of complete TAs, and S-GW service areas may overlap each other. There is no one to one relationship between an MME area and an S-GW service area.

Tracking area includes one or several E-UTRAN cells. The network allocates a list with one or more TAs to the UE. In certain modes of operation, the UE may move freely in all TAs in the list, without updating the MME, via EMM TA update procedures. A single physical cell may belong to more than one TA.

5.2.5 Other Network Elements

Other network elements are central to EPS functionality—for example, the HSS, SGSN, PCRF, HeNB subsystem, DeNB, etc. See 23.002 [31] and 23.401 [1] for more details.

5.2.6 Reference Points (Interfaces)

This section describes the reference points (interfaces) in the EPS and their key functions in the network.

5.2.6.1 S1 Interfaces

The S1 interface connects the E-UTRAN to the EPC. More specifically, the S1 interface supports the exchange of signaling information between the eNB and EPC. From a logical standpoint, the S1 is a point-to-point interface between an eNB within the E-UTRAN and an MME in the EPC. A point-to-point logical interface should be feasible even in the absence of a physical direct connection between the eNB and MME.

The S1 interface supports the following:

- Procedures to establish, maintain, and release E-UTRAN radio access bearers (RABs)
- Procedures to perform intra-LTE handover and inter-RAT handover
- The separation of each UE on the protocol level for user specific signaling management
- The transfer of non access stratum (NAS) signaling messages between the UE and EPC
- Location services by means of transferring requests from the EPC to E-UTRAN, and location information from E-UTRAN to EPC
- Mechanisms for resource reservation for packet data streams.

Two types of S1 interface are defined: the S1 control plane interface (S1-MME) between the E-UTRAN and MME, and the S1 user plane interface (S1-U) between the E-UTRAN and S-GW. See Figures 5.2-5 and 5.2-6. Their functions are described briefly below; see 36.410 [32] and 36.413 [33] for more information.

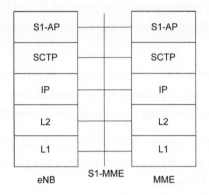

Figure 5.2-5. S1-MME reference point between E-UTRAN and MME

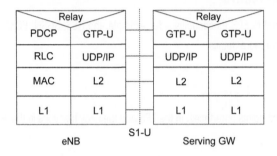

Figure 5.2-6. S1-U reference point between E-UTRAN and S-GW

S1-MME Interface

S1-MME is defined between the eNB and the MME for the control plane protocol, S1-AP, in 36.413 [33]. The EPS NAS, defined in 24.301 [29], is transported transparently over S1-AP, terminating at the MME. The stream control transmission protocol (SCTP) transport layer provisions the network for guaranteed S1-AP delivery. See Figure 5.2-5.

S1-MME Functions

Context management is responsible for setting up, modifying, and releasing the S1 UE context and the associated E-UTRAN radio and S1 resources for one or more enhanced radio access bearers (E-RABs). These are used for user data transport in both the serving MME and the eNB. Establishment and modification of UE context is initiated by the MME. The release of the context is triggered by the MME either directly or following a request received from the eNB.

E-RAB management is responsible for setting up, modifying, and releasing E-UTRAN radio and S1 resources for one or more E-RABs that are used for user data transport. Procedures for establishment and modification of E-UTRAN resources are initiated by the MME and require respective QoS information to be provided to the eNB. Release of E-UTRAN resources is triggered by the MME either directly or following a request received from the eNB.

Mobility functions for UEs in LTE-ACTIVE state enable the following handovers:

- **Intra-LTE**, a change of eNBs within the EPS via the S1 interface with EPC involvement. This is also known as an S1-initiated handover. The MME and S-GW can be relocated through this procedure.
- **Inter-3GPP RAT**, a change of RAN nodes between different RATs via the S1 interface with EPC involvement. This is also known as an inter-3GPP-RAT handover.
- **Mobility to cdma2000 system**, a change of RAN nodes between LTE and non-3GPP cdma2000 systems.

Handover signaling functions consist of the following:

- **Handover preparation**, initiated by the source eNB to the serving MME to request resource preparation at the target cell.
- **Handover resource allocation** for E-UTRAN handover operation to the E-UTRAN, initiated by the serving MME to reserve resources at the target cell.
- **Handover notification**, initiated by the eNB to notify the MME that the UE has arrived at the target cell and the S1 handover has been completed successfully.
- **Path switch request**, initiated by the eNB to request the switch of a downlink GTP tunnel towards a new GTP tunnel endpoint.
- **Handover cancellation** can be used by the source eNB to cancel an ongoing handover, after which handover procedures are terminated in the source eNB and the MME. The MME also releases the associated resources allocated for this handover in the target system (target eNB, S-GW, etc.).
- **eNB status transfer**, which transfers the uplink PDCP-SN and HFN receiver status and the downlink PDCP-SN and HFN transmitter status from the source to the target eNB via the MME during an intra-LTE S1 handover for each respective E-RAB for which PDCP-SN and HFN status preservation applies.
- **MME status transfer**, which transfers the uplink PDCP-SN and HFN receiver status and the downlink PDCP-SN and HFN transmitter status from the source to the target eNB via the MME during an S1 handover for each respective E-RAB for which PDCP-SN and HFN status preservation applies.

Paging procedures enable the EPC to page the UE through eNBs that have one or more cells belonging to the TA in which the UE is registered.

NAS signaling transport is used to transfer NAS signaling related information between the UE and serving MME. An S1 UE context may be established as part of this functionality if one does not already exist. NAS protocol information is further discussed in 24.301 [29].

S1 interface management

- **Error indication** procedures are used by the MME and eNB to indicate self-detected errors, which may or may not be related to the UE S1 context. The S1AP IDs are tagged with error indication messages if the errors are related to the UE S1 context.

- **Reset procedures** are used by the MME and eNB to request that all or part of the S1 interface context be reset. For a partial reset, the initiating node indicates the S1AP IDs for the connections to be reset.

- **S1 setup** procedures are used by the MME and eNB to exchange application level data for S1 interoperability. S1 setup is initiated by the eNB, which sends the global eNB ID, name, and supported TAs while the MME responds with the list of served PLMNs and globally unique MME identities (GUMMEIs). The relative MME capacity used for MME load balancing at the eNB is also transferred.

Overload procedures are used by the MME to tell the eNBs to reduce the signaling load. The MME stops the overload control operation via an overload stop procedure. Overload actions include (1) rejecting RRC connection requests for non-emergency mobile originated data transfer, (2) rejecting all new RRC connection requests for signaling, or (3) permitting RRC connection establishments for emergency causes only.

S1 setup procedures are used by the MME and eNB to exchange application level data for S1 interoperability. S1 setup is initiated by the eNB, which sends the global eNB ID, name and supported TAs while the MME responds with the list of served PLMNs and globally unique MME identities (GUMMEIs). The relative MME capacity used for MME load balancing at the eNB is also transferred.

eNB and MME configuration update procedures are used by the MME and eNB to update application level data for S1 interoperability, similar to S1 setup procedures, but can be initiated by nodes over the S1-MME interface. Note that the global eNB ID cannot be updated using this procedure.

UE capability information indication is used to provide UE radio capability information when received from the UE by the MME, as initiated by the eNB.

Trace is used to configure UE trace procedures for UEs in the ECM-CONNECTED state. Each activated trace is associated with a trace reference, interfaces to be traced, and trace depth for each of the traced interfaces. Trace activation and deactivation is initiated by the serving MME. UE trace procedures are further described in 32.421 [34], 32.422 [35], and 32.423 [36].

Location reporting enables the MME to request that the eNB for a given UE either (1) directly report the current serving cell or (2) report or stop reporting upon change of the serving cell.

S1 cdma2000 tunneling is used to carry cdma2000 signaling between the UE and cdma2000 RAT over the S1 interface. This function comprises signaling for (1) UE preregistration with the cdma2000 HRPD network, (2) UE preregistration and paging with the cdma2000 1xRTT network and (3) handover preparation signaling for handovers from the E-UTRAN to cdma2000 HRPD/1xRTT. See 23.402 [2] for more details on inter-RAT non-3GPP handover procedures.

Warning message transmission provides the means to start and overwrite warning message broadcasts. The MME merely forwards this procedure as received from the cell broadcast center (CBC) over the SBc interface. See 29.168 [37] for more details.

S1-U Interface

S1-U is defined between the eNB and the S-GW. See Figure 5.2-6. The S1-U interface is used for user plane (per bearer) GTP user (GTP-U) tunneling and inter-eNB path switching during S1 and X2-initiated handover. The UDP transport layer provisions the network for non-guaranteed S1AP delivery. User plane tunneling information (GTP-U IP addresses and TEIs) between the S-GW and eNB is coordinated by the MME as part of the S1AP E-RAB management procedures.

An end marker message, which is a GTP-U message, is initiated by the S-GW and sent across the S1-U and X2-U interface for each of the GTP-U tunnels, marking the termination of these tunnels for user data transfer. Any G-PDU (carrying user plane data) received after the end marker message are deemed invalid and discarded silently by the receiving nodes.

Multiple S1-U logical interfaces may exist between the eNB and the S-GW. The S-GW selection function is performed by the MME. See 29.281 [38] for more details on GTP-U functions, noting that GTP-U remains on GTPv1 (a variant of the GPRS tunneling protocol) and is common across user plane interfaces.

5.2.6.2 X2 Interfaces

The X2 interfaces provide interconnection of two eNB components within the E-UTRAN architecture. The architecture is open, allowing interconnection of eNBs from different manufacturers. The X2 supports the exchange of signaling between two eNBs as well as the forwarding of PDUs to the respective endpoints, allowing continuation of the E-UTRAN services offered via the S1 interface. The specifications require a clear separation between the radio network layer and the transport layer; thus the network signaling and X2 data streams are separated from the data transport resource and traffic handling as shown in Figure 5.2-7. The radio network layer, which consists of a radio network control plane and a radio network user plane, defines procedures related to the interaction between eNBs. The transport layer provides services for the user plane and signaling transport. The protocol structure of the X2 is detailed in 36.420.

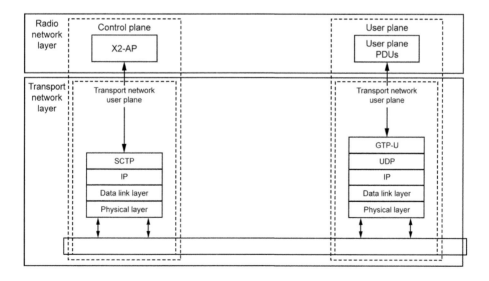

Figure 5.2-7. X2 interface protocol structure

X2-C Interface

The X2 control (X2-C) interface is defined between the eNBs for transport of the X2 application part (X2-AP) control plane protocol, in 36.423 [39]. The E-UTRAN RRC defined in 36.331 [27] is transported over X2-AP for handover information transfer. The SCTP transport layer provisions the network for guaranteed X2-AP delivery.

X2-C Functions

Mobility management for UEs in the ECM-CONNECTED state enables intra-LTE handovers within the EPS (also known as X2-initiated handovers) via the X2 interface without EPC involvement, up to the handover completion phase, when the S1 path switch is required for the target eNB and S-GW to sync up to the new S1-U user path. S-GWs can be relocated through the mobility management function..

Handover preparation is initiated by the source eNB to the target eNB to request resource preparation at the target cell. The source eNB specifies the E-RABs and associated parameters to be established for RAB management functionality in the eNB. The source eNB informs the target eNB of the S1-U tunnel parameters of the serving S-GW for each E-RAB to be established. The source eNB also indicates the E-RAB status for DL forwarding. The eNB stores the UE security capabilities and handover restriction list, which specifies the supported encryption and integrity algorithms in the UE, and the restricted roaming areas and access restrictions, respectively. Access stratum security information is used by the target eNB to derive the AS security configuration, outlined in 33.401 [18]. Indication of the UE and MME single radio voice call continuity (SRVCC) support status is stored and used in the eNB. For more details on SRVCC, see 23.216 [40].

UE trace and location reporting, if active for the current UE context in the source eNB, is also activated if supported in the target eNB. UE trace procedures are further described in 32.421 [34], 32.422 [35], and 32.423 [36]. The target eNB performs the required resource allocation and sends the handover preparation status. For each E-RAB that the source eNB has proposed for downlink data forwarding, the target eNB may include the DL X2-U tunnel endpoint for forwarding of DL PDUs. In addition, the UL X2-U tunnel endpoint may be included for each E-RAB target that the eNB requests for UL PDU forwarding. The RRC handover information is sent to the UE transparently over the X2-AP with a handover request acknowledge message.

Status transfer is sent from the source eNB to the target eNB to transfer the uplink receiver and downlink transmitter PDCP status for each E-RAB for which PDCP sequence number (SN) and hyper frame number (HFN) status preservation applies during X2-initiated handover. After the transfer status is sent, PDCP status in the source eNB is considered frozen and any out of sequence packets received for the E-RABs will be forwarded to the target eNB if UL forwarding is enabled or else be discarded. In cases in which status preservation does not apply, packets will be sent to the S-GW.

UE context release is sent from the target eNB to the source eNB to indicate successful handover and trigger release of resources at the source eNB. The source eNB releases the radio and control plane resources including the S1-MME related to the UE context.

Handover cancel can be used by the source eNB to cancel an ongoing handover. The target eNB removes all associated references and releases any resources previously reserved as part of the handover preparation phase.

Inter-cell interference coordination is used by the eNB to transfer load and inter-cell inftererence coordination information between the intra-frequency neighbor eNBs. This allows receiving eNBs to make radio resource assignments to reduce interference. Information includes interference levels, sensitivity experienced by the sending eNB, and downlink power restriction status at the cell resource block level.

Load management allows the exchange of overload and traffic load information between eNBs so that the eNBs can control the traffic load appropriately. This information may be sent spontaneously to selected neighbor eNBs or reported in response to a neighbor eNB request.

X2 Interface Management Functions

Error indication procedures are used by the eNB to indicate to its neighbors any detected errors.

Reset procedures are used by the eNB to indicate to its neighbors that a failure has occurred, and all active X2 interface contexts are to be reset and related resources removed.

X2 setup procedures are used by the eNB to exchange application level data with its neighbors for X2 interoperability.

eNB configuration update procedures are used by the MME and eNB to update application level data for X2 interoperability, similar to the X2 setup procedures.

Trace is used to configure UE trace procedures for UEs in the ECM-CONNECTED state. Each trace activated is associated with a trace reference, interfaces to be traced, and trace depth for each interface. Trace activation and deactivation are initiated by the serving MME. UE trace procedures are further described in 32.421 [34], 32.422 [35], and 32.423 [36]. Applicability to X2 is initiated by the source eNB during the handover preparation procedure, if trace activations exist for the UE context to be handed over to the target eNB.

Data exchange for self-optimization allows two eNBs to exchange information in order to support self-optimization functionality.

X2-U Interface

X2-User (X2-U) interface is defined between the eNBs that provide user plane per bearer GTP-U (29.281 [38]) tunneling for data forwarding during X2-initiated handover. The UDP transport layer provisions the network for non-guaranteed S1AP delivery. User plane tunneling information between the source and target eNBs is coordinated as part of the X2-AP mobility management procedures. See 29.281 [38] for more details on GTP-U function.

5.2.6.3 Other Interfaces

In addition to the S1 and X2 interfaces, the following interfaces are defined for the EPC:

S3 is the reference point between the MME and SGSN. It enables the exchange of user and bearer information for inter-3GPP access network (UTRAN/GERAN) mobility in idle or active state. S3 is based on GTP version 2-Control plane (GTPv2-C) defined in 29.274 [41].

S4 is the reference point between the S-GW and SGSN. Two S4 paths—S4-C and S4-U—are based on GTPv2-C defined in 29.274 [41] and GTPv1-U defined in 29.281 [38], respectively. S4 provides control and mobility support between the GPRS core and the 3GPP anchor function of the S-GW. It also provides user plane tunneling between the EPC and SGSN if direct tunneling is not enabled through the S12 interface.

S5/S8 are reference points between the S-GW and the PDN GW, providing user plane tunneling and tunnel management. S8 is the inter-PLMN variant of the S5 interface. S5/8-C and S5/8-U versions are based on GTPv2-C defined in 29.274 [41] and GTPv1-U defined in 29.281 [38], respectively, for the GTP variant.

S6a is the reference point between MME and HSS used to transfer subscription and authentication data for authenticating and authorizing user access to EPS services. It is based on the Diameter application with required extensions defined in 29.272 [42].

S10 is the reference point between MMEs used for user information transfer and MME relocation support. S10 is based on GTPv2-C defined in 29.274 [41].

S11 is the reference point between the MME and S-GW used to support mobility and bearer management. S11 is based on GTPv2-C defined in 29.274 [41].

S12 is the reference point between the S-GW and UTRAN used for direct user plane tunneling during E-UTRAN and UTRAN handovers. S12 is based on GTPv1-U defined in 29.281 [38].

S13 is the reference point between the MME and the equipment identity register (EIR) used for UE identity validation. S13 is based on the Diameter application with required extensions defined in 29.272 [42].

SBc is the reference point between the MME and the CBC used for warning-message delivery and control functions. The SBc-AP protocol is defined in 29.168 [37].

Gx is the reference point between the PCRF and the policy enforcement and charging function (PECF), providing transfer of policy and charging rules from PCRF to PECF in the PDN GW in the EPS. Gx is based on the Diameter application with required extensions defined in 29.212 [43].

Rx is the reference point between the authentication framework (AF) and PCRF defined in 23.203 [13]. Rx is based on the Diameter application with required extensions defined in 29.214 [44].

SGi is the reference point between the PDN and PDN-GW.

5.3 Quality of Service in EPS

Quality of service (QoS) in the EPS is based on the EPS bearer service layered architecture shown in Figure 5.3-1.

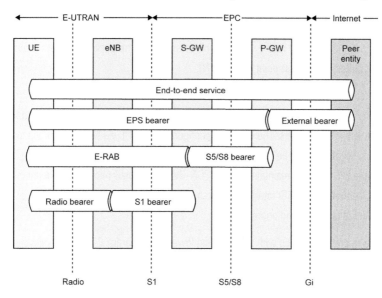

Figure 5.3-1. EPS bearer service architecture (from 36.300 [10] Fig. 13.1-1)

5.3.1 EPS Bearer

The EPS bearer provides PDN connectivity service to the UE via EPS access. The EPS bearer defines the bearer level QoS control that can be specified in the EPS, which means that all traffic through a particular EPS bearer has to use the same set of QoS parameters. Different QoS requirements for bearer controls would require additional EPS bearers to be established.

There are two EPS bearer types, default and dedicated, described in Table 5.3-1. EPS bearer contexts are handled by EPS session management (ESM) procedures, as detailed in 24.301 [29].

Table 5.3-1. Characteristics of EPS bearer types

EPS Bearer	Default	Dedicated
Traffic flow template	No (match all packet filters)	Yes
Quantity per UE	One per PDN	Zero or more per PDN
Resource type	Non-GBR	GBR or non-GBR
Required for EMM-REGISTERED state	Yes	No

A default EPS bearer must be established with the EPC for the UE to be in the EMM-REGISTERED state and hence able to access EPS services. Additional default bearers can be established with additional PDNs if simultaneous access to multiple PDNs is required.

Dedicated EPS bearers are defined as additional EPS bearers established with any PDN after the default EPS bearer has been established with that PDN. Dedicated EPS bearers are linked to the default bearer via the linked EPS bearer identity (LBI) at the NAS ESM layer.

The traffic flow template (TFT) consists of one or more packet filters, each identified with a unique packet filter identifier. Each packet filter has an evaluation precedence index, which is unique among all the packet filters associated with PDN contexts that share the same PDN address and access point name (APN).

TFTs are not configured for default EPS bearers; however, a TFT is associated with each dedicated EPS bearer. Association between an UL TFT and the EPS bearer is maintained in the UE. The UE uses the packet filters defined in the UL TFT to route packets in the uplink to different EPS Bearers. The UL TFT is configured in the UE via NAS ESM procedures. Typically, packets that do not match any of the UL TFT will be delivered to the EPS via the default bearer (match all). Similarly, association between the DL TFT and the EPS bearer is maintained in the PDN-GW, which uses packet filters defined in the DL TFT to route packets in the downlink (received on the SGi interface) to associated EPS bearers. See 23.060 [45] for more details on TFT, packet filtering, and associated attributes and operations.

The EPS bearer has a one-to-one relationship with the following lower layer bearers: E-RAB, a composite of the S1 bearer and data radio bearer (DRB); and the S5/S8 bearer.

5.3.1.1 E-UTRAN Radio Access Bearer

The E-RAB is used to transport packets of an EPS bearer between the UE and the EPC, and has a one-to-one relationship with a single EPS bearer. The E-RAB is defined as the concatenation of the S1 bearer and the corresponding DRB (described below).

5.3.1.2 S5/S8 Bearer

The S5/S8 bearer is used to transport packets of an EPS bearer between the S-GW and the PDN-GW, and has a one-to-one relationship with the EPS bearer.

5.3.1.3 S1 Bearer

The S1 bearer is used to transport packets of an EPS bearer between the eNB and the EPC, specifically the S-GW over S1-U interface. The S1 bearer has a one-to-one relationship with the E-RAB.

5.3.1.4 Data Radio Bearer

The DRB is used to transport packets of an EPS bearer between the eNB and the UE, and has a one-to-one relationship with a single EPS bearer. The eNB and UE map the EPS bearer QoS received from the MME via the S1-MME and NAS ESM, respectively, to the radio bearer QoS, which in turn is used to configure the related E-UTRAN and UE resources. For more details see 36.331 [27].

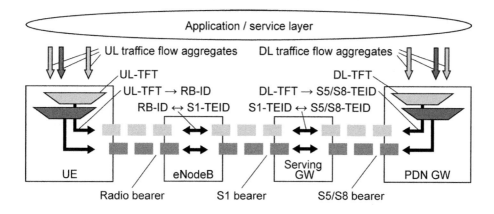

Figure 5.3-2. Layered architecture and binding relationships for various elements of EPS bearers
(from 23.401 [1] Fig. 4.7.2.2-1)

Figure 5.3-2 depicts the binding relationships between the different layer bearers with GTP-based S5/S8. See 23.401 [1] for a detailed explanation and 23.402 [2] for information on the EPS bearer with PMIP-based S5/S8.

5.3.2 QoS Profile and Parameters

The EPS bearer and E-RAB QoS profile includes service level QoS parameters defined by the following:

- QoS class identifier (QCI)
- Allocation and retention priority (ARP)
- Guaranteed bit rate (GBR), applicable only for GBR bearer
- Maximum bit rate (MBR), applicable only for GBR bearer.

5.3.2.1 QoS Class Identifier

QCI is a scalar value. It is used as a reference to node-specific parameters that handle bearer level packet forwarding behavior in order to meet a set of performance characteristics required by applications and services for which these bearers are utilized. QCI provides a consistent QoS identification used extensively in various ways through every node along the path within the EPS and with the UE.

Table 5.3-2 shows the standardized QCI values and standardized characteristics specified in 23.203 [13]. The objective of this standardization is to ensure QoS requirement interoperability between multi-vendor nodes, shared networks, and roaming scenarios, independent of access systems

Table 5.3-2. Standardized QCI characteristics (adapted from 23.203 [13] Table 6.1.7)

QCI	Resource type	Priority	Packet delay budget	Packet error loss rate	Example services
1	GBR	2	100 ms	10^{-2}	Conversational voice
2	GBR	4	150 ms	10^{-3}	Conversational video (live streaming)
3	GBR	3	50 ms	10^{-3}	Real time gaming
4	GBR	5	300 ms	10^{-6}	Non-conversational video (buffered streaming)
5	Non-GBR	1	100 ms	10^{-6}	IMS signaling
6	Non-GBR	6	300 ms	10^{-6}	Video (buffered streaming) TCP-based (e.g., www, e-mail, chat, ftp, p2p file sharing, progressive video, etc.)
7	Non-GBR	7	100 ms	10^{-3}	Voice Video (live streaming) Interactive gaming
8	Non-GBR	8	300 ms	10^{-6}	Video (buffered streaming) TCP-based (e.g., www, e-mail, chat, ftp, p2p file sharing, progressive video, etc.)
9	Non-GBR	9	300 ms	10^{-6}	Video (buffered streaming) TCP-based (e.g., www, e-mail, chat, ftp, p2p file sharing, progressive video, etc.)

The following performance characteristics are specified along with each of the QCI values:

- Resource type, which differentiates between GBR and non-GBR service data flows (SDFs)
- Priority, which is associated with every QCI, with 1 as the highest value
- Packet delay budget, which supports the configuration of scheduling and link layer functions by defining the maximum amount of time that packets are allowed to be delayed between the UE and PDN-GW
- Packet error loss rate, which defines the upper bound for packets lost through non-congestion related causes. By definition these are packets processed by the link layer protocol or the sender but not successfully received at the upper layer of the receiver. This specification supports the appropriate link layer protocol configurations (e.g., L1 HARQ and RLC ARQ).

5.3.2.2 Allocation and Retention Priority

ARP contains information that is used for bearer admission control in a resource-level network:

- Priority level used to differentiate the relative importance between different bearers, especially during setup, modification, or replacement decisions (potentially replacing other bearers).
- Pre-emption capability, indicating whether the bearer should be allowed to preempt other bearers.
- Pre-emption vulnerability, indicating whether the bearer can be preempted (i.e., eligible for a request to be replaced by other bearers).

Note that ARP has no effect on run-time QoS control of the EPS bearers and the associated lower layered bearers.

5.3.2.3 Guaranteed Bit Rate

GBR indicates the guaranteed EPS bearer and E-RAB bit rates for a GBR type bearer, defined for uplink and downlink directions.

5.3.2.4 Maximum Bit Rate

MBR indicates the maximum EPS bearer and E-RAB bit rates for a GBR type bearer, defined for uplink and downlink directions.

5.3.2.5 Other Bit Rate Parameters

In addition to the QoS parameters just described, 3GPP defines parameters for bit rates of data per groups of bearers.

UE-AMBR sets the limit for the aggregate bit rate that can be expected to be provided across all non-GBR bearers per UE. Enforcement of the UE-AMBR in both the uplink and downlink is performed in the eNB.

APN-AMBR sets the limit for the aggregate bit rate that can be expected to be provided across all non-GBR bearers over all PDN connections of the same PDN. Enforcement of the APN-AMBR in the uplink is performed in the UE and PDN-GW, while downlink is performed in the PDN-GW.

See 23.401 [1] and 23.203 [13] for more information.

5.4 Security in the Network

Service access in the PS domain requires a security association to be established between the UE and the PLMN. A separate security association must be established between the UE and the IMS core network subsystem (IMS CN SS) before access can be granted to multimedia services hosted.

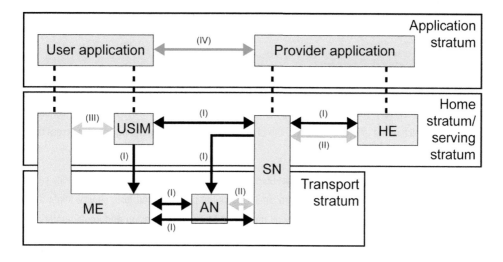

Figure 5.4-1. Security features in the network (from 33.401 [18] Fig.4-1)

As shown in Figure 5.4-1, five security feature groups are defined in 33.401 [18].

(I) **Network access** provides users with secure access to services and protects against attacks on the access interfaces.

(II) **Network domain** enables nodes to securely exchange signaling data and user data, and protects against attacks on the wire line network.

(III) **User domain** provides secure access to mobile stations.

(IV) **Application domain security** enables applications in the user and provider domains to securely exchange messages.

(V) **Visibility and configurability of security** allow the user to learn whether a security feature is in operation or not and whether the use and provision of services should depend on the security feature.

A more detailed description of security features in the EPS is available in 33.401 [18]. Security architecture for non-3GPP accesses to EPS is covered in 33.402 [19].

5.5 Services

For an end user the EPS will in many aspects behave much like current fixed-access networks. The services deployed on the EPS will not be limited to a single access technology such as E-UTRAN but will more likely be part of a larger overall service framework. As the EPS and E-UTRAN access have been designed for much more cost-effective operation, it is anticipated that much wider use of data services will be made in the future, not limited to but including an ample amount of machine-to-machine communication possibly without the direct interaction of a human end user.

This section examines some common services offered on today's networks to understand how they can be maintained in an EPS. Indeed, a significant portion of this section is dedicated to the handling of traditional circuit-switched voice calls, a technical challenge that is the focus of much activity today.

5.5.1 End-to-End Service Concept

What constitutes end-to-end service is very much in the eye of the beholder. In other words, what are considered the end points of a communication can vary greatly depending on who the observer is. Figure 5.5-1 depicts the concept, presenting as part of the overall QoS picture the different aspects of the EPS that are involved in delivering services across the end-to-end chain.

Depending on what service is used and what bearer is set up over the EPS, the resulting end user experience can vary greatly in different instances, even when network conditions are similar. Thus the price charged for a service might depend on the QoS achieved. End-to-end service concepts must be well understood by network design and implementation teams and by the teams that develop and deploy new services. A solid understanding of the underlying technology is perhaps more important than ever to develop proper marketing, including pricing of the different services that can be deployed on the EPS. Without such an understanding, implementing a service such as circuit-switched voice via circuit-switched fallback (CSFB), for example, could result in a poor end-user quality of experience (QoE). All parties involved in delivering the service must know the tradeoffs, and not only on an engineering level. The International Telecommunications Union (ITU) has established conventions about the relationship between QoS and QoE that includes non-technical factors such as the end user's subjective expectations of service quality.

Figure 5.5-1 shows how QoE is built from both technical and non-technical QoS attributes. These merge with the customer expectations that are derived from many external factors. The contributions of network and terminal performance will be considered here.

Section 5.3 described how the EPC and E-UTRAN categorize different high level services by QoS class identifier (QCI). Each of these services could then be managed properly. The MAC process discussion of Chapter 4 suggested that the scheduler could effectively manage a real time service and deliver a high level of QoS because the traffic is fully deterministic and resources can be granted proactively rather than by request only. By taking care to properly manage both services and the QCI allocation of services, an improved end-user QoE based on the highly efficient allocation and use of resources can be achieved.

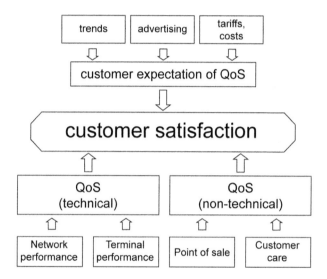

Figure 5.5-1. Relationship between customer satisfaction (QoE) and QoS

5.5.2 Circuit-Switched Services for EPS

One aspect of the EPS that was not fully defined in Release 8 is the management of traditional circuit-switched services, voice in particular. As the long term solution, a majority in the industry appear to favor the use of voice over IP (VoIP) and an IP multimedia subsystem (IMS) in the core network to deliver voice over LTE (VoLTE). Much effort is being put into developing VoLTE technology and at the time of this writing various industry groups are reporting some success in trials. However, although IMS exists in regular IP networks today, it is not yet widely developed (standardized and optimized) for use in mobile networks.

Another factor to consider is the time it will take to deploy LTE on a large scale. Operators are aiming for 100% LTE coverage, but experience tells us that goal is a long way off and the need to fall back to legacy formats will exist for some time. Thus, for those networks that do not support IMS or for the instances when a call is outside the LTE coverage area, various fallback options have been defined or proposed:

- Circuit-switched fallback (CSFB), which is a single radio approach
- Dual radio via legacy network (also called "dual standby" and "simultaneous voice and data LTE")
- Single radio voice call continuity (SRVCC) using VoLTE with circuit-switched backup
- Voice over LTE generic access (VoLGA)
- "Over the top" options such as Skype.

Of these options, CSFB and dual radio rely solely on legacy infrastructure for voice coverage and do not use any form of voice over LTE, while SRVCC relies on LTE within an LTE coverage area and transfers the call to the legacy network when LTE coverage is no longer available. Test considerations for these approaches are covered in this book in Section 6.12.5.

5.5.2.1 Voice over IMS

IMS is the framework for delivery of multimedia services and was standardized in Release 5 for delivery of internet services on GPRS. The specification was later updated and expanded to cover CDMA and wireless LAN. In 2010, the Global System for Mobile Communications Association (GSMA) adopted VoLTE with IMS in the core network as the preferred delivery system (see http://www.gsma.com/technicalprojects/volte/), and the industry seems to be coalescing around this approach.

Within the IMS framework, the session initiation protocol (SIP) is used for call setup and connection control of services. SIP does more than just call control, however; it also provides extended or supplementary services such as call hold, multi-party calls, SMS delivery, video calling, etc.

Today LTE networks and devices that support voice through IMS must grapple with interoperability issues between non-3GPP and LTE, which were never designed to work together, and between 3GPP legacy systems and LTE. Thus operators with different legacy systems have to get to the global LTE end point through different routes.

The 3GPP specifications for various aspects of IMS for telephony and supplementary services are found in 22.173 [46], 23.228 [47], 23.292 [48], 24.173 [49], and 24.229 [50].

5.5.2.2 CS Fallback in EPS

CSFB in EPS enables the provisioning of voice and other circuit-switched domain services by reuse of the circuit-switched infrastructure when the UE is served by the E-UTRAN and the E-UTRAN coverage overlaps those of 3GPP legacy access networks (GERAN/UTRAN) or non-3GPP cdma2000 1xRTT. This means that if CSFB is implemented, the E-UTRAN will not support circuit-switched service and will hand over any circuit-switched call initiated or received to either a GERAN/UTRAN or a non-3GPP cdma2000 1xRTT access. A CSFB-enabled terminal may establish one or more CS-domain services via these legacy access networks.

CSFB can coexist with the IMS-based services in an operator's network. To ensure the proper handling of a priority CSFB call to or from a priority service subscriber (government or emergency response personnel, for example), the multimedia priority service (MPS) is used to indicate the subscriber's priority status, which is then provided to the MME. See 23.401 [1] and 23.272 [51] for more information. CSFB to 1xRTT is covered in 36.300 [3].

5.5.2.3 Dual Radio

Simultaneous voice and LTE (data) uses two UE radios connected at the same time to two completely different networks. One radio is used for voice on 1xRTT and the other is connected to the LTE or enhanced high rate packet data (eHRPD) network for data. No interoperability between the two radios is required. Although this is what has become the first implementation of LTE voice, the approach has significant drawbacks including the cost of two systems, the resulting device complexity, and the battery drain from increased power consumption. Although the voice side can be considered separately, the data radio still has to handle handovers between LTE and eHRPD.

5.5.2.4 SRVCC

Single radio voice call continuity is an LTE functionality that allows a VoIP/IMS (VoLTE) call in the LTE packet domain to be moved to a legacy voice domain (GSM/UMTS or CDMA 1xRTT) when necessary. SRVCC allows operators with a legacy cellular network to deploy VoIP/IMS-based services in conjunction with the rollout of an LTE network, providing subscribers with coverage over a much larger area than would typically be available at this stage of the rollout. For more on SRVCC, see 23.216 [40].

5.5.3 Conclusion

Because circuit-switched services are still dominant and they are not covered by Release 8, they have been the focus of this section, which ends with some concluding remarks about the possible voice service options. In general, CSFB will enable delivery of CS services in an E-UTRAN environment but the user will be restricted to the services of the traditional UTRAN or GERAN. Call setup will be delayed compared to what is experienced today in the UTRAN and GERAN. In other words, with CSFB at best the user will get a service similar to what is available today.

The main issue related to solutions running voice over IMS is the cost structure of the IMS subsystem and the uncertainty about its scalability. IMS uptake so far has been limited, and thus only limited real world experience is available to tell us whether IMS-based solutions will be effective over E-UTRAN and EPS. Voice over IMS technology is readily applicable to other access technologies including the UTRAN and GERAN. It is expected that some variant will become the final solution to the overall circuit-switched problem, although it may take some time to become fully stabilized and integrated.

It is highly likely that some operators and service providers will launch E-UTRAN and EPS networks using non-3GPP CS solutions. However, as with the Internet of today, solutions can be found to all these problems that are more or less stable, scalable, maintainable, and cost-effective. The ever increasingly competitive landscape makes it difficult to predict whether the adaptation of a solution that has not been standardized by 3GPP would allow for a significant first mover advantage. One possibly significant obstacle to proprietary solutions is the security requirement to provide full support for lawful interception, including adaptation to possible local governmental regulations. But it does seem likely that some non-standard solutions will find their way onto the connections of end users as they have done in the past.

5.6 References

[1] 3GPP TS 23.401 V11.2.0 (2012–06) General Packet Radio Service (GPRS) enhancements for Evolved Universal Terrestrial Radio Access Network (E-UTRAN) access

[2] 3GPP TS 23.402 V11.3.0 (2012–06) Architecture enhancements for non-3GPP accesses

[3] 3GPP TS 36.300 V11.2.0 (2012–06) Evolved Universal Terrestrial Radio Access (E-UTRA) and Evolved Universal Terrestrial Radio Access Network (E-UTRAN); Overall description; Stage 2

[4] 3GPP TS 22.278 V12.1.0 (2012–06) Service requirements for the Evolved Packet System (EPS)

[5] ETSI TS 102 250-1 V1.1.1 (2003–10) Speed Processing, Transmission and Quality Aspects (STQ); QoS aspects for popular services in GSM and 3G networks: Part 1: Identification of Quality of Service aspects

[6] 3GPP TS 22.101 V12.1.0 (2012–06) Service aspects; Service principles

[7] 3GPP TS 23.261 V10.2.0 (2012–03) IP flow mobility and seamless Wireless Local Area Network (WLAN) offload; Stage 2

[8] 3GPP TR 25.913 V9.0.0 (2009–12) Requirements for Evolved UTRA (E-UTRA) and Evolved UTRAN (E-UTRAN)

[9] 3GPP TS 22.115 V11.5.0 (2011–12) Service aspects; Charging and billing

[10] 3GPP TS 23.122 V11.2.0 (2012–06) Non-Access-Stratum (NAS) functions related to Mobile Station (MS) in idle mode

[11] 3GPP TS 43.022 V10.0.0 (2011–03) Functions related to Mobile Station (MS) in idle mode and group receive mode

[12]   3GPP TS 25.304 V10.5.0 (2012–06) User Equipment (UE) procedures in idle mode and procedures for cell reselection in connected mode

[13]   3GPP TS 23.203 V11.6.0 (2012–06) Policy and charging control architecture

[14]   3GPP TS 32.240 V11.4.0 (2012–06) Telecommunication management; Charging management; Charging architecture and principles

[15]   3GPP TS 33.106 V11.1.1 (2011–12) Lawful interception requirements

[16]   3GPP TS 33.107 V11.2.0 (2012–06) 3G security; Lawful interception architecture and functions

[17]   3GPP TS 33.108 V11.3.0 (2012–06) 3G security; Handover interface for Lawful Interception (LI)

[18]   3GPP TS 33.401 V11.4.0 (2012–06) 3GPP System Architecture Evolution (SAE); Security architecture

[19]   3GPP TS 33.402 V11.4.0 (2012–06) 3GPP System Architecture Evolution (SAE); Security aspects of non-3GPP accesses

[20]   3GPP TS 33.210 V12.0.0 (2012–06) 3G security; Network Domain Security (NDS); IP network layer security

[21]   3GPP TS 25.401 V10.2.0 (2011–06) UTRAN overall description

[22]   3GPP TS 26.114 V11.4.0 (2012–06) IP Multimedia Subsystem (IMS); Multimedia Telephony; Media handling and interaction)

[23]   3GPP TS 23.251 V11.2.0 (2012–06) Network Sharing; Architectural and functional description

[24]   3GPP TS 23.271 V10.2.0 (2012–03) Functional stage 2 description of Location Services (LS)

[25]   3GPP TS 22.368 V11.5.0 (2012–06) Service requirements for Machine-Type Communications (MTC); Stage 1

[26]   3GPP TS 22.153 V11.1.0 (2011–06) Multimedia priority service

[27]   3GPP TS 36.331 V11.0.0 (2012–06) Evolved Universal Terrestrial Radio Access (E-UTRA); Radio Resource Control (RRC); Protocol specification

[28]   3GPP TS 22.011 V11.2.0 (2011–12) Service accessibility

[29]   3GPP TS 24.301 V11.3.0 (2012–06) Non-Access-Stratum (NAS) protocol for Evolved Packet System (EPS); Stage 3

[30]   3GPP TS 36.401 V10.4.0 (2012–06) Evolved Universal Terrestrial Radio Access Network (E-UTRAN); Architecture description)

[31]   3GPP TS 23.002 V11.3.0 (2012–06) Network architecture

[32]   3GPP TS 36.410 V10.3.0 (2012–06) Evolved Universal Terrestrial Radio Access Network (E-UTRAN); S1 general aspects and principles

[33]   3GPP TS 36.413 V11.0.0 (2012–06) Evolved Universal Terrestrial Radio Access (E-UTRA) ; S1 Application Protocol (S1AP)

[34]   3GPP TS 32.421 V11.3.0 (2012–06) Telecommunication management; Subscriber and equipment trace; Trace concepts and requirements

[35]   3GPP TS 32.422 V11.4.0 (2012–06) Telecommunication management; Subscriber and equipment trace; Trace control and configuration management

[36]   3GPP TS 32.423 V11.1.0 (2012–06) Telecommunication management; Subscriber and equipment trace; Trace data definition and management

[37]   3GPP TS 29.168 V11.3.0 (2012–06) Cell Broadcast Centre interfaces with the Evolved Packet Core; Stage 3

[38]   3GPP TS 29.281 V11.3.0 (2012–06) General Packet Radio System (GPRS) Tunnelling Protocol User Plane (GTPv1-U)

[39] 3GPP TS 36.423 V11.1.0 (2012–06) Evolved Universal Terrestrial Radio Access Network (E-UTRAN); X2 Application Protocol (X2AP)

[40] 3GPP TS 23.216 V11.5.0 (2012–06) Single Radio Voice Call Continuity (SRVCC); Stage 2

[41] 3GPP TS 29.274 V11.3.0 (2012–06) 3GPP Evolved Packet System (EPS); Evolved General Packet Radio Service (GPRS) Tunnelling Protocol for Control plane (GTPv2-C); Stage 3

[42] 3GPP TS 29.272 V11.3.0 (2012–06) Evolved Packet System (EPS); Mobility Management Entity (MME) and Serving GPRS Support Node (SGSN) related interfaces based on Diameter protocol

[43] 3GPP TS 29.212 V11.5.0 (2012–06) Policy and charging control over Gx reference point

[44] 3GPP TS 29.214 V11.5.0 (2012–06) Policy and charging control over Rx reference point

[45] 3GPP TS 23.060 V11.2.0 (2012–06) General Packet Radio Service (GPRS); Service description; Stage 2

[46] 3GPP 22.173 V12.1.0 (2012–06) IP Multimedia Core Network Subsystem (IMS); Multimedia Telephony Service and supplementary services; Part 1

[47] 3GPP TS 23.228 V11.5.0 (2012–06) IP Multimedia Subsystem (IMS); Stage 2

[48] 3GPP TS 23.292 V11.3.0 (2012–06) IP Multimedia Subsystem (IMS) centralized services; Stage 2

[49] 3GPP TS 24.173 V11.2.0 (2012–03) IMS multimedia telephony communication service and supplementary services; Stage 3

[50] 3GPP TS 24.229 V11.4.0 (2012–06) IP multimedia call control protocol based on Session Initiation Protocol (SIP) and Session Description Protocol (SDP); Stage 3

[51] 3GPP TS 23.272 V11.1.0 (2012–06) Circuit Switched (CS) fallback in Evolved Packet System (EPS); Stage 2

Links to all reference documents can be found at www.agilent.com/find/ltebook.

Chapter 6

Design and Verification Challenges

6.1 Introduction

The material in this chapter was the motivation for writing this book. Agilent is actively involved in developing the design and measurement tools the LTE industry needs to efficiently turn LTE concepts into deployed and operational systems. Although the process of developing the radio equipment for a new standard is complex and no one model captures everything, Figure 6.1-1 is an attempt to define the product development lifecycle and provides the outline for this chapter.

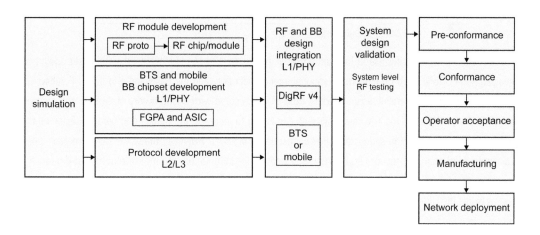

Figure 6.1-1. Example development lifecycle

This chapter is divided into fifteen major sections, shown in Table 6.1-1, each addressing different aspects of the lifecycle. Section 6.2 begins with design simulation. The complexity of communication systems is continuing to rise, and correspondingly the value of simulation early in the lifecycle is increasingly important. The ability to integrate electronic design automation (EDA) tools with instruments to stimulate, measure, and emulate hardware subsystems extends the reach of simulation tools further into the design lifecycle. This section offers numerous examples of how Agilent's EDA tools and other instruments can be used to overcome design and verification challenges in the earliest phases of product development.

LTE and the Evolution to 4G Wireless: Design and Measurement Challenges, Second Edition, Edited by Moray Rumney.
Copyright Agilent Technologies, Inc. 2013. Published by John Wiley & Sons, Ltd.

Table 6.1-1. Design and measurement sections

Section	Title
6.2	Simulation and Early R&D Hardware Testing
6.3	Testing RFICs With DigRF Interconnects
6.4	Transmitter Design and Measurement Challenges
6.5	Receiver Design and Measurement Challenges
6.6	Receiver Performance Testing
6.7	Testing Open- and Closed-Loop Behaviors of the Physical Layer
6.8	Design and Verification Challenges of MIMO
6.9	Beamforming
6.10	SISO and MIMO Over The Air Testing
6.11	Signaling Protocol Development and Testing
6.12	UE Functional Testing
6.13	Battery Drain Testing
6.14	Drive Testing
6.15	UE Manufacturing Test

Section 6.3 addresses the increasing use of high-speed serial interconnect between subsystems within user equipment (UE) and the evolved node B (eNB or base station). Traditional interconnect between the baseband and IF or RF subsystems has been analog, with some use of parallel digital systems, both of which come with a long history of probing and measurement techniques. With the peak data rates in wireless systems continuing to increase, the analog and parallel digital interconnect methods cannot keep pace and high speed serial interconnect is taking over. However, the use of serial interconnect, which can include embedded control, presents an entirely new domain for stimulation and measurement that demands a new generation of test equipment. Section 6.3 describes the development of the DigRF standard for interconnecting baseband and RF subsystems within the UE.

The design and measurement challenges of LTE transmitters are covered in Section 6.4. This section begins with some of the simpler RF measurements that can be made with general purpose analog signal analysis techniques and then covers the rich variety of measurements based on digital demodulation that are necessary to fully analyze the highly complex and flexible signals that make up the LTE air interface. With the introduction of digital interfaces, transmitter development now involves mixed signal analysis with a digital interface on the input to the transmitter module and an RF interface on the output.

Following the discussion of transmitters is a similar discussion of receivers in Section 6.5. This section describes the basic RF characteristics of the receiver including blocking, selectivity, spurious emissions, and reference sensitivity. Section 6.6 next covers receiver performance testing under faded RF channel conditions. This section looks at the most complex of receiver performance verification challenges: closed-loop analysis of a MIMO receiver in a faded channel, which requires real time feedback of the channel conditions to enable adaptive modulation control and frequency-selective scheduling, in addition to the use of incremental redundancy for damaged packets and retransmission for lost packets. The section finishes with an overview of methods for analyzing performance at the application layer. This includes throughput testing as well as channel state information (CSI) testing for the channel quality indicator (CQI), precoding matrix indicator (PMI), and rank indication (RI).

Section 6.7 takes a closer look at the open and closed-loop behaviors of the physical layer using the demodulation capabilities of the Agilent 89600 VSA software. This level of lower layer testing provides essential insight into whether the UE is correctly responding to the dynamic radio environment, which might otherwise be missed by higher level tests such as end-to-end throughput.

Section 6.8 focuses on the RF challenges presented by multi-antenna systems including MIMO. The theoretical gains possible from such systems are well documented; however, the practical gains that will be seen in realistic conditions are influenced by many factors that involve new methods for analyzing antenna design, the channel propagation conditions, and the received signals.

Section 6.9 investigates MIMO beamforming from the perspective of the eNB. Beamforming is a very powerful technique but for it to work effectively requires precise knowledge of the transmit phase of each of the eNB antenna ports. This section explains the theory of beamforming and then provides a test solution for verifying beamforming performance for an eight-antenna system.

The radiated "over-the-air" (OTA) performance of the UE is the subject of Section 6.10, which covers existing SISO test methods and the ongoing developments in the area of MIMO OTA. The latter is a particularly difficult test challenge and at the time of this writing, 3GPP and CTIA have not yet concluded on a single test approach. There are seven different methods being studied and each is described.

The chapter changes focus in Section 6.11 by moving from RF aspects to consider the challenges of signaling protocol development. The different phases from early development testing through conformance testing to interoperability testing are discussed along with the tools available to facilitate this aspect of the product lifecycle. A more formal discussion of conformance testing for both RF and signaling is provided in Chapter 7.

Continuing up the signaling protocol stack is Section 6.12, which discusses UE functional testing. This section takes a more user-centric view of UE functionality and includes the network elements and servers necessary to test the UE in an environment much closer to a real, operational network. Tests include voice functionality and inter-RAT handover performance as well as end-to-end throughput testing at the application layer.

Section 6.13 looks at battery drain testing. With the ever-increasing demands being put on high-end mobile devices, power consumption is often a limiting factor. This section takes a look at some of the tools available to help measure and optimize battery current drain.

When networks are deployed, drive test tools are often used as part of the process. Section 6.14 describes the different types of drive test tools and the kinds of information they provide to the installation and maintenance engineer.

Finally Section 6.15 covers the unique test requirements of UE manufacturing. Historically UE manufacturing tests were often just a subset of conformance tests executed using signaling with a base station emulator. However, the demands for high-speed testing have caused the use of signaling to be dropped in favor of non-signaling approaches. Such approaches offer significant time savings but present new challenges in the area of DUT control, which is now based on proprietary mechanisms closely associated with the chipset chosen for the UE design.

6.2 Simulation and Early R&D Hardware Testing

LTE presents a number of design and test challenges for system designers who are

- Evaluating existing designs, algorithms, and hardware (for example, those based on W-CDMA, HSPA, or WiMAX) to determine if they can be reused in LTE systems
- Defining design requirements and specifications for new hardware
- Developing and verifying new algorithms as the wireless specification evolves
- Designing, verifying, and testing the performance of RF and mixed-signal designs and hardware independent of the baseband hardware and firmware (for example, coded bit error ratio (BER) or block error ratio (BLER) designs in which baseband coding functionality such as turbo coding impacts RF BER and BLER performance)
- Verifying the interoperability of LTE designs with multiple signal formats (including WiMAX, WLAN, W-CDMA, and HPSA).

This section discusses physical layer simulation to assist in the design of 3GPP LTE and LTE-Advanced RF systems and circuits. It also covers early R&D testing of these hardware designs, which is facilitated by combining simulation software tools with test equipment in an integrated RF workflow environment.

Agilent SystemVue design software and LTE and LTE-Advanced Wireless Libraries provide a rich set of LTE FDD and TDD downlink and uplink models and simulation examples, which are described in this section to illustrate simulation benefits. Examples show how these software tools can be used for LTE design requirement partitioning, algorithm development, and RF and mixed signal transmitter and receiver design. The effect of local oscillator (LO) phase noise on OFDMA BER performance—a key area of interest—is examined by comparing LTE and Mobile WiMAX OFDMA BER in a simulated receiver design. Also covered are the benefits and technical considerations of combining simulation with test equipment to address test needs during product development. Several LTE hardware BER test examples are included.

The complexities of LTE, along with time-to-market pressures, make it critical to minimize system integration risk whenever possible to avoid costly design turns and product delivery delays. The concepts and simulation techniques presented in this section can help minimize integration risk throughout the product development lifecycle, beginning with the initial design and extending through R&D hardware testing.

6.2.1 Physical Layer Simulation Tools

6.2.1.1 Agilent SystemVue

Modeling complex LTE systems requires powerful tools. Agilent SystemVue is an electronic design automation (EDA) environment for electronic system level (ESL) simulation, verification, implementation, and measurement. It enables wireless communication system architects and algorithm developers to create model-based designs of the physical layer (PHY). SystemVue can complete a working PHY layer communication system that integrates real-world baseband and RF using any combination of software, hardware, simulation, and measurements.

As a dedicated platform for ESL design and signal processing realization, SystemVue replaces general-purpose digital, analog, and math environments. It provides essential RF capability, cuts PHY development and verification time in half, and connects to the developer's mainstream EDA flow. Figure 6.2-1 shows a high level view of SystemVue functionality, which is focused in four areas: (1) polymorphic baseband algorithms and IP; (2) reference IP and applications, including reference model libraries for all major wireless communication standards; (3) accurate RF models and channel effects; and (4) hardware connectivity with interfaces to Agilent measurement hardware.

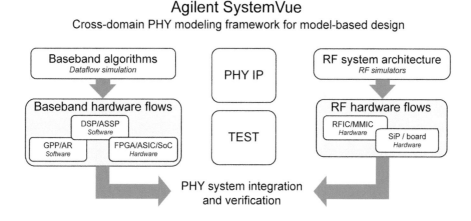

Figure 6.2-1. SystemVue high level functionality

In the SystemVue environment, developers can use C++, MATLAB, a math language, VHSIC hardware description language (VHDL), or Verilog to do baseband algorithm research, development, and implementation. Figure 6.2-2 is a flowchart showing a typical baseband algorithm development process using a tool such as SystemVue. Typically a developer will study the design requirements (such as performance and functionality), research industry and company materials, and then decide how best to meet the requirements. The baseband algorithm is developed in SystemVue using C++ or MATLAB and simulation and evaluation are carried out. If the requirements are met, the developer can implement and document the algorithm. The implementation in turn must be tested and any necessary modifications made until the code is ready for use.

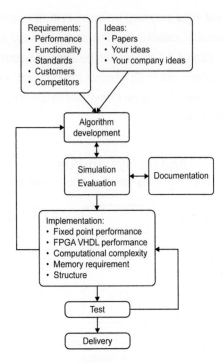

Figure 6.2-2. Baseband algorithm development flowchart

6.2.1.2 SystemVue 3GPP LTE and LTE-Advanced Libraries

The Agilent LTE and LTE-Advanced Wireless Libraries provide SystemVue with preconfigured support for the physical layer of 3GPP Releases 8, 9, and 10 with standards-based simulation models and examples. The block set, reference designs, and test benches in the Wireless Libraries assist in the design and verification of communication systems by providing configurable physical layer waveforms and data for Release 8 (LTE) and Release 10 (incorporating LTE-Advanced). The Libraries are useful for simulation-based exploration of complex algorithms, helping ensure that the physical layer meets or exceeds 3GPP real-world performance requirements.

The LTE-Advanced Baseband Verification Library, for example, provides measurement-hardened "golden reference" models that accelerate the 3GPP LTE PHY design and verification process. The Library puts reliable measurement know-how at the front of the design process, where it can help improve system design instead of merely characterizing the system's nonconformity after the fact. The Library can be used as a parameterized reference design to create internal test vectors at the block level, or to fill in gaps to complete a fully coded working PHY so that system-level performance can be continuously monitored.

The LTE Baseband Exploration Library unlocks access to algorithmic source code. System developers can explore the 3GPP LTE-Advanced standard interactively, probe inside algorithms with a debugger, and modify the design in order to precisely test any level of abstraction. Agilent provides a high-quality, independent reference that works with RF and offers a seamless transition into test.

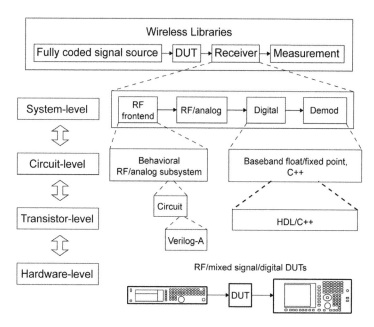

Figure 6.2-3. Simulation conceptual overview

In addition to simulation models, the LTE and LTE-Advanced Wireless Libraries include a number of preconfigured examples for simulating LTE and LTE-Advanced metrics such as error vector magnitude (EVM) and throughput on the MIMO fading channel. To highlight Agilent's simulation design and verification capabilities in these areas, Figure 6.2-3 shows how an LTE system can be modeled and tested.

Radio frequency system designs are constructed using parameterized behavioral block models (components) such as amplifiers, filters, and mixers. Component performance is specified using parameters such as gain, 1 dB compression point, and third order intercept point. These parameters can be varied until the system design meets specification; then they serve as the basis for the circuit design requirements.

As the design cycle progresses, the parameterized behavioral models are replaced by transistor-level circuit designs. Co-simulation is performed with the Agilent Circuit Envelope simulator so that circuit-level effects are included in the system-level simulation. Verifying a system design with the circuit effects included in the simulation helps minimize risk as the design cycle moves from the system level to the circuit level. Greater modeling fidelity is provided as the design process moves from concept to implementation.

In the baseband domain, floating and fixed point behavioral models are used to construct baseband algorithm functionality. Co-simulation with external simulators can be used to replace these blocks with custom algorithms written in hardware description language (HDL). Custom models defined by the user also can be written and simulated. To help mitigate RF/baseband integration risks and unexpected behavior late in the system testing phase, RF and baseband designs are simulated and verified together using the parameterized behavioral models, co-simulating with transistor-level circuit designs and with external simulators for algorithms written in HDL.

As the design cycle transitions to the hardware testing stage, Agilent's design software can be combined with test equipment to verify LTE system-level performance. With this approach, some portions of the design are modeled in simulation and other portions are represented as hardware devices under test (DUTs). This approach allows simulated signals to be downloaded to an arbitrary waveform signal generator to turn simulated signals into "real world" physical test signals. The test signals are then used as stimuli for hardware DUTs during R&D testing. The DUT outputs are captured with signal analyzers and read back into the design software for simulation post-processing.

The power of this approach is demonstrated in examples that show how LTE BER measurements can be made on RF and mixed-signal hardware using simulated baseband capability to represent missing baseband hardware functionality. The use of simulation enables the RF and mixed-signal hardware to be tested independent of the baseband hardware. It also helps isolate issues between the baseband and RF sections if problems arise later in the hardware integration phase.

6.2.3 LTE-Advanced Downlink Simulation Example

Part of the LTE-Advanced Wireless Library downlink channel coder block is shown in Figure 6.2-4. The double lines on the channel coder symbol indicate that it is hierarchical; moving down a level reveals the algorithm blocks. The transport block processing for the DL-SCH, PCH, and MCH transport channels is shown in the lower part of the figure, and the corresponding algorithm blocks for the LTE-Advanced Wireless Library downlink channel coder are shown in the upper part. The channel coding procedures of the DL-SCH consists of the following: adding CRC to the transport block, segmenting the code blocks and adding code block CRC, adding 1/3 rate turbo code with trellis termination, matching circular buffer rates, and concatenating code blocks.

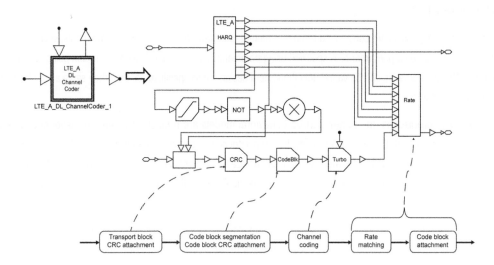

Figure 6.2-4. Downlink channel coder of transport block
(transport block processing of DL-SCH, PCH, and MCH from 36.212 [1] Figure 5.3.2-1)

Figure 6.2-5 shows to how to simulate downlink physical channel processing in a SystemVue LTE-Advanced source. An overview of LTE-Advanced downlink physical channel processing according to 36.211 [2] is shown in the upper part of the figure, and the corresponding LTE Wireless Library simulation blocks are shown in the lower part. The models shown with double arrows at the input and output are used to denote the support of MIMO with up to 8 antennas. The physical layer modulation procedures illustrated in the figure include scrambling of coded bits in each of the codewords, modulation (BPSK, QPSK, 16QAM, and 64QAM) of scrambled bits, layer mapping, precoding for support of multi-antenna transmission, and generation of complex-valued time-domain OFDM signals for each antenna port.

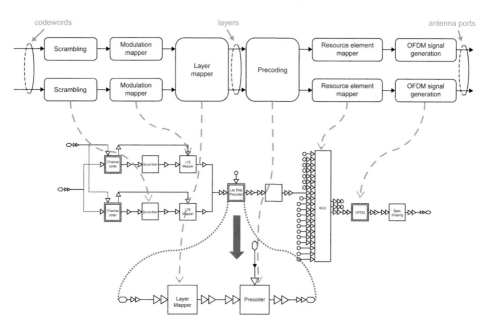

Figure 6.2-5. Part of LTE-Advanced Wireless Library downlink source
(downlink physical channel processing diagram from 36.211 [2] Figure 6.3-1)

Figure 6.2-6 shows the LTE-Advanced Wireless Library downlink channel decoder for the DL-SCH, PCH, and MCH. The DL-SCH channel decoding procedures consist of circular buffer rate de-matching, turbo decoding, code block concatenation, and addition of CRC decoder-to-output data bits and parity bits. The parity bits are fed back to the HARQ controller block, which is used to control HARQ transmission for downlink transport channels to generate HARQ control signals such as transport block size and retransmission number using the HARQ ACK/NACK feedback from the receiver. The output pin HARQ_Bits will be fed back to LTE-Advanced downlink MIMO source to do closed-loop HARQ simulation.

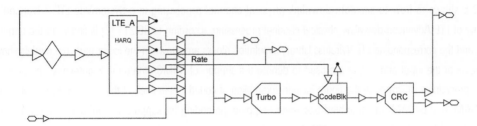

Figure 6.2-6. Downlink channel decoder of DL-SCH

The LTE-Advanced downlink MIMO receiver block diagram is shown in the top half of Figure 6.2-7. The lower part of the figure is the SystemVue schematic implementation. The receiver performs the operations of time and frequency synchronization, channel estimation, maximum likelihood (ML), minimum means squared error (MMSE), and zero forcing (ZF) decoding for spatial multiplexing and channel decoding. The first operation of the downlink receiver is timing and frequency synchronization, which includes initial synchronization and time-domain carrier frequency offset (CFO) compensation, joint estimation of residual CFO, and sampling frequency offset (SFO) after FFT operation. The CFO and SFO estimates are filtered with a loop filter and accumulated to compensate CFO/SFO in the time and frequency domains, respectively. The second operation is the FFT, which transforms the time domain representation of the received signal into the frequency domain. Following this transformation it is possible to extract the resource element mapping for the PDSCH of each UE. MIMO channel estimation can then be performed for each UE to extract the channel impulse response (CIR) from which the channel state information (CSI) can be estimated for reporting on the uplink. The received frequency-domain symbols for each layer are decoded in the MIMO receiver and then de-mapped with a soft decision de-mapper, which determines probability metrics for each bit. The bit metrics are sent to the channel decoder for de-interleaving and turbo decoding to decode the final data sequence.

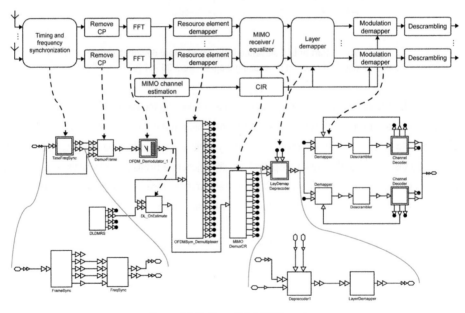

Figure 6.2-7. Downlink MIMO receiver

6.2.4 LTE-Advanced Uplink Simulation Example

SystemVue LTE and LTE-Advanced Libraries provide top-level uplink baseband multiple antenna sources and multiple antenna receivers.

There are two major parts to the SystemVue LTE-Advanced uplink source: the UL-SCH channel coding and the SC-FDMA modulator. Figure 6.2-8 shows the uplink channel coder of the UL-SCH. The double lines on the channel coder symbol indicate that it is hierarchical; moving down a level reveals the algorithm blocks. The transport block processing for the UL-SCH transport channel is shown on the left, and the corresponding algorithm blocks for the LTE-Advanced Wireless Library uplink channel coder are shown on the right. The LTE-Advanced UL-SCH channel coder has two more functions (multiplexing data and control information and channel interleaver) than the DL-SCH channel coder.

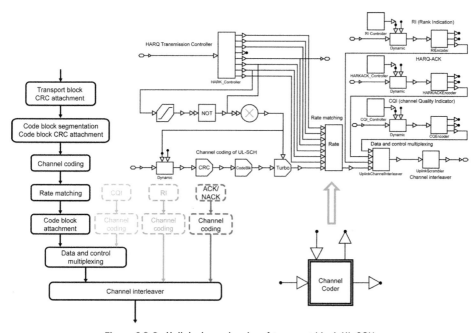

Figure 6.2-8. Uplink channel coder of transport block UL-SCH
(transport block processing of UL-SCH from 36.212 [1] Figure 5.2.2-1)

Unlike the LTE Release 8 uplink, the Release 10 LTE-Advanced uplink has been updated to support up to four transmit antennas and up to four layers (spatial streams). Figure 6.2-9 shows part of schematic of an LTE-Advanced uplink MIMO source. The upper part of Figure 6.2-9 is a diagram showing uplink physical channel processing. Its steps are scrambling of coded bits, modulation (QPSK, 16QAM, and 64QAM) of scrambled bits, mapping of the complex-valued modulation symbols onto one or several transmission layers, transformation of precoding to generate complex-valued symbols, mapping of precoded complex-valued symbols to resource elements, and generation of complex-valued time-domain SC-FDMA signals for each antenna port. The lower part of Figure 6.2-9 is a schematic of PUSCH physical channel processing in the SystemVue LTE-Advanced Wireless Library.

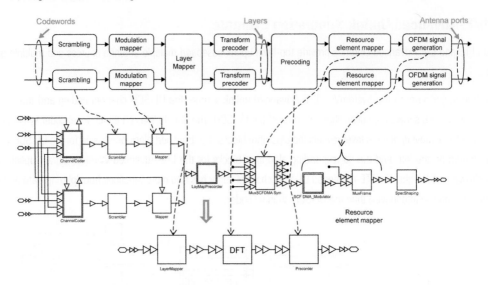

Figure 6.2-9. Part of LTE-Advanced Wireless Library uplink source
(uplink physical channel processing from 36.211 [2] Figure 5.3-1)

In the upper part, only the UL-SCH channel coding and PUSCH modulation are introduced. PUCCH, SRS, and PRACH generation and modulation are not covered here. The uplink receiver, including channel decoding of UL-SCH and physical channel demodulation, are covered in this section.

Figure 6.2-10 shows the channel decoder of the UL-SCH in SystemVue. The channel decoding operations of the UL-SCH consist of channel de-interleaving, circular buffer rate de-matching, turbo decoding, code block concatenation, and decoding of CRC to output data bits and parity bits. The parity bits are fed back to the HARQ controller block, which is used to control HARQ transmission for the uplink transport channels to generate HARQ control signals such as transport block size and retransmission number using the HARQ ACK/NACK feedback from the receiver. The output pin HARQ_Bits will be fed back to the LTE-Advanced downlink MIMO source to do closed-loop HARQ simulation.

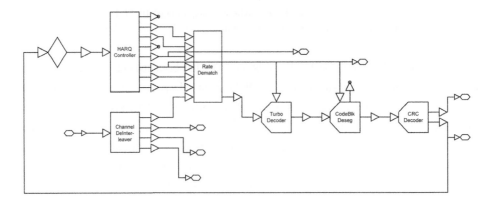

Figure 6.2-10. Uplink channel decoder of UL-SCH

Figure 6.2-11 shows part of the LTE-Advanced uplink receiver. The LTE-Advanced uplink MIMO receiver structure is on the upper part of the figure, and the SystemVue schematic is on the lower part. The uplink MIMO receiver is similar to the downlink MIMO receiver structure in Figure 6.2-7 with the exception of the inverse discrete Fourier transform (IDFT) blocks, which convert the signal from the frequency domain to the time domain prior to constellation de-mapping (see Figure 2.3-4). Despite the similarities in the high level uplink and downlink receiver structures, the algorithms in each block are quite different because the uplink adopts SC-FDMA and the downlink uses OFDMA. For example, the P-SS, S-SS, and reference signals are used for time and frequency synchronization in the downlink, and the PRACH is used for time and frequency synchronization in the uplink. For MIMO channel estimation, the downlink uses UE-specific reference signals and channel impulse response (CIR) interpolation in the frequency domain, and the uplink uses a demodulation reference signal (DMRS) in the time-domain first, and then transforms the time domain CIR into the frequency domain CIR by using an FFT prior to interpolation in the frequency domain.

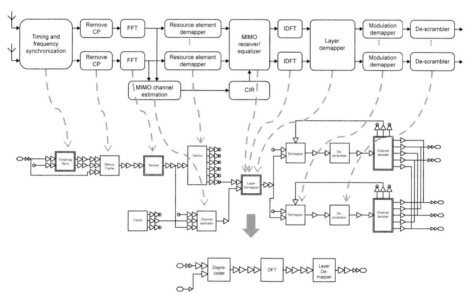

Figure 6.2-11. Part of LTE-Advanced uplink MIMO receiver

6.2.5 MIMO Channel Model Simulation

SystemVue provides full support for the spatial channel model extended (SCME) radio channel propagation models defined for RF performance requirements in 36.101 [3] for the UE and 36.104 [4] for the eNB. These models apply from Release 8 onwards. See Section 6.6.2 for more details. In addition, SystemVue also supports the new generation of geometry-based channel models developed by the ITU-R as part of the evaluation process for IMT-Advanced technologies (see Section 8.3). The advantage of geometry-based models over correlation-based models is that they can separate the properties of the antennas from those of the propagation channel. The basis for the IMT-Advanced channel model is the WINNER II model [5]. The IMT-Advanced models are specified in ITU-R M [IMT.EVAL] [6] and are further referenced by 3GPP in 36.814 [7], the LTE-Advanced physical layer technical report.

In January 2012 the ITU-R formally approved 3GPP's LTE-Advanced as a technology meeting the requirements of IMT-Advanced. However, at the time of this writing, performance requirements for LTE-Advanced continue to be based on the SCME models from Release 8. The IMT-Advanced channel models are not yet incorporated into the 3GPP specifications although they are being used as part of the evaluation of MIMO over-the-air (OTA) test methods.

With the continued development of MIMO technology through Release 11 and Release 12, it remains important for system designers to evaluate their designs against the ITU-R channel models even though these are not currently being used for conformance testing. SystemVue provides the necessary support for that purpose. Figure 6.2-12 shows the principle behind geometry-based models.

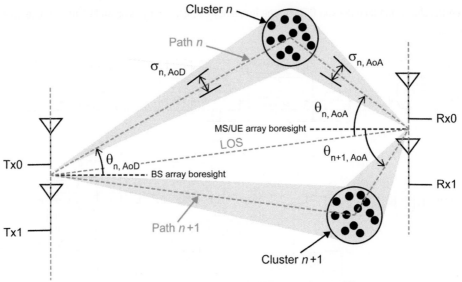

Figure 6.2-12. Geometry-based MIMO channel model [8]

The model comprises one or more clusters, each representing typically 20 localized scatterers. The parameters that define the clusters are randomly chosen from tabulated distribution functions. The first to be chosen are called large scale (LS) parameters and they define shadow fading, angular spreads, and delay. Then, small scale (SS) parameters are similarly chosen at random including power, angles of arrival (AoA), and angles of departure (AoD). The model does not explicitly locate the scatterers but rather implies their location based on the chosen parameters. Once the LS and SS parameters have been chosen, the only remaining freedom is the initial phase of each reflected signal. The overall model is realized through the concept of "drops." A drop is a single realization of the model for a fixed time. Unless short term time variability is introduced for some parameters, many drops are required before a statistically valid model is generated.

IMT-Advanced models were defined for the following deployment scenarios: indoor hotspot, urban microcell (Umi), urban macrocell (Uma), and rural macrocell. The Umi and Uma models are being used in the evaluation of MIMO OTA test methods (see Section 6.10.6.2). The SystemVue MIMO Channel Builder is an optional block set that provides both WINNER II and IMT-Advanced (LTE-Advanced) MIMO channel models for system performance simulation. Figure 6.2-13 shows an LTE-Advanced MIMO channel model in SystemVue complying with 36.814 [7] Annex B.

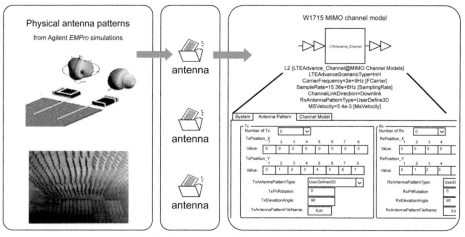

Figure 6.2-13. LTE-Advanced MIMO channel model in SystemVue

The LTE-Advanced MIMO channel model icon and its graphical user interface are shown on the right of Figure 6.2-13. On the left are the physical antenna pattern parameters including orthogonal polarizations—typically linear theta (Θ) and phi(ϕ) polarizations—or far-field radiation patterns, which can be generated for instance by Agilent EMPro antenna modeling software or some other electromagnetic (EM) simulation tool. Ideal antenna patterns such as an omni-directional pattern, 3-sector pattern, and 6-sector pattern are also supported options. SystemVue is capable of supporting the most complex examples for LTE-Advanced. Figure 6.2-14 shows a schematic for a downlink 8x8 LTE-Advanced system.

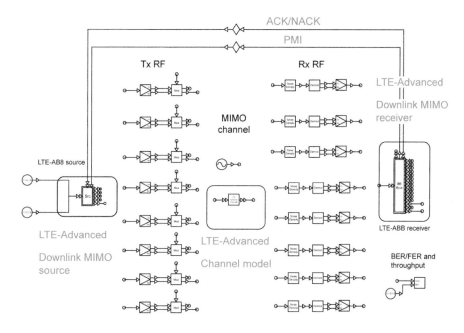

Figure 6.2-14. Schematic of downlink 8x8 LTE-A scenario, with fading and closed-loop throughput measurement incorporating active HARQ feedback

243

Although 8x8 MIMO has been defined, it is challenging to implement in the UE and 8-layer spatial multiplexing is more likely to be used to enable multi-user MIMO. See Section 2.4.4.9.

The primary advantage of any simulation is an early indication of system performance. Figure 6.2-15 demonstrates how a simulation can predict 2x2 MIMO antenna performance well ahead of the availability of real hardware. In this example a Rayleigh radio environment was produced in an anechoic chamber by using a multi-probe antenna and channel emulator. Then, a two-dipole antenna array with half-wavelength separation was placed at the center of a turntable, and the table oriented at different angles to produce different AoA effects.

The trace with the circles in Figure 6.2-15 shows the experimental results obtained from the real antenna hardware, which reflects the real world directionality of the antenna array. To measure this a fully coded downlink LTE MIMO signal was introduced into the anechoic chamber using probe antennas and the received signals from the dipole antennas were captured using the Agilent 89600 VSA software. The data file was then fed into a simulated reference receiver in SystemVue to generate the throughput results. The trace with the squares in Figure 6.2-15 represents a complete simulation of the same experiment. A measured antenna pattern file for the dipole antennas was loaded into SystemVue and the MIMO Channel Builder models were used to predict the faded wireless channel with directional MIMO antenna effects. The same simulated reference receiver that was used to model the real hardware was also used to decode the results.

The fully simulated MIMO throughput is remarkably similar to the experimental results considering the small vertical scale. This fact demonstrates that accurate 3D electromagnetic simulations of the industrial design of the UE can be made very early in the development cycle. The 3D models can include measured or simulated antenna pattern data with optional loading effects from specific anthropomorphic mannequin (SAM) head and hand phantoms. They can be combined with early UE and eNB reference algorithms and early RF transceiver models to obtain reliable link-level performance data, months in advance of having actual working hardware. By integrating these early simulation models with test equipment when working UE hardware becomes available, additional insights about real world performance can be gained.

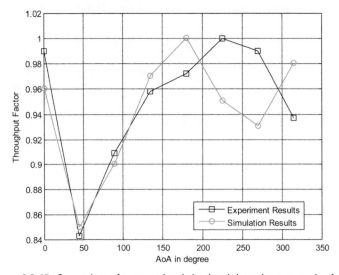

Figure 6.2-15. Comparison of measured and simulated throughput vs. angle of arrival

6.2.6 Mixed-Signal Design Challenges

Today's system-level designs are mixed-signal designs that combine baseband/digital, analog, and RF signals to meet a system-level design specification such as throughput, BER, or EVM. A system engineer typically has to make tradeoffs among various RF and mixed-signal impairments—phase noise, I and Q imbalance, 1 dB compression point of the power amplifier, etc.—when deciding on an overall system performance budget such as the one illustrated in Figure 6.2-16.

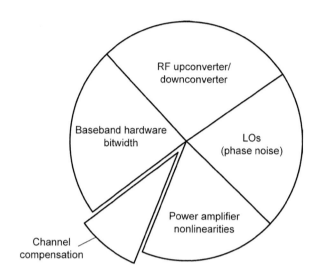

Figure 6.2-16. Example impairments for a system performance budget

A simplified RF/mixed-signal system design is shown in Figure 6.2-17.

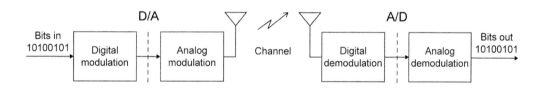

Figure 6.2-17. RF/mixed-signal system design

6.2.6.1 RF Transmitter Example

An example of how simulations are used in making RF transmitter design tradeoffs is shown in Figure 6.2-18.

The LTE-Advanced uplink signal source is used to generate an LTE uplink signal. The output of the uplink source is a modulated IF signal that is up-converted to RF using a mixer and an LO source. The LO Phase noise, defined in dBc/Hz, is defined for various frequency offsets. A nonlinear power amplifier is modeled by specifying the gain in dB and the output 1 dB compression point in dBm. The output signal is then filtered and EVM is measured using the LTE uplink EVM sink.

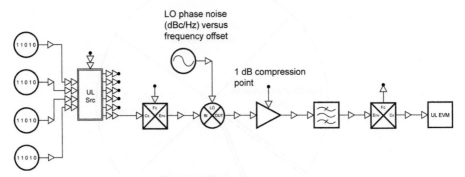

Figure 6.2-18. Transmitter baseband/RF design tradeoffs vs. phase noise and power amplifier 1 dB compression point

In the simulation, the power level of the LO phase noise at the 1 kHz frequency offset is swept relative to the LO power level from −105 dBc/Hz to −55 dBc/Hz. Also, the 1 dB compression point of the power amplifier is swept from 13 dBm to 19 dBm. The resulting plots of EVM versus the swept design impairments are shown in Figure 6.2-19. These types of simulation enable system engineers to gain insight into design sensitivities so that informed decisions and tradeoffs can be made for design requirements.

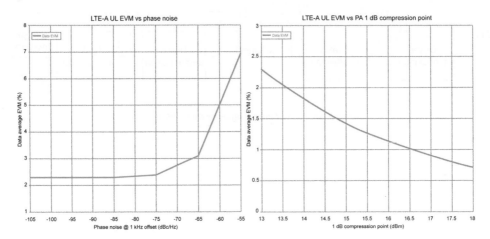

Figure 6.2-19. EVM vs. LO phase noise and 1 dB compression point

Simulations similar to the example here are valuable early in the design cycle to determine design requirements. Once development advances to the detailed design phase, simulation can be used to verify overall system-level RF and mixed-signal performance.

6.2.2.3 Receiver Mixed-Signal Design Challenges

Similar to transmitter design, receiver design presents a set of mixed-signal design challenges. Typically, wireless standards define receiver performance in terms of BER or BLER at a minimum received RF input power level. In addition, the receiver's BLER is specified with impairments such as fading, added noise, and adjacent channel interference.

To measure BLER requires that baseband coding and decoding functionality be available both for simulated designs and for verifying actual hardware. Baseband and RF designs often need to progress in parallel to meet the objectives of development schedules, which makes it difficult to measure receiver performance early in the development cycle.

Baseband simulation can be used to enable parallel development so that RF receiver test activities can begin earlier. For example, BER simulation using the LTE-Advanced Wireless Library to evaluate receiver LO phase noise requirements is shown in Figure 6.2-20. Phase noise is of key interest in OFDMA systems because the close subcarrier spacing can make OFDMA system performance susceptible to phase noise and frequency error. An RF receiver design is inserted between the LTE-Advanced downlink simulation signal source and downlink simulation receiver, and the signal to noise ratio (SNR) into the RF receiver is swept, along with the down-converter LO phase noise.

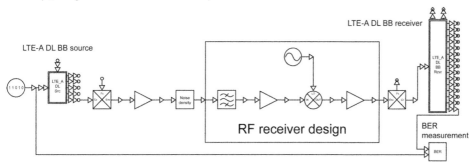

Figure 6.2-20. Schematic to measure BER of RF receiver vs. swept SNR and phase noise

The SNR is swept from −1 dB to 1 dB in 0.2 dB steps for QPSK, and from 8 dB to 10 dB for 64QAM. The LO phase noise at 10 kHz offset is set to −60 dBc/Hz, −70 dBc/Hz, and −80 dBc/Hz. Note that the 10 kHz frequency offset is within the LTE 15 kHz subcarrier spacing. The simulation results are shown in Figure 6.2-21.

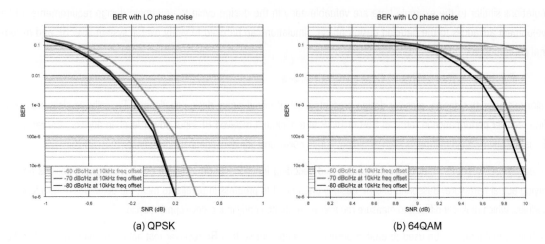

Figure 6.2-21. Coded BER simulation results for an RF receiver

Figure 6.2-21 shows that the LO phase noise impacts BER performance more significantly for 64QAM than for QPSK. This result is expected, as the noise margin is lower when higher order modulation is used. This is just one example of the many tradeoffs that must be considered in designing for higher order modulation and the associated higher data throughputs.

6.2.7 Carrier Aggregation

LTE Release 8 provides extensive support for deployment in a variety of channel bandwidths, ranging from 1.4 MHz to 20 MHz, in both paired (FDD) and unpaired (TDD) spectrum. Beyond 20 MHz, the only reasonable way to achieve the target peak-throughput for LTE-Advanced is to increase the transmission bandwidth relative to the 20 MHz limit specified in Release 8. Therefore, LTE-Advanced specifies spectrum allocations of up to 100 MHz. Since it is very rare to find available contiguous spectrum of that bandwidth, the ITU allows the use of carrier aggregation, wherein multiple component carriers (CCs) are aggregated (combined) to provide the necessary bandwidth. Carrier aggregation is discussed more fully in Section 2.1.11.

The 3GPP Release 9 Technical Report 36.815 [9] identifies twelve deployment scenarios for study in later releases. The scenarios include contiguous and non-contiguous aggregation. Figure 6.2-22 is a SystemVue schematic for Scenario 7, which is a triple carrier FDD non-contiguous example for bands 1, 3, and 7 with 10MHz, 10MHz, and 20MHz bandwidths, respectively. The resulting spectrum is shown in Figure 6.2-23.

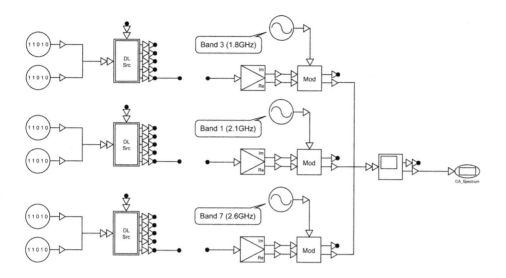

Figure 6.2-22. SystemVue schematic of Scenario 7, multiband carrier aggregation FDD downlink at Bands 3, 1, and 7

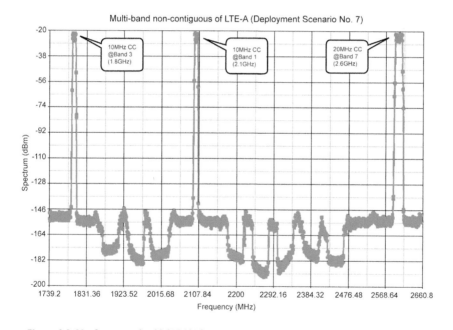

Figure 6.2-23. Spectrum for 36.815 [9] Scenario 7 triple non-contiguous carrier aggregation

Three branches are used in Figure 6.2-22 to generate three LTE-Advanced downlink carriers and up-convert them to 1.8 GHz, 2.1 GHz, and 2.6 GHz, respectively. The SystemVue Data Flow Block SignalCombiner combines the carriers into one RF signal. The output of the SignalCombiner block is the non-contiguous carrier aggregated signal. The SignalCombiner block can combine multiple input signals with different sample rates, different frequencies, and different bandwidths into a single signal at the specified frequency and sample rate. In this multiband non-contiguous carrier aggregation example, the bandwidth of the first two component carriers is 10 MHz and the sampling rate is 122.88 MHz (15.36 x 8 MHz). The bandwidth of the third component carrier is 20 MHz and its sampling rate is 245.76 MHz (30.72 x 8 MHz).

This example creates a signal with more than 800 MHz (2.6 GHz–1.8 GHz) bandwidth. The sampling rate of carrier aggregation must be larger than 800 MHz. SignalCombiner's output sampling rate is set to 921.6 MHz (30.72 x 30 MHz), which corresponds to Figure 6.2-22 (266.8–1739.2 = 921.6 MHz). Very wide bandwidth aggregated carriers are likely to be generated by separate signal generators and combined at RF but baseband solutions continue to reach new limits.

6.2.8 Multi-Standard Radio

Along with carrier aggregation, another mechanism that creates multicarrier signals is the multi-standard radio (MSR) feature introduced in Release 9. (See Sections 2.1.10, 6.4.7, and 8.2.6). In Release 9, MSR was restricted to a single band but in Release 11 it was extended to multiband (sometimes referred to as non-contiguous MSR). Figure 6.2-24 (a) through (d) shows a SystemVue simulation for a dual-band MSR signal comprising three carriers in the 850 MHz band (GSM + W-CDMA + GSM) and one LTE carrier at 2.3 GHz.

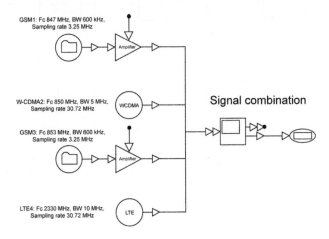

(a) SystemVue schematic of dual-band MSR signal

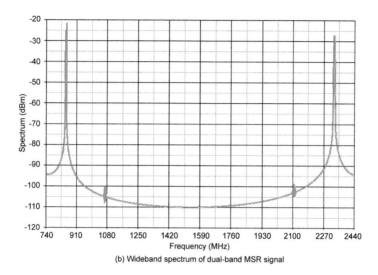

(b) Wideband spectrum of dual-band MSR signal

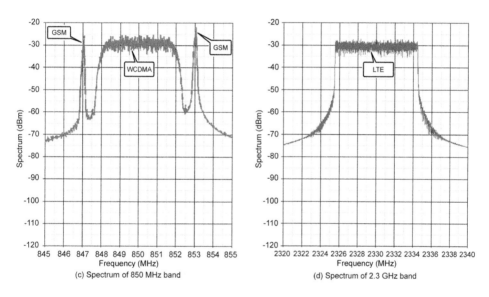

(c) Spectrum of 850 MHz band

(d) Spectrum of 2.3 GHz band

Figure 6.2-24. SystemVue dual-band MSR simulation

The schematic in Figure 6.2-24 (a) combines two GSM carriers, one W-CDMA carrier, and one LTE carrier into a single signal with the first three carriers in the 850 MHz band and the fourth carrier in the 2.3 GHz band. Figure 6.2-24 (b) shows the spectrum of the combined MSR signal. Figure 6.2-24 (c) is the spectrum of the 850 MHz band and Figure 6.2-24 (d) is the spectrum at 2330 MHz.

6.2.9 Crest Factor Reduction for LTE-Advanced

The higher peak-to-average power ratio (PAPR) that comes from carrier aggregation places a greater demand on the linearity of the power amplifier. In order to limit the adjacent channel leakage and other distortion effects, it is necessary for the amplifier to operate in its linear region; that is, below its saturation point. High linearity requirements for the power amplifier lead to low power efficiency and therefore to high power consumption (Class A amplifier). In the base station, power consumption itself is not a primary concern as it is in mobiles, but the heat generated in the eNB by high power consumption is a concern. Digital clipping can be used to digitally distort the signal so that the peak power is reduced at the loss of some signal quality.

Some form of crest factor reduction (CFR) is generally necessary for the LTE downlink due to the high PAPR created by the OFDM modulation. The LTE uplink, however, uses SC-FDMA modulation, which has a lower PAPR. The UE also operates at much lower power and so CFR for the LTE uplink is not necessary. With the development of LTE-Advanced, the need for CFR for both the downlink and uplink increases. In the downlink, carrier aggregation further increases the potential PAPR of the signal, and in the uplink, the PAPR is increased by simultaneous transmission of the control and data signals, the use of clustered SC-FDMA, and the use of carrier aggregation.

6.2.9.1 CFR for Release 8 LTE Downlink Single Component Carrier

There exist numerous methods to achieve CFR including amplitude clipping, selected mapping techniques, coding schemes, tone reservation, and nonlinear commanding transform.

An example of CFR using polar amplitude clipping is shown in Figure 6.2-25. This is based on the algorithm discussed in the IEEE paper "Constrained Clipping for Crest Factor Reduction in Multiple-User OFDM" [10]. The clipping is followed by digital filtering, which limits the degradation on the wanted signal EVM and nulls out-of-channel components.

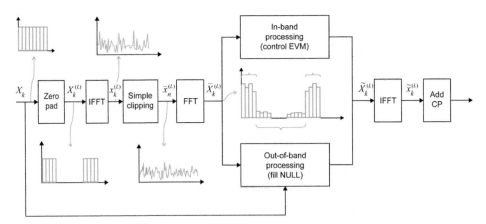

Figure 6.2-25. Simplified block diagram of CFR algorithm of LTE-Advanced downlink OFDMA signal

The input signals X_k in Figure 6.2-25 are OFDMA symbols in the frequency domain. The first step in the CFR process is to pad zeros in the frequency domain to up-convert the data using perfect frequency domain interpolation. An IFFT transforms the frequency domain signal into the time domain. The time domain signal is then clipped using a polar clipping technique.

This process reduces the peak power of the signal but also introduces unwanted in-channel and out-of-channel distortions. To control these distortions the clipped signal is transformed back into the frequency domain. The in-channel processing block ensures that the EVM (in-channel) does not exceed a specified limit, which is based on the modulation depth. The out-of-channel processing block nulls the out-of-channel components generated by the clipping process.

Polar clipping reduces the input signal $x_n^{(L)}$ to some level A_{max} to create

$$\overline{x}_n^{(L)} = \begin{cases} x_n^{(L)}, & \left| x_n^{(L)} \right| \le A_{max} \\ A_{max} e^{j \angle x_n^{(L)}}, & \left| x_n^{(L)} \right| > A_{max} \end{cases}$$

where $\left| x_n^{(L)} \right|$ is the amplitude of the nth sample before clipping and is the maximum permissible amplitude. The purpose of the in-channel processing block is to ensure that the EVM does not exceed a specified limit. The EVM for the single-subcarrier case is defined as

$$EVM\left(\overline{x}_n^{(L)}\right) = \frac{1}{S_{max}} \sqrt{\frac{1}{N} \sum_{k=0}^{N-1} \left| E_k \right|^2}$$

where S_{max} is the maximum amplitude of the symbol constellation, $\left| E_k \right|^2$ is the error vector power on the kth subcarriers and $E_k = \overline{X}_k^{(L)} - X_k^{(L)}$. Table 6.2-1 lists S_{max} values for the LTE modulation depths.

Table 6.2-1. S_{max} value for LTE

Modulation scheme	S_{max}
QPSK	1
16QAM	$\sqrt{\frac{18}{10}}$
64QAM	$\sqrt{\frac{98}{42}}$

The in-channel processing unit is needed only when $EVM\left(\overline{x}_n^{(L)}\right) > Th$, where Th is the target EVM threshold. The value of Th would normally be set at or lower than the system requirement shown in Table 6.2-2.

Table 6.2-2. EVM requirements (36.104 [4] Table 6.5.2-1)

Modulation scheme for PDSCH	Required EVM [%]
QPSK	17.5%
16QAM	12.5%
64QAM	8%

If the clipping is light enough that the EVM requirement is already met by the simply clipped signal—i.e., $EVM\left(\overline{x}_n^{(L)}\right) \le Th$—then no in-channel processing is necessary and $\widetilde{X}_k^{(L)} = \overline{X}_k^{(L)}$, $k \in [0, N-1]$. If, however, $EVM\left(\overline{x}_n^{(L)}\right) > Th$, the in-channel processing block first sorts E_k in ascending order.

If the root mean square (rms) average of the M smallest E_k values is less than or equal to $Th \cdot S_{max}$, but the rms average of the M + 1 smallest E_k values is greater than $Th \cdot S_{max}$, the value M is recorded and the subcarrier indices k that correspond to the M smallest E_k values in a set M are collected. The in-channel modified vector $\widetilde{X}_k^{(L)}$ is obtained according to

$$\tilde{X}_k^{(L)} = \begin{cases} \bar{X}_k^{(L)}, & k \in M \in I \\ X_k + Th \cdot S_{max} e^{j\angle E_k}, & k \in (I \backslash M) \end{cases}$$

Figure 6.2-26 shows a vector diagram to illustrate the in-channel processing algorithm. This figure illustrates the equation above when $k \in (I \backslash M)$.

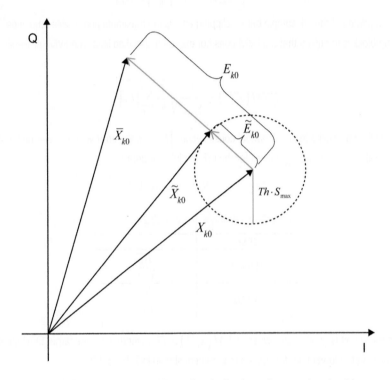

Figure 6.2-26. Vector diagram illustrating the in-channel processing algorithm

In SystemVue the out-of-channel subcarriers are nulled as follows: $\tilde{X}_k^{(L)} = 0$, $k \in O$. The OFDMA multiple access scheme in the LTE downlink allows several users to simultaneously access services. Each user (represented by one PDSCH) can have a different modulation scheme (QPSK, 16QAM, or 64QAM). This requires three EVM thresholds to be maintained: *EVM_Threshold_QPSK*, *EVM_Threshold_16QAM*, and *EVM_Threshold_64QAM*.

Apart from the PDSCH, which as noted can have three different modulation levels, the LTE downlink also has many control physical channels such as the PBCH, PDCCH, PCFICH, and PHICH. For the PBCH, a QPSK threshold is used in the in-channel processor block. For the PDCCH, PCFICH, and PHICH, the original signals are regenerated to replace these corrupted signals in their corresponding subcarrier position: $\tilde{X}_k^{(L)} = X_k$, $k \in \{subcarriers\ of\ PDCCH,\ PCFICH,\ and\ PHICH\}$.

The clipping process also distorts the downlink reference signals. Clipped signals are replaced in the in-channel processor unit using the following equation: $\tilde{X}_k^{(L)} = X_k$, $k \in$ *{subcarriers of physical signals}*. Figure 6.2-27 shows the schematic implementation of CFR in SystemVue, which has the same blocks as in Figure 6.2-25.

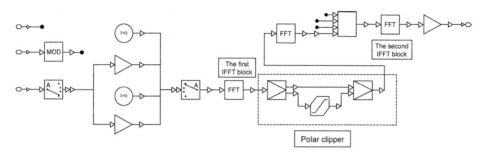

Figure 6.2-27. Schematic of CFR with single level of clipping and filtering

Figure 6.2-28 shows the spectrum of a 20 MHz FDD downlink signal with all resource blocks assigned to one PDSCH with QPSK modulation. The spectra with and without CFR are almost identical because the out-of-channel subcarriers were nulled in the CFR algorithm.

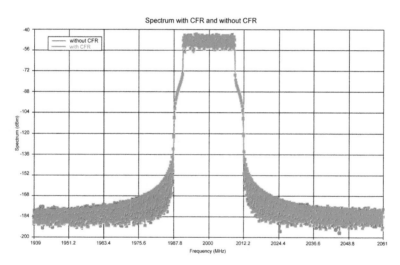

Figure 6.2-28. Spectra (CFR and without CFR) of FDD LTE-A 20 MHz system with QPSK modulation

Figure 6.2-29 (a) shows the complementary cumulative distribution function (CCDF) of the composite signal, indicating that a CFR reduction from 9 dB to 6.8 dB has been achieved. Figure 6.2-29 (b) shows the PDSCH constellation confirming that only QPSK modulation is present. The EVM threshold used for clipping the QPSK signal is 12.5%. Figure 6.2-29 (c) shows the EVM of the different physical channels and signals. The EVM values of the PBCH and PDSCH QPSK are 10.24% and 9.24%, respectively. Both values are less than the 12.5% QPSK EVM requirement. The EVM values of P-SS, S-SS, and RS are 0.164%, 0.146% and 0.180%, evidence that the CFR process did not corrupt the physical signals when the original signals were regenerated and inserted into the composite signal.

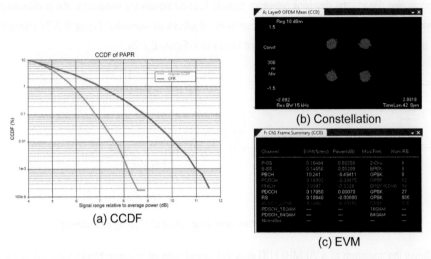

Figure 6.2-29. FDD 20 MHz QPSK signal showing (a) CCDF with and without CFR, (b) QPSK constellation, and (c) EVM of each physical channel and physical signal after CFR

Figure 6.2-30 shows CCDF, EVM, and constellation with three PDSCHs, each with a different modulation scheme. The first 30 resource blocks are assigned to PDSCH 1 with QPSK, the second 30 resource blocks are assigned to PDSCH 2 with 16QAM, and the last 40 resource blocks are assigned to PDSCH 3 with 64QAM modulation. The EVM thresholds for QPSK, 16QAM, and 64QAM are 10%, 8%, and 6%, respectively.

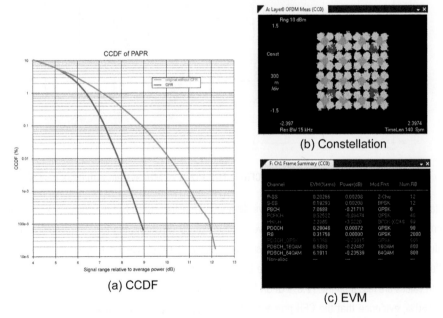

Figure 6.2-30. FDD 20MHz QPSK/16QAM/64QAM signal (a) CCDF with and without CFR (b) Constellation showing three modulation formats and (c) EVM of each physical channel and physical signal after CFR

Figure 6.2-30 (a) is the CCDF with and without CFR. The PAPR at 0.1% probability without CFR is 8.9 dB, which improves to 7.2 dB with CFR. Figure 6.2-30 (b) shows the constellations of the three PDSCH after CFR. Figure 6.2-30 (c) shows the EVM for each physical channel after CFR. The EVM values of the PBCH and QPSK PDSCH are 7.1% and 6.1%, respectively, both less than the chosen 10% QPSK EVM CFR threshold. The 16QAM PDSCH EVM is 6.6%, which is less than the chosen 8% 16QAM CFR EVM threshold. The PDSCH 64QAM EVM is 6.2%, which is just above the chosen 6% 64QAM CFR EVM threshold.

Figure 6.2-31 shows the CCDF of a 20 MHz 16QAM TDD 20 MHz signal. The uplink-downlink configuration is Configuration 0 [D S U U U D S U U U] (see Table 3.2-1) and the special subframe configuration is Configuration 0 [DwPTS: 3 GP: 10 UpPTS: 1] (see Table 3.2-2). The PAPR of the signal is 8.8 dB without CFR and 7 dB after CFR.

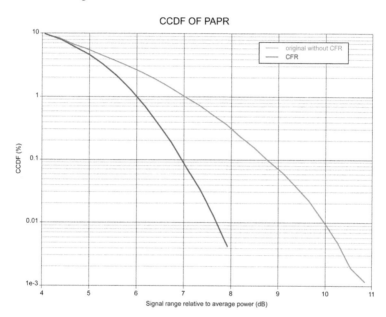

Figure 6.2-31. CCDF with and without CFR of 20 MHz 16QAM TDD LTE-A 20MHz signal with 16QAM modulation

6.2.9.2 CFR for LTE-Advanced Carrier Aggregation

LTE-Advanced supports up to five aggregated component carriers in order to achieve 100 MHz occupied bandwidth for higher data rates. A single downlink component carrier has about 9 dB PAPR at 0.1% probability when all resource blocks are allocated to one PDSCH, and the uplink has about 6 dB PAPR when all resource blocks are allocated. In the worst case a two-carrier aggregation will increase PAPR by 3 dB.

The LTE-Advanced base station and UE both support the following three component carrier aggregation scenarios:

- Intra-band contiguous aggregation
- Intra-band non-contiguous aggregation
- Inter-band aggregation (by definition non-contiguous).

The frequency spacing between the carriers leads to a number of different implementation options for the UE as shown in Figure 6.2-32 (a) through (d).

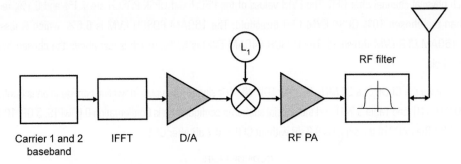

(a) Single baseband and radio, suitable for intra-band contiguous aggregation

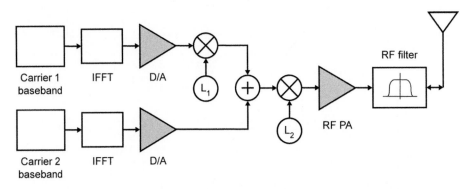

(b) Separate basebands, one with first stage mixer with low power combiner, suitable for intra-band non-contiguous aggregation

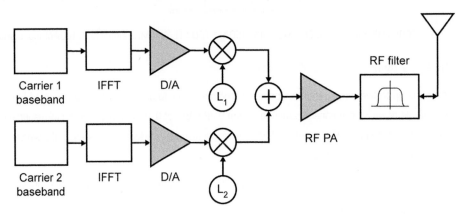

(c) Multiple baseband and first stage mixers with low power RF combiner single radio, suitable for intra-band non-contiguous aggregation

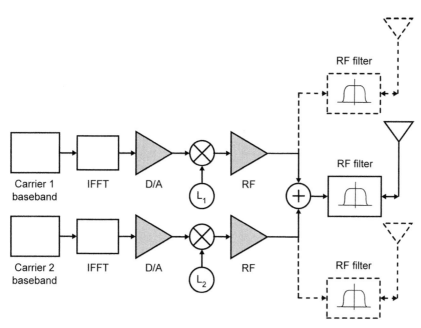

(d) Multiple baseband and radios with high power combiner, suitable for inter-band non-contiguous aggregation

Figure 6.2-32. Possible UE carrier aggregation implementations

Figure 6.2-32 illustrates various transmitter architectures based on where the component carriers are combined; i.e., at digital baseband, in analog waveforms before RF mixing, after the mixer but before the power amplifier, or at high power after the power amplifier. For contiguous intra-band aggregation the UE and particularly the eNB will likely have one power amplifier. Connected to the power amplifier can be a single RF chain of a zero-IF mixer, wideband digital-to-analog converter, and wideband IFFT as in Figure 6.2-32 (a). Figure 6.2-32 (b) combines the analog baseband waveforms from the component carriers first (e.g., via a mixer operating at an IF of roughly the bandwidth of the other component carrier for the two-carrier example shown). The resulting wideband signal is then up-converted to RF. Figure 6.2-32 (c) uses zero IF up-conversion of each component carrier before combining into a single power amplifier. Figure 6.2-32 (d) is for inter-band aggregation with wide frequency spacing and uses multiple RF chains and multiple power amplifiers after which the high-power signals are combined and fed into a single or multiple antenna.

Figure 6.2-32 (d) can use simple CFR per carrier but the other three options all require more advanced solutions. There are two common CFR methods used in such circumstances. With Method 1, each component carrier does CFR separately and then all component carriers are summed. With Method 2, clipping is applied for the composite signal and then each component carrier is filtered separately and summed. Figure 6.2-33 shows this approach.

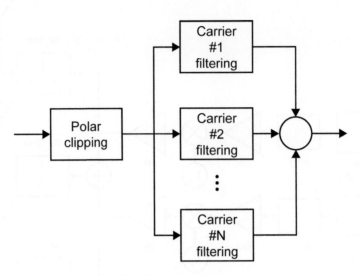

Figure 6.2-33. Clipping carrier aggregation signals

The two different carrier aggregation CFR methods will now be compared. Figure 6.2-34 is an example of Method 1 for a 2 x 20 MHz signal using high power combining at RF.

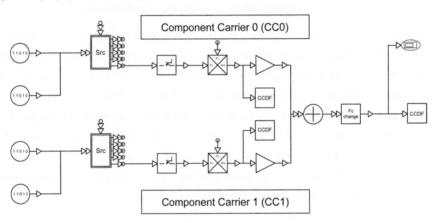

Figure 6.2-34. Schematic of 2 x 20 MHz intra-band contiguous carrier aggregation with CFR

All resource blocks are assigned to one 16QAM PDSCH in CC0 and one QPSK PDSCH in CC1, respectively. Both carriers apply CFR independently. The EVM thresholds for QPSK and 16QAM are set to 12% and 10%. CC1 can clip a little more heavily because QPSK modulation allows higher EVM.

Figure 6.2-35 shows CCDF curves of CC0, CC1, and the combined 2 x 20 MHz signal. The PAPR values of CC0 and CC1 at 0.1% probability are 7.2 dB and 6.7 dB. The PAPR of the combined 2 x 20 MHz signal is 8.2 dB.

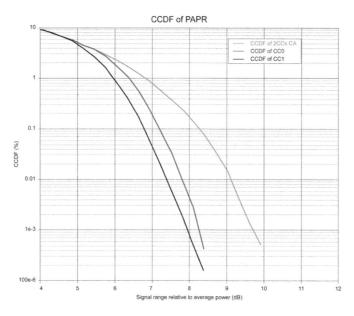

Figure 6.2-35. CCDF of individual and combined carriers with CFR using method 1

Figure 6.2-36 shows the analysis for the combined 2 x 20 MHz signal using method 1. The CC0 16QAM PDSCH EVM is 8.54% and the CC1 QPSK PDSCH EVM is 11.11%. These EVM values are within the limits defined by the clipping thresholds.

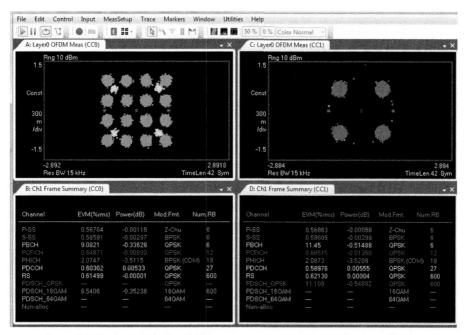

Figure 6.2-36. Analysis for 2 x 20 MHz contiguous carrier aggregation using method 1

Figure 6.2-37 is the schematic for CFR using method 2.

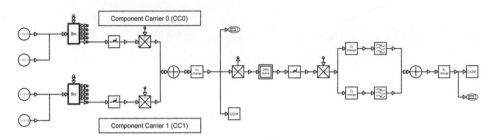

Figure 6.2-37. Schematic of 2 x 20 MHz contiguous carrier aggregation with clipping and filtering

The PDSCH configuration with CC0 and CC1 is similar to the previous example but this time CFR is not applied at the carrier level but later, after the carriers are combined.

Figure 6.2-38 shows the CCDF for the aggregated carriers with and without CFR. The addition of CFR improves the PAPR at 0.1% probability from 9 dB to 7.4 dB, which is 0.8 dB better than the results using method 1.

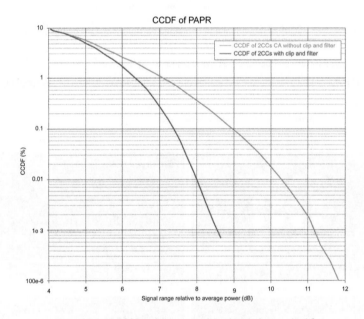

Figure 6.2-38. CCDF of aggregated carriers using method 2

Figure 6.2-39 shows the analysis for the combined 2 x 20 MHz signal using method 2. The CC0 16QAM PDSCH EVM is 7.8% and the CC1 QPSK PDSCH EVM is 8.8%, both within the correct limits. The EVM values of the P-SS, S-SS, and RS are > 6.5% in both CC0 and CC1. Unlike the physical signals in method 1, P-SS, S-SS, and RS in this case are corrupted after clipping since they cannot be regenerated. Also, CC0 and CC1 have almost the same EVM values for 16QAM and QPSK since only one clipping level could be chosen for the composite signal, and this level had to be based on the more sensitive 16QAM content of CC0. In summary, analysis shows that neither method is better overall and tradeoffs have to be made.

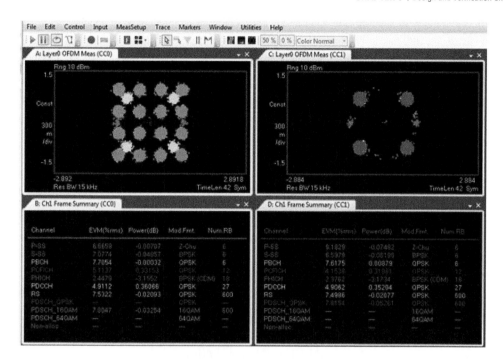

Figure 6.2-39. Analysis for 2 x 20 MHz contiguous carrier aggregation using method 2

6.2.10 Digital Predistortion of LTE-Advanced

The power amplifiers used in communication systems are inherently nonlinear at the top end of their operating range. This nonlinearity generates spectral regrowth, which leads to adjacent channel interference and violations of the out-of-band emission requirements mandated by regulatory bodies. It also causes in-band distortion, which degrades the signal quality. To reduce the nonlinearity, the power amplifier can be operated at a lower power; however, this results in very low efficiencies, typically less than 10%. With > 90% of the DC power lost and turning into heat, the amplifier's performance, reliability, and ongoing operating expenses are all degraded.

An alternative to operating the power amplifier below its maximum power is to apply linearization to it using techniques such as feedback, feed-forward, and digital predistortion (DPD). Typical performance indices of the basic linearization techniques are illustrated in Table 6.2-3.

Table 6.2-3. Comparison of linearization techniques

Technique	Intermodulation distortion cancellation	Bandwidth	Power added efficiency	Additional circuitry
Feedback	Good	Narrow	Medium	Medium
Feedforward	Good	Wide	Low	Large
Predistortion	Medium	Wide	High	Small

Among these linearization techniques, DPD is one of the most cost-effective. Compared to feedback and feed-forward linearization techniques, DPD has several advantages: excellent linearization capability, preservation of the overall power amplifier efficiency, and ability to use existing digital signal processors and converters. The principle behind DPD shown in Figure 6.2-40 is to add a digital predistorter in the baseband to create a nonlinearity that is complementary to the compressing characteristic of the power amplifier. By cascading the predistorter with the power amplifier compression, the result becomes linear with predistortion. The power amplifier can be used up to its saturation point while still maintaining good signal linearity, thereby significantly increasing its efficiency.

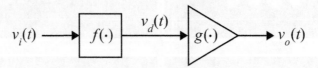

Figure 6.2-40. Digital predistortion principle

The DPD output is $v_d(t)$ and the power amplifier output is expressed by

$$v_0(t) = g\left(\left|f\left(\left|v_i(t)\right|^2\right)v_i(t)\right|^2\right)f\left(\left|v_i(t)\right|^2\right)v_i(t).$$

The power amplifier is linearized when

$$G = g\left(\left|f\left(\left|v_i(t)\right|^2\right)v_i(t)\right|^2\right)f\left(\left|v_i(t)\right|^2\right).$$

There are two classes of DPD, the memoryless model and the memory model. The memoryless model is used for power amplifiers that have a memoryless nonlinearity; i.e., the output depends only on the instantaneous input through a nonlinear mechanism. This instantaneous nonlinearity is usually characterized by the AM/AM and AM/PM responses of the power amplifier, in which the output signal amplitude and phase deviation of the power amplifier output are given as functions of the amplitude of the instantaneous input. Both polynomial algorithms and lookup tables (LUT) are used for memoryless models.

When the signal bandwidth gets very wide, power amplifiers begin to exhibit memory effects. This is especially true for the high power amplifiers used in the eNB. The causes of the memory effects can be attributed to thermal constants of the components in the biasing network that have frequency dependent behaviors. As a result, the output of the power amplifier depends not only on the instantaneous input, but also on past input values. In other words, the power amplifier becomes a nonlinear system with memory. For such a power amplifier, memoryless predistortion can achieve only limited linearization; therefore, DPD with memory effect modeling must be used.

The most important algorithm for models with memory for DPD implementation is the Volterra series and its derivatives. However, the large number of coefficients of the Volterra series makes it unattractive for practical applications. For that reason, Volterra's derivatives including Wiener, Hammerstein, Wiener-Hammerstein, parallel Wiener structures, and memory polynomial models are used instead. The memory polynomial is interpreted as a special case of a generalized Hammerstein model and is further elaborated by combining with the Wiener model.

There are two approaches to construct digital predistorters with memory structures. The first approach is to directly identify the inverse of the power amplifier response. This approach is called the direct learning architecture (DLA). However, obtaining the inverse of a nonlinear system with memory is generally a difficult task. The second approach is to use the indirect learning architecture (IDLA) to design the predistorter. The advantage of IDLA is that it eliminates the need for modeling and parameter estimation of the power amplifier.

Figure 6.2-41 shows IDLA, which is the predistortion technique used in SystemVue W1716 DPD simulation software. The feedback path labeled "Predistorter Training" (block A) has $y(n)/G$ as its input, where G is the intended power amplifier small signal gain, and $\hat{z}(n)$ as its output. The actual predistorter is an exact copy of the feedback path (copy of A); it has $x(n)$ as its input and $z(n)$ as its output. Ideally, $y(n) = Gx(n)$, which renders $z(n) = \hat{z}(n)$ and the error term $e(n) = 0$. Given $y(n)$ and $z(n)$, this structure enables the parameters of block A to be found directly, which yields the predistorter. The algorithm converges when the error energy $\|e(n)\|^2$ is minimized.

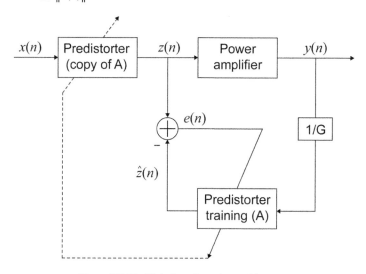

Figure 6.2-41. Digital predistortion architecture

6.2.10.1 Memory Polynomial Predistortion

Figure 6.2-41 illustrates the DPD architecture where $x(n)$ is the input signal to the predistortion unit, whose output $z(n)$ feeds the power amplifier to produce output $y(n)$. The most general form of nonlinearity with a finite memory of length $Q+1$ is described by the Volterra series, which consists of a sum of multidimensional convolutions. In the training branch of Figure 6.2-41, the Volterra series predistorter can be described by

$$z(n) = \sum_{k=1}^{K} z_k(n) \tag{1}$$

where

$$z_k(n) = \sum_{m_1=0}^{Q} \cdots \sum_{m_k=0}^{Q} h_k(m_1, \cdots, m_k) \prod_{l=1}^{k} y(n - m_l) \tag{2}$$

and where Z_k is the k-dimensional convolution of the input with Volterra kernel h_k.

This is a generalization of a power series representation with a finite memory of length Q+1. The $z(n)$ also can be written as follows:

$$z(n) = h_0 + \sum_{m_1=0}^{Q} h_1(m_1) y(n-m_1) + \sum_{m_1=0}^{Q} \sum_{m_2=0}^{Q} h_2(m_1, m_2) y(n-m_1) y(n-m_2) + ... \tag{3}$$

A memory polynomial predistorter uses the diagonal kernels of the Volterra series and can be viewed as a generalization of the Hammerstein predistorter. The memory polynomial predistorter is used to linearize power amplifiers with memory effects. The predistorter is constructed using indirect learning architecture, thereby eliminating the need for model assumption and parameter estimation of the power amplifier. Compared to the Hammerstein predistorter, the memory polynomial predistorter has slightly more terms. However, it is much more robust and its parameters can be easily estimated by way of least-squares. In the training branch of Figure 6.2-41, the memory polynomial predistorter can be described by

$$z(n) = \sum_{k=1}^{K} \sum_{q=0}^{Q} a_{kq} y(n-q) \left| y(n-q) \right|^{k-1} \tag{4}$$

where $y(n)$ and $z(n)$ are, respectively, the input and output of the predistorter in the training branch, and a_{kq} are the coefficients of the predistorter.

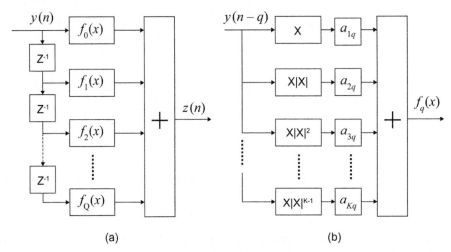

Figure 6.2-42. (a) The structure of the memory polynomial and (b) the structure of the nonlinear polynomial

The structure of the memory polynomial and the structure of the Qth polynomial are shown in Figure 6.2-42. If $Q=0$, the structure in Figure 6.2-42 (a) is the memoryless polynomial. Since the model in equation (4) is linear with respect to its coefficients, the predistorter coefficients a_{kq} can be directly obtained by least-squares.

By defining a new sequence

$$u_{kq}(n) = \frac{y(n-q)}{G} \left| \frac{y(n-q)}{G} \right|^{k-1}, \tag{5}$$

at convergence

$$z = Ua \tag{6}$$

where

$$z = \left[z(0), z(1), \ldots, z(N-1) \right]^{T},$$

$$U = \left[u_{10}, \cdots, u_{K0}, \cdots, u_{1Q}, \cdots, u_{KQ} \right],$$

$$u_{kq} = \left[u_{kq}(0), u_{kq}(1), \cdots, u_{kq}(N-1) \right]^{T}, \text{ and}$$

$$a = \left[a_{10}, \cdots, a_{K0}, \cdots, a_{1Q}, \cdots, a_{KQ} \right]^{T},$$

the least-squares solution for $z = Ua$ is

$$\hat{a} = (U^{H}U)^{-1}U^{H}z \tag{7}$$

where $(\bullet)^{H}$ denotes complex conjugate transpose.

After the memory polynomial coefficients $\hat{a} = \left[\hat{a}_{10}, \cdots, \hat{a}_{K0}, \cdots, \hat{a}_{1Q}, \cdots, \hat{a}_{KQ} \right]^{T}$ are obtained, they are loaded into the nonlinear filter of Figure 6.2-42 (b).

6.2.10.2 Example implementation of DPD

Ideally the toolset for implementing a DPD algorithm will be accurate, will avoid dependence on a vendor-specific chipset or hardware implementation for the initial modeling, and will integrate with the rest of the baseband processing subsystem to minimize additional costs. Connectivity with a range of other tools for hardware verification is also important.

To illustrate the DPD concept an example using SystemVue W1716 DPD software will now be described. This software features a wizard-based user interface that enables modeling of typical memory effects in both low and high-power amplifiers. The tool can also be applied to transceiver and automatic gain control modules.

The SystemVue simulation approach is complimented by readily available test equipment that can be used to quickly assess the extent to which components within the system can be linearized.

The behavior of the predistorter or DUT can be experimentally measured using the instantaneous complex baseband waveform approach. A typical setup is presented in Figure 6.2-43. The digital baseband waveform is downloaded into a vector signal generator (VSG) that will feed the DUT. The output of the DUT is then down-converted and sampled by the vector signal analyzer (VSA). The sampled input and output data are captured and used to extract behavioral models—both the DPD model and the power amplifier (PA) model—for the DUT.

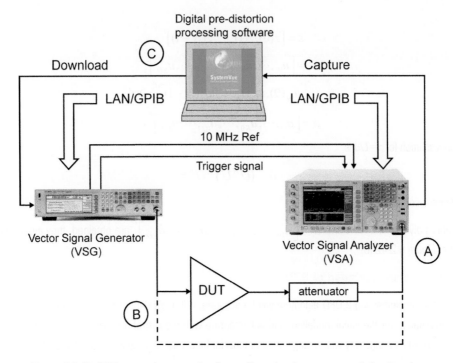

Figure 6.2-43. DPD system incorporating SystemVue, signal generator, and signal analyzer

The choice of signal analyzer should take into account the need to cover the 3rd and 5th order harmonics of the signal bandwidth. The process for DPD measurement follows the five-step process in Figure 6.2-44. Two methods to extract the DPD model are possible depending on the choice of signals. Method (a) in Figure 6.2-44 uses the PA input and PA output (points B and A in Figure 6.2-43). Method (b) in Figure 6.2-44 uses the baseband input and PA output (points C and A in Figure 6.2-43).

The five-step process for testing the DUT with DPD is as follows:

1. Create the DPD stimulus signal. The DPD stimulus waveform (e.g., 802.11ac, LTE/LTE-A, WCDMA, or user-defined) is created and downloaded via LAN or GPIB into the VSG.
2. Capture the PA output. The PA's response—both input and output (points B and A in Figure 6.2-43) or just PA output (point A in Figure 6.2-43)—is captured from the VSA. The PA output is captured by inserting the PA between the VSG and VSA with appropriate signal calibration, including any signal padding with attenuators.
3. Generate the DPD model. This step is to extract DPD models based on PA input and PA output. This step includes time delay estimation and adjustment. The propagation delay through the DUT will introduce a mismatch between the data samples used to calculate the instantaneous AM/AM and AM/PM characteristics of the DUT. This mismatch will translate into dispersion in the AM/AM and AM/PM characteristics that can be misinterpreted and considered as memory effects.
4. Apply predistortion to the stimulus signal. The DPD+PA response is captured by applying stimulus to the extracted DPD model and downloading the DPD output waveform into the VSG. The PA output waveform is then captured from the VSA.
5. Verify DPD performance. The DPD+PA response is verified and the performance improvements possible with DPD can be shown.

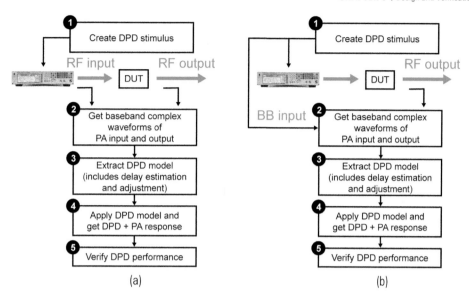

Figure 6.2-44. Model extraction procedure (a) with both PA input and output, (b) with both baseband input and output

Step 2 in method (b) needs additional coarse time-delay estimation. This is performed at the sample level to synchronize between the real PA output waveform and the simulated baseband waveform because a trigger signal is used to capture the signal in the VSA whereas the baseband waveform is the ideal waveform and has no time-delay. The coarse time delay is first estimated as the cross-covariance between the input sequence and the output sequence of the PA. The equation is as follows:

$$corr(n) = \sum_{k=1}^{N} x_{BB}(k)y^{H}(k-n)$$

where N is the number of captured samples. The results are shown in the following figures.

Figure 6.2-45 shows 2 x 20 MHz contiguous carrier aggregation. The lowest trace is the stimulus (reference), the upper trace is the power amplifier response without DPD, and the middle trace is the power amplifier response with DPD. The improvement in adjacent channel performance is evident.

Figure 6.2-46 shows an LTE-Advanced downlink signal with 3 x 20 MHz contiguous carrier aggregation.

Figure 6.2-47 shows an LTE-Advanced downlink signal with 2 x 20 MHz + 20 MHz non-contiguous carrier aggregation.

Figure 6.2-48 shows an LTE-Advanced uplink signal with 20 MHz + 20 MHz non-contiguous carrier aggregation.

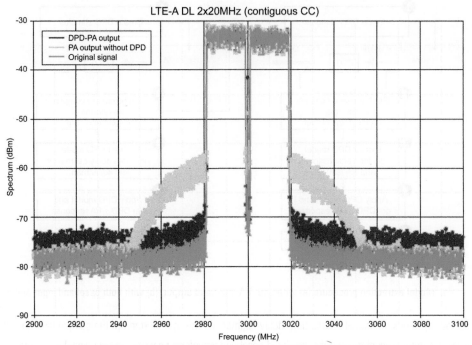

Figure 6.2-45. DPD performance for 2x20 MHz contiguous carrier aggregation

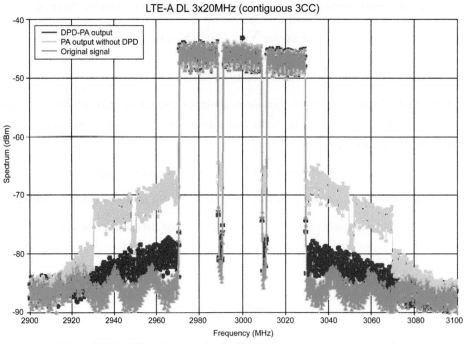

Figure 6.2-46. DPD performance for 3 x 20 MHz contiguous carrier aggregation

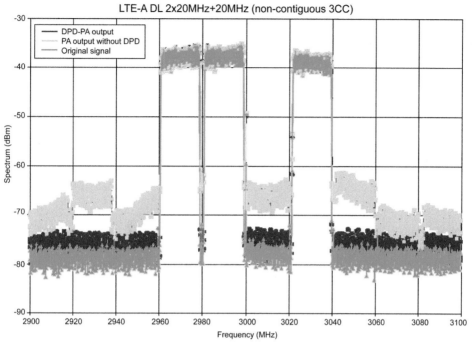

Figure 6.2-47. DPD performance for 2 x 20 MHz + 20 MHz non-contiguous carrier aggregation

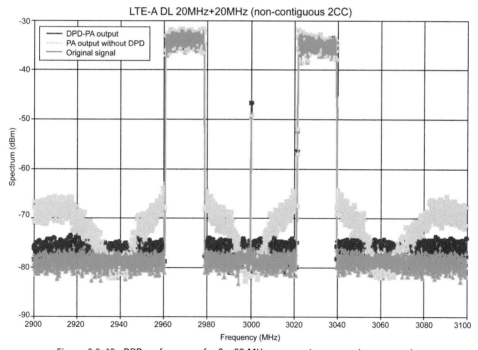

Figure 6.2-48. DPD performance for 2 x 20 MHz non-contiguous carrier aggregation

Crest factor reduction supplements and improves the effectiveness of DPD. Figures 6.2-49 and 6.2-50 show a 20 MHz 16QAM downlink signal with DPD and CFR applied versus the uncorrected response. A 200 W LDMOS Doherty power amplifier is tested in this example. Table 6.2-4 shows the EVM and ACLR results.

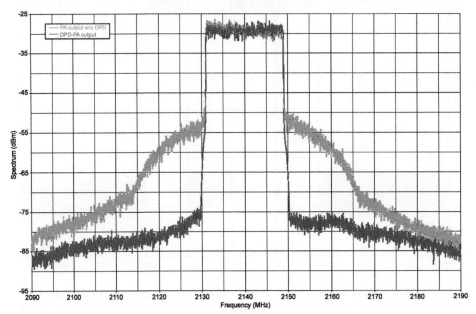

Figure 6.2-49. Power amplifier output spectra with DPD and without DPD

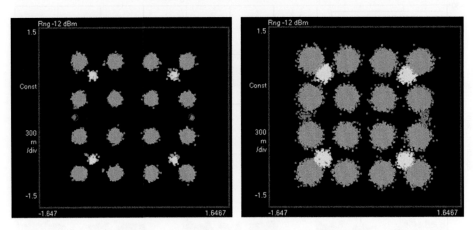

Figure 6.2-50. DPD + CFR performance for 20 MHz downlink 16QAM carrier:
constellation of power amplifier output with DPD (left);
constellation of power amplifier output without DPD (right)

Table 6.2-4. ACLR and EVM performance for 20 MHz downlink carrier

ACLR	Lower alternate channel	Lower adjacent channel	Upper adjacent channel	Upper alternate channel	EVM (%)
System requirement	45 dB	45 dB	45 dB	45 dB	12.5
Stimulus signal	58.6	55.1	57.1	58.0	5.3
Uncorrected power amplifier response	47.6	29.3	28.2	48.7	10.1
Power amplifier with DPD and CFR	55.0	51.0	49.1	43.4	5.5

6.2.11 Combining Simulation and Test Hardware Development

By combining simulation with test instruments, system engineers can extend the test functionality of their equipment, adding pre-processing and post-processing of test signals using flexible simulation models.

This section introduces an integrated approach to model-based design for advanced communication systems called "Design-Verify-Test." Design-Verify-Test is a hybrid approach to development which uses stimulus/response test equipment to bring real-world effects into the design of communication architectures and physical layer algorithm development. SystemVue is used to integrate these domains into a single, versatile development platform.

Traditionally, RF and DSP designers have worked separately to produce architectures within their respective domains. However, today's evolving communication systems require cross-domain tradeoffs to optimize the final design.

There are several benefits to the Design-Verify-Test approach. System engineers can gain practical performance insight from measurement-based validation of initial algorithmic designs, helping them accomplish the following:

- Diagnose and solve cross-domain problems early during development
- Reduce excess design margin in both baseband/DSP and RF transceiver architectures
- Create an accelerated model-based design methodology that sits above traditional RF and baseband hardware design flows and connects to them, making them more powerful.

System development for layer 1 is traditionally split between baseband/DSP and RF/analog disciplines. While baseband/DSP and RF components may initially be designed together and even verified together as a hardware prototype, there are intermediate states of component development that are difficult to validate until all components are ready for cross-domain integration.

Impairments in the radio channel are often not accounted for early in development. These are the combined effects of radio propagation, noise, and interference from other users. Functional gaps also pose a challenge: development groups may concentrate on one component of an overall system and, perhaps due to vendor or intellectual property constraints, lack access to complementary parts to create a virtual system. Figure 6.2-51 and Table 6.2-5 show where the Design-Verify-Test approach can address these early development challenges.

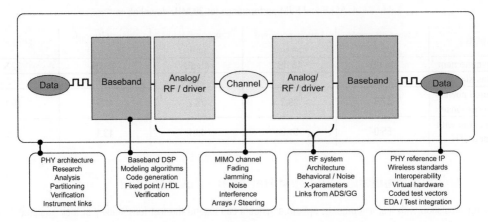

Figure 6.2-51. Example of areas in which the SystemVue Design-Verify-Test approach enables system engineers to continuously evaluate the link-level performance in a cross-domain communication design

Table 6.2-5. Summary of design and verification challenges addressed by a "Design-Verify-Test" approach based on Agilent SystemVue

Design needs addressed	Verification needs addressed
Simulation of missing transmit or receive hardware, to complete a virtual system	Early pre-commercial standards support for test instruments
Simulation of channel impairments	Platform for structured verification scripting and test plan execution
Continuous verification of systems that are partially implemented or are mixed-hierarchy prototypes	
Improved RF/analog modeling fidelity	Consistency of approaches and measurements for verification of algorithms vs. hardware
Unified approach to RF and baseband/DSP	MIMO hardware multiplicity for early R&D validation

The Design-Verify-Test approach is able to "virtualize" the missing pieces of the complete system, including the RF. This allows both the component designer and the system engineers to maintain a link-level perspective of performance from the first day of a project. At any point in the mutual baseband/RF design flow an individual algorithm, hardware block, or analog subsystem can be represented using either simulations or measurements, and the performance can be evaluated both at a detailed physical signaling level (e.g., digital debug using a logic analyzer) and at the link level (e.g., system throughput or BER/FER).

With the Design-Verify-Test approach, wideband stimulus and response test equipment is represented as software-defined instrumentation within the design process, increasing the fidelity of the simulations at the earliest possible phases of design. An advantage of the Design-Verify-Test approach is that the design environment transitions seamlessly from early simulation through to later hardware verification, thus promoting consistency and high levels of reuse for assets such as scripting and IP. This process is illustrated in Figure 6.2-52. The Design-Verify-Test approach can be applied to multimode system-on-chip (SoC) and application-specific integrated circuit (ASIC) design; to field programmable gate array (FPGA), embedded DSP, and application-specific standard product (ASSP) development; to board-level integration; and to RF technologies. For example, using SystemVue's LTE-Advanced reference library an SoC algorithm designer can experience an instrument-grade verification environment for BER and throughput testing before the first tape out or prototype FPGA.

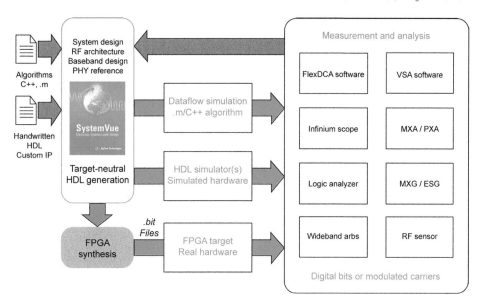

Figure 6.2-52. The SystemVue cross-domain, model-based design platform

Figure 6.2-53 shows the many way in which the simulation environment can connect with the outside world.

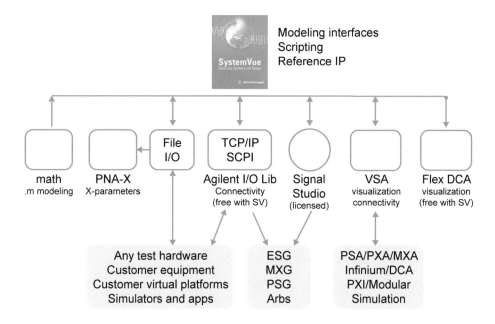

Figure 6.2-53. SystemVue connectivity to the outside world

Figure 6.2-54 shows a typical example of mixed signal test connectivity for a transceiver. The first step of testing is to stimulate the appropriate point in the block diagram and the second step is to capture the desired response at a later point. There is no need for the stimulus and response to be in the same domain.

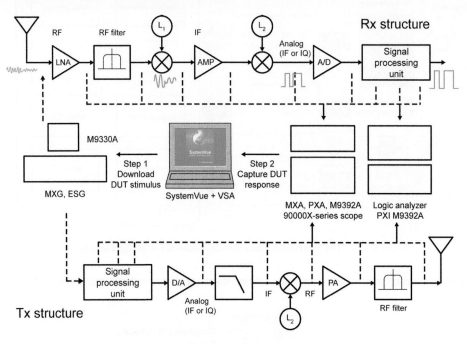

Figure 6.2-54. Combining simulation and test to measure EVM, BER/FER, and throughput at various stages along the RF/mixed-signal transmitter and receiver chain

The remainder of this section will show Design-Verify-Test in action using a 5 MHz + 5 MHz intra-band non-contiguous design example. The steps are the following:

- Creating and downloading signals for test
- Capturing waveforms using a vector signal analyzer
- Using math language to control swept simulation
- Downloading and capturing MIMO signals
- Using math language to control instruments
- Making mixed-domain hardware measurements.

6.2.11.1 Creating and Downloading Signals for Test

Figure 6.2-55 is a SystemVue schematic for an LTE-Advanced intra-band non-contiguous 5 MHz + 5 MHz signal. The model includes a Gaussian noise source to control the SNR of the signal. The right-most part of the schematic is the signal sink used to capture the waveform for subsequent download to a signal generator.

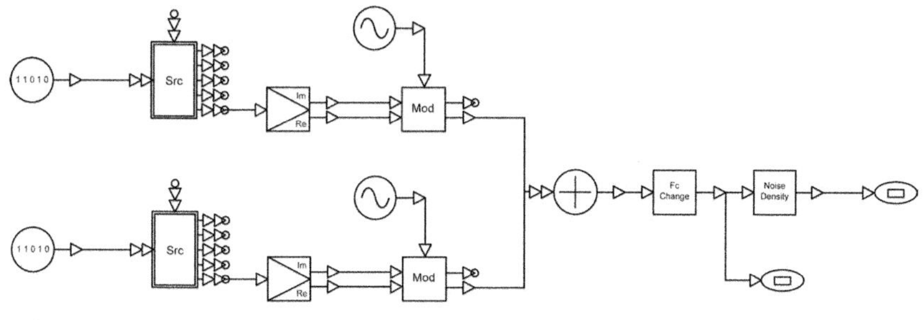

LTE-A baseband source of component carrier 1

LTE-A baseband source of component carrier 2

Figure 6.2-55. Schematic for an LTE-A intra-band non-contiguous 5 MHz + 5 MHz signal with additional Gaussian noise

The signal sink is a model available in the SystemVue Algorithm Library. This model contains the parameters necessary to capture the simulated signal and download it to the chosen signal generator.

6.2.11.2 Waveform Capture Using Vector Signal Analysis

The following examples describe how signals are captured using a variety of instruments. Figure 6.2-56 shows the 5 MHz+5 MHz signal being captured by the 89600 VSA software by means of suitable hardware. The captured signal is filtered and the component carrier 1 signal is then demodulated and a BER measurement made.

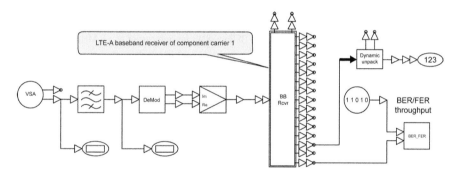

Figure 6.2-56. SystemVue schematic for signal capture, filtering and demodulation with BER measurement

The choice of receiver hardware for the signal capture depends on the domain of the signal. If the signal is at RF, then a suitable analyzer such as an Agilent PXA or M9392A signal analyzer is appropriate. For baseband analog IQ signals an oscilloscope can be used and for digital signals a logic analyzer can be used. The source on the left-hand side of Figure 6.2-56 (VSA) is a library component that contains all the parameters required to interface the 89600 VSA software performing the signal capture to the chosen instrument.

6.2.11.3 Using Math Language to Perform Swept Simulation

So far the simulation examples have been for single configurations but it is often necessary to repeat a simulation using different parameters such as BER vs. SNR. Figure 6.2-57 is an example of BER vs. SNR for the CC1 signal of the 5 MHz + 5 MHz signal. In each iteration, the SNR changes and the BER measurement is repeated.

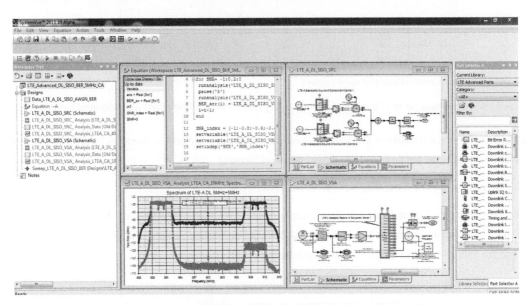

Figure 6.2-57. SystemVue workspace for test of BER vs. SNR for one carrier of a 5 MHz + 5 MHz signal

Figure 6.2-58 shows the output BER vs. SNR curve of CC1 for both QPSK and 16QAM modulation formats.

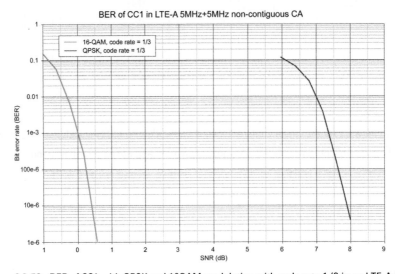

Figure 6.2-58. BER of CC1 with QPSK and 16QAM modulation with code rate 1/3 in an LTE-Advanced non-contiguous 5 MHz + 5 MHz carrier aggregation signal

6.2.11.4 MIMO Signal Download and Capture

The preceding examples have been examples of SISO. For simulation and verification with MIMO systems it is possible to use the SystemVue standard library components to simultaneously download and capture multiple streams of data.

Agilent offers three types of analyzer with specific support for MIMO signal capture:

1. Two channel N9010A EXA and N9020A MXA signal analyzers
2. Four channel Infiniium 90000 X-Series oscilloscopes
3. Eight channel N7109A multichannel signal analyzer.

These analyzers can be used to capture MIMO signals for playback in simulation. Figure 6.2-59 shows the 89600 VSA software's MIMO source module with parameter support for multiple antennas.

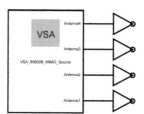

V1 {VSA+89600B_MIMO_Source@Data Flow Models}
VSATitle='Simulation source
RepeatData=Reacquire
Pause=NO
SetupFile=
SampleRateOption=FromTraces
NumberOfAntennas=4 Antennas
Antenna1_Trace=A
Antenna2_Trace=B
Antenna3_Trace=C
Antenna4_Trace=D
RecordingFile=
AutoCapture=NO
ShowAdvancedParams=YES
DefaultHardware=NO
SetFrequencySpan=NO
SetCenterFrequency=NO
SetRange=NO
RecordingLength=0s
GapOut=NO

Figure 6.2-59. The 89600 VSA software configured as a MIMO simulation source (up to 4 channels) that can inject either recorded or live MIMO measurement waveforms into the simulation

6.2.11.5 Using Math Language to Control Instruments

SystemVue's built-in math language capability offers another option for instrument connectivity. Using a TCP/IP object, SystemVue can connect to instrumentation and communicate SCPI commands or ASCII .m scripting files. The example in Figure 6.2-60 shows a trace capture from an MXA analyzer using .m code, which is seen on the right side of the screen. The use of .m code allows logic to be added to the download model.

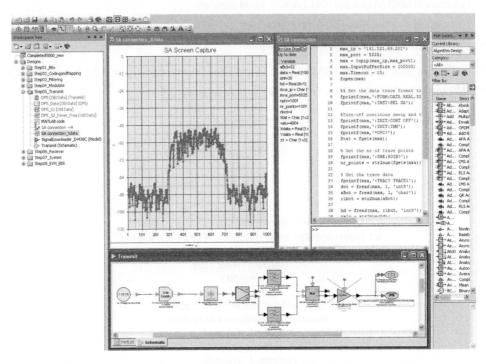

Figure 6.2-60. A signal trace captured manually from an Agilent MXA signal analyzer, which uses .m code

A complementary capability is the use of MATLAB from within SystemVue in conjunction with SystemVue's own math language parser. Using MATLAB with SystemVue's instrument control "toolbox" offers an expanded set of instrument interfacing capability. SystemVue provides examples of IQ and screen trace data acquisition (from a spectrum analyzer) using SystemVue's native math language capability. Direct MATLAB integration into SystemVue provides additional plotting capability.

6.2.11.6 Mixed-Signal/RF Hardware Measurement Example Using Design-to-Test FPGA Flow

An RF transmitter with FPGA hardware and a DAC can be used to illustrate the challenges of mixed-signal/RF transmitter hardware measurements. Baseband, analog, and RF issues can be particularly difficult to diagnose and debug during the system integration phase. Potential hardware integration issues can begin early in the design phase but not necessarily be identified right away due to disparate design flows and tools used by FPGA and RF engineers. SystemVue enables baseband designs and RF designs to be designed and evaluated together at a system level in one design environment, which helps the system engineer identify potential system integration issues early in the design phase.

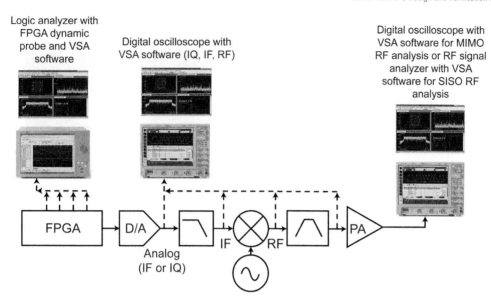

Figure 6.2-61. Troubleshooting mixed-signal/RF FPGA digital baseband, analog IQ, IF, and RF issues along a transmitter chain using a logic analyzer and digital oscilloscope with 89600 VSA software

Figure 6.2-61 shows a test methodology using the 89600 VSA software with logic analyzers, digital oscilloscopes, and RF signal analyzers to probe various points along the mixed-signal/RF transmitter chain. While logic analyzers are more commonly used in digital baseband applications to measure test vectors consisting of 0s and 1s, here the logic analyzer is being used to perform EVM measurements at digital baseband. This can help the system engineer debug potential FPGA problems and evaluate the RF EVM performance across the DAC boundary by comparing the EVM at the DAC input measured with a logic analyzer to the EVM at the DAC output measured with an oscilloscope.

Using the logic analyzer's dynamic probe capability with the 89600 VSA software enables an additional level of testing to diagnose and debug issues within an FPGA design. An example is measurement of the fixed point precision of an FPGA filter, which impacts the RF spectral performance and EVM of a signal. To see the effect of filter precision on signal quality, the RF system engineer can directly measure the spectral performance and EVM of the digital IF or digital IQ in the FPGA hardware using a logic analyzer to capture the digital data and the 89600 VSA software to measure the EVM. In addition, as shown by the vertical dashed lines representing the logic analyzer inputs in Figure 6.2-61, FPGA dynamic probing with the logic analyzer and VSA software can be performed at multiple points within the FPGA implementation, such as before and after the filter inside the FPGA.

To further illustrate this capability, a fixed-point LTE IQ modulator is designed in an FPGA using SystemVue as shown in Figure 6.2-62. The simulated EVM spectrum and the 64QAM constellation show minor waveform impairments resulting from the fixed-point design, which results in a 0.9% simulated EVM.

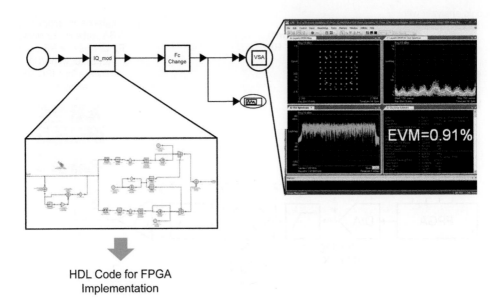

HDL Code for FPGA
Implementation

Figure 6.2-62. Simulation of fixed-point design and HDL code generation

After verifying the functionality of the IQ modulator, HDL code is generated by SystemVue for the FPGA implementation. The HDL code is used to generate a .bit file to program the FPGA. The resulting FPGA hardware is shown being tested in Figure 6.2-63. The logic analyzer on the left is being used with dynamic probing to probe at two test points within the FPGA, and the 89600 VSA software running in the logic analyzer is used to demodulate the 64QAM waveform.

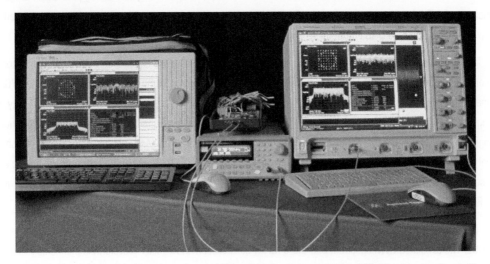

Figure 6.2-63. Logic analyzer with dynamic probe and VSA software used to debug an FPGA implementation (left); mixed-signal digital oscilloscope used at the DAC output to evaluate the performance across the mixed-signal DAC boundary (right)

The two test points are shown in Figure 6.2-64. They are (left) after filtering (EVM= 0.52%) and (right) at digital IF (EVM= 0.9%). In this example the EVM performance at the filtered IQ and digital IF are within the expected performance limits; however, if a problem existed within the FPGA implementation—for example, a bit reversal affecting EVM performance—it would be identified and diagnosed. This type of issue is identified by comparing the EVM performance at the input and output of the filter.

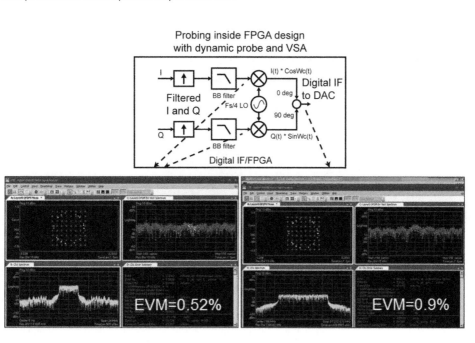

Figure 6.2-64. Logic analyzer dynamic probe measurements with 89600 VSA software at two probe points within an FPGA IQ modulator implementation

The EVM measured at the digital IF (input to the DAC) is 0.9%, which agrees with the simulated digital IF result of 0.91% shown in Figure 6.2-62. In addition, the output of the DAC is also measured with the analog input of a high performance digital oscilloscope to evaluate the impact of the DAC. This is shown in Figure 6.2-65. The EVM is measured at 1.4% at the output of the DAC versus 0.9% at the digital IF as shown in Figure 6.2-64. This small difference in EVM indicates negligible error contributions from the DAC for this particular example. If an issue were occurring across the DAC boundary, this measurement approach would help to quantify the error contributions of the DAC relative to the overall EVM at the system level.

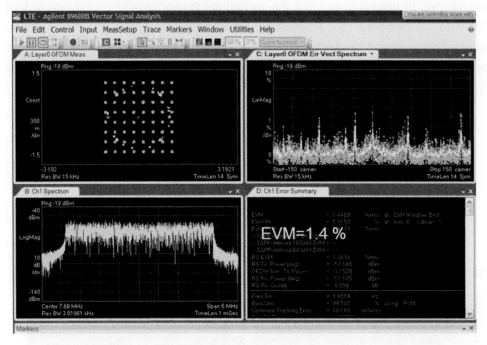

Figure 6.2-65. Measurement at the output of the DAC using the test setup in Figure 6.2-63

6.2.12 Summary

The simulation concepts and examples described in this section can be applied to accelerate development and minimize risk throughout the LTE and LTE-Advanced product development cycle. The synergy that is achieved by combining simulation with test provides even greater power and flexibility for hardware testing. Using the 89600 VSA software in simulation along with logic analyzers, digital oscilloscopes, and RF signal analyzers provides a common test methodology with a consistent user interface to help diagnose issues along the mixed-signal/RF transmitter/receiver chain (baseband, analog IQ, IF, RF). This powerful capability can be used to identify potential issues earlier in the cycle, when they are easier and less costly to fix.

6.3 Testing RFICs With DigRF Interconnects

6.3.1 The Mobile Industry Processor Interface Alliance and DigRF

The introduction of LTE brings architectural change to virtually every subsystem in the wireless network, from IP-based infrastructure to the chipsets used in mobile handsets. The Mobile Industry Processor Interface (MIPI™) Alliance is driving several of the changes at the chipset level. The MIPI Alliance has developed several standards that affect both baseband (BB) and RF chipsets used in mobile cellular devices. The MIPI standards work is in the following areas:

- Battery interface working group defining a robust, cost-efficient communication interface specification for smart batteries and low cost batteries.
- Camera working group developing the camera serial interface (CSI) imaging protocol.
- Display working group developing the display serial interface (DSI) packet protocol carrying the display command set (DCS) application layer support for display management.
- Low-speed multi-point link working group developing audio and control interfaces.
- Low latency interface (LLI) working group producing a specification for an interface with sufficient performance to allow sharing a DRAM memory between two chips.
- System power management working group developing the system power management interface (SPMI) using a master/slave approach to handle the increasing challenge of power management in complex mobile devices.
- DigRF working group developing the DigRF v4 standard for BB to RFIC connectivity.
- Unified protocol working group developing the unified protocol (UniPro) standard, which is a transport layer protocol for interconnecting application processors, modems, and peripherals.
- Physical layer working group developing the underlying physical layer interconnect that carries the higher level application protocols. The current D-PHY standard runs at up to 1 Gbps with up to four lanes giving 500 MB per second, which explains the D in the name coming as it does from the Roman numeral for 500. Recently the specification has been updated to support speeds up to 1.5 Gbps. The D-PHY standard is aimed mainly at the DSI and CSI protocols. The next development is the M-PHY standard, which will run at up to 6 Gbps per lane and be able to support DigRF v4, LLI, and UniPro as well as the next-generation CSI and DSI protocols.
- RF front end (RFFE) working group defining control interface solutions for RF front-end modules and components.

TThe efforts of the MIPI Alliance will bring to the mobile device industry the type of subsystem interchangeability that has existed in the computer industry for many years.

Early in 2007 the DigRF standardization activity was incorporated into the MIPI Alliance. There are three existing versions of the DigRF standard. DigRF v4 is the newest and most relevant for LTE as the link bandwidth is appropriate for the expected data rates. However, the DigRF standard is a proprietary standard available only to MIPI Alliance members. Therefore the details of the standard will not be covered here. Instead, this section examines the challenges of integrating and testing DigRF v4 in LTE designs.

6.3.1.1 Why DigRF?

DigRF has been designed to address several issues. By making the baseband to RF link fully digital, baseband IC design can now become a less expensive, digital-only process. DigRF v4 specifies a serial link, reducing pin count, and further simplifying designs. DigRF v4 also incorporates low-power modes that can reduce standby power consumption. With a standard baseband to RFIC interface, it now becomes theoretically possible to have interchangeable subsystems allowing a mobile cellular device vendor to "mix and match" baseband and RFICs based on the performance attributes desired. As a practical matter, true interchangeability of ICs will likely await further standardization efforts to ensure that all relevant physical and protocol standards exist.

6.3.1.2 Testing Baseband and RFICs With DigRF

The multi-gigabit DigRF v4 standard removes the potential inter-chip communication bottleneck resulting from the higher data rates implicit in LTE. However, this fourth version of the DigRF standard introduces multiple levels of design and test complexity due to changes such as a new link protocol and the high-speed serial link.

From a test perspective, the integration of DigRF into the baseband IC is relatively straightforward. This is because the baseband IC is now completely digital and can be tested with standard digital test tools. Straightforward does not imply easy, however, as the DigRF v4 interface is a gigabit-speed interface requiring special care and test to integrate. This section will focus on the baseband testing issues of the RFIC interface. The wider issues of the RFIC are covered elsewhere.

6.3.2 New Challenges in Building and Testing RFICs

The move to a digital baseband interface introduces a number of new challenges to designing and testing the RFIC. These challenges exist in the physical and protocol layers of the digital and RF domains and they include the following:

- The formerly analog baseband to RFIC communication link is now a high-speed serial digital interface that requires special design and test techniques.
- Because analog sources can no longer be used to stimulate the RFIC on the baseband interface, testing requires different equipment and a different methodology than used in previous generation chipsets.
- Unlike other serial interfaces, the DigRF has a protocol stack encapsulated within the digital interface and this "dual protocol stack" complicates characterization and validation.
- The information flowing on the interface between the baseband and RFICs includes data and control traffic.
- Information transfers must comply with strict time constraints (time determinism).

These challenges and the associated considerations for testing will now be examined more closely.

6.3.2.1 Digital Serial Link

To provide the data rates needed for LTE implementations, the DigRF interface runs at gigabit per second rates. Under these conditions, high-speed analog effects can impair signal quality and degrade link bit error ratio (BER).

Similar to other high-speed serial buses, the current DigRF standards employ data-encoding mechanisms along with embedded clocks and state machines for link and transaction operations. To analyze the data, the embedded clock must be extracted before the data can be decoded and fully analyzed. A clear understanding of clock recovery, encode/decode protocols, and real-time measurements is critical for success. Most general-purpose measurement tools cannot understand signals encoded for DigRF v4 and information encoding so therefore provide only raw information about the digital data.

6.3.2.2 RFIC Test Methodology

Digital IQ data and control information is now packetized and transferred between the baseband and RFICs over the DigRF interface. Previously, RF engineers and validation teams used analog sources to stimulate the RFIC on the IQ analog interface. Now that the analog interface is gone, new tools are needed to enable the same RF physical measurements through the DigRF digital serial interface.

6.3.2.3 Dual Protocol Stack

Like many other serial buses, the DigRF interface is described as a stack of multiple layers in which each layer has a specific function and mode of operation. These layers grow from the physical layer to the application or software layer, and they include the link layer (or mode of operation of the bus), the data-encoding scheme, the frame structure, the flow control, the error handling mechanism, and others.

Ensuring that an RFIC properly interoperates with a baseband IC requires verification that all layers of the digital protocol stack are designed and operate in accordance with the DigRF v4 specification. A suitable test environment must provide analysis and stimulus capabilities on the entire digital protocol stack.

DigRF is designed to be used in mobile devices and is unlike most serial buses because there is another protocol stack "encapsulated" within the digital interface—this stack representing the mobile device's wireless protocol (e.g., GSM, W-CDMA, LTE, WiMAX). As with other protocols, this stack extends from the RF physical layer to the user application layer.

Things can become confusing as the physical and protocol layers of the two stacks are intermixed. Because the RF physical-layer information is encapsulated in the payload of DigRF frames, it is seen as existing "above" the DigRF protocol layer. This is different from typical layering schemes in which the physical layer resides at the bottom of the stack.

The characterization and validation of a DigRF-enabled RFIC requires separate measurement of each layer of the digital and RF protocol stacks. This in turn requires a test infrastructure that provides insight to all layers (Figure 6.3-1). From a stimulus point of view, the ability to test an RFIC under real-world conditions requires a test environment that can encapsulate IQ stimulus data within DigRF traffic to create the required dual-stack stimulus model.

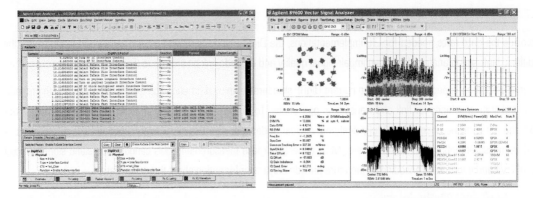

Figure 6.3-1. Agilent DigRF test dual interface presenting information from the digital protocol layers (bit-and packet-level view, left) and the RF physical information (modulation analysis, right) from the same measurement

6.3.2.4 Mixed Traffic

The information flowing between RF and baseband ICs consists of IQ data encapsulated in DigRF frames as well as control information that is sent from the baseband IC to the RFIC. This control information stays within the RFIC. Here are some examples of control traffic:

- Turning on and configuring the RFIC in loopback mode
- Changing transmit output power level
- Changing the RF output frequency.

When an RFIC is tested, proper validation requires test equipment that can configure and create this mix of control and data traffic. To test an RFIC, the stimulus environment must be able to insert RFIC control information within the IQ data flow.

6.3.2.5 Time Determinism

Information transfers between the RF and baseband ICs must comply with strict time constraints. Therefore, it is important for the test environment to precisely measure when each frame is sent from one IC to the other and provide real-time detection of time-constraint violations.

6.3.3 Addressing the Cross-Domain Issues

To address the need to test in both the digital and RF domains, Agilent has combined traditional RF measurement tools with digital and protocol analysis and stimulus tools in a complete test environment for cross-domain RFIC test.

This new test platform enables DigRF protocol debugging as well as comprehensive stimulus and analysis across the digital and RF domains to enable and to accelerate the turn-on, validation, and integration of DigRF v4-based devices.

The DigRF test environment incorporates testing of both the transmitter and receiver paths. For transmit path testing, the typical environment includes three major elements:

- A signal analyzer for modulation analysis on the antenna side of the RFIC. Advanced signal-analysis tools, such as the Agilent 89600 VSA software, enables detailed signal analysis including digital demodulation of the IQ data according to the radio protocol being transmitted.
- A DigRF exerciser/analyzer provides digital and RF stimulus on the baseband side of the RFIC.
- Signal generation software, such as Agilent design software or Signal Studio, generates the IQ data to be sent to the RFIC. See Section 6.2 for more information on how the design software is used to generate signals. The signal generation software is hosted on the DigRF exerciser as an analog signal source and cannot be used on the DigRF interface.

For receiver path testing, the typical environment includes four key elements:

- A vector signal generator provides an appropriate RF signal on the antenna side of the RFIC.
- Signal generation software, such as Agilent design software or Signal Studio, provides baseband modulated data for the vector signal generator.
- A DigRF exerciser/analyzer enables analysis of received DigRF packets.
- Signal analysis software such as the 89600 VSA software, which enables analysis of the resulting IQ data carried by the DigRF interface.

The 89600 VSA software can run inside the DigRF exerciser. With the DigRF analyzer and 89600 VSA software, both the digital and RF paths can be examined and analyzed in parallel from the same measurement.

Since RF designers are used to connecting test equipment directly to the IQ port between the RF and baseband ICs, the test methodology and tools are now changed to connect with the digital interface. Fortunately, many of the tools can still be used; they are just hosted on different hardware. See Figure 6.3-2 for a before and after illustration of the test configuration for TX testing. For TX testing of the RFIC, the Agilent Radio Digital Cross Domain (RDX) test system connects to the DigRF interface and hosts the signal creation software which acts as a kind of "virtual signal generator." For RX testing, the RDX tester hosts the 89600 VSA software acting as a "virtual signal analyzer." The RDX tester inserts and extracts the IQ data to and from the DigRF data stream while interleaving any control or timing packets as part of the DigRF control stream.

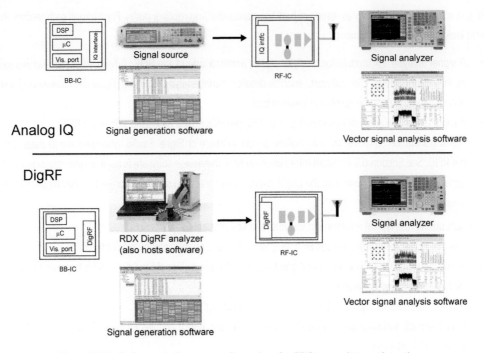

Figure 6.3-2. Before and after test configurations for RFIC transmitter path testing

6.3.4 Stepping Through the Layers

A typical design process will progress from left to right as shown in Figure 6.3-3. Digital designers are accustomed to working with one set of tools in the digital domain; RF designers are used to working with another set of tools in the RF domain. With DigRF in the picture, a unified set of tools will benefit both digital and RF designers.

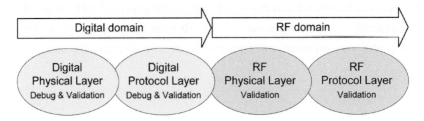

Figure 6.3-3. Working through the four layers is an efficient way to characterize and validate an LTE device.

6.3.4.1 Digital Physical Layer

Because the baseband-side interface to the RFIC is digital, it is necessary to ensure proper DigRF operation before testing the RF sections of the IC. The RFIC design may integrate "legacy" inputs with analog IQ inputs that are separate from the DigRF interface. This offers multiple ways to access the RF subsections. In such cases, it might be possible to simultaneously test the DigRF interface and the RF subsections.

Because the DigRF v4 interface is a high-speed, multi-lane, bidirectional link—with edge rates operating at sub-nanosecond intervals—the signal's rise time, pulse width, timing, jitter, and noise must all be measured and controlled. Specialized tools for testing high-speed serial interfaces include high-performance oscilloscopes and bit-error-ratio testers (BERTs). To minimize possible signal disruption, specialized probing solutions are also needed for testing high-speed serial links. The combination of a high-performance oscilloscope and an advanced probing system is the primary tool used to test the transmit lines of the DigRF interface. A multi-lane BERT is the primary tool used to test the receive lines of the DigRF interface.

6.3.4.2 Digital Protocol Layer

Testing in this layer involves seven distinct activities:

- Active and passive testing of the DigRF link
- "Stateful" exercising of the DigRF link
- Exercise and analysis of bus-mode transitions
- Testing of 8b/10b encoding mechanisms
- Verification of RFIC responses
- Testing of RFIC responses to DigRF errors
- Checking of the RFIC initialization sequence.

Test considerations for each of these activities are discussed next.

Active and Passive Testing of the DigRF Link

There are two ways to test a DigRF v4 link: active and passive. During the turn-on of an IC, the test environment must emulate a peer device communicating with the device under test (DUT). This environment is called an active tester because it is an active "citizen" of the link.

When an RFIC is integrated with a baseband IC, it is important to understand the behavior of the link—though with minimal intrusion on the signal and link—in order to understand the root causes of any interoperability issues. In this situation, the test equipment is called a passive tester because it does not participate in the operation of the link.

To minimize the possibility of the instrument disrupting the DigRF signals, it is important to minimize the stub effect with tip resistors and also ensure that the capacitive loading from the probes is extremely low. Active probing systems (not to be confused with active testing) are the most efficient way to ensure reliable measurements and minimize signal degradation.

Because the root cause of link problems may be either a physical or protocol defect, the use of a common probing system with oscilloscopes and DigRF analyzers can minimize the chances of misleading results. Figure 6.3-4 shows a common set of tip resistors that can be connected to both types of instruments. The active probe adds just 150 fF of capacitive loading.

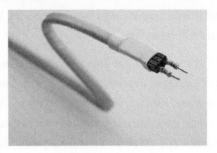

DigRF v4 analyzer probe

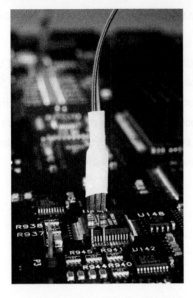

Plug-on socket connection for E2678A differential probe head

Figure 6.3-4. This set of tip resistors helps to minimize the stub effect and can be connected to either an oscilloscope or a DigRF analyzer.

Stateful Exercising of the DigRF Link

To perform the DigRF link layer test of an RFIC requires an active test system that mimics a peer device connected to the DigRF port of the DUT. The nature of active test equipment can be divided into two categories: "stateless" and "stateful" test devices.

The stateless test environment generates a stimulus with limited or no knowledge of the protocol state machines of the DUT.

The stateful test environment (or "exerciser") incorporates the DigRF protocol state machines and acts much more like a real device. A typical example in which a retry sequence is tested illustrates the benefits of an exerciser. Most digital protocol stacks include a packet resend mechanism: A receiver can require the sender to retransmit a packet if the packet was not received properly the first time it was sent. A stateful test platform will recognize the request to resend the packet and act according to the resend sequence definition. A stateless device, however, will simply ignore the request.

A stateful test environment allows testing of the following bus modes:
- Transitions from sleep mode to active mode
- Retry sequences
- Flow-control mechanisms that require the sender to slow down or speed up the traffic
- Emulation of the dynamic physical characteristics of a bus that may change in response to protocol events (e.g., termination or voltage level).

Exercise and Analysis of Bus-Mode Transitions

To optimize both power consumption and performance, the DigRF v4 bus has been designed to operate in multiple high speed and low power modes. When no data is being sent, the bus shifts into a sleep mode that requires very little power. When data must be transferred, the bus can quickly wake up and start transferring data.

The DigRF test environment should support power management features and the associated bus transitions. From a stimulus point of view, the test platform must support these modes deterministically so that it can check DUT mode transitions and verify specification-compliant execution.

From an analysis point of view, it is necessary to perform two types of measurements with one analysis module: track the transitions and then capture the data between the transitions, especially just after the bus has woken up. In this application, instrumentation lock time is critical because the test needs to capture data from the embedded clock. To reliably measure the DUT behavior and capture the initial data transmission while the bus is waking up, the tester's lock time must be faster than that of the DUT. The Agilent RDX tester includes a multipath clock recovery mechanism to capture data during very fast bus mode transitions.

Testing of 8b/10b Encoding Mechanisms

The 8b/10b coding converts a byte-wide data stream of random ones and zeros into a DC-balanced stream of ones and zeros. The code also provides sufficient signal transitions to enable reliable clock recovery. The average number of ones and zeros in the serial stream must be maintained at nearly-equal levels.

It is important to ensure that the DUT properly encodes and decodes the data. It is also essential to properly identify disparity errors to determine how the DUT detects and recovers from such exceptions. Disparity errors tend to occur infrequently and the analyzer triggering mechanisms must include real-time detection of such errors.

Verification of RFIC Responses

At this point in the development process, it is important to confirm that the RFIC is responding properly to commands. To validate RFIC responses—and to activate loopback mode—it is necessary to send a customized command sequence to the RFIC. The exerciser graphical user interface (GUI) shown in Figure 6.3-5 illustrates commands that can be used to build one or more custom frames to enable verification of the RFIC response.

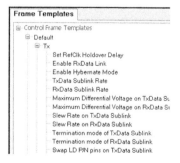

Figure 6.3-5 Control commands can be generated from a template list to manage the state of the RFIC.

Testing of RFIC Responses to DigRF Errors

It is useful to create errors in order to validate (1) how well the DUT detects each error and recovers from it and (2) whether the error recovery mechanism is compliant with the specification. Testing with known errors also increases the test coverage of the protocol state machines by analyzing exceptional transitions between each state. There are multiple error categories, including low level errors such as disparity or symbol errors, and higher level errors such as cyclic redundancy check (CRC) errors in a packet. Figure 6.3-6 shows how errors can be selected from a template list and added into the main stimulus file.

Figure 6.3-6. One or more frames with errors can be created and sent to the DUT to analyze how well it recovers from them.

Checking of the RFIC Initialization Sequence

Before each operation, the RFIC must be initialized with a sequence of control commands that set up its internal registers and configure its mode of operation. This sequence is executed only once, after which the desired operation is initiated. If the initialization sequence is short—for example, to put the DigRF port in loopback mode—the custom frame GUI can be used to configure the initialization sequence. In some cases this sequence includes a large number of configuration parameters that cannot be created manually. Figure 6.3-7 shows how the stimulus software can get a large command sequence from a file and execute it once before the test.

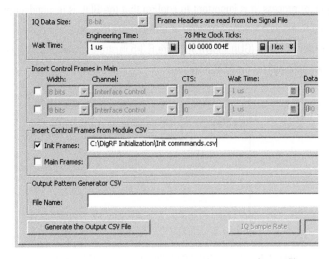

Fig 6.3-7. Inserting a complex command sequence from a file

6.3.4.3 Transitioning From Digital to RF Domains

Once the DigRF link itself appears to be working properly, testing of the RF transmit path can be started. Data is sent to the RFIC by embedding digitized IQ data into DigRF frames. One of the advantages of the Agilent RDX tester is its ability to insert IQ data generated by standard tools into the DigRF data stream.

Conversely, to test the RF receive path, the RDX tester has the ability to separate the command and data frames so the data can be routed to the 89600 VSA software. This allows detailed analysis and characterization of the received IQ data without the need for a connected baseband IC or special IQ port that bypasses the DigRF interface (though a direct IQ port can also be very useful for debug). Figure 6.3-1 showed simultaneous displays of the 89600 VSA software's RF physical analysis along with packet analysis showing digital protocol information. Other sections in this chapter discuss the actual RF physical layer tests appropriate for LTE designs.

6.3.4.4 RF Physical Layer

The RF physical layer tests are performed through the antenna interface and either the DigRF interface or a special IQ port on the baseband side of the RFIC. To perform tests via the DigRF interface, a test instrument must be able to insert IQ data into the DigRF bitstream (for transmitter test) or extract IQ information from the bitstream (for receiver test). This is a key feature of the Agilent RDX tester.

A variety of tools can be used to create and generate the IQ waveforms. For example, design packages such as Agilent design software make it possible to generate waveforms based on design models. Another common tool is Agilent's Signal Studio software, which provides waveform creation capabilities for a variety of modulation formats including LTE. MATLAB or other analysis tools (or programming languages) can also be used to create custom waveforms.

Tools such as Signal Studio for 3GPP LTE can create the standards-based LTE signals needed to verify RF uplink and downlink performance. These signals can include transport channel coding for receiver testing using block error ratio (BLER) measurements.

TThe RDX tester for DigRF includes a signal inserter tool to translate IQ data into the appropriate DigRF-compliant bitstream. It also allows insertion of control frames, time-accurate strobe messages, and status messages. The end result is a stimulus file that contains four items:

- The RFIC initialization sequence (control traffic)
- Data frames containing digitized IQ information
- On-the-fly control frames
- Time-accurate strobe (TAS) messages.

Debug and qualification of the RF subsystems of the RFIC involves extensive testing and verification of the transmitter, receiver and control paths. Amplifier linearity, control algorithms, modulation quality, and a host of other items must also be checked. These issues are discussed in the later sections of this chapter.

6.3.4.5 RF Protocol Layer

Testing this layer generally begins later in the process of designing and integrating the chipset and the overall system. As a result, it has relatively little bearing on the DigRF subsystem and so is not covered here.

6.3.5 Conclusion

The multi-gigabit DigRF v4 standard is rapidly emerging as the next-generation serial interface between mobile baseband and RFICs because it removes the inter-chip communication bottleneck. However, it also creates measurement challenges that exist in—and span—the physical and protocol layers of the digital and RF domains. For more information on DigRF test, see www.agilent.com/find/digrf.

6.4 Transmitter Design and Measurement Challenges

From the perspective of the RF engineer, LTE promises a dauntingly wide range of design and measurement challenges, arising from a number of factors:

- The requirement to handle six channel bandwidths from 1.4 to 20 MHz
- The use of different transmission schemes for the downlink (OFDMA) and uplink (SC-FDMA)
- Flexible transmission schemes in which the physical channel configuration has a large impact on RF performance—much more so than in CDMA systems
- Specifications that include both FDD and TDD transmission modes. While the examples in this section focus on FDD, the general principles described apply equally to TDD, with the added complexities of "time-sliced" signal generation and analysis.
- Challenging measurement configurations resulting from the spectral, power, and time variations due to traffic type and loading
- Further challenges resulting from the need to support multi-antenna techniques such as TX diversity, spatial multiplexing (MIMO), and beamsteering
- The need for making complex tradeoffs between in-channel, out-of-channel, and out-of-band performance.

In addition,

- Release 9 introduced multi-standard radio (MSR) base station transmitter requirements, which are covered in Section 6.4.7.
- Release 10 introduced "LTE-Advanced" requirements including carrier aggregation (CA) for both downlink and uplink, as well as uplink enhancements for partially allocated PUSCH (multi-clustered SC-FDMA) and simultaneous transmission of control and data, which is discussed briefly in the Section 6.4.8.

As with the development of other modern communication standards, the design task involves troubleshooting, optimization, and design verification with an eye toward conformance and interoperability testing. Since the number of possible measurements and operating permutations is virtually infinite, this section will focus on representative examples of impairments and the associated measurements, along with a discussion of the essential aspects of measurement setup and interpretation.

6.4.1 General Design and Verification Challenges for LTE Transmitters

This section discusses general challenges of transmitter design and some basic verification techniques, starting with basic characteristics and then moving on to LTE-specific aspects.

6.4.1.1 Output Power and Power Control

Accurate average power measurement of time-invariant signals is not a major challenge for LTE. Accurate broadband power measurements can be made using power meters, signal and spectrum analyzers, or vector signal analyzers. However, due to the nature of the downlink and uplink signal characteristics, the more typical case for LTE involves output power measurements that are much more specific. These involve measurement all the way down to the resource element (RE) level, which is one OFDMA or SC-FDMA symbol lasting 66.7 μs on one subcarrier. For such measurements a power meter is of no value, but a spectrum analyzer, signal analyzer, or vector signal analyzer is essential. In particular, power measurements associated with specific portions of the signal often require the digital demodulation capabilities of vector signal analyzers, described further in Section 6.4.6.

6.4.1.2 Out-of-Channel and Out-of-Band Emissions

Out-of-band emissions are regulated to ensure compatibility between different radio systems. The primary requirement is for control of spurious emissions from very low (9 kHz) frequencies to very high (12.75 GHz) frequencies. LTE in this respect is no different than any other radio system and spurious emissions will not be discussed further. LTE gets more interesting at the band edges, however, where the signal has to meet the out-of-channel requirements as well as the out-of-band requirements, which are often tighter. With LTE supporting channel bandwidths up to 20 MHz, and with many bands too narrow to support more than a few channels, a large proportion of the LTE channels are also at the edge of the band.

Controlling transmitter performance at the edge of the band requires careful filter design to trade off the required out-of-band attenuation without affecting the in-channel performance of the channels near the band edge. This tradeoff must also consider costs (in terms of financial cost, power or power efficiency, physical space, etc.), which must be balanced with optimization of the in-channel and out-of-band performance and the location in the transmitter block diagram where this tradeoff is achieved. Requirements for out-of-channel emissions are covered by adjacent channel leakage ratio (ACLR) and spectrum emission mask (SEM) measurements, as was the case for UMTS. These measurements are generally made with spectrum or signal analyzers using built-in routines for ACLR and SEM. The measurements can be done using either swept analysis in a signal or spectrum analyzer or using FFT analysis in a vector signal analyzer. The swept approach offers higher dynamic range and faster measurements.

6.4.1.3 Power Efficiency

Power efficiency is a critical design factor for both eNB and UE transmitters. A battery design must meet power consumption targets while ensuring that the transmitter meets the output power, modulation quality, and emission requirements. There are no formal requirements for power efficiency although this may change in the future with increased environmental awareness. Instead, power efficiency remains an ever-present design challenge to be met through design choices and optimization. The subject of battery drain testing is covered in Section 6.13.

6.4.1.4 Strategies for Handling High Peak Power

As discussed in Section 2.2, OFDMA signals can have a high peak to average power ratio (PAPR) and eNB power amplifiers must have a high degree of linearity to avoid producing out-of-channel distortion products. Power amplifiers with high linearity for the eNB are expensive and modest in their power efficiency. Two complementary methods exist to counteract this challenge: Crest Factor Reduction (CFR), which attempts to limit the signal peaks, and predistortion, which attempts to match the signal to the non-linear characteristics of the amplifier. Both methods are DSP-intensive, with predistortion being the more advanced method with the best performance but also the more difficult method to implement. CFR is discussed further in Section 6.2.9.

For the UE it is still necessary to control PAPR, but the use of SC-FDMA in the uplink rather than OFDMA means that the PAPR of the signal is no worse than that of the underlying modulation depth for QPSK or 16QAM. These are, respectively, 4 dB and 2 dB better at 0.01% probability than the Gaussian peaks typical of OFDMA. The original intent to define 64QAM performance requirements for the uplink is no longer a priority. However, the introduction in Release 10 of carrier aggregation, clustered SC-FDMA, and simultaneous PUSCH and PUCCH increases the need to manage uplink PAPR.

Crest Factor Reduction

CFR was first widely used with CDMA signals and is also an important technique for LTE, although the specifics of the implementation will be somewhat different. CFR is distinct from predistortion in that it attempts to limit the peaks in the signal before it reaches the amplifier rather than shaping the input signal to compensate for amplifier nonlinearity. As such, CFR is a general technique that can be applied to any amplifier design. CFR improves headroom at the cost of degraded in-channel performance. OFDM signals without CFR have RF power characteristics similar to that of additive white Gaussian noise (AWGN), with peak power excursions more than 10 dB above the average power. It is impractical to design and operate power amplifiers with this level of headroom. Careful use of CFR can substantially reduce peak power requirements while maintaining acceptable signal quality.

All CFR techniques involve balancing PAPR improvements to alleviate out-of-band distortion vs. the detrimental impact on in-channel distortion and the cost and power associated with the increased baseband processing overhead. Perhaps the simplest type of CFR is clipping or signal compression, in which intermittent RF power peaks are either removed (by means of a simple clipping algorithm) or scaled (in the case of compression). Because of the dramatic effects of clipping and compression on signal quality and the availability of increased processing power, more advanced techniques are also used.

An example of a more sophisticated CFR technique is tone reservation, which is possible only on the downlink since the eNB can control the resource allocation for every RB. A dummy transmission is allocated on the reserved tones, which cancels out the peaks generated by the composite signal. Although tone reservation promises to accomplish CFR without degrading signal quality, it has several disadvantages that need to be considered, including a loss of spectral efficiency (due to the loss of some subcarriers), loss of useful power, and the increased computational overhead.

The effectiveness of CFR can be evaluated using the complementary cumulative distribution function (CCDF) applied to a series of instantaneous power measurements. This measurement technique is described in Section 6.4.5.2 with examples of LTE uplink SC-FDMA signals and in Section 6.4.8.1 with examples of LTE-Advanced enhanced-uplink signals.

The CCDF can be measured after the CFR operation and compared with either the actual or predicted CCDF of the signal without CFR. The CCDF measurement will yield a family of direct measurements of the reduction in PAPR, along with an associated probability of a specific PAPR. Measuring the power CCDF of a signal can be misleading if the signal is not stable; for instance, if some or all of the signal is not continuously transmitted. In such cases, CCDF measurements should be made time-specific.

The positive consequence of applying CFR is to increase the headroom available within the amplifier. This headroom can be used to drive the amplifier harder or to improve the out-of-channel performance as measured by ACLR or SEM, or some combination of the two. To choose the correct operating point for CFR, it is essential also to measure the in-channel quality (error vector magnitude or EVM, etc.), since in all cases this quality will be degraded.

Predistortion

Predistortion enables the use of amplifier technologies that are both more power-efficient and less costly, although predistortion also adds design and operational complexity. Predistortion is a more advanced power management technique than CFR because it requires tight coupling to a specific amplifier design. Predistortion maintains the in-channel performance while operating in the non-linear region of the amplifier. This minimizes signal compression so that out-of-channel performance does not degrade at the higher operating level. A number of analog and digital predistortion techniques are available, from analog predistortion to feed-forward techniques and full adaptive digital predistortion. These techniques operate with varying degrees of effectiveness over varying bandwidths, and the eNB design may use a combination of techniques including CFR to optimize overall cost and performance. DPD is discussed further in Section 6.2.10.

6.4.1.5 Phase Noise

Optimizing designs for sufficient phase noise performance is particularly challenging in OFDM systems for two reasons. First, excessive phase noise degrades the orthogonality of the closely spaced subcarriers causing frequency domain inter-carrier interference (ICI) leading to impaired demodulation performance. This is discussed in Section 2.2 and in "Vector Modulation Analysis and Troubleshooting for OFDM Systems" [11]. Second, phase noise reduction can be expensive in terms of system cost and power efficiency. These costs are, relatively speaking, a larger issue for the UE as opposed to the eNB.

6.4.1.6 In-Channel Signal Quality

Many factors influence in-channel signal quality, from baseband signal processing to modulation, filtering, up-conversion, CFR and predistortion techniques, and the power amplification process. For LTE one of the most significant of these factors is filtering. Section 2.1.6.1.1 introduced the fact that the LTE specifications do not define a transmit filter for either the UE or the eNB. In UMTS, an RRC filter is specified for both the transmitted signal and for its measurement, giving the designer an exact target at which to aim. For the LTE designer there is no standardized filter. This fact can be viewed both as an opportunity to optimize the in-channel and out-of-channel performance and as a challenge because there is no longer a fixed target.

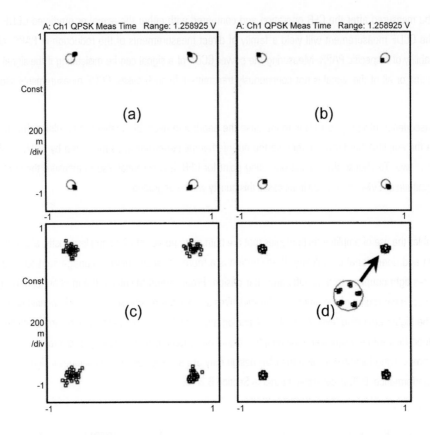

Figure 6.4-1. Constellation diagram examples of single carrier impairments in the IQ plane

One of the most useful indicators of signal content and some types of in-channel distortion is the constellation diagram. From this it is possible to tell if the demodulator is locked to the signal, what signals and modulation types are present, and whether specific types of distortion are indicated. There are seven primary types of distortion and most have a distinct constellation pattern when viewed with single carrier modulation. The distortions are white noise, phase noise, IQ gain imbalance, quadrature error, phase error, AM to AM (linear compression), and AM to PM (non-linear compression). Figure 6.4-1 shows examples of four impairments. They are (a) 1 dB IQ gain imbalance, (b) 5 degree quadrature error, (c) in-channel interference from a frequency spur 36 dB below the carrier, and (d) general state spreading that would be characteristic of problems such as white noise, improper filtering, small amounts of compression or a symbol rate error.

A signal generator can be used to generate a single carrier digitally modulated signal at the appropriate bandwidth to test for component or subsystem IQ impairments. A skilled engineer can look at an IQ constellation and quickly determine which distortion mechanisms are present. If more than one distortion is present, it becomes harder to identify the individual components. At the other extreme, a numeric EVM measurement will indicate only the amount of distortion, giving no indication of the mechanism that created it. Thus while EVM is necessary to evaluate signal quality for conformance testing, the IQ constellation can be an essential tool for troubleshooting the source of distortion, particularly when a simple multi-level stimulus such as single carrier 16QAM is used.

In a multicarrier system such as OFDMA, the uncorrelated relative phase relationships of the subcarriers turn many of these distortions into state spreading, masking the underlying error. State spreading is a term used to describe the spreading of energy from the ideal constellation point. Some impairments do produce a distinctive constellation with multicarrier signals, but most just result in state spreading. Further reading on this subject can be found in [11] and "Effects of Physical Layer Impairments on OFDM Systems" [12].

Impairments such as AM to AM and AM to PM are sometimes best measured with dedicated distortion analysis tools such as Agilent's Distortion Suite software. Such general purpose tools offer the advantage of not requiring LTE-specific stimulus and demodulation. These methods can be used on components or subsystems and provide an estimate of EVM contribution.

6.4.2 A Systematic Approach for Verifying LTE Transmitter Performance

When complex digitally modulated signals are verified and optimized, it is tempting to go directly to advanced digital demodulation measurements such as those described in Section 6.4.6 using vector signal analysis. However, it is usually more productive and sometimes necessary to follow a verification sequence that begins with basic spectrum measurements and continues with vector measurements (combined frequency and time) before switching to digital demodulation and modulation analysis. This sequence is shown in Figure 6.4-2 and will be used as a way of partitioning the measurements discussed in the remainder of Section 6.4.

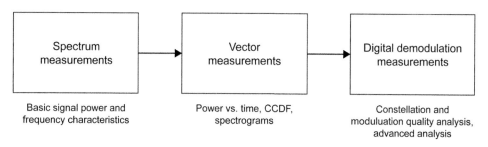

Figure 6.4-2 Sequence for verification of LTE transmitter performance

This sequence is useful because it improves the chance that underlying signal problems will be identified at the earliest stages of design before possibly being masked when complex signals are measured. In particular, the frequency and time domain measurements at the beginning of the sequence enable the verification of many signal parameters (including some associated with the digital modulation itself) without the need to perform digital demodulation. This is an advantage in early development when fully coded signals may not be available or are in some way questionable. The use of simpler measurements also means that development progress can be made using less sophisticated test equipment than is needed for the more advanced measurements.

6.4.3 Measuring Signals at Different Locations in the Transmitter

Before discussing the process of verifying transmitter performance it is necessary to address the growing use of digital interfaces, which are starting to replace traditional analog interfaces between modules within the UE and eNB. This trend leads to the need for new probing techniques and mixed domain signal analysis. Many transmitter measurements are a straightforward matter of connecting the transmitter RF output directly to an RF signal analyzer input and measuring signal characteristics and content. Some measurements, however, will require connecting, probing, and measuring at early or intermediate points in the transmitter signal chain. Figure 6.4-3 shows a typical UE block diagram and the possible ways in which signals can be injected or probed at different points.

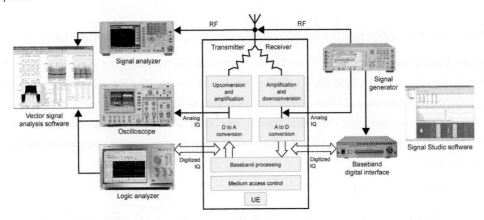

Figure 6.4-3. Stimulus and analysis of different points in the UE block diagram

6.4.3.1 Measuring the RF Output at the Antenna Connector

The RF output or antenna connector is the traditional point for signal analysis. For the critical maximum power requirements defined at the antenna connector, some modern spectrum and signal analyzers now have an accuracy approaching that of power meters [13], although achieving high accuracy depends on careful selection and use of cables, connectors, and adapters. Testing the high output power of an eNB requires the use of external attenuators to protect the analyzer inputs. For more information on best practices for RF signal analysis, see Agilent application notes AN-150 [14] and AN-1303 [15].

Spectrum and signal analyzers have the added benefits over power meters of time selectivity and frequency selectivity, often found in combination. With appropriate software this time and frequency selectivity can be extended into digital demodulation, which allows individual OFDM subcarriers or groups of subcarriers to be measured separately, and individual symbols or groups of symbols to be measured separately as well. These selective measurements are invaluable for troubleshooting and system optimization.

RF analysis at the antenna connector is used to evaluate the complete transmission system, but RF analysis is also used to evaluate subsystems and components such as upconverters, filters, local oscillators, and power amplifiers, that require probing at different parts of the system. The increasing use of DSP techniques such as CFR and predistortion in the baseband stages of transmitters adds complexity to the interpretation of RF measurement results, whether they involve spectrum or time domain measurements or digital demodulation analysis. For evaluation of components or subsystems an incremental approach is often used in which the transmitter system is stimulated with a signal generator or its own baseband/IF, and successive measurements are made at later points in the signal chain to quantify the distortion, spurious emissions, and other impairments contributed by individual stages and elements of the transmitter.

6.4.3.2 Measuring IF Signals

Signal analysis at the IF is used to determine the quality of signals after the first stage of up-conversion or to determine the quality of an IF signal generated directly from baseband. IF analysis also provides a baseline for understanding the relative contributions of up-conversion and RF amplifier circuits. Quantifying the individual distortions that make up the overall transmitter signal quality allows the system designer to isolate problems and make necessary design tradeoffs for power, cost, and board real estate.

In some transmitters the IF signal is generated directly by an all-digital baseband section feeding a single digital-to-analog converter (DAC), whose output is the analog IF signal. While this analog IF signal can be measured directly, it may be useful to also understand the quality and content of the input to the DAC. Indeed, it may be beneficial to compare the signal quality of the DAC output to its digitized input. For this type of mixed signal analysis, a product such as the Agilent 89600 VSA software can be used with logic analyzers to capture digital rather than RF input data prior to analysis using the same algorithms and user interface as traditional RF measurements. Mixed signal analysis is discussed in Sections 6.2.2 and 6.3.3.

6.4.3.3 Measuring Analog and Digital Baseband Signals

The process of measuring analog and digital baseband IQ (BBIQ) signals is similar to that for the analog and digital IF signals described above. Signal analysis begins with data acquisition and the method depends on the system design.

For systems using analog BBIQ, the signals may be in dual single-ended or dual differential form, and accurate analysis of these signals often requires probing along with probe calibration and compensation. Modern spectrum and signal analyzers such as the Agilent X-Series analyzers offer the option of analog BBIQ inputs in single-ended and differential form, along with software to assist in probe identification, calibration, and compensation. A discussion of these techniques is beyond the scope of this book, but they are discussed in "Analysis of Baseband IQ Signals" [16].

Modern designs increasingly use digital rather than analog baseband approaches and, as with the digital IF described above, analysis starts by acquiring the digital signal using an appropriate tool such as a logic analyzer. The baseband signal can then be analyzed with VSA software as if the signal had been captured as an analog or RF waveform. The ability to capture a signal in analog or digital form at any point in the block diagram and then analyze it with the same software provides the ability to precisely analyze the quality of the signal as it progresses through the transmit chain, making design optimization and troubleshooting more productive and less complicated.

Such measurement requires generation of a digital baseband or IF signal to stimulate later sections of the transmitter for signal analysis. Digital signal generation solutions are available such as the Agilent MXG Series signal generators, which offer a digital output connector driven by the same signal generation software as is used for analog signal generation.

The digital connector, known as a digital signal interface module (DSIM), provides flexible data formats, clocking, and physical interfaces for IQ, serial, and parallel data streams including digital IF signals. To match circuit requirements, DSIMs may also perform resampling and provide adjustable clock phase and skew.

6.4.4 Spectrum Measurements

6.4.4.1 Channel Power, Amplitude Flatness, Center Frequency, and Occupied Bandwidth

The following are basic spectrum measurements that can be made on an LTE signal:

- Channel power
- Amplitude flatness
- Center frequency
- Occupied bandwidth (OBW).

The center frequency and amplitude flatness measurements made using basic spectrum analysis are indicative results since the formal definition of these measurements requires digital demodulation of the signal.

As with most digitally modulated signals, LTE signals are significantly noise-like in their power and spectral characteristics. Traditional spectrum analyzers are not designed for noise-like and burst-like signals, although with care spectrum analyzers can be used to make useful measurements. Fortunately many modern spectrum and signal analyzers as well as vector signal analyzers are designed to accurately measure signals such as these, implementing true RMS detection and appropriate averaging rather than the traditional video bandwidth filtering. For more information see "Spectrum Analyzer Detectors and Averaging for Wireless Measurements" [17].

When the signal frequency span of interest stays within the available sampling bandwidth of a digitizer, the best way to measure the power of an LTE signal is, in most cases, to use the power vs. time calculation available in demodulation applications. This feature integrates power over a specific period of time. The alternative is to use a non-swept FFT-based spectrum analysis method with power integration over a particular frequency span. The time synchronization and time selectivity of demodulation (selecting which part of the frame to measure) are important for ensuring that the desired portion of the signal is measured. In addition, demodulation provides a way to separately measure the relative power of the LTE channels and signals as discussed in Section 6.4.6.1. If time synchronization is not required or if timing is available from a trigger or a power on/off event, power can be accurately measured without demodulation by averaging spectrum or RF envelope measurements and using band power markers to select the signal of interest. Signal power may also be measured using traditional swept-tuned spectrum analysis and dedicated LTE measurement applications as part of the ACLR and SEM measurements. An important requirement for all of these measurement approaches is the use of consistent or equivalent signals (signal content) because LTE IF and RF power levels vary with signal content and other parameters. The non-swept FFT and swept-tuned methods will now be described.

Non-swept FFT-based Spectrum Analysis Method

For the FFT techniques used in vector signal analyzers there is an inverse relationship between settings for frequency span and length of the maximum time record. In measurements such as in-channel carrier power of a 5 MHz signal over a 10 ms frame it is necessary to have both a 5 MHz frequency span and a relatively long (10 ms) time record. This results in a very large FFT block size of 50,000 points or more. The frame of a 20 MHz signal would require a 200,000 point FFT time record.

Figure 6.4-4 shows an example measurement of a fully allocated 5 MHz downlink signal made with a large FFT block size to provide a high resolution spectrum measurement with averaging of 100 traces. The trace provides an accurate power measurement and any significant amplitude flatness errors can be seen. It is also possible to approximate the center frequency and the OBW. Band power markers are used to integrate and measure the power over the desired bandwidth, which is typically the same as the occupied bandwidth of the signal. Note, however, that the spectrum measurement in this example is not time-selective and indicates only the power of the averaged spectra. The center frequency can be calculated from the centroid of the occupied bandwidth markers, but this calculation may be of questionable accuracy because the subcarrier allocation is not always symmetrical about the center frequency.

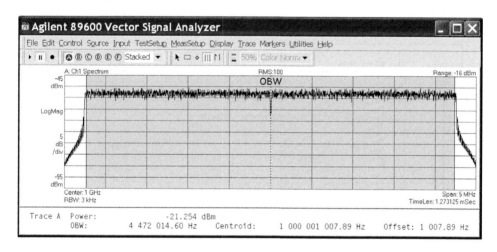

Figure 6.4-4. Spectrum of 5 MHz downlink showing power, OBW, and center frequency, measured with non-swept FFT spectrum analyzer

A more accurate measurement of center frequency that is not affected by asymmetry in the subcarrier allocation can be made for the downlink by identifying the spectral minimum associated with the central subcarrier, which is not transmitted. This subcarrier is used to accommodate any LO leakage caused by impairments in the transmitter, such as IQ offset, so that this distortion does not degrade the wanted part of the signal. The accuracy using this method is significantly better than 1 ppm. The required frequency accuracy for the downlink is ±0.05 ppm and is measured using digital demodulation as part of the EVM measurement process. For the uplink, the central subcarrier is not reserved and so cannot be used to estimate the center frequency.

Swept-Tuned Spectrum Analysis Method

In contrast to the FFT-based method, the swept-tuned analyzer does not have a limitation of its sweep span until the maximum frequency of the analyzer is reached. Thus it is capable of covering a wider span than the digitizer supports. For example, an ACLR measurement for a 20 MHz downlink signal requires a 100 MHz span to cover the adjacent and alternative 20 MHz channels on either side of the carrier. This can be achieved in one measurement by swept-tuned analysis but would require several FFT measurements if the digitizer has a bandwidth of less than 100 MHz.

Figure 6.4-5 shows a swept-tuned carrier power measurement of a 5 MHz TDD downlink signal. The measurement is made using a swept spectrum method of frequency vs. time with a 100 kHz resolution bandwidth (RBW). The upper trace shows the non-constant carrier power vs. time for TDD uplink/downlink configuration 3 (see Table 3.2-1) , and the lower trace shows the power versus frequency using a time-gated sweep so that the LO sweeps through the carrier only during periods when the carrier is switched on.

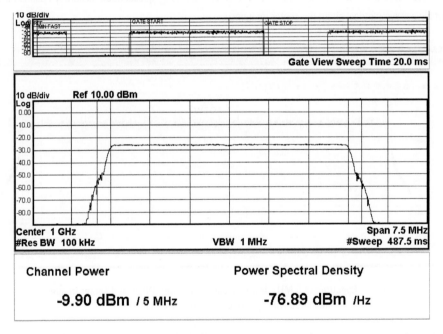

Figure 6.4-5. Spectrum of a TDD 5 MHz downlink showing carrier power measured using a swept-tuned spectrum analyzer with a 6.5 ms time gate, 100 kHz RBW, and 487.5 ms sweep time

Note that although OFDM signals may look like random noise, the variation in power that appears in the signal in Figure 6.4-4 during the power-on burst is a result of the specific locations of reference signals (RS) and control channels, as well as the user resource block allocations. Therefore, when a signal under test is a repetitive signal—e.g., one frame of an E-UTRA test model (see Section 7.2.10.1)—increased averaging may not reduce the variance of the measured power. The periodic triggering of such measurements at a particular point in each time-repetitive frame results in the same trace through the frequency vs. time plane. Small movements of the gate sweep start and stop time can cause changes in the result on the order of 1 dB.

To make the measurement more stable although slightly slower, the sweep time can be set such that the RBW sweeps through the entire frequency vs. time domain of the power burst in each frame. Assuming that the number of points in the trace is large enough (i.e., points ≥ span/RBW), the sweep time should be more than the gate length span/RBW. This is illustrated in Figure 6.4-6. The sweep time setting used in Figure 6.4-6 is an example of the recommended setting, which is 6.5 ms gate length x 7.5 MHz span/100 kHz RBW = 487.5 ms. With this slower sweep time the power result is stable enough so that averaging is not required and the averaging count can be set to 1.

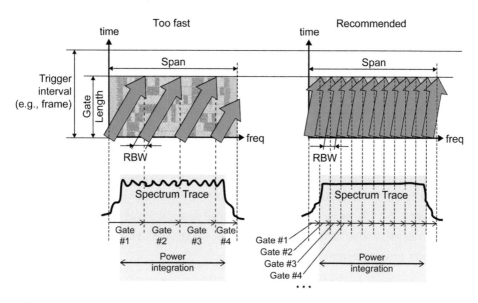

Figure 6.4-6. Swept spectrum analysis showing (left) the consequence of a sweep time that is too fast, leading to unstable results, and (right) the recommended sweep time, which is stable

6.4.4.2 Filtering and Spectral Shape

Some fine details of the signal spectrum can provide insight into system optimization, although they are not essential to conformance testing. For example, different choices of baseband filter parameters may optimize in-band flatness vs. computational efficiency, or load vs. signal-to-noise ratio (SNR). An example is shown in Figure 6.4-7.

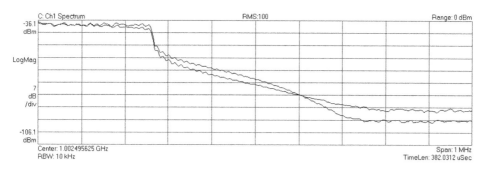

Figure 6.4-7. Comparison of different baseband filter choices on OFDMA signal roll off

A high resolution averaged spectrum measurement can show the effect of different baseband filter choices on the signal's transition band and stop band. The formal implications of such choices have to be evaluated using the conformance measurements for in-channel and out-of-channel performance.

6.4.5 Vector (Frequency and Time) Measurements

6.4.5.1 Power vs. Time Measurements

Since RF signal power varies due to factors including signal content, power control, and discontinuous transmission (DTX) and since power affects signal quality, design, and operational choices (such as UE power control), accurate measurements of power versus time (PVT) and statistical power behavior are essential. Power measurements can be made without demodulation and are generally scaled in time (seconds). With the benefit of the signal timing references that demodulation provides, these demodulation measurements allow power to be analyzed in terms of symbols, slots, or frames and are especially useful in correlating power to signal traffic and control actions. Examples are discussed in Section 6.4.6.

Basic PVT measurements can be made through direct analysis of the RF envelope using a spectrum analyzer in a zero-span mode, although the accuracy of this approach is limited due to the fact that the RBW filters are not rectangular channel filters and may not have the bandwidth (up to 20 MHz) to measure wider LTE signals. PVT measurements and power statistics measurements (see Section 6.4.5.2) are more typically made with vector signal analyzers, which can implement wide, flat channel filters for accurate measurements. The 89600 VSA software also provides band power markers, which are useful in making some time-specific power measurements.

PVT measurements made without demodulation can be performed simply and quickly, without the setup and processing needed for demodulation. Such measurements can be performed even on signals that have not been successfully demodulated. The measurements provide confirmation of absolute power levels and some visual indications of signal structure but typically not frame timing. In making PVT measurements the 89600 VSA software digitizes the entire channel bandwidth (as its FFT span) and reports the power in this span as log magnitude vs. time. Band power markers are used to integrate the power over a particular region in time. Depending on the signal structure and power behavior it can be useful to measure over periods ranging from one or more frames to a single slot. A PVT measurement example of the half frame (5 ms) of a downlink signal is shown in Figure 6.4-8. Band power markers are used to indicate the average power over a 1 ms interval.

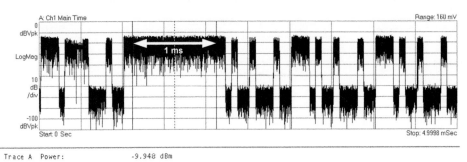

Figure 6.4-8. PVT measurement of one half frame of a downlink signal
with band power markers to indicate average power over 1 ms

When a long duration measurement is made, the average power of any portion can be easily measured using band power markers.

6.4.5.2 Power Statistics CCDF Measurements

CCDF measurements provide a broad quantitative measurement of signal power behavior and are useful in evaluating the operating points and efficiencies of transmitters. CCDF measurements are particularly important for evaluating OFDMA (and to a lesser extent SC-FDMA) systems in which high PAPR makes it impractical to amplify the signal in unmodified form due to the excessive headroom required.

An example of a CCDF power measurement is given in Figure 6.4-9. The X-axis of the CCDF measurement is log power (dB) referenced to the average power of the signal during the selected measurement interval. The Y-axis of the measurement is a logarithmic scale of probability, commonly from 0.001% to 100%. A point on the CCDF curve indicates the probability (as a percentage of individual power measurements) that the signal will reach a given peak level above the measured average power level.

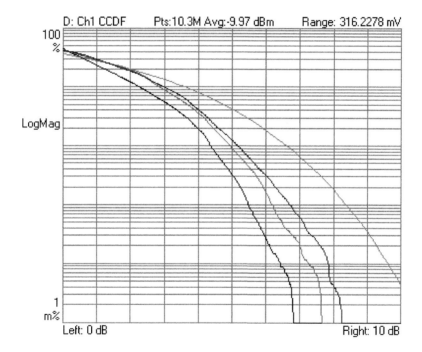

Figure 6.4-9 CCDF measurements of uplink signals: from left to right, QPSK, 16QAM, and 64QAM

So that CCDF measurements will not be misinterpreted, they must be made with knowledge of the signal under test. For example, removing power from the signal by turning it off for periods of time (discontinuous transmission) can dramatically increase the apparent level of the peaks since the average signal power is lower. However, from the perspective of the power amplifier, such a signal is not any harder to amplify since the absolute level of the peaks has not changed.

Provided care is taken not to measure CCDF during the periods when the signal is off, CCDF is a relatively straightforward measurement to interpret because it is automatically normalized for average amplitude, and modern analyzers can quickly compile the statistics for a large number of power measurements. This provides accurate average power and low variance for the statistics of the relatively rare peak power events.

The example shown in Figure 6.4-9 is from an uplink signal containing a three RB PUSCH in a 10 MHz system bandwidth with SC-FDMA precoding and three different modulation formats. The data is encoded using pseudo-random bit sequence PN15 since PN9 is not long enough for evaluating CCDF of higher order modulation formats. The three curves in the measurement were obtained using time gating to isolate each modulation format and represent, from left to right, QPSK, 16QAM, and 64QAM. Note that for all modulation formats this SC-FDMA signal shows significantly lower peaks than the AWGN reference curve that is furthest to the right. Downlink LTE signals prior to any CFR are generally very close to the AWGN curve and more challenging to amplify. Note that although uplink 64QAM is specified at baseband, RF requirements are not specified.

Section 6.4.1.4 discusses how CCDF measurements can be used along with in-channel and out-of-channel measurements to evaluate the effectiveness of CFR techniques. Section 6.4.8.1 shows some additional CCDF examples for Release 10 enhanced-uplink signals.

6.4.5.3 Spectrum vs. Time: The Spectrogram

One of the most useful general spectrum measurements for LTE signals is the spectrogram. The spectrogram provides a means of interpreting the signal structure at a glance. Specifically, the spectrogram can simultaneously present hundreds of contiguous or overlapping spectra revealing underlying detail of the frame structure.

Spectrograms are made up of a sequence of ordinary spectrum measurements, each of which is compressed to a height of 1 pixel row on the display with the amplitude values of the spectra encoded in color. This produces a display of spectrum vs. time containing hundreds or even thousands of spectrum measurements. The spectrogram enables an entire LTE uplink or downlink frame to be shown. All of the spectrum data is preserved and individual spectrum measurements can be selected from the spectrogram for more detailed analysis using a spectrogram marker.

The spectrogram allows easy visual recognition of the major signal characteristics in the downlink including channel types and symbol transitions based on subcarrier allocation. These measurements all occur without the need for digital demodulation. For uplink signals the frame structure is simpler, but the spectrogram remains a very useful tool to show the position in time and frequency of any dynamic RB allocation, and it can provide an indication of the images created by IQ errors. An uplink example is shown in Section 6.4.6.9.

To show the LTE frame structure the spectrogram should be made with the analyzer operating in real time mode—that is, the spectra for the measurement should be produced from time records in which no data from the input signal is missing from the spectrum calculations, and the time records for each spectrum overlap. A typical overlap figure for these measurements is 75 to 95%.

In order to make such heavily overlapped, real time measurements of signals that can be up to 20 MHz wide, it is necessary to use time capture, a feature of many vector signal analyzers. During time capture, the sampled data is streamed directly to memory, without gaps and without performing analysis that would interrupt gap-free recording. After capture, spectrum/ spectrogram measurements are made during playback of the recording, and the overlap percentage is varied to adjust both the effective "speed" of the playback and the time increment represented by each new spectrum in the display. Figure 6.4-10 shows a spectrogram of part of an LTE downlink frame in which the channel allocations are clearly visible. Also seen are the fine horizontal lines indicating the spectral spreading that occurs at each symbol transition where the collective phase and amplitude of the subcarriers change abruptly, resulting in a discontinuity in the transmitted signal.

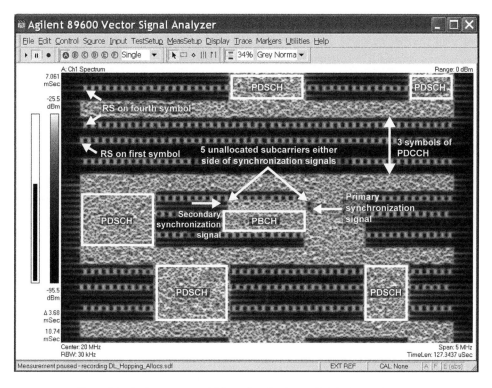

Figure 6.4-10. Spectrogram of 3.7 ms duration from a 5 MHz downlink signal

In this spectrogram the Y-axis is time, progressing from top to bottom over a time period of approximately 3.7 ms, which is just over 50 symbols. The X-axis is frequency, and signal power is represented in shades of gray, although on the actual instrument the power variations are represented in color. The most obvious patterns in the signal are the RS subcarriers, which are allocated to the first and fourth symbol of every seven symbol slot. The RS are allocated every sixth subcarrier, which creates a ladder effect across the entire channel bandwidth. The definition of the RS subcarrier allocation for this single antenna example (see also Figures 2.4-3 and 3.2-9) is that for the first symbol of each slot the RS starts from the first subcarrier, whereas on the fourth symbol of the slot the RS starts from the fourth subcarrier. A left-right alternating pattern is thus created that can be seen clearly in the spectrogram. The first RS allocations appear near the top of the spectrogram with no power at the lowest subcarrier, indicating that this symbol is the fourth symbol of the slot rather than the first.

The first five symbol periods of the spectrogram show allocations in the central part of the channel and a block at the upper end. These allocations are likely to be the physical downlink shared channel (PDSCH) carrying the user data. This allocation is followed by three symbol periods in which the entire channel bandwidth is allocated. The periodicity of this allocation, which occurs every 14 symbols (one subframe), suggests that it is the physical downlink control channel (PDCCH), as seen in Figure 3.2-9. The next period of activity is at the center of the channel, which shows the pattern created by the primary and secondary synchronization signals followed by the physical broadcast channel (PBCH). Recall from Figure 3.2-9 that the synchronization signals occupy only the central 62 subcarriers and that the five subcarriers on either side are unallocated. This small gap can clearly be seen in the spectrogram. The next symbol contains the PBCH, which occupies 72 subcarriers along with the RS across the rest of the channel. The following three symbols for the central 72 subcarriers contain the rest of the PBCH. Any other allocation in the spectrogram is most likely to be more PDSCH.

The time capture recordings used for spectrograms can also be used for signal demodulation. For troubleshooting purposes it can be useful to compare both spectrum and demodulation measurements of the same captured signal.

6.4.6 Analysis of Signals After Digital Demodulation

By performing digital demodulation according to the radio specifications using a vector signal analyzer it is possible to analyze the structure and quality of signals in the domain for which they were designed. This analysis provides a low level interoperability test. In addition to these basic tasks it is also possible to perform more advanced analysis. Here are some examples:

- Separately measure the characteristics of individual components of the signal right down to the resource element (RE) level of one symbol for one subcarrier.
- Understand how modulation quality varies over time to identify periodic or single shot effects.
- Measure subtle effects of different operating conditions (average signal power, battery voltage and impedance, operating temperature, number of transmitters driven, etc.).
- Isolate the source of signal impairments.
- Evaluate tradeoffs of modulation quality vs. design or operational parameters (CFR or predistortion techniques, amplifier operating points, component choices, etc.).
- Validate correct coding of physical signals and physical channels.

The procedure for basic demodulation will be discussed first before moving on to the more advanced demodulation techniques that reveal the full nature of the signal structure. Digital demodulation and subsequent signal analysis is generally carried out using vector signal analysis software such as the 89600 VSA software coupled with some form of signal capture, which can be from a variety of sources such as those described in Section 6.4.3. An alternative is to use a dedicated signal analyzer with built-in measurement applications such as the X-Series signal analyzers with LTE measurement application. The digital demodulation examples and associated measurements used in the remainder of this section are illustrated using the 89600 VSA software. The power of this tool is not fully conveyed here due to the limitations of two color printing. Color is used in the 89600 VSA software to great effect, highlighting aspects of the complex LTE signal structure for purposes of identifying and isolating specific signal and channel effects and impairments.

6.4.6.1 Procedure for Basic Digital Demodulation

Figure 6.4-11 shows a typical display of a digitally demodulated signal. The 89600 VSA software user has complete flexibility to choose (1) at what point in the demodulation process the signal will be analyzed and (2) the format and number of simultaneous displays. This particular example is a downlink LTE signal with a 5 MHz channel bandwidth. Each of the six displays shows an orthogonal view of the signal, and many more views are possible depending on the purpose of the analysis. Before detailed analysis can begin, however, the first priority is to configure the analyzer correctly to obtain demodulation lock to the signal.

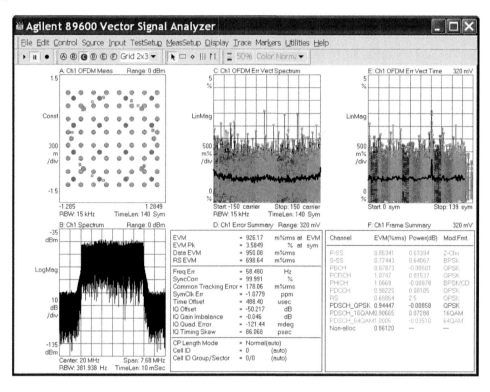

Figure 6.4-11. Example analysis of a digitally demodulated 5 MHz LTE downlink signal using Agilent 89600 VSA software

The spectrum and vector measurements described earlier in this section are relatively easy to make since they require few analyzer parameter settings: once the signal frequency, bandwidth, and input range (level) have been established, not much can go wrong. However, to successfully perform digital demodulation requires many more settings. Fortunately, most of these can be configured automatically using preset demodulation parameters.

For situations in which the signal frequency is not known precisely or its stability is suspect, the frequency lock range of the analyzer can be extended. This extended lock range increases the amount of data processed and reduces the measurement update rate with no reduction in accuracy. After demodulation lock has been achieved, a precise frequency error measurement as reported in the Figure 6.4-11 Trace D error summary will be available to set the analyzer center frequency more accurately.

After frequency and input range, the next important demodulation parameter that must be set is the uplink or downlink selection. This selection is critical because unlike some other systems, the LTE uplink and downlink synchronization reference signal structure and modulation formats are quite different. It would be nearly impossible to make sense of a downlink signal using uplink demodulation parameters and vice versa. Similarly the choice of FDD or TDD frame structure is important because again the differences in synchronization reference signal structure between the two are large.

The next important parameter is the channel bandwidth, and LTE has six to choose from: 1.4 MHz, 3 MHz, 5 MHz, 10 MHz, 15 MHz, and 20 MHz. For analysis of IQ signals obtained directly from supported front-end acquisition hardware, selecting the appropriate channel bandwidth from the demodulation properties will also preset an appropriate front-end hardware acquisition span and IQ sample rate. While span is typically thought of as a display parameter, the 89600 VSA software acquisition span is closely related to the signal acquisition bandwidth. If the bandwidth is too wide, unwanted noise and interference could impact measurement accuracy. Wider spans also required higher sample rates, which could negatively affect measurement speed, without providing any benefit. For analysis of signals coming from a simulation environment or a recording file, it is also necessary to specify the correct IQ sample rate. In this case the IQ sample rate will be used as part of the 89600 VSA software recording file header information, enabling the IQ samples to be interpreted correctly.

The final parameter to consider is the appropriate sync type, which determines the specific downlink or uplink reference signal component to be used for initial synchronization of the 89600 VSA software to the LTE signal being measured. For downlink analysis, valid choices include primary synchronization signal (P-SS) or RS. (The acronym P-SS is not formally defined in the 3GPP specifications but is used for convenience in the 89600 VSA software.) For uplink analysis, valid sync type choices include PUSCH demodulation reference signal (DMRS), PUCCH DMRS, sounding reference signal (SRS), or physical random access channel (PRACH). Selecting between different synchronizing methods allows fuller verification of the signal being tested. For example, coding errors in the downlink P-SS may not prevent measurements using the RS, but would create problems during system operation. Note that for uplink signals some additional, higher layer configuration parameters are required in order to achieve reliable 89600 VSA synchronization. These parameters can be configured under the demodulation properties profile allocation editor within the 89600 VSA software.

Depending on where in the up-conversion or down-conversion process the signal is being probed, the frequency spectrum may be reversed. There is a setting within the demodulator properties that takes this into account so that demodulation is still successful. The mirrored spectrum is corrected before demodulation by performing a complex conjugate on the time-domain representation of the signal. This setting can also compensate for swapped I and Q signals.

If these steps are followed, it should be possible to view the demodulated signal. Each of the traces in Figure 6.4-11 will now be described as an introduction to basic digital demodulation. This particular set of displays is not the default for the 89600 VSA software but has been chosen to represent a broad range of basic capabilities.

Description of Basic Digital Demodulation Traces

Figure 6.4-11 Trace B (bottom left) shows the classic spectrum analyzer view of signal power vs. frequency, calculated from a single FFT over the whole time interval being used for demodulation. This view, although it does not show digital demodulation, is often comforting during the early stages of the digital demodulation process because it confirms that the two most important

314

parameters of frequency and input range are consistent with the signal. From the trace it can be determined that the occupied bandwidth is around 4.5 MHz, which indicates a fully allocated 5 MHz downlink signal. The power spectral density for the 380 Hz resolution bandwidth is around −45 dBm. This power spectral density should not be confused with the aggregate signal power, which can be measured using band power markers across the entire 4.5 MHz occupied bandwidth.

The spectrum analyzer view establishes that the signal is generally at the correct frequency and power. The next important trace to consider is the error summary shown in Trace D (bottom middle). Important metrics to verify initially include SyncCorr, which indicates the observed initial synchronization correlation quality level. Ideally this level would approach 100%. The SyncCorr metric is a useful, quick indicator of how well the 89600 VSA software is locked to the received signal's previously selected sync type parameter signal component. Other useful, quick checks include EVM, frequency error, auto-detected cyclic prefix (CP) length, and cell ID metrics.

After the initial synchronization quality check and other summary metric checks are made, the next important trace to examine is Trace A (top left). This trace shows an IQ constellation of the demodulated signal in which, for OFDMA, each point represents the amplitude and phase of one subcarrier in the frequency domain and one symbol in the time domain. From this trace it should be immediately apparent whether the analyzer has successfully locked to and demodulated the signal. If the trace has a regular pattern, then lock has almost certainly been achieved. If the pattern is unstable or just shows noise, then lock has not been achieved and a check of the demodulation setup and signal quality will be required to trace the cause. Trace A also reveals the different modulation depths detected in the signal. In this example the most obvious is 64QAM, followed by 16QAM and then by QPSK, all of which are allocated for the physical downlink shared channel (PDSCH) used in this example. The QPSK modulation points also represent subcarrier allocations for the physical broadcast channel (PBCH), physical control format indicator channel (PCFICH), physical downlink control channel (PDCCH), and reference signal (RS). A closer look reveals BPSK modulation points representing the secondary synchronization signal (S-SS). (As with P-SS, S-SS is not an official 3GPP acronym.) Additional points on the unity circle represent the P-SS modulated with a Zadoff-Chu phase sequence. The signal in this example was chosen because it represents simultaneously all the modulation types available on the downlink.

The dots in Trace A, representing each physical channel or physical signal, are color-coded with the same colors used in the other displays. The key to the color coding is provided in Trace F (bottom right), the frame summary display. This display lists all the physical signals and physical channels in the signal. For each entry there is an assigned color, and four measurement results are displayed: EVM, detected signal or channel power relative to the un-boosted RS power level (0 dB unity circle), the allocated RB, and modulation format.

Trace D (bottom middle) is the error summary. The top section provides quality statistics for the composite signal including overall EVM, peak EVM, EVM of the PDSCH data, and EVM of the RS pilot signals. In the middle section of Trace D is the frequency error (which is a residual error subtracted from the signal prior to calculating EVM), and various statistical metrics that indicate the types of error that make up the overall EVM. These include the IQ gain imbalance, IQ quadrature error, and IQ timing skew. Apart from the frequency error and the IQ offset, the other error terms do not have definitions in the specifications. Rather, to the skilled designer, they are troubleshooting tools that can help pinpoint the underlying errors that make up the overall EVM. The bottom section of Trace D shows the auto-detected OFDMA symbol CP length, as well as the auto-detected cell ID and the cell ID group or sector derived from demodulation of the P-SS and S-SS.

Further insight into the nature of the signal errors can be found in the two remaining traces. Trace C (top middle) is the error vector spectrum, which shows the EVM as a function of subcarrier in the frequency domain. The value of analyzing EVM in this domain is to determine if there are frequency-specific elements in any observed error. Typical errors that are discernable in this trace are spurious signals—which could elevate the EVM of subcarriers located near the spur—as well as baseband modulation errors and IF or RF filter responses, both of which can shape the error spectrum. In any practical signal it is expected that the EVM will degrade at the channel edges due to effects of channel and band filters. Maintaining the optimal balance between in-channel and out-of-channel performance is one of the biggest challenges for the LTE transmitter designer, and tools such as 89600 VSA software can be invaluable for that purpose. This particular signal has an almost flat EVM spectrum with only a slight increase at the channel edges. The heavy dark trace is the average EVM per subcarrier over the measurement period, which for this example is one radio frame (10 ms or 140 symbols). There is a third dimension to this trace that explains the solid nature of the data. For any one subcarrier there will be one EVM result for every measured symbol. In this case 140 data points will be overlaid horizontally on the vertical axis for each of the 301 subcarriers. From visual inspection it can be seen that the peak EVM is just over 3% with most of the peaks around 2% and the average just under 1%. This aligns with the rms EVM and peak EVM statistics in the error summary of Trace D. It is also possible to change the X-axis units to RBs rather than subcarriers; this averages the data for each set of 12 adjacent subcarriers to derive a single rms EVM value per RB.

The final view of this signal is error vector time, in Trace E (top right). This trace shows EVM in the time domain as a function of the OFDMA symbol. In this example, the errors are generally evenly spread in time, although there is a noticeable jump in the average EVM at the end of the sixth subframe, which could indicate a problem somewhere that needs further investigation. The color coding is more evident here than in Trace C and shows that the allocation of the different PDSCH modulation formats is sequential in time. As with Trace C, there is an additional dimension to this trace. For each of the 140 OFDMA symbols in time, 301 data points will be overlaid on the vertical axis, indicating the EVM of each subcarrier during that symbol period. It is also possible to configure the X-axis in units of slots rather than symbols; this averages the data for each set of seven adjacent OFDMA symbols (assuming normal CP) within a given slot to derive a single rms EVM value per slot. Common errors visible in this trace include amplifier gain instability, particularly for TDD signals in which thermal settling can be a factor. This trace can also be useful for identifying other problems that are temporal in nature, such as timing glitches or interference from logic signals. Large single-symbol errors could indicate amplifier clipping due to a symbol interval with particularly high peak amplitude.

It should be evident from this introductory view of digital demodulation just how powerful a vector signal analyzer can be for understanding the structure and quality of LTE signals.

6.4.6.2 Coupling Measurements Across Domains

One of the most powerful 89600 VSA software tools used in modulation analysis is the coupling of markers across different measurements, traces, and domains. This tool is particularly effective for analyzing LTE signals, which can have a high symbol content (large number of symbols per frame, burstiness, etc.) and a large number of different signal elements (channels, reference signals, OFDM subcarriers, etc.). Coupled markers allow the user to understand the identity and characteristics of a symbol simultaneously in time, frequency, power, and error. To make this analysis even more specific, coupled markers can be combined with the selection or de-selection of individual LTE physical signals and physical channels for display and analysis.

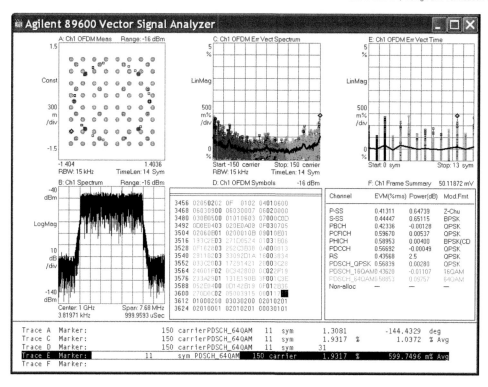

Figure 6.4-12. Example of using coupled markers across four different measurement domains

An example of the use of coupled markers across multiple measurement displays is shown in Figure 6.4-12. This downlink demodulation measurement takes advantage of the 89600 VSA software's ability to simultaneously show multiple color-coded displays and a complete marker table. With markers coupled and all LTE physical signals and physical channels selected for analysis, a peak search in either the error vector spectrum or error vector time mode indicates the largest error during the measurement interval of one subframe (14 symbols for normal CP). The exact symbol associated with this error can now be understood in terms of time domain OFDMA symbol index, frequency domain subcarrier number, IQ magnitude and phase values, physical channel type, encoded modulation format, decoded symbol table bit value, and other parameters. Note that in this case the highest detected error involves a 64QAM PDSCH physical channel symbol from OFDMA symbol index 11, located on the outermost edge subcarrier index 150 and modulated using one of the highest power outer IQ constellation states, which was decoded as 64QAM state bit value 0x1F (see the Trace A, C, D, and E marker values at the bottom of Figure 6.4-12). This example identifies the important fact that the highest error is located at the edge of the channel spectrum, which is consistent with the rest of the error vector spectrum display in Trace C. Although this characteristic of the signal error appears obvious in the error vector spectrum display, it would not be clear from examination of either the error vector time or error summary trace displays. Therefore a flexible measurement tool that provides many different views into the measured signal across different domains is extremely valuable in troubleshooting, providing a much fuller understanding of the underlying causes of observed impairment. The Trace C error vector spectrum shows a more pronounced degradation (increase in EVM) at the channel edges than was the case with the equivalent Trace C of the previous example in Figure 6.4-11.

6.4.6.3 Measuring With and Without Equalization

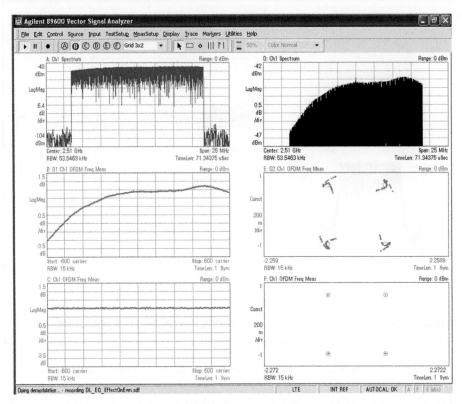

Figure 6.4-13. Example of impact of equalization on RS power

Figure 6.4-13 shows a simple QPSK downlink signal. Trace A (top left) is the non-demodulated spectrum, which reveals a 20 MHz channel bandwidth. Some slight unflatness is evident. Trace D (top right) is the same measurement but with an expanded scale to show the detail of the unflatness, which is about 3 dB peak to peak. The 89600 VSA software's ability to display the same measurement in different views with different display attributes is especially convenient.

Trace B (middle left) shows the demodulated downlink RS subcarrier power observed in the third last symbol of slot index 1. In this example, slot index 1 (the second slot of radio frame) contains no data allocations, so only the RS subcarriers are active, and their magnitude follow the same shape as the RF spectrum in Traces A and B. Trace E (middle right) shows the constellation of the RS subcarriers. The shape of the distortion is simple enough that it can be mapped visually onto the unflatness in Trace B. The lower frequencies start at the lower power nearer the center of the constellation, and then the power rises with nearly constant phase. This rise is followed by a small counterclockwise phase error, which changes direction with the power and then remains nearly constant to the highest frequency. In Trace F (bottom right) the 89600 VSA software's equalizer has been turned on and the constellation is now corrected to the ideal values, and the RS power and phase response are now normalized. This is seen in Trace C (bottom left), where the RS power is now constant across the channel.

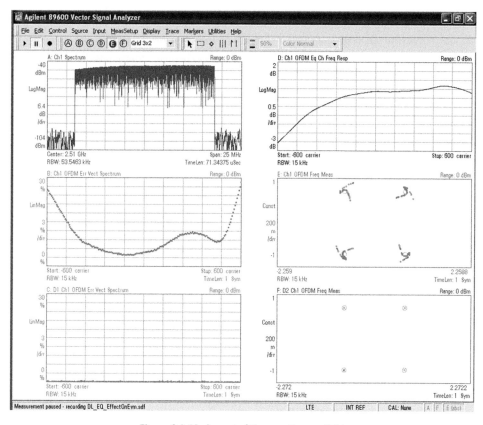

Figure 6.4-14. Impact of the equalizer on EVM

Figure 6.4-14 shows further analysis of the same signal. Trace A (top left) is the same spectrum shown in Figure 6.4-13. Trace D (top right) is the frequency response of the equalizer coefficients. This matches the pattern of the RS power in Trace B of Figure 6.4-13. It would also be possible to display the phase or group delay component of the equalizer coefficients (not shown in this example). Trace E (middle right) shows the same unequalized constellation that appears in Figure 6.4-13. Trace B (middle left) shows the unequalized EVM spectrum. Note how EVM rises at the channel edges and reaches around 27%. Without equalization the EVM spectrum response reflects not only the un-compensated frequency magnitude response of Trace D, but also the corresponding un-compensated phase or group delay distortion component (not shown in this example). This explains why EVM at the channel edges is similar in value even though the amplitude error at the upper frequencies is small. It is the phase error at the higher frequencies that contributes this EVM. The requirements for UE spectral flatness in 36.101 [3] Subclause 6.5.2.4 apply only for the amplitude component. This example indicates why phase is just as important as amplitude error. When the equalizer is applied to the signal, the constellation in Trace F (bottom right) returns to the ideal points and Trace C (bottom left) shows how the equalized EVM spectrum is flat and reduces to nearly zero.

6.4.6.4 Measuring EVM at Different Points in the CP

As discussed in Section 2.1.6.1.1, the EVM definition requires that measurements be made at two different points in time during the CP. The reasons are discussed extensively in Section 2.1.6.1.1 and will not be repeated here other than to say that EVM is an important aspect of transmitter design that needs analysis beyond a simple pass-fail result. The 89600 VSA software makes EVM measurements at the required positions within the CP and returns the largest result. However, so that LTE designers can investigate how much of the CP has been used up by time domain distortions in the signal, it is possible to modify the default window lengths used to make EVM measurements. Figure 6.4-15 shows some of the flexibility built into the 89600 VSA software to allow evaluation of EVM at different points in the CP. There are also other parameter selections for turning on and off the equalizer along with various RS pilot tracking options to allow in-depth analysis of the source of errors in the signal.

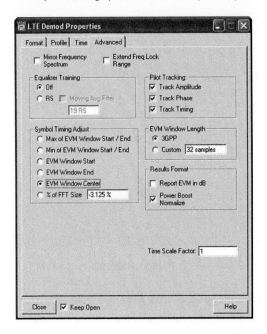

Figure 6.4-15. Advanced demodulation properties

6.4.6.5 Time-Based Measurements Across the Radio Frame

An example of three different time-domain views for one frame of a 5 MHz downlink signal is shown in Figure 6.4-16. Trace B (bottom left) shows the detected allocations vs. time, and is a demodulated version of the spectrogram shown in Figure 6.4-10. Note that time is now on the X-axis rather than the Y-axis. Figure 6.4-16 can be compared easily to Figure 3.2-8. The regular PDCCH three-slot allocation is evident across the entire channel at the beginning of every subframe, with the RS showing up in between.

Within the first subframe the primary and secondary synchronization signals as well as PBCH can all be clearly seen occupying the central 62 and 72 subcarriers (see Section 3.2.5.1). The remainder of the first subframe allocation consists of a 16QAM PDSCH. This can be identified by the consistent color coding used across the demodulation traces, including the color key within the frame summary Trace D (bottom right). The frame summary also reports that 16QAM PDSCH is the only active data modulation format present in the measured radio frame (10 ms or 140 symbols), occupying a total of 150 allocated RB. From Traces A, B, and C we can clearly see that these PDSCH RB allocations occur in subframes 0, 3, and 6 with RB allocations extending across the entire 5 MHz transmission BW. The 16QAM PDSCH is allocated for 6 of the 10 slots during the 10 ms radio frame, with each allocation being the full 25 RB in the frequency domain. This gives a total of 150 RB as reported in the frame summary.

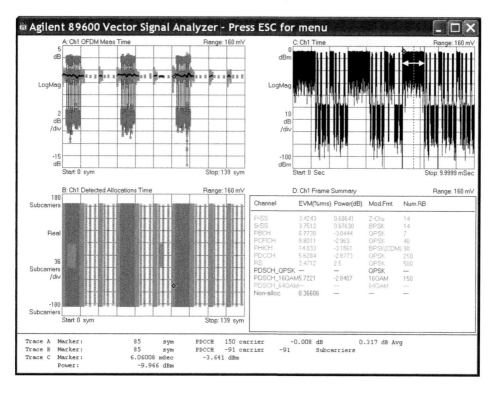

Figure 6.4-16. Example of time-based measurements across the frame using coupled markers linking demodulated symbol power, frequency allocation, and RF envelope power

In Trace A (top left) the demodulated IQ subcarrier power is displayed on a Y-axis scale of 2 dB per vertical division for each of the 140 measured OFDMA time symbols along the X-axis. The first three time symbols show very little variation in power since this is the QPSK PDCCH along with some power boosted RS subcarrier activity in the first time symbol. The next eleven symbols represent the 16QAM PDSCH and there is a much larger power spread of around 10 dB. The power is clustered in three horizontal bands, which represent the three different ideal power levels present in a 16QAM signal. Since this is a 301 subcarrier signal, for every symbol in time on the X-axis there will be 301 vertical data points representing the power of every subcarrier. The average of this power for each subcarrier across the measurement interval is shown by the horizontal dark line through the middle. The final observable pattern is the RS, which shows up as a tight burst of power on the first and fourth symbol of each slot. The variation in the RS power is very low since this is the reference used to equalize the signal.

The results in Trace A and Trace B (bottom left) are linked by coupled markers to Trace C (top right), which shows the RF envelope power vs. time. A peak search in Trace C identifies the peak power at 6.06 ms, while a band power calculation indicated by the white arrow for 11 symbols centered on the dotted line at 6.6 ms calculates the average power of −10 dBm for that particular PDSCH allocation. This particular signal uses full allocation across the channel so the peak power in Trace C is constant for all allocated symbols. If the allocations had been for less than the whole channel bandwidth, Trace C would have shown appropriate step changes in the peak power.

The detected allocations and symbol power vs. time traces can be used to understand the effect of RB allocation on power and to establish appropriate and consistent conditions for the measurement of peak and average power. Without a full understanding of the symbol structure in the frame, the power measurements needed for design and optimization will be inconsistent and potentially misleading.

6.4.6.6 Controlling the Measurement Interval

To view the signal with more precision, measurement offsets and intervals can be set in the time domain. Any time-specific portion of the signal can be isolated for analysis, as shown in the demodulator setup example in Figure 6.4-17.

The terms used to describe the timing control are as follows:

- Result length, which determines the length (in slots) of the acquired IQ time-continuous capture buffer aligned to the start of the user-specified analysis start boundary event.
- Measurement offset, which determines the delay offset (in slots and symbols) from the start of the user specified result length from where the measurement interval processing starts.
- Measurement interval, which determines the measurement interval duration (in slots and symbols) starting from the user specified measurement offset to be used for the measurement analysis.

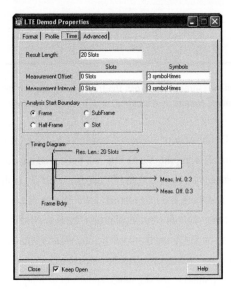

Figure 6.4-17. Configuration of measurement interval and offset within the capture interval (result length)

By definition, the measurement offset plus the measurement interval cannot exceed the result length. The following section uses this ability to isolate a particular part of the signal for more detailed analysis.

6.4.6.7 Measuring Power Across the Subframe

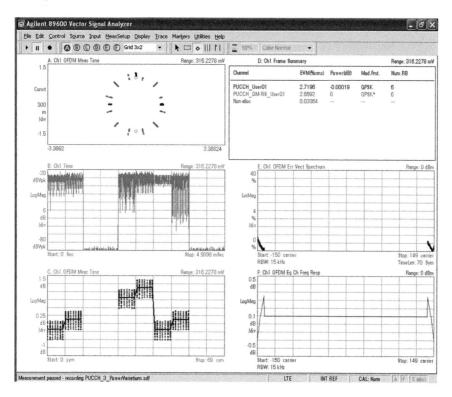

Figure 6.4-18. Hopping PUCCH Format 1b showing amplitude error across the channel

Figure 6.4-18 is a 5 MHz uplink signal with only the PUCCH allocated. The PUCCH format chosen is Format 1b in which the PUCCH is allocated for one RB at the lowest frequency followed by one RB at the highest frequency. See Figures 3.2-12 and 3.2-13. Three different PUCCH subframes (indexes 0, 2, and 3) are allocated with a gap of one subframe. Therefore in the time domain the first subframe will have one slot at the lowest RB followed by one slot at the highest RB. The next subframe is empty and then the pattern of the first subframe is repeated twice. Each slot consists of two PUCCH symbols followed by three DMRS symbols followed by two more PUCCH symbols as was shown in Figure 3.2-12.

PUCCH and DMRS Structure

Before describing the power dynamics in this example, the constellation in Trace A (top left) containing the PUCCH and DMRS requires explanation. The measurement interval of five subframes contains three active subframes. Each subframe transmits just one useful PUCCH Format 1b control data useful QPSK symbol. In this example each subframe transmits a different PUCCH useful QPSK symbol value and these are represented by three clusters of dots at the left, top, and right of the 45-degree rotated QPSK grid. Each subframe's PUCCH control data useful QPSK symbol value is carried by eight PUCCH symbols: two on

323

either side of the three DMRS symbols in the first slot and the same again in the second slot. Each PUCCH symbol comprises 12 subcarriers, each with a phase offset. The unmodulated base subcarrier phases are calculated from a cyclically shifted base reference sequence with further scrambling and orthogonal spreading. The result is a subcarrier sequence of 12 phase values on a 30 degree raster. The base reference sequence component varies depending on the slot index, and the cyclic shift component varies depending on the symbol index within the slot. One subframe contains eight PUCCH symbols, each with 12 subcarriers, resulting in a basis function of length 96 subcarriers per subframe. The PUCCH control data useful QPSK symbol value is then modulated onto this basis function to determine the actual subcarrier phase to be transmitted per subcarrier.

Rather than display the complexity of the underlying signal structure, the 89600 VSA software removes the nominal reference rotation introduced by the basis function. As a result all 96 points distributed around the 30 degree raster are de-mapped onto one of the four rotated QPSK points in the grid, revealing the original PUCCH control data useful QPSK symbol value. The purpose of this complex PUCCH coding is (1) to introduce robustness to the PUCCH in the presence of interference and (2) to deal with the fact that the PUCCH RB is a shared resource, and several UEs can be allocated to the same frequency and time. The orthogonal coding included in the basis function allows the eNB receiver to separate out each individual UE.

Should any one of the underlying subcarrier amplitudes or phases contain an error, it would be seen in Trace A as a point lying outside the nominal rotated QPSK grid. Trace A shows clearly that the 96 points, which represent the PUCCH control data useful QPSK symbol values, are tightly clustered, indicating that the signal contains few errors.

The remaining 12 states that do not lie on the rotated QPSK constellation points represent the PUCCH Format 1b DMRS symbol subcarrier magnitude and phase values. The phase values for the DMRS subcarriers are defined using methods similar to those used to generate the basis function used for the PUCCH, the main difference being the use of a different orthogonal sequence. Each slot contains three DMRS symbols, each symbol comprising 12 subcarriers. With two slots per subframe the DMRS are represented in Trace A by 72 data points per subframe. Unlike the PUCCH symbols, the DMRS symbol data points are displayed exactly as they are transmitted without removing the basis function rotation of each subcarrier.

Analyzing the Power Variation Across the Frame

The constellation in Trace A shows amplitude distortion, which creates a starburst effect. This is because both the 89600 VSA software receiver equalizer and pilot amplitude tracking settings are both disabled, resulting in the observed IQ constellation magnitude variations. These variations are due to an uncompensated 1 dB power boost applied to the middle subframe and a 0.3 dB channel flatness error revealed by the frequency hopping.

Trace B (middle left) shows the unsynchronized power vs. time of the signal. Small variations can be seen between the slots, but at 6 dB per division the unflatness is not obvious. Trace C (bottom left) shows the demodulated power vs. symbol, which is a much more sensitive measurement than the unsynchronized power vs. time in Trace B. This highlights the limitations of general purpose measurements for LTE. In Trace C the six transmitted slots can clearly be identified. Looking first at the average power (solid line) per symbol, four discrete levels can be identified. The first slot is at −0.20 dB and the second slot is about 0.3 dB higher. Then a gap of one subframe occurs, and the next slot is at +0.8 dB followed by another about 0.3 dB higher. The next subframe repeats the first. This measurement is made without any pilot amplitude tracking, so the observed 0.3 dB power steps are due to the uncompensated level unflatness of the transmitted signal when it hops from the lowest RB to the highest RB.

The average power per symbol measurement is also made without any equalization, resulting in a vertical spread of subcarrier power at each symbol. The PUCCH is a single RB so it has 12 subcarrier points per symbol. The magnitude spread is about 0.5 dB peak to peak and can also be seen in Trace A. Note that this 0.5 dB magnitude spread is directly attributable to the uncompensated equalizer frequency response shown in Trace F (bottom right) for both the low frequency and high frequency PUCCH RB. There is equalizer information only for the outer PUCCH RBs since nothing is transmitted anywhere else in the frequency domain. By applying the equalizer to the signal, it is possible to remove almost all the amplitude spreading error.

Trace E (middle right) shows the unequalized EVM vs. subcarrier. Note again that no information is shown except for the outer RB since the rest of the signal is unallocated. The dark line for the lower RB starting at 6% and dropping to 1% represents the lower PUCCH, which gives an average EVM of around 3%. The upper PUCCH has an average EVM closer to 2%. This is summarized in Trace D (top right) in which the PUCCH average EVM is 2.7%. A similar analysis can be done for the DMRS.

6.4.6.8 Measuring Power Across the Symbol

The 89600 VSA software can be used to look in even more detail at the signal. Figure 6.4-19 is an analysis of just two adjacent symbols in an SC-FDMA uplink. The first symbol is a PUCCH followed by a PUCCH DMRS. The IQ constellation is shown in Trace A (top left) and follows the same rules explained in detail in Section 6.4.6.7.

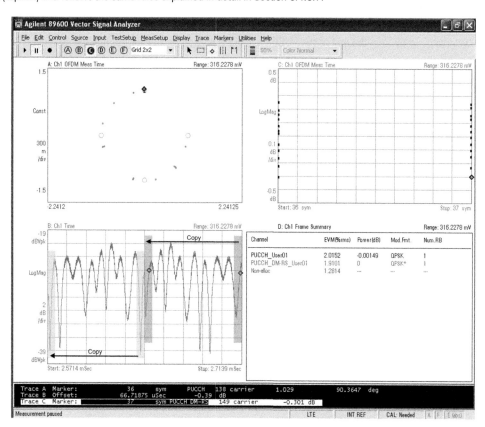

Figure 6.4-19. Analyzing power across the symbol showing the CP

Trace C (top right) plots the amplitude vs. time of the constellation points shown in Trace A. The X-axis shows the measurement interval to be two symbols with an offset of 36 symbols, which means only two sets of data will be displayed, representing symbol index 36 and 37 (or the 37th and 38th symbols) within the captured signal. The results on the left of Trace C show the amplitude of the 12 subcarriers that make up symbol index 36, which is a PUCCH symbol. The spread is just over 0.5 dB, so this symbol is well-behaved. On the right hand side are the 12 subcarrier amplitudes that make up the adjacent PUCCH DMRS symbol.

Trace A has an uncoupled marker enabled and placed on one of the symbol index 36 PUCCH control data useful QPSK symbol points. Note that the Trace A marker readout (bottom of screen) clearly identifies the selected symbol point as belonging to the PUCCH. Trace C also has an uncoupled marker enabled and placed on one of the symbol index 37 PUCCH DMRS symbol points. Similarly, the Trace C marker readout (bottom of screen) clearly identifies the selected symbol point as belonging to the PUCCH DMRS.

Trace B (bottom left) is the time domain waveform of the PUCCH and adjacent PUCCH DMRS symbols. The most interesting aspect of this trace is that it shows the CP prepended to each symbol. Note that two time markers have been placed across the tail end of the PUCCH DMRS symbol period. From the Trace B marker readout (bottom of screen), it can be observed that the two time markers are actually separated by the nominal 66.7 μs symbol period. A cyclic prefix copy of the final 4.69 μs of the PUCCH DMRS symbol is inserted at the start of the same second symbol period located in the middle of the trace. Similarly, a cyclic prefix copy from the end of the PUCCH symbol is inserted at the start of the first symbol period located on the left of the trace.

6.4.6.9 IQ Image and LO Leakage

Uplink signals with changing allocations can serve to demonstrate the effects of IQ errors as described in Section 6.4.2.8, along with demodulation measurements that provide an alternative view. Consider first the spectrogram of an UL signal shown in Figure 6.4-20.

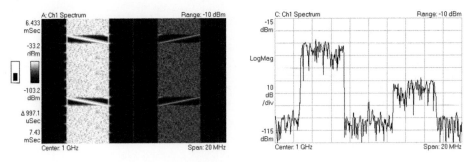

Figure 6.4-20. Spectrogram and spectrum IQ images

The spectrogram on the left shows the channel allocations over a period of about 1 ms. Clearly visible are the mirror images around the center frequency of the desired carriers—the result of IQ errors. The spectrum trace on the right by default reflects the most recent measurement appearing at the bottom of the spectrogram; however, a "trace select" marker from the spectrogram display can be used in the trace on the right for analysis of any single spectrum from the spectrogram buffer. In that spectrum, trace band power markers can be used to compare the power of the allocated carriers and their images.

The spectrogram clearly reveals the symmetrical signal-duplicating effect of IQ distortion about the center frequency without the need for demodulation. These measurements can also indicate the suppression of unwanted subcarrier energy of more than 30 dB for this signal.

A more accurate analysis of IQ image requires demodulation of the signal. Figure 6.4-21 analyzes three adjacent symbol periods from a 5 MHz uplink 16QAM PUSCH. The three symbols chosen for analysis are a PUSCH DMRS symbol in the center of the slot along with a single PUSCH data symbol on either side. In order to demonstrate typical IQ image and LO leakage distortions, the signal was deliberately impaired using the features in the MXG signal generator. To emulate LO leakage a 0.1% IQ offset was added, and to emulate the IQ image a 1 dB IQ gain imbalance was added. Given the richness of the information in Figure 6.4-21, this figure will be used to provide further explanation of the uplink signal structure prior to explaining how the IQ image and LO leakage measurements are made.

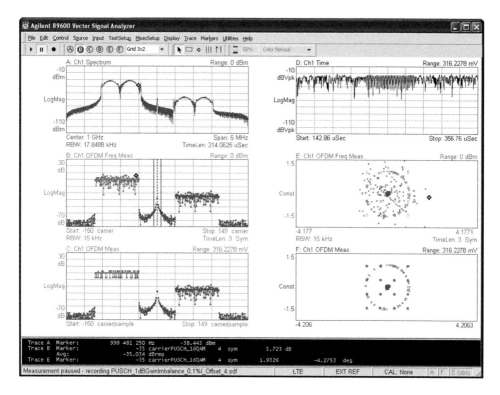

Figure 6.4-21. Example of IQ image and LO leakage distortion on the uplink signal

Unsynchronized Frequency Spectrum

Figure 6.4-21 Trace A (top left) shows the frequency spectrum of the signal in a 5 MHz span. From this trace the approximate allocation bandwidth of 1 MHz and the presence of an image about 30 dB below the wanted signal can be determined. An LO leakage spike at the channel center frequency can be seen as well. The high level of spectral spreading of the wanted signal almost engulfs the LO leakage spur. This spreading is generated by the discontinuity at the SC-FDMA symbol boundaries and was discussed in Section 6.4.5.3. To see the signal in the same way that an uplink receiver sees it, each symbol, excluding the symbol boundary transients, must be analyzed separately. Figure 6.4-22 shows such an analysis. On the left is the spectrum of one SC-FDMA symbol carrying PUSCH data and on the right is the spectrum of one symbol of DMRS. Since the symbol transitions have been excluded, the LO leakage is no longer partially obscured by the spectral spreading. Note how the spectra are very different and the reversed spectrum of the image is very clear. The distinctive spectrum in Figure 6.4-21 Trace A is the combination of two symbols of SC-FDMA plus one symbol of DMRS.

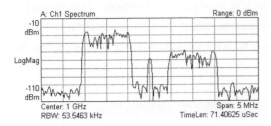 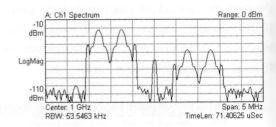

Figure 6.4-22. Spectrum of SC-FDMA symbol (carrying PUSCH data) and PUSCH DMRS symbol. Both exclude spectral spreading due to inter-symbol discontinuities.

Synchronized Subcarrier Amplitude vs. Time

Figure 6.4-21 Trace B (middle left) shows an FFT of the signal that is fully synchronized to sample only during the central part of the symbol, excluding symbol boundaries. The spectral spreading seen in the unsynchronized view of Trace A is noticeably absent in Trace B. The X-axis units of Trace B are subcarrier numbers, and it can be verified that this is a 6 RB allocation of 72 subcarriers with a bandwidth of 1.08 MHz. Note the span is 300 subcarriers (4.5 MHz), which represents the maximum number of subcarriers that can be allocated in a 5 MHz channel whereas Trace A shows the full 5 MHz channel..

Another difference between Trace A and Trace B is the representation of the LO leakage. LO leakage is a CW tone at the channel center frequency that, when viewed in an unsynchronized analog spectrum, is represented as a single tone with a shape matching the RBW of the analyzer. The CW nature of LO leakage is partially obscured in Figure 6.4-21 Trace A due to the inclusion of symbol discontinuities, but is more evident in Figure 6.4-22.

In contrast, the synchronized FFT spectrum of Figure 6.4-21 Trace B shows the LO leakage energy spread across many subcarriers. The reason is that the raster for the subcarrier frequencies in the uplink is offset by one half subcarrier (7.5 kHz) from the channel center frequency. The spreading seen in Figure 6.4-21 Trace B is an artifact of the FFT process. Any energy falling on a frequency that is not exactly aligned with the FFT bin spacing will result in energy being spread (or leaked) into adjacent bins according to the $\sin(x)/x$ function roll off described in Section 2.2.1.

The Y-axis in Figure 6.4-21 Trace B represents the amplitude of each subcarrier in the frequency domain. For each subcarrier in the X-axis there will be as many data points displayed as there are symbols in the chosen measurement interval. In this example three symbols are being analyzed so for the allocated part of the signal there are $72 \times 3 = 216$ discrete subcarrier data points. Two distinct patterns are visible in the allocated part of the signal. There is a line of constant amplitude data points around 0 dB and another set of what look like noisy data points above and below 0 dB.

Synchronized Subcarrier Constellation (Frequency Domain)

It is easier to see what is going on in Figure 6.4-21 Trace B by looking at Trace E (middle right). This is the same subcarrier amplitude data as Trace B but with phase replacing frequency to form an IQ constellation display. Each data point represents the amplitude and phase of one subcarrier for the duration of each measured 66.7 μs symbol period. A cluster of points around the unity circle can clearly be seen (magnitude level 1.0 equivalent to 0 dB). This represents the Zadoff-Chu phase modulation of the DMRS subcarriers. On Trace B these subcarriers are represented by the flat line of data points around 0 dB in the allocated part of the signal.

Also represented in Trace E are the amplitude and phase of the unallocated subcarriers. These are at very low levels and form a noisy ball at the center of the display. This energy represents the unwanted LO leakage and image responses. The remainder of the data points in Trace E can be considered random, having no observable pattern. These data points represent the subcarriers carrying the PUSCH data encoded with SC-FDMA modulation. The reason these subcarriers have no regular pattern is explained in the definition of SC-FDMA, given in Section 2.3.1 and specifically in Figure 2.3-1. During the period of each SC-FDMA symbol, there are as many PUSCH data time symbols transmitted as there are subcarriers in the allocation. In this example there are 72 allocated subcarriers.

On the downlink, the mechanism for transmitting 72 PDSCH data symbols within one 66.7 μs OFDMA symbol is to assign each data symbol to its own subcarrier and transmit all data symbols simultaneously, which creates a high PAPR composite signal that is difficult to transmit. However, on the uplink with SC-FDMA the 72 PUSCH data symbols are transmitted sequentially in time using one broadband carrier. In this example each PUSCH data symbol is represented as one point on a 16QAM constellation. Within the period of one 66.7 μs SC-FDMA symbol, a time domain waveform is generated by transitioning from one PUSCH data time symbol to the next. This single carrier waveform, which has a PAPR no worse than 16QAM, is then converted from the time domain to the frequency domain using a discrete Fourier transform (DFT) prior to subcarrier allocation mapping.

By sampling at the correct rate this transformation results in 72 DFT subcarrier bins with a spacing of 15 kHz, each with its own amplitude and phase held constant for the duration of the SC-FDMA symbol. The fact that each subcarrier is constant for 66.7 μs explains why it is then possible to insert the 4.69 μs CP on each subcarrier, even though the PUSCH data time symbols encoded by the 72 subcarriers are 72 times shorter in length and are not constant during the SC-FDMA symbol period. By using SC-FDMA modulation, the relationship between the original PUSCH data time symbols and each individual allocated subcarrier is lost. Only when the entire 72 subcarriers are summed in the time domain within the eNB receiver is it possible to recreate the original transmitted PUSCH data time symbol in the 16QAM constellation.

Trace E shows another example of using coupled markers. One of the outlying subcarriers on the right hand side is identified with a marker. By coupling the markers between Trace E and Trace B it is possible to identify which subcarrier this outlying point in the constellation represents in the frequency domain. In this example the outlying subcarrier identified in Trace E is seen in Trace B as being near the high end of the allocation.

It is worth summarizing key points regarding the definition of OFDMA and SC-FDMA. On the downlink, all data, control, and reference symbols have a period of 66.7 µs, and a 1:1 relationship to the 66.7 µs OFDMA symbols is allocated to each subcarrier. On the uplink, the same frequency domain modulation and mapping technique is used to generate PUSCH reference symbols and PUCCH control and reference symbols. The SC-FDMA modulation technique is used only on the uplink for the PUSCH data symbol generation process. For an SC-FDMA allocation of M subcarriers, M PUSCH data time symbols are transmitted sequentially during the 66.7 µs SC-FDMA symbol period. The period of a PUSCH data time symbol is therefore 66.7 µs/M. A PUSCH data time symbol is defined by all its M subcarriers that make up the SC-FDMA symbol. An individual subcarrier in an SC-FDMA modulated signal has no meaning.

Time Domain

The amplitude component of the time domain waveform for the three symbol periods is shown in Figure 6.4-21 Trace D (top right). During the first SC-FDMA symbol period the waveform appears random, but in fact it is the amplitude component of a precisely controlled trajectory around the points in a 16QAM constellation representing the PUSCH data time symbols. By simply changing the displayed data from log magnitude (LogMag) to phase, a similar trace can be shown for the phase component of the time domain waveform. The phase trace would also appear to be random, but like the amplitude waveform; the phase is also entirely deterministic. The middle symbol representing the DMRS looks much more ordered. The DMRS consists of 72 subcarriers, each with a specified phase following a sequence across the frequency domain.

The time domain waveform for the DMRS has a very distinctive pattern comprising some 36 discrete peaks during the 66.7 µs period leading up to the right hand end of the symbol. There are additional peaks at the start of the symbol, which represent the 4.69 µs of the CP. The 36 peaks are to be expected, because the summation of 72 discrete subcarriers at the same amplitude and different phase creates a 72nd order time domain waveform that has 36 peaks.

The third symbol period is a further example of the SC-FDMA modulated PUSCH, except that the PUSCH data time symbol values and hence the details of the resulting time domain waveform are naturally different from the first symbol.

Full Uplink Digital Demodulation Constellation (Frequency and Time)

The full picture of the signal is most powerfully described in Figure 6.4-21 Trace F (bottom right). This constellation represents the information content of the signal as observed after the full digital demodulation process has been performed by the uplink receiver. Since the uplink modulation contains both frequency domain (OFDMA) and time domain (SC-FDMA) components, this constellation is also mixed domain containing both frequency and time domain elements. The frequency domain DMRS unity circle in Trace E is repeated in Trace F. The big difference between Trace E and Trace F is the transformation of Trace E's apparently random SC-FDMA subcarriers into the 16QAM time domain constellation of Trace F. The 16QAM constellation is the result of performing an inverse discrete Fourier transform (IDFT) on the PUSCH frequency domain subcarriers to recreate the original 16QAM time domain representation of 144 transmitted PUSCH data time symbols.

A final view of the signal is shown in Figure 6.4-21 Trace C (bottom left). This view is an alternative representation of the same signal in Trace F, but the phase information has been replaced with frequency information. If Trace B is compared to Trace C, the DMRS information remains unchanged, similar to the unity circle remaining unchanged between Trace E and Trace F.

Also note that the non-allocated subcarrier regions including both the IQ image and LO leakage distortion energy impairments remain unchanged. A new thing to note in the Trace C mixed domain results is that the apparently random allocated subcarriers from the Trace B frequency domain are now replaced by a regular pattern of three discrete levels. These levels represent the three discrete magnitude state sets in the 16QAM constellation of Trace F.

Note that the X-axis has now changed from being a "carrier" in Trace B frequency domain to a "carrier¦sample" in Trace C mixed domain. Thus the allocated region 16QAM amplitude data points can be read from left to right as the time domain sequence of the original 72 PUSCH data symbols. By coupling a marker in Trace C to a marker in Trace F, it is possible to scroll from left to right through the individual PUSCH data symbols in Trace C, matching them to their corresponding constellation points in Trace F. A tabulated demodulation display of the signal is available and is discussed in Section 6.4.6.10.

Measuring LO Leakage (Relative Carrier Leakage Power)

The definition of LO leakage is given in Sections 2.1.6.1.2 and 2.1.6.1.3 and is based on 36.101 [3] Subclause 6.5.2.2. The first reference defines requirements for LO leakage from the perspective of wanted signal quality and the second reference considers IQ impairments including LO leakage in terms of how they impact other users when the allocation does not occupy the full channel bandwidth. Many names are used to describe the LO leakage phenomenon including IQ component, relative carrier leakage power (RCLP), IQ offset, and IQ origin offset. There is also a class of requirements in 36.101 [3] Subclause 6.5.2.3 for "in-band emissions" (which are actually in-channel emissions) and these include a "DC component," which is the same thing as LO leakage. To use the formal term, the RCLP requirement for signal power above 0 dBm is −25 dBc. At lower powers the requirement is relaxed as described in Section 2.1.6.1.2.

The measurement of RCLP is calculated as part of the EVM process according to 36.101 [3] Annex F. The RCLP (IQ offset) error is removed from the signal prior to measuring EVM but the quantity removed must meet the requirement limits for RCLP. If the center of the signal has not been allocated, it is possible to measure RCLP using band power markers. Figure 6.4-23 Traces A and B are used to illustrate such a measurement, with rms band power markers set to encompass both the allocated PUSCH RBs and the unused center RB, respectively. The Trace A rms band power marker result shows that the mean allocated subcarrier power is −0.2 dB, measured over the full 6 RB PUSCH frequency domain allocation. The Trace B rms band power marker result shows that the equivalent mean unused subcarrier power is −34.7 dB, measured over just the center unused RB. From these two results the RCLP (IQ offset) is calculated to be

$$(-34.7) - (-0.2 + 10 \log(6)) = -42.3 \text{ dB}.$$

Note that the 10 log(6) total power adjustment factor accounts for the 6 RB PUSCH allocation used in this example.

331

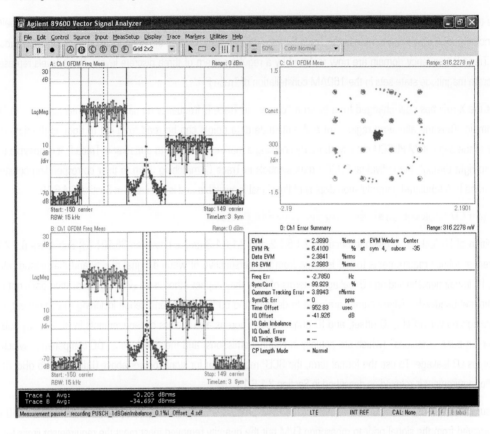

Figure 6.4-23   Example of IQ image and LO leakage distortion on the uplink signal

The requirements for RCLP apply to all signal allocations, and if the center of the signal has been allocated, then the only way to measure RCLP is as part of the EVM process. The measured IQ offset is also provided as a result metric in the error summary display in Trace D. As shown the reported IQ offset metric value of −41.9 dB closely matches the RCLP result calculated above using the rms band power marker method.

Measuring In-Band Emissions (DC component and IQ Image)

The definition of the DC component for in-band emissions in 36.101 [3] Subclause 6.5.2.3 differs slightly from the RCLP definition. The in-band requirements are defined for a partially allocated signal so that the DC component and IQ image falling on unallocated subcarriers can be assessed in terms of their interference with other users. The measurement of in-band emissions is performed directly at the output of the receiver FFT step, prior to any equalization being applied.

The DC component is defined as the power of the RB containing the center frequency divided by the total power of the allocated RBs. If the maximum number of RBs for the channel bandwidth is even rather than odd, then the DC component is measured for the RB on either side of the center frequency, excluding any allocated RB.

The IQ image requirement is −25 dBc and is defined as the power in a non-allocated image RB relative to the average power of all the allocated RBs. In Figure 6.4-23 Trace B, this can be estimated as being near the limit of −25 dBc. This image is due to the 1 dB IQ gain imbalance impairment that was added using the MXG signal generator IQ adjustment feature to emulate real distortion in the UE. The IQ image measured value can be verified more accurately by further use of band power markers.

The 0.1% IQ offset applied to this signal has created an LO leakage 17 dB better than the requirements, while the 1 dB IQ gain imbalance has created an image level that is close to failing the requirements. These results give some indication of the level of IQ impairments that can be tolerated.

For the downlink, the LO leakage is not a problem since there is one subcarrier intentionally left unused. This was not an option for the uplink because the use of SC-FDMA requires a continuous transmission bandwidth and will always incorporate the LO leakage when the allocation exceeds half of the available RB. Given this situation, it was decided to offset the uplink subcarrier raster from the center frequency by a one half subcarrier so that the LO leakage would fall between the subcarriers and be spread out accordingly.

6.4.6.10 Verifying Demodulated Symbol Values

The demodulation measurements discussed in this section have focused on modulation quality results in various forms. Another important demodulation result is the symbol table trace data, which displays the values of the demodulated symbols. The symbol table includes only the user-selected physical signals and physical channels within the specified measurement interval, and thus it can be used to narrow the analysis to as little as a single symbol. For example, synchronization problems can be caused when the RS are incorrectly coded or when the coding (the symbol sequence) is generated according to one version of the standard while the demodulation is performed according to another. In the two symbol tables shown in Figure 6.4-24, the analyzer is set to display only the complex QPSK-like RS symbol values observed on every sixth subcarrier from a single downlink antenna port. A previously stored set of good symbol table trace data (stored in register D1) can be compared to the currently measured data in which the downlink RS sync is not found due to an error.

A: Ch1 OFDM Symbols			Range: 0 dBm		C: D1 Ch1 OFDM Symbols			Range: -16 dBm
			SYNC NOT FOUND					
0 00	00	02	01		0 00	02	01	00
24 03	03	00	02		24 01	02	03	00
48 01	01	03	01		48 02	01	03	00
72 00	03	03	02		72 02	02	00	02
96 00	01	03	02		96 01	03	01	00
120 00	01	03	00		120 03	03	03	01
144 01	01	01	02		144 00	02	03	02
168	00	03	01	00	168	01	03	01	00
192	01	00	03	00	192	03	03	03	01
216	01	02	02	01	216	00	02	03	02
240	01	03	02	01	240	03	01	00	00
264	03	01	02	01	264	03	03	03	01
288	02	02			288	00	02		
312								
336								
360								

Figure 6.4-24. Example of RS demodulation

With the 89600 VSA software set to display only the RS symbol values, the mismatch between the symbol table for the current measurement (left) and a stored measurement (right) indicates a possible cause for synchronization failure and can be used to diagnose the failure's root cause and to fix it.

6.4.6.11 Digital Demodulation Summary

The range of possible digital demodulation measurements is far greater than those described here. More examples and LTE-specific information are provided in Section 6.7 and references [18], [19], [20], and [21] at the end of this chapter.

6.4.7 Multi-Standard Radio Base Station Transmitters

This section discusses the multi-standard radio (MSR) requirements first introduced in Release-9 in 37.104 [22] and the associated conformance tests in 37.141 [23]. Modern base station transmitters have evolved to support wider bandwidths enabling not only support for multiple carriers of a single radio format but also multiple formats simultaneously in one transmitter. As an example, GSM, W-CDMA, and LTE carriers can now be simultaneously transmitted from a single MSR base station. The MSR standard also refers to multiple radio access technologies (multi-RATs). The ability to support multiple radio formats in a cellular network is important to reducing both base station size, cost, and number of sites to be deployed. MSR base stations are expected to enable smooth and seamless network migration from the current 2G/3G networks to LTE and LTE-Advanced technologies.

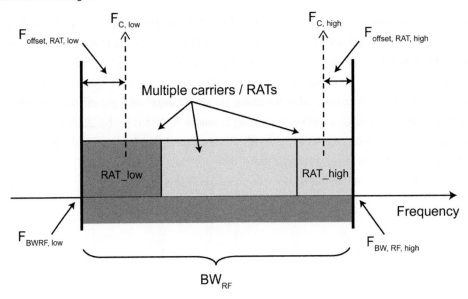

Figure 6.4-25. RF bandwidth related symbols and definitions for MSR base station (37.104 [22] Figure 3.2-1)

Figure 6.4-25 illustrates the overview of the MSR base station in the frequency domain. The term BWRF is the bandwidth in which a base station transmits and receives multiple carriers and/or RATs simultaneously. For the detailed definitions of other symbols in the figure, refer to 37.104 [22] Section 3.2. This model was further extended in Release 10 to account for non-contiguous carriers. The MSR receiver conformance test requirements for the most part adopt the requirements of each individual RAT such as GSM, W-CDMA, and LTE. The receiver test configurations use one target carrier of either W-CDMA or LTE positioned at one edge of the BWRF channel, plus one GSM carrier acting as interference at the opposite end of the BWRF channel.

n contrast, the MSR transmitter conformance tests use multi-RAT configurations including channel power, modulation quality (EVM), frequency error (derived from EVM), spurious emissions, and operating band unwanted emissions such as spectrum emission mask (SEM). The exceptions that require single RAT configurations only are ACLR, occupied BW, and time alignment between transmitter branches. However, manufacturers may want to test some of these exceptions under MSR configuration to create more realistic usage scenarios and to cover more of the formats that the base station under test supports.

6.4.7.1 Spectrum Measurements

Similar to the single-RAT cases, the spectrum measurements for multi-RATs can be made using either swept-tuned analysis in a signal or spectrum analyzer or FFT analysis in a vector signal analyzer. Swept analysis is generally appropriate for the out-of-band (or out-of-channel) testing such as spurious emissions, ACLR, and SEM because the required frequency span is wider than the measurements for a single carrier typically made using FFTs.

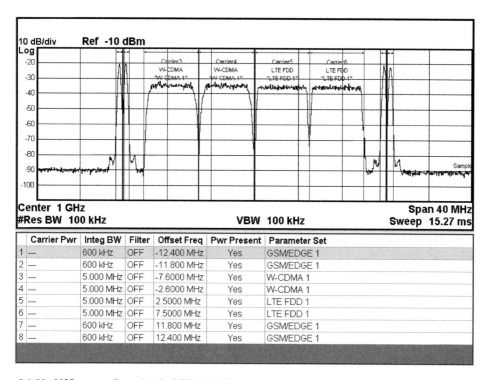

	Carrier Pwr	Integ BW	Filter	Offset Freq	Pwr Present	Parameter Set
1	---	600 kHz	OFF	-12.400 MHz	Yes	GSM/EDGE 1
2	---	600 kHz	OFF	-11.800 MHz	Yes	GSM/EDGE 1
3	---	5.000 MHz	OFF	-7.6000 MHz	Yes	W-CDMA 1
4	---	5.000 MHz	OFF	-2.6000 MHz	Yes	W-CDMA 1
5	---	5.000 MHz	OFF	2.5000 MHz	Yes	LTE FDD 1
6	---	5.000 MHz	OFF	7.5000 MHz	Yes	LTE FDD 1
7	---	600 kHz	OFF	11.800 MHz	Yes	GSM/EDGE 1
8	---	600 kHz	OFF	12.400 MHz	Yes	GSM/EDGE 1

Figure 6.4-26. MSR test configuration 4c (UTRA, E-UTRA, GSM multi-RAT) with base station RF bandwidth = 25 MHz

Figure 6.4-26 shows an example of swept spectrum analysis using the Agilent PXA Signal Analyzer with N9083A MSR Measurement application for the multi-RAT, multicarrier-active configuration defined in 37.141 [23] Subclause 4.8.4.3. This example, which is called "Test Configuration 4c (TC4c)," requires an RF bandwidth (BWRF) of 25 MHz, with two pairs of GSM carriers at each end of BWRF, two W-CDMA carriers left of the band center frequency, and two LTE FDD 5 MHz carriers right of center. The conformance test requirements define many test cases that are dependent on which combinations of multi-RATs the base station supports.

6.4.7.2 Digital Demodulation Measurements

In evaluating signal modulation qualities such as the EVM of each carrier in a multi-RAT, multicarrier-active configuration, a key consideration will be how to best acquire all the active carriers throughout the wide bandwidth that the MSR base station supports.

For transmitter conformance testing, measurements are made using test models, which in this case are repetitive waveforms. The conformance test specification defines several multicarrier test configurations, and the test requirements are defined per RAT and per carrier, so it is not necessary to capture all the signals simultaneously. The fact that all signals are being transmitted is sufficient to create the conditions whereby any inter-carrier or inter-RAT distortions will be present.

Each carrier in the multicarrier-active condition can be analyzed individually using an acquisition input bandwidth wide enough to capture each carrier in turn. This narrowband analysis approach is deemed simpler and more cost-effective than using a more expensive wideband analyzer to capture all the carriers at once, which would also create very large waveform samples for computing EVM. The widest cellular carrier bandwidth is the LTE 20 MHz carrier. This remains the case in Release 10, which can support up to 100 MHz bandwidths by aggregating up to five carriers; however, each carrier is still analyzed individually.

Simultaneous capture all of the active carriers of interest requires more expensive hardware, but this may be justified by the significant improvement in test efficiency realized not only in single carrier analysis but also in MSR device verification and troubleshooting tasks, including functional design validation and real-world operating tests. Figure 6.4-27 compares the two acquisition approaches for MSR modulation analysis: the sequential acquisition of each carrier using narrow bandwidth hardware (a) and the simultaneous acquisition of all carriers using wide bandwidth hardware (b).

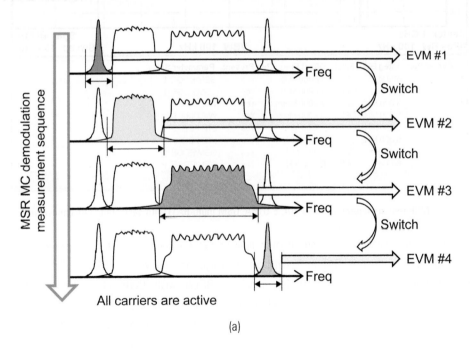

(a)

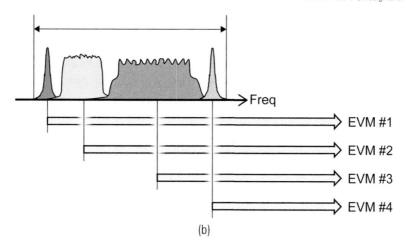

Figure 6.4-27. MSR modulation analysis comparing sequential narrowband acquisition of each carrier (a) and simultaneous broadband capture (b)

Note that regardless of the approaches used, whether narrow bandwidth or wide bandwidth, an appropriate receiver filter is required for each carrier of interest. The filter rejects adjacent carrier power interference so that the analyzer can achieve good-enough synchronization and the demodulation robustness necessary for each carrier in the multicarrier-active condition. In the case of W-CDMA, the requirement is a 3.84 MHz root raised cosine filter with a roll-off factor of 0.22. For other RATs such as GSM and LTE, the receive filter is not explicitly defined. However, an adjacent carrier rejection filter will be needed with an appropriately steep roll-off factor to avoid inter-carrier interference on the measurement.

6.4.8 Release 10 LTE-Advanced Transmission

This section discusses additional transmitter functionality defined in Release 10 for LTE-Advanced. The following important enhancements are covered:

- Carrier aggregation (CA) in the downlink and uplink
- Multi-clustered SC-FDMA in the uplink
- Simultaneous transmission of PUCCH and PUSCH in the uplink.

These enhancements further complicate a challenging radio environment for transmitters in terms of higher PAPR and more complicated spurious management within and between component carriers (CCs).

6.4.8.1 Peak-to-Average Power Ratio

For LTE-Advanced eNB power amplifiers, downlink CA is not a major new challenge because the concept of multicarrier is central to the development of MSR described in Section 6.4.7. The CCs in LTE-Advanced are required to be backward-compatible with Release 8 and Release 9 carriers. The situation for the enhanced uplink is not so straightforward, because all of the Release 10 new features can increase the PAPR over that of Release 8 that was shown in Figure 6.4-9.

Some examples of CCDF power measurements for a 10 MHz LTE-Advanced uplink signal are given in Figure 6.4-28. The X- and Y-axis profiles are the same as in Figure 6.4-9 for Release 8. All the modulation formats of the PUSCH in Figure 6.4-28 are QPSK with PN15. Each PUSCH cluster consists of three RBs. There are five curves in the measurement. From left to right they are (1) non-clustered PUSCH with SC-FDMA precoding, which is the baseline release 8; (2) multi-clustered PUSCH with SC-FDMA precoding; (3) simultaneous transmission of PUCCH and non-clustered PUSCH; (4) simultaneous transmission of PUCCH and multi-clustered PUSCH; and finally (5) the AWGN reference curve.

The use of multi-clustered SC-FDMA as well as simultaneous PUCCH and PUSCH transmission increases the PAPR above that of the legacy non-clustered SC-FDMA, reducing the original advantage of SC-FDMA but not by as much as would be the case with full OFDM, which can exceed the PAPR of AWGN.

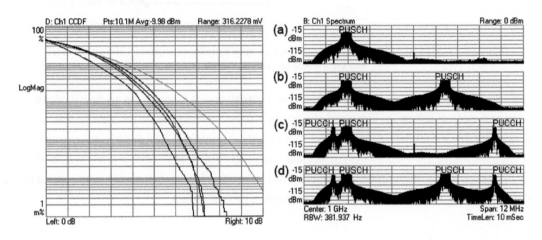

Figure 6.4-28. CCDF and spectrum measurements of various LTE-Advanced uplink signals vs. Release 8 PUSCH

Uplink CA is another factor that can increase the PAPR of a UE power amplifier (PA) when a single PA is used to cover both carriers.

6.4.8.2 Intermodulation Products

Both multi-clustered SC-FDMA and simultaneous PUCCH with PUSCH create multiple carriers within the channel bandwidth. Any nonlinearity within the PA will result in intermodulation products falling within and just outside the wanted channel. Figure 6.4-29 shows some examples of FFT spectrum measurements on the same four signals seen in Figure 6.4-25 but with 50% IQ clipping applied to emulate PA compression.

It can be seen that as the number of carriers within the channel increases, the number of intermodulation products also increases. This increases pressure to either make the PA more linear or back off the maximum power in order to meet the modulation quality requirements.

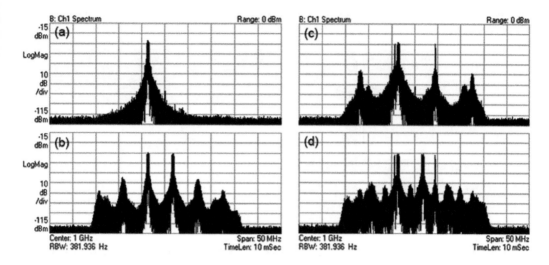

Figure 6.4-29. FFT spectrum measurements of various LTE-Advanced uplink signals vs. Release 8 PUSCH

6.4.8.3 Digital Demodulation Measurements

For multicarrier CA signals the fundamental techniques of demodulation are not different from those discussed in Section 6.4.6 provided each carrier is analyzed separately. However, even though the modulation quality requirements are defined separately for each CC, it is important to measure the CCs under the conditions when they are all active.

Figure 6.4-30 shows examples of the side-by-side modulation analysis across LTE-Advanced multiple CCs with the 89600 VSA software. The signal being analyzed is a contiguous aggregation of two 10 MHz and one 5 MHz carriers.

The traces A, B, and C show various demodulation results of the first CC at −10 MHz offset from the center frequency; traces D, E, and F show the results of the second CC at 0 Hz offset; traces G, H, and I show the results of the third CC at +7.5 MHz offset; and trace J shows the spectrum of all three carriers.

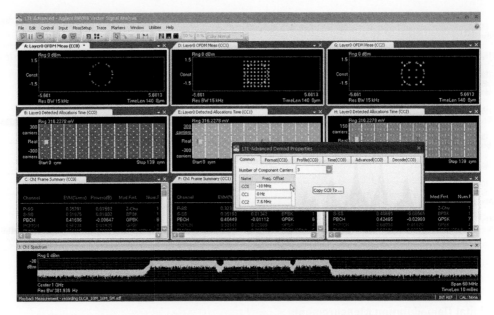

Figure 6.4-30. Digital demodulation results for three carrier uplink contiguous aggregation

6.5 Receiver Design and Measurement Challenges

6.5.1 Introduction

This section focuses on the design issues and challenges associated with testing LTE receivers from RF through baseband. The principles apply to both FDD and TDD access modes although the examples here are FDD. This section takes a systematic approach to testing the various receiver components in order to facilitate testing of the fully assembled receiver.

Open-loop receiver testing is discussed first. In this case the DUT does not send feedback information to the source. Open-loop testing is sufficient to test the fundamental characteristics of the individual components in the receiver and also is a first step in validating the demodulation algorithms in the baseband section. However, full verification of the overall receiver performance in real world conditions requires closed-loop testing through a faded channel, a method that is discussed later in Section 6.6. In closed-loop testing lost packets are retransmitted using incremental redundancy based on real-time packet acknowledgement feedback from the DUT. The modulation and coding used for transmission are similarly based on real-time feedback from the DUT. This feedback may be optimized for subbands within the overall channel bandwidth to enable frequency selective scheduling.

A block diagram of a classical receiver, shown in Figure 6.5-1, provides the basis for the discussions in this section. Modern transmitters and receivers utilize the same blocks shown here; however, today there is a higher degree of integration with single components such as RFICs performing multiple functions. The concepts embodied in the classical block diagram still apply although there may be fewer places at which signals can be injected or observed for testing.

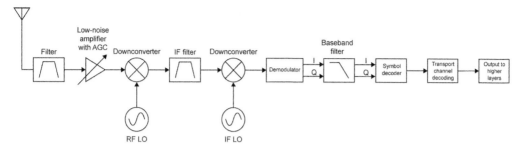

Figure 6.5-1 Classical receiver block diagram

6.5.2 Verifying the RF Receiver

The main objective of receiver testing is to make performance measurements on the entire receiver. However, many factors can influence receiver performance, so the basic receiver sub-blocks are considered first and uncertainty contributions eliminated or quantifiably reduced.

6.5.2.1 Amplitude Flatness

One of the unique aspects of LTE is that it supports six channel bandwidths ranging from 1.4 MHz to 20 MHz. To simplify system operation and roaming, the UE is mandated to support all of these bandwidths, even though actual deployment in any one area may be restricted to fewer bandwidths. The LTE 20 MHz bandwidth is significantly wider than the maximum bandwidths of today's other cellular systems; therefore, special attention to phase and amplitude flatness is required during receiver design. Filters, amplifiers, and mixers in particular now have to operate correctly over multiple channel bandwidths. One benefit of the LTE signal structure is that the reference signals (RS) are spread in both frequency and time over the entire LTE signal. The UE and eNB receivers can use these signals along with digital signal processing (DSP) techniques to compensate for amplitude and phase linearity errors in the receiver and radio channel. Relying on the receiver to correct these errors based on the RS ultimately comes at the price of degrading the signal to interference plus noise ratio (SINR), so a well-behaved frequency response in the receiver is still important. Channel flatness needs to be tested across each supported bandwidth and band, particularly at the band edges where the duplex filter attenuates the edge of the signal.

A simple way to verify receiver flatness is to use a broadband signal at the front of the receiver and measure the response at later stages. An actual LTE signal, such as one of the downlink test models, can be used. Care must be given to the signal configuration, however, because the flexibility of LTE means that the signal may not be fully occupied at the same power level over the entire bandwidth. An example of a simpler test signal is band-limited noise, which most signal generators can create. Averaging several hundred samples in a spectrum analyzer gives a good approximation of the receiver amplitude flatness. Caution is necessary, though, because the results are dependent on the flatness of the signal generator. This flatness varies with frequency and the signal should be checked prior to measuring the DUT. Typical signal generator flatness over 20 MHz is a few tenths of a dB but is better at the narrower LTE bandwidths. A technique for correcting flatness errors in the test equipment is described later in this section.

Another way of verifying receiver flatness is to create a multitone signal with a sufficient number of tones spaced over the 20 MHz bandwidth. In the downlink this can be simulated using an LTE signal composed of only the RS; however, since the RS are not present in every symbol, the bursty nature of the signal makes it harder to measure. A good representation of the RF front end frequency response can be made with 20 or 30 continuous wave (CW) tones spaced over the LTE bandwidth being tested. More tones can be added to increase the resolution of the test results. There are solutions available today that use a signal generator and spectrum analyzer in conjunction with software to automatically create, measure, and correct a user-defined multitone signal for testing amplitude flatness and intermodulation distortion products. As already indicated, a signal generator is flat to within a few tenths of a dB over a 20 MHz bandwidth without the use of correction techniques; however, after corrections are applied, the signal generator output will be flat to within a few hundredths of a dB.

Figure 6.5-2 shows a multitone test system using a signal generator and spectrum analyzer in conjunction with Agilent Signal Studio for Multitone Distortion software. A benefit of this solution is that any part of the measurement setup included between the signal generator and spectrum analyzer RF path—for example, cables, connectors, or amplifiers—can be corrected for both amplitude flatness and intermodulation distortion products. Figure 6.5-3 shows the amplitude flatness of tones spaced over a 100 MHz bandwidth before and after corrections. Although this example uses a bandwidth that is wider than the 20 MHz bandwidth supported for LTE, from Release 10 onwards it is possible to aggregate carriers up to a maximum bandwidth of 100 MHz, which may be contiguous. Note in Figure 6.5-3 that the scale per division in the multitone traces is 0.2 dB per division.

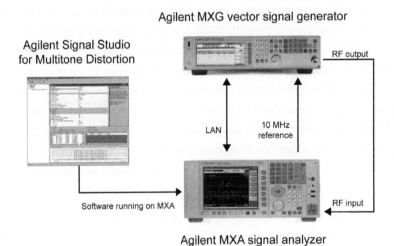

Figure 6.5-2. A calibrated, multitone system to measure amplitude flatness and intermodulation distortion

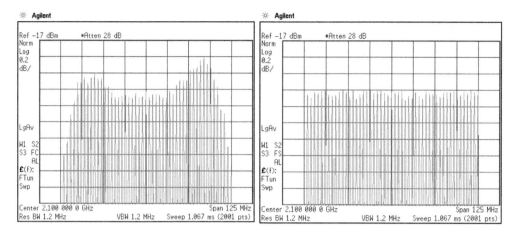

Figure 6.5-3. Amplitude flatness of tones spaced over a 100 MHz bandwidth before and after correction

6.5.2.2 Phase Linearity

The phase linearity of the receiver front end is just as important as the amplitude flatness. The receiver can compensate for phase linearity in the received signal using the RS in a manner similar to that used for correcting the amplitude flatness. Poor phase response results in degraded error vector magnitude (EVM) and lowers the margin of error for determining the correct location on the constellation diagram of a given symbol. The phase linearity of a system cannot be measured directly with a traditional signal generator and spectrum analyzer. A network analyzer can be used if there are directly accessible components in the RF front end; however, this approach may be hard to realize with the high degree of integration in designs today. Another approach is to measure the phase by creating modulated LTE signals and then demodulating them in a vector signal analyzer, which has the added benefit of providing the amplitude response at the same time. How to create, demodulate, and analyze these signals is addressed in the following sections.

Trace A (top left) and Trace B (bottom left) in Figure 6.5-4 show amplitude flatness and phase linearity of a 5 MHz downlink LTE signal. The error vector time and error vector spectrum of the signal are shown in Trace C (top middle) and Trace D (bottom middle). Trace E (top right) and Trace F (bottom right) show the constellation diagram and frame summary of the signal as another way to identify and isolate problems.

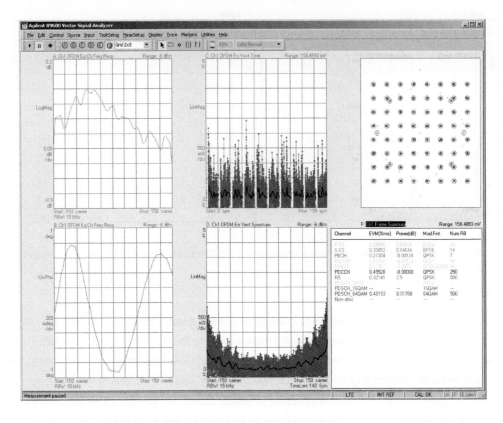

Figure 6.5-4. Amplitude flatness and phase linearity measurements of an LTE signal derived from digital demodulation of reference signals

6.5.2.3 Automatic Gain Control

The receiver's RF front end contains automatic gain control (AGC) circuitry to maintain the proper signal levels at the various receiver stages. AGC is necessary since the operating range for the received signal is around 75 dB. Traditional analog receivers measure the average power level of the received signal and adjust gain blocks and the output of variable gain amplifiers to maintain the signal at a constant level for processing by following stages. The LTE radio system is more sophisticated, making use of the reference signals (rather than the average power) in conjunction with the RF AGC circuitry to set the gain. The task of the UE is limited to receiving transmissions from one eNB at a time; however, the eNB receives signals from multiple UEs arriving at different levels. The eNB therefore has to set the receiver input level for the strongest signal and use a wide dynamic range to capture the weaker signals.

Basic RF AGC operation can be tested with a band-limited additive white Gaussian noise (AWGN) signal by varying the input over the expected range of amplitudes and then monitoring the output of the AGC for the correct level using a spectrum analyzer. The overall AGC operation must be tested with an LTE signal that has been modulated with the reference signals to determine if the baseband processing is providing the correct feedback to the AGC circuitry. Creating modulated LTE signals will be addressed later in this section.

The downlink PAPR is similar to Gaussian noise and can exceed 10 dB at 0.001% probability; however, in the eNB the peaks are typically limited to approximately 7 dB. This must be accounted for when the gain of the AGC is set so that there is enough head room in the amplifier to handle the signal peaks and prevent distortion from occurring.

The receiver may have multiple gain stages that can be switched in or out of the circuit to compensate for large variations in signal strength. Although the AGC will maintain the signal at the desired level, the phase of the receiver may change when gain stages are enabled or disabled. This effect should be quantified, as phase changes can seriously affect demodulation quality and any switching needs to be minimized with hysteresis to prevent oscillation around the switch point levels. There are no requirements for receiver phase, and all the performance tests are carried out at mid-range levels or at reference sensitivity, so phase performance remains a desirable goal of good design rather than a formal requirement.

6.5.2.4 Noise Figure

Each of the sub-blocks in the RF front end should be evaluated for gain and noise to assure proper signal-to-noise ratio (SNR) through the various stages. Properly placed amplifiers and filters help optimize SNR and overcome conversion losses, attenuation, and added noise. Each component in the front end adds noise to the received signal, and the assumption from the technical report 36.942 [24] on LTE RF scenarios is that the UE receiver has a noise figure of 9 dB with the eNB being significantly better at 5 dB. The entire front end or individual components can be measured using a noise figure meter or a spectrum analyzer with a noise figure application and noise source.

6.5.2.5 Receiver Error Vector Magnitude

Error vector magnitude is a measure of signal impairment normally applied to transmitted signals. However, it is equally useful in verifying receiver performance although no specific EVM requirements are defined in the specifications for receivers. Impairments in the RF front end—such as a non-ideal amplitude and phase response, noise, IQ imbalances, phase noise, and compression—can cause the symbols to be in the wrong location relative to the ideal IQ constellation. Such errors become more important with higher-order modulation such as 64QAM, in which the vector distance between symbols is reduced, lowering any margin for error. The margin for error in the receiver is further reduced in real world conditions in which fading and interference are present.

The EVM is a singular metric that provides a gauge of how well a given signal matches the ideal signal and it gives good insight into the ways each component in the system may be distorting the received signal. However, as a singular metric, EVM masks the cause of the errors created by individual components. Therefore, detailed IQ analysis is usually required to identify the root cause. IQ analysis is covered in Section 6.4.1.6. Although there are no requirements in the specifications that call for IQ measurements in receivers, it is still useful from a design and troubleshooting standpoint to understand the characteristics and performance of individual receiver components and also the overall RF front end.

EVM measurements are made by demodulating the signal in a vector signal analyzer, which requires a modulated signal as the stimulus to the input of the receiver or individual component. EVM can be measured with many types of signals; however, a modulated LTE signal should ultimately be used to verify performance. For a discussion on digital demodulation see Section 6.4.6. A useful source of standard downlink test signals (test models) is found in the eNB conformance test specification 36.141 [25]. These may be of some use for testing the UE receiver, but it should be noted that the purpose of these signals is to test the eNB transmitter performance, and these signals do not directly represent configurations that are used for the UE receiver tests, which are based on reference measurement channels. Test signals are discussed further in Section 6.5.4.

6.5.3 Verifying the Baseband Receiver

6.5.3.1 Analog-to-Digital Converter (ADC)

Traditionally, the analog signal from the RF section can be demodulated into the I and Q components using analog techniques, normally with an IQ demodulator. However, in an age of software-defined radios, the down-converted IF signal is usually digitized by an analog-to-digital converter (ADC) and then fed to the baseband section for demodulation and decoding. It is possible to test the baseband receiver independent of the RF section by generating an appropriate IF signal. An IF signal can be created easily by simply setting the signal generator output frequency to the desired IF frequency.

Measuring the output of the ADC poses a challenge because the output is now in digital format and standard spectrum analyzers make measurements on analog signals. One solution is to analyze the digital bits from the ADC with a different instrument, and a logic analyzer is a natural choice for capturing this digital data. The difficult part of this solution is processing the captured data into a meaningful result since most logic analyzer applications are not focused on generating RF metrics. Using the Agilent 89600 VSA software is one way to solve this challenge. The 89600 VSA software is commonly run in Agilent spectrum analyzers to demodulate various modulation formats; however, the software can be run in the logic analyzer, too. The software offers a unique way to analyze the ADC performance by making traditional RF measurements directly on digital data. This approach gives designers the ability to quantify the ADC contribution to overall system performance by comparing the digital results with the RF measurements made earlier in the block diagram using the same measurement algorithms.

The output of the ADC may be converted to a high speed serial interface such as the common public radio interface (CPRI) in the eNB or DigRF in the UE. In some cases the topology of the receiver may allow access to the digital data only via one of these industry standard buses, complicating the data analysis process. However, industry-standard solutions are available that accept these high speed serial data streams. For example, Agilent offers a DigRF analyzer that can either analyze the digital data (as a logic analyzer would) or pass the data to the 89600 Vector Signal Analysis software, enabling vector measurement of the demodulated data. Refer to Section 6.3 for a more detailed discussion of DigRF.

6.5.3.2 Baseband Demodulation

The digitized signals from the ADC are transferred to the baseband section where FPGAs or ASICs perform signal demodulation. Up to this point the measurements have been relatively straightforward because test equipment has been used for both the generation and analysis of the LTE signals. Now the receiver in either the eNB or the UE must demodulate the signal and indicate the result.

One challenge associated with testing the baseband section of the receiver is how to physically deliver the test signals to the DUT. Depending on where the receiver is in the development cycle, the test signal could be injected into the receiver as an RF, IF, analog IQ, or digital IQ signal in the baseband section. Most signal generators can create signals directly for testing each of the different sections of a receiver. The digital outputs of signal generators are generally raw I and Q samples with highly configurable physical characteristics including logic type, numeric format, number of bits, bit order, sample rate, and clock options. However, most baseband LTE radio designs are expected to use a dedicated industry-standard digital interface such as CPRI in the eNB or DigRF in the UE. As noted above, solutions are available today for analyzing these industry-standard buses as well as stimulating them. See Section 6.3 for more details. The rest of this section focuses on different test signals and how they can be used and assumes that the signal is physically delivered to the receiver in the appropriate format.

Signal generators can provide timing signals or accept trigger signals to facilitate synchronizing to either the UE or eNB receivers. To facilitate synchronization, information about the LTE signal being generated can be preprogrammed into the DUT; for example, when a UE is tested, the UE can be forced via preprogramming to use the physical layer cell ID group and sector generated by the signal generator. It may be helpful to configure a vector signal analyzer to demodulate the same signal that the receiver is configured to demodulate. The demodulation and decoding algorithms in the baseband section can be verified with physical layer coded LTE signals, which can be easily configured using a signal generator with an application such as Signal Studio for 3GPP LTE. The RBs configured at the channel and band edges are of particular interest because band and channel filters are likely to distort and attenuate part of the signal. Although test signals are easily measured and interpreted using a vector signal analyzer with its multiple displays, the interface to a real LTE receiver is likely to be a simple terminal interface with proprietary commands and results. Thus a useful feature of a receiver is the ability to write the demodulated data from each channel to a file for post analysis to ensure that the received bits match the transmitted bits. The payload data in Signal Studio can be set to be a pseudorandom sequence, a regular pattern, or a user-defined file.

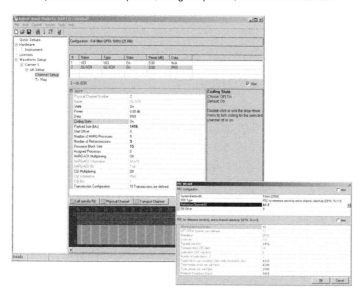

Figure 6.5-5. Example from Agilent Signal Studio of a QPSK R=1/3 uplink FRC used for making eNB sensitivity and in-channel selectivity measurements

An example from Signal Studio of an uplink FRC for testing an eNB is shown in Figure 6.5-5. Basic demodulation is verified at the subframe level. Once this step is complete, the next step is to check the transport channel decoding. The specifications define fixed reference channels (FRCs) that are used as reference configurations for defining receiver requirements. These signals are a good starting point for initial verification of the transport channel decoding algorithms.

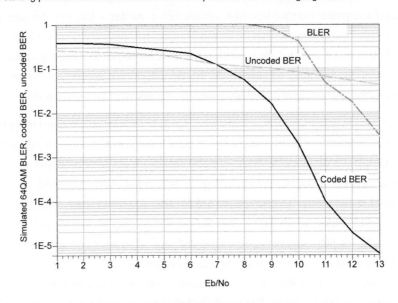

Figure 6.5-6. Simulation of UE receiver performance for 64QAM PDSCH showing the relationships between uncoded BER, coded BER, and BLER

Once the receiver has been verified and is correctly demodulating and decoding the signal, bit error ratio (BER) and block error ratio (BLER) measurements can be performed. Figure 6.5-6 shows a simulation of UE receiver demodulation performance for a 64QAM downlink PDSCH as a function of E_b/N_0, which is the energy per bit divided by the noise power spectral density. This gives a general idea of the relationships between uncoded BER, transport channel coded BER, and BLER. The uncoded BER is a measure of the bit error ratio at the physical layer before transport channel decoding, while the coded BER is a measure of the bit error ratio after the transport channel decoding. Uncoded BER is a more sensitive measure of receiver performance than BLER or coded BER and is useful during early phases of receiver characterization. The receiver requirements use BLER as the metric for performance, which is expressed in the form of a throughput relative to the maximum throughput of the FRC. The transport channel coding with forward error correction clearly improves BER performance. The simulation was performed using Agilent's system design software, which is discussed in Section 6.2.

BER measurement requires a pseudorandom sequence to be configured for the payload data in the signal generator and for that sequence to be made known to the receiver, which can then auto-correlate to it and calculate BER. Some signal generators can calculate the BER if the demodulated and decoded signal can be routed back to the signal generator as a transistor-to-transistor logic (TTL) or complementary metal oxide semiconductor (CMOS) signal. Unlike UMTS and earlier systems, LTE has no requirements based on BER, and the loopback mechanisms defined for measuring UE BER are not supported in LTE. All receiver tests for both UE and eNB are based on BLER, and BER testing remains an R&D tool.

The eNB conformance tests require that the eNB calculate and report its own BLER. For the UE this is an option for proprietary testing but the conformance tests calculate BLER independent of the device by counting acknowledgement (ACK) and negative acknowledgement (NACK) reports transmitted by the UE on the uplink. The UE and eNB BLER requirements (expressed in terms of throughput) are based on hybrid automatic repeat request (HARQ) retransmission and therefore cannot be fully measured using an open-loop signal generator. Closed-loop receiver testing is discussed in Section 6.6.

6.5.3.3 Open-Loop HARQ Functional Testing

The performance of HARQ processing requires closed-loop testing; however, testing of basic open-loop functionality at the physical layer can eliminate problems later in the design cycle. Section 3.4.1 discusses the definition of HARQ. The signal generator can be configured with the number of retransmissions that is required to generate a signal based on hypothetical NACK responses from the DUT. Additionally, for each of the HARQ processes, a different modulation type, redundancy version (RV) index, and resource block allocation can be configured, testing the receiver's ability to correctly decode data from multiple HARQ. The parameter variation from one subframe to the next can be reduced for troubleshooting when problems are discovered. Figure 6.5-7 shows an example using Signal Studio. The signal is configured to transmit a single DL-SCH with transport channel coding. The first transmission is in subframe 0 and the retransmission is in subframe 8 with different coding due to a simulated NACK response from the DUT.

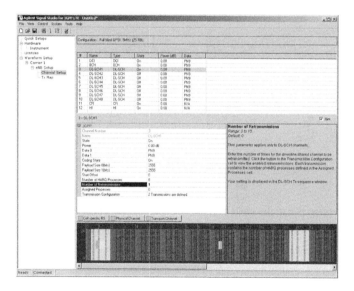

Figure 6.5-7. Example from Agilent Signal Studio showing the original transmission (subframe 0) and the retransmission (subframe 8) in response to a simulated NACK response from the DUT

6.5.4 Receiver Performance Under Impaired Conditions

The receiver test discussions up to this point have focused on verifying and testing receiver performance using highly accurate reference signals created by a signal generator. However, it is essential to be able to degrade these near-perfect signals in a known and controlled way to gain additional insight into the characteristics of the receiver after it has been shown to work with the unimpaired signals.

6.5.4.1 Phase Noise Impairments

The first impairment that will be examined is phase noise, and it can be used to investigate several receiver performance factors. LTE uses OFDMA and SC-FDMA modulation schemes based on either 7.5 kHz (downlink only) or 15 kHz subcarrier spacing. This makes receiver demodulation performance very sensitive to phase noise impairments. A signal generator can be used as a substitute for the local oscillator (LO) in the receiver to investigate demodulation performance. The default phase noise of a signal generator typically is much better than that of the DUT receiver; however, by degrading the phase noise, the performance of the receiver can be evaluated over a range of different operating conditions. This ability to emulate different phase noise levels is useful in determining the optimum LO phase noise requirements for a given receiver design. Figure 6.5-8 shows a screen image of the Agilent N5182A MXG vector signal generator with adjustable phase noise capability. The start and stop frequencies of the "pedestal" region can be set along with the level of the pedestal itself.

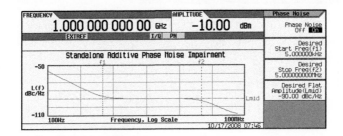

Figure 6.5-8. Agilent N5182A MXG vector signal generator showing phase noise profile for receiver evaluation

Figure 6.5-9 shows a phase noise plot of the MXG with and without additional phase noise added to the signal. Figure 6.5-10 shows the spectrum of two CW tones separated by 15 kHz, which is the normal subcarrier spacing (7.5 kHz is used only in the downlink for broadcast applications). Notice the two levels of phase noise when the pedestal region is varied from 125 dBc/Hz to −90 dBc/Hz. When modulation is applied to the subcarriers and the energy spreads in the frequency domain, the −90 dBc/Hz phase noise leads to significant inter-carrier interference. Phase noise can also be added directly to the signal that the receiver is trying to demodulate. This tests the ability of the receiver to demodulate signals from a transmitter in which the bulk of the EVM budget is represented by phase noise, an impairment more likely to occur in the UE transmitter than in the eNB transmitter. Figure 6.5-11 shows the EVM performance of a downlink signal with phase noise at −90 dBc/Hz in the pedestal region. Note that the EVM performance has degraded from less than 0.5% to more than 6%. The upper figure is close to the entire 64QAM EVM requirement, which indicates that the −90 dBc/Hz phase noise level is above what can be used for this modulation depth.

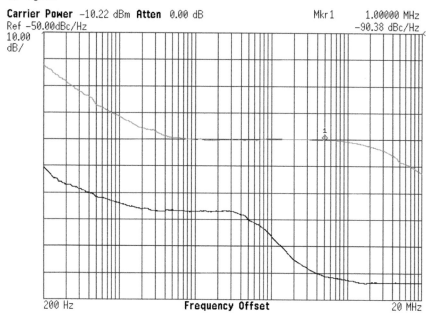

Figure 6.5-9. MXG phase noise with and without phase noise impairment

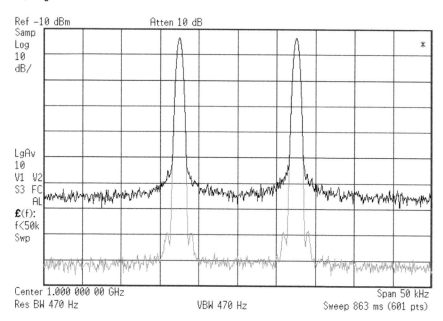

Figure 6.5-10. MXG with two tones spaced at 15 kHz with and without the phase noise impairment

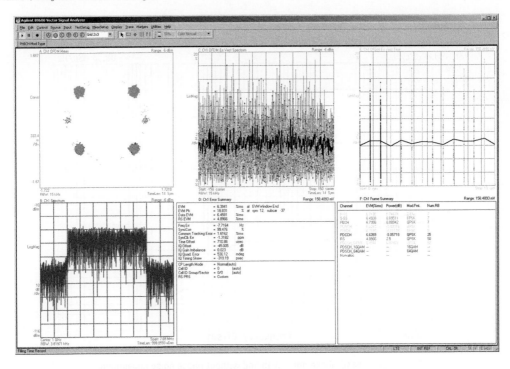

Figure 6.5-11. Modulation analysis of the MXG with phase noise at −90 dBc/Hz showing EVM performance around 6%

Note the interesting constellation in Trace A (upper left). It might be expected that phase noise would produce a rotation of the states rather than what looks here like Gaussian noise. Each subcarrier represents a particular amplitude and phase for one symbol, but the phase noise causes a loss of frequency orthogonality such that energy from the adjacent subcarriers leaks into the wanted subcarrier frequency. Even if the symbols on adjacent subcarriers are of the same amplitude, as in this QPSK case, the phase of the interference will vary depending on the adjacent symbol value. The resulting phase and amplitude error projected on the wanted symbol creates noise rather than just phase rotation.

6.5.4.2 AWGN Impairments

Most receiver and performance requirements are specified with AWGN, which is added to the signal in a controlled way to create a known SINR to represent broadband interference. Table 6.5-1 shows a typical eNB test requirement from 36.141 [25] in which AWGN interference is specified.

Table 6.5-1. Wide area base station dynamic range requirement for 10 MHz channel bandwidth (36.141 [25] Table 7.3-1)

E-UTRA channel bandwidth [MHz]	Reference measurement channel	Wanted signal mean power [dBm]	Interfering signal mean power [dBm]/BW_{config}	Type of interfering signal
10	FRC A2-3 in Annex A-2.	−69.9	−79.5	AWGN

This test requires the signal to be 9.6 dB above the AWGN interference power as shown in Figure 6.5-12.

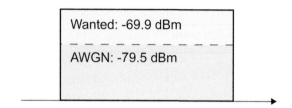

Figure 6.5-12. Wanted and AWGN power requirement for eNB dynamic range test

Most modern signal generators have the ability to generate multiple carriers and AWGN. Using a single signal generator to generate both the wanted and interfering signals makes configuring tests requiring AWGN simpler, more accurate, and less expensive than using separate equipment to generate the signal and noise. However, configuring a single signal generator correctly requires careful consideration.

The output power of a signal generator is defined as the total power of the signal in a wide bandwidth. If multiple signals are combined, such as the wanted signal and AWGN, it is necessary to calculate the combined total power of the waveforms. In addition, "AWGN power" as it is defined in the test specification and as it is implemented in the signal generator may not be exactly the same, so it is important to understand how the wanted and actual AWGN power is calculated.

The AWGN power specified in Table 6.5-1 is defined in a bandwidth, denoted as BW_{config}, which is the transmission bandwidth configuration of the wanted E-UTRA signal. BW_{config} is always less than the channel bandwidth BW_{ch}. (See Section 3.2.4 for a definition of bandwidths). For a channel bandwidth of 10 MHz, BW_{config} is 9 MHz.

To ensure the optimum AWGN flatness across the desired channel bandwidth, it is normal to generate an AWGN signal that is wider than the channel. In this case, it is important to take care when setting the AWGN and signal power to create a known SINR since the power bandwidths of the signal and AWGN may be slightly different.

Figure 6.5-13 shows how a wanted signal and AWGN of a different bandwidth are combined in a way such that the desired SINR is achieved over the transmission bandwidth configuration of the wanted signal. The AWGN flat noise bandwidth, BW_{flat}, has been set to be the same as the 10 MHz channel bandwidth, which is wider than the 9 MHz transmission bandwidth configuration.

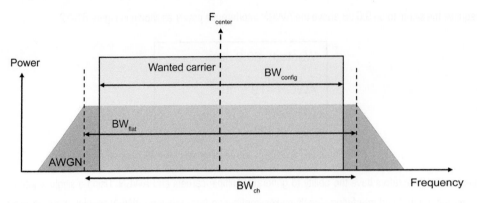

Figure 6.5-13. Wanted signal and AWGN interferer bandwidth

Note from Figure 6.5-13 that the AWGN power looks like a trapezoid and is wider than $\mathrm{BW}_{\mathrm{flat}}$. To calculate the total power of the AWGN, the power falling within $\mathrm{BW}_{\mathrm{flat}}$ has to be multiplied by a factor that is specified by the signal generator model. For the MXG signal generators this factor is 1.25.

The AWGN power defined in Table 6.5-1 for $\mathrm{BW}_{\mathrm{config}}$ can now be related to the total AWGN power using the equation

$$\mathrm{AWGN}_{\mathrm{total}} = \mathrm{AWGN}_{\mathrm{BW}_{\mathrm{config}}} + 10\mathrm{Log}_{10}\left(\frac{\mathrm{BW}_{\mathrm{flat}} \times 1.25}{\mathrm{BW}_{\mathrm{config}}}\right).$$

Using this 10 MHz channel bandwidth example and an AWGN $\mathrm{BW}_{\mathrm{flat}}$ of 10 MHz to cover the wanted signal, the total AWGN power can be calculated as

$$\mathrm{AWGN}_{\mathrm{total}} = -79.5 + 10\mathrm{Log}_{10}\left(\frac{1.25 * 10^7}{9 * 10^6}\right) \cong -78.07 \text{ dBm}.$$

The total power of the wanted signal plus AWGN is

$$\mathrm{P}_{\mathrm{total}} = 10\mathrm{Log}_{10}\left(10\frac{\mathrm{Wanted}}{10} + 10\frac{\mathrm{AWGN}_{\mathrm{total}}}{10}\right) = -69.28 \text{ dBm}.$$

The MXG has been designed with the ability to specify the individual components of the combined signal, including direct entry of the required C/N ratio in dB. From these parameters the appropriate AWGN power is calculated by the instrument. See Figure 6.5-14.

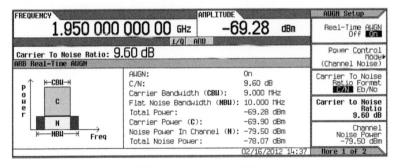

Figure 6.5-14. AWGN setup menu on MXG signal generator

6.5.4.3 IQ Impairments

An IQ demodulator is the most common method of demodulating data from the RF carrier. The demodulator can introduce a number of errors for which the baseband will have to compensate. Gain imbalances in the demodulator can cause unwanted image responses. These unwanted images can also exist when the phase relationship between the I and Q paths is not maintained in quadrature. Additionally, the I and Q paths can have small, unwanted DC offsets that will cause local oscillator leakage (LO feedthrough) to appear at the output. Yet another error is IQ skew (time delay) between the I and Q signals caused by physical path length differences of circuit traces. All of these errors degrade the received signal quality and ultimately the demodulation performance of the receiver. A signal generator can add calibrated IQ distortions to the signal to emulate errors in the transmitter or receiver to determine the impact on demodulation performance. Figure 6.5-15 shows the effect of a 5 ns delay between the I and Q signals on a downlink signal with QPSK modulation. In Trace A (top left) the QPSK constellation states are just starting to become spread due to the timing error, although the consequence of the timing error is more apparent when EVM versus subcarrier is plotted in Trace C (top middle) as a function of subcarrier frequency. The characteristic "V" shape is the result of the timing error progressively affecting carriers farther from the central frequency. The EVM is shown to be around 2% in Trace D (bottom middle).

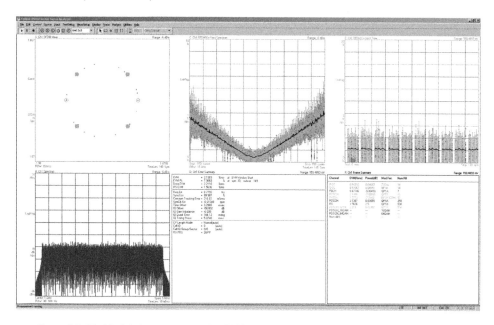

Figure 6.5-15. Modulation accuracy using 89600 VSA software for a 5 MHz downlink signal with 5 ns of timing skew between the I and Q paths

6.6 Receiver Performance Testing

6.6.1 Introduction

Section 6.5 covered basic receiver characteristics under static (unfaded) channel conditions at reference sensitivity. Also covered were the interference, blocking, and adjacent channel selectivity performance. This section covers the UE and eNB receiver performance requirements for the wanted channel including the effects of co-channel interference—often modeled as additive white Gaussian noise (AWGN)—and an impaired or faded downlink radio channel. These conditions are representative of the conditions typical in a loaded network. Section 6.6.2 provides an introduction to fading. Section 6.6.3 covers the requirements for the UE and Section 6.6.4 covers the eNB.

6.6.2 Channel Propagation Impairments

This section describes the models used to characterize the radio channel. There are several levels of these. The primary models, which are typically associated with Rayleigh fading, are based on the spatial channel model, extended (SCME). These were derived from a series of field experiments that demonstrate that MIMO throughput depends not only upon multipath, but also upon the angle of arrival of the signals at the receiver and the angular spread of those signals. These characteristics influence the degree of correlation between the MIMO signals seen by the receiver: the higher the correlation, the lower the throughput. For the sake of standardization, rather than including all these angular definitions in the conformance tests, it was decided for LTE to define fixed correlation matrices (for high, medium and low correlation) that can be used for conducted tests without showing how these correlation values might be computed from specific assumptions about the propagation channel or the transmit and receive antennas at either end. These matrices cover all the common use cases and can be extended to other use cases by formula. These models cover a delay range of 5 μs and Doppler spread up to 300 Hz.

In addition to the primary models, there are secondary fading models. The first of these is for beamsteering. This model adds both a beamsteering matrix with time-variable phase and the precoding matrix W to the correlation matrix calculation.

For channel quality indicator (CQI) tests, a second special fading model is used that varies the amplitude of the channel matrix at a rate defined by the Doppler frequency. For this case, the channel is identical between every pair of Tx/Rx.

There is an multicast/broadcast over single frequency network (MBSFN) channel model which is quite similar to SCME with the exception that it includes very long path delays (delay spread) up to 28.5 μs. This model is necessary for the kinds of deployment envisaged for multimedia broadcast multicast services (MBMS) services.

Finally, there is the high speed train channel model, which is not really fading at all, but a precisely defined pattern of Doppler shifting. For this case the Doppler extends to 750 Hz.

6.6.2.1 Receiver Testing With Channel Propagation Impairments

The impairments affecting the receiver characteristics defined in 36.101 [3] Section 7 are located within the receiver itself or result from of out-of-channel interfering signals. In any real mobile radio environment, the dominant distortion mechanism that the receiver has to deal with is distortion of the wanted signal resulting from impairments in the radio channel propagation conditions. These distortions, commonly known by the more generic term "fading," are the result of changing reflections and frequency shifts caused by fluctuations in the distance between the transmitter and receiver. The radio specifications define specific propagation conditions for use in the receiver performance tests, the general concepts of which are briefly introduced here.

In free space, objects in the environment such as mountains, buildings, and vehicles can reflect, refract, and block transmitted signals. These objects can be at different locations from the receiver with some objects close and some far away. As a result of this variation in distances, a large number of time-delayed copies of the transmitted signal can arrive at the receiver antenna. These time-delayed copies have different phase relationships that cause both constructive and destructive addition at the receiver.

Small variations in the phase relationships occur because of the movement of objects in the environment—for example, pedestrians and vehicles. If the UE is moving, then the phase variations can be large with the rate of change being a function of the UE speed. Empirical data shows that the received signal level fluctuates up and down; occasionally dropping to very low levels when nearly complete signal cancellation occurs. A moving transmitter or receiver also can cause the signal to undergo a Doppler (frequency) shift. Since the time-delayed copies of the signal arrive from different directions relative to the direction of motion, some of the signals are shifted to higher frequencies and some are shifted to lower frequencies. This causes particular difficulty in OFDM systems since simple frequency shifting cannot be used to remove the inter-subcarrier interference.

As the number of paths increases, the superposition of multiple copies of the transmitted signal at different amplitudes, timings, frequencies, and phases results in a stochastic (non-deterministic) signal that follows a Rayleigh distribution. This distribution is known to accurately represent the amplitude fluctuations and frequency variations (also known as Doppler spreading) typical in urban environments. This type of multipath propagation condition is used in specifying receiver performance requirements. For the LTE UE, propagation conditions are specified in 36.101 [3] Annex B, and a similar set of conditions is specified for the eNB in 36.104 [4] Annex B.9.

Propagation conditions are made up of three components: a multipath delay profile; a Doppler spread; and, for multi-antenna requirements, a set of correlation matrices defining the correlation between the transmitting and receiving antennas. An example of a delay profile for a typical pedestrian environment is given in Table 6.6-1.

Table 6.6-1. Extended Pedestrian A model (EPA) (36.101 [10] Table B.2.1-2)

Excess tap delay [ns]	Relative power [dB]
0	0.0
30	−1.0
70	−2.0
90	−3.0
110	−8.0
190	−17.2
410	−20.8

For LTE, three delay profiles are defined: one for pedestrian use shown in Table 6.6-1, one for vehicular use, and one for typical urban use. A delay profile includes the number of delay taps, the delay, and the attenuation. The delay profile is further defined by a root mean square (rms) delay spread. The three delay profiles for LTE are shown in Table 6.6-2.

Table 6.6-2. Delay profiles for E-UTRA channel models (36.101 [10] Table B.2.1-1)

Model	Number of channel taps	Delay spread (rms)	Maximum excess tap delay (span)
Extended Pedestrian A (EPA)	7	45 ns	410 ns
Extended Vehicular A (EVA)	9	357 ns	2510 ns
Extended Typical Urban (ETU)	9	991 ns	5000 ns

In addition to delay profile, it is necessary to specify a maximum Doppler frequency shift. Three different frequencies are used to represent low, medium, and high speeds of 5 Hz, 70 Hz, and 300 Hz, respectively. At 2 GHz these frequencies translate to UE velocities of 2.7 km/h, 37.8 km/h, and 162 km/h. The maximum Doppler frequencies are combined with the delay profiles to produce the combinations that are used to define the receiver performance requirements. Although there are three delay profiles and three Doppler frequencies, only five of the possible nine combinations are used. These are shown in Table 6.6-3.

Table 6.6-3. Channel model parameters (36.101 [10] Table B.2.2-1)

Model	Maximum Doppler frequency
EPA 5 Hz	5 Hz
EVA 5 Hz	5 Hz
EVA 70 Hz	70 Hz
ETU 70 Hz	70 Hz
ETU 300 Hz	300 Hz

The pedestrian profile is defined for the low speed 5 Hz Doppler frequency only; the vehicular profile is defined for 5 Hz and 70 Hz only; and the typical urban profile excludes the 5 Hz case. One additional delay profile has been defined specifically for high speed trains. This definition is based on the velocity of a train and the distances between base stations located at a specified distance from the track. The result is a Doppler shift with a nearly square wave profile.

An important difference exists between the methods of specifying propagation conditions for LTE and for UMTS. In UMTS, the delay profiles are specified for different UE velocities. Thus for performance testing, each frequency band requires a different set of Doppler frequencies. In the early days of UMTS, when only one band was defined, this requirement was not a problem. In Release 11, however, 28 bands are defined (26 by 3GPP and 2 by ETSI; see Section 2.1.1), and this has created a huge growth in test configurations. To reduce the test burden for LTE, it was decided to fix the Doppler frequencies for the extended pedestrian A (EPA), extended vehicular A (EVA), extended typical urban (ETU), and high speed train (HST) profiles and apply these to all frequency bands. As a result the effective UE velocity has now become variable as a function of the test frequency, and for the HST case, the inter-site distances have also become variable. This approach is not as pure as the fixed velocity approach for UMTS; nevertheless, fixing the Doppler frequencies is expected to provide sufficient test coverage with the benefit of significant test simplification.

Figure 6.6-1 shows amplitude fluctuations over a period of one second for one path of Rayleigh fading at a Doppler spread of 10 Hz (5.4 km/h at 2 GHz). The dynamic range of the signal is contained mainly within the top 10 dB although much deeper fades can be seen.

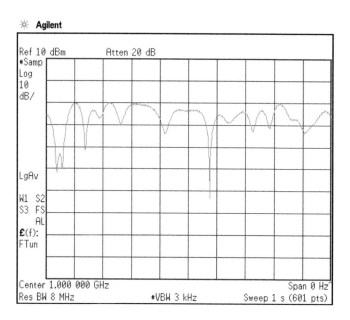

Figure 6.6-1. Time domain profile of one path of Rayleigh fading applied using the Agilent PXB MIMO receiver tester

The constellation diagram in Figure 6.6-2 shows how one path of Rayleigh fading affects a QPSK modulated signal. Although the signal in this simple example is not an LTE signal with multicarrier OFDMA or broadband SC-FDMA modulation, it illustrates the problem that Rayleigh fading presents to the receiver and the need to properly verify the receiver under a variety of propagation conditions. The trajectory of the constellation can be seen to change in both amplitude and phase as is expected with Rayleigh fading. It is interesting to note that the constellation approaches the origin during deep fades and the Doppler spreading causes the trajectory to migrate from one quadrant to the next. The propagation conditions defined for LTE use up to nine paths, creating a much more complex signal than the ones shown in Figures 6.6-1 and 6.6-2.

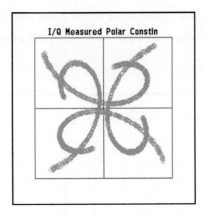

Figure 6.6-2. Constellation diagram of a single carrier QPSK signal with one path of Rayleigh fading

The third element of the propagation conditions consists of the antenna correlation matrices. A radio system with two transmit antennas and two receive antennas will have four different propagation channels. This then requires four separate fading emulators to properly emulate the radio channel. An example of how to generate multi-channel signals with fading using the Agilent N5106A PXB MIMO receiver tester is shown in Figure 6.6-3. This figure shows the fading profiles for a downlink 4x2 MIMO case. A more complete description of channel emulation can be found in the application note "MIMO Channel Modeling and Emulation Test Challenges" [26]. Correlation matrices are further discussed in Sections 6.6.3 and 6.8.

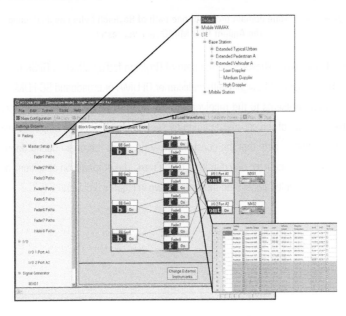

Figure 6.6-3. Agilent N5106A PXB MIMO receiver tester configured for 4x2 channel emulation
showing the LTE downlink MIMO fading profiles

6.6.3 UE Receiver Performance Characteristics

This section examines some of the UE receiver performance characteristics for data throughput in faded channel conditions. Also covered is the UE's ability to correctly report channel state information (CSI), which is required by the eNB to dynamically adjust the downlink signal to the varying propagation and interference conditions. These techniques are used to optimize throughput.

6.6.3.1 Testing Throughput Performance in a Faded Channel

Throughput testing is performed using an eNB emulator (also known as a system simulator) consisting of a source and receiver under the control of a real time protocol stack. An example of an eNB emulator is the Agilent E6621A wireless test set. This product is capable of setting up a connection with the UE and maintaining and controlling that connection in real time. By accurately setting downlink power and noise conditions, and by using uplink UE reports such as the UE CSI and hybrid-automatic repeat request (HARQ) information to adapt to the downlink channel conditions, the eNB emulator is capable of emulating real world conditions. For receiver performance testing, it is also necessary to add a fading emulator to the downlink for both single channel and multiple channel (MIMO) configurations to fully exercise the UE under a wide range of radio link conditions.

To test the receiver performance characteristics defined in 36.101 [3] Section 8 requires the addition of a fading propagation channel for the downlink only. The uplink remains unfaded (ideal radio conditions) in order that the performance of the receiver in the eNB emulator does not influence the measurements made of the UE receiver. Testing of the full system with bi-directional fading is outside the scope of the 3GPP requirements but is something that is carried out by network equipment and UE vendors as well as network operators (see Section 7.5).

The term fixed reference channel (FRC) is used to describe the signal definition. An FRC implies that the underlying reference measurement channel (RMC) does not include any of the dynamic parameters, such as coding rate or modulation depth, that are necessary if adaptive modulation and coding (AMC) is applied. Some of the CSI test requirements make use of the UE's reported values; in these cases, the eNB emulator must vary the RMC accordingly. In real network operation AMC is active and the format of the signals and allocations varies in a much more complex manner.

During the development of UMTS the possibility of defining requirements for using variable reference channels (VRCs) was discussed but never implemented. These would have tested closed-loop receiver performance with AMC enabled. The LTE requirements from Release 8 are also defined only for FRC and hence the terms FRC and RMC may appear interchangeable. VRCs have been used in the application layer performance tests defined in Release 11. These tests do not have performance requirements defined in 36.101 [3]. This topic is covered in Section 7.4.5. Referring to a receiver test condition using the term FRC rather than RMC makes it clear that AMC is not enabled in this scenario.

The FRCs for the receiver requirements have no specific names, but those for the performance requirements are named numerically: R.3 FDD, R.4 FDD, R.7 TDD, etc. A full discussion of the many receiver and performance requirements is beyond the scope of this chapter, so one example is described here in detail to show how the various elements including the FRC, channel propagation conditions, and other parameters combine to create the right conditions for ensuring that the UE can deliver the required performance.

6.6.3.2 Throughput Performance Example With Spatial Multiplexing

The performance requirement chosen for discussion is a requirement for multi-layer spatial multiplexing (a.k.a. MIMO) as specified in 36.101 [3] Subclause 8.2.1.4. The downlink transmission mode for this case is transmission mode 4 (TM4) "closed-loop spatial multiplexing"; see Section 2.4 for more details. The term "closed-loop" refers to the use of dynamic precoding with precoding matrix indicator (PMI) feedback. Otherwise this test is an open-loop test as far as the RMC is concerned, since AMC is disabled and an FRC is used. The requirement is complex and involves many advanced features of LTE. The purpose of the requirement is to define the throughput performance for two spatial layers with wideband and frequency selective precoding. Table 6.6-4 defines the general parameters for the requirement.

Table 6.6-4. Test parameters for multi-layer spatial multiplexing (FRC) 36.101[3] Table 8.2.1.4.2-1

Parameter		Unit	Test 1-2
Downlink power allocation	ρ_A	dB	−3
	ρ_B	dB	−3 (Note 1)
N_{oc} at antenna port		dBm/15kHz	−98
Precoding granularity		Physical resource block (PRB)	50
PMI delay (Note 2)		ms	8
Reporting interval		ms	1
Reporting mode			PUSCH 3-1
CodeBookSubsetRestriction bitmap			110000

Note 1: $\rho_B = 1$.
Note 2: If the UE reports in an available uplink reporting instance at subframe SF#n based on PMI estimation at a downlink SF not later than SF#(n−4), this reported PMI cannot be applied at the eNB downlink before SF#(n+4).

Table 6.6-5 provides the specific performance requirements for several configurations of channel bandwidth, modulation and coding scheme (MCS), reference channel, channel propagation conditions, and MIMO channel correlation matrix for 2x2 and 4x2 antenna configurations.

Table 6.6-5. Minimum performance for multi-layer spatial multiplexing (FRC) (36.101 [3] Table 8.2.1.4.2-2)

Test number	Bandwidth	Reference channel	OCNG pattern	Propagation condition	Correlation matrix and antenna configuration	Reference value		UE category
						Fraction of maximum throughput (%)	SNR (dB)	
1	10 MHz	R.35 FDD	OP.1 FDD	EPA5	2x2 low	70	18.9	2–5
2	10 MHz	R.11 FDD	OP.1 FDD	ETU70	2x2 low	70	14.3	2–5

The performance requirement is the signal-to-noise ratio (SNR) at which 70% throughput is achieved. The SNR definition for the 2x2 MIMO case is

$$SNR = \frac{\hat{E}_s^{(1)} + \hat{E}_s^{(2)}}{N_{oc}^{(1)} + N_{oc}^{(2)}}$$

where $\hat{E}_s$ represents the received energy per resource element (RE) during the useful part of the symbol (excluding the cyclic prefix) averaged across all allocated resource block (RB) and normalized to the subcarrier spacing. The denominator N_{oc} is the power spectral density of a white noise source, averaged per RE and normalized to the subcarrier spacing. The superscripts refer to the receiver antenna number.

There are two ways to devise a test based on the SNR requirement. The first is to determine the SNR at which 70% throughput is reached, and the second is to fix the SNR and determine the throughput. For practical reasons the LTE conformance tests fix the SNR and measure the throughput because this can be achieved in a single step. However, in an R&D environment the variable SNR method involving iteration is useful in determining how many dB of margin the receiver has relative to the minimum requirement. Table 6.6-6 shows the parameters for the required FRC, in this case R.11 FDD (explained below).

Table 6.6-6. Fixed reference channel with two antenna ports (from 36.101 [3] Table A.3.3.2.1-1)

Parameter	Unit	Value					
		R.10 FDD	R.11 FDD	R.11-2 FDD	R.11-3 FDD	R.30 FDD	R.35 FDD
Reference channel							
Channel bandwidth	MHz	10	10	5	10	20	10
Allocated resource blocks		50	50	25	40	100	50
Allocated subframes per radio frame		10	10	10	10	10	10
Modulation		QPSK	16QAM	16QAM	16QAM	16QAM	64QAM
Target coding rate		1/3	1/2	1/2	1/2	1/2	1/2
Information bit payload							
For subframes 1, 2, 3, 4, 6, 7, 8, 9	Bits	4392	12960	5736	10296	25456	19848
For subframe 5	Bits	n/a	n/a	n/a	n/a	n/a	n/a
For subframe 0	Bits	4392	12960	4968	10296	25456	18336
Number of code blocks per subframe (Note 3)							
For subframes 1, 2, 3, 4, 6, 7, 8, 9	Bits	1	3	1	2	5	4
For subframe 5	Bits	n/a	n/a	n/a	n/a	n/a	n/a
For subframe 0	Bits	1	3	1	2	5	3
Binary channel bits per subframe							
For subframes 1, 2, 3, 4, 6, 7, 8, 9	Bits	13200	26400	12000	21120	52800	39600
For subframe 5	Bits	n/a	n/a	n/a	n/a	n/a	n/a
For subframe 0	Bits	12384	24768	10368	19488	51168	37152
Max. throughput averaged over one frame	Mbps	3.953	11.664	5.086	9.266	22.910	17.712
UE category		1–5	2–5	1–5	1–5	2–5	2–5

The FRC definition provides all the parameters necessary to calculate the maximum throughput averaged across one 10 ms frame. The reason for the averaging is that within the frame structure the number of data symbols varies in each subframe, but over the length of one frame the number of data symbols is constant. See Section 3.2.6 for more details on the frame structure. R.11 FDD is used for the requirement being described. It is a fully allocated RMC (all available RBs are used) giving a maximum fully allocated throughput of 12.960 Mbps. However, since no data is allocated on subframe 5 during these tests (to prevent mixing of data and SIBs), the maximum throughput averaged over the frame is 11.664 Mbps. The eNB emulator therefore has to be capable of disabling data transmission in subframe 5. The target performance for the configurations given in Table 6.6-5 would be 70% of this figure. The other RMCs defined in Table 6.6-6 are used for different test requirements. The coding rate refers to the amount of redundancy in the channel coding; the lower figures represent more repetition with corresponding lower throughput and better resilience to noise and interference. The maximum throughput for R.11 FDD is three times that of R.10 FDD due to the use of higher order modulation and a higher coding rate. This higher throughput indicates that R.10 FDD is a more robust signal and is used for requirements based on much lower SNR than requirements using R.11 FDD.

Table 6.6-5 specifies the required MIMO correlation matrix and antenna configuration. The details for these are shown in Table 6.6-7. The definition of the correlation matrices is given in terms of the eNB correlation α and UE correlation β as defined in 36.101 [3] Annex B.2.3. See also Section 6.8.5.2. For the 2x2 case the channel correlation matrix R is defined for the eNB and UE as

$$R_{eNB} = \begin{pmatrix} 1 & \alpha \\ \alpha^* & 1 \end{pmatrix} R_{UE} = \begin{pmatrix} 1 & \beta \\ \beta^* & 1 \end{pmatrix}.$$

When α and β are zero, the channel can be described as if it were two completely independent channels with no leakage from one to the other. As α and β approach 1, the channel starts to look like a single input, single output (SISO) channel in which the transmitted signals add together in the channel and the signals at both receivers are identical, thus reducing any potential spatial multiplexing gain to zero. For further discussion of channel correlation refer to Sections 2.4 and 6.8.

Table 6.6-7. MIMO correlation matrices (36.101 [3] Table B.2.3.2-1)

Low correlation		Medium correlation		High correlation	
α	β	α	β	α	β
0	0	0.3	0.9	0.9	0.9

The low correlation scenario is essentially a perfect MIMO channel in which the paths are completely de-correlated. For the 2x2 case this means a potential doubling of throughput compared to the case of a SISO channel. Very low correlation is an impractical and unlikely real-world scenario; however, it does enable like-for-like comparison with the equivalent performance requirements in UMTS, which are also based on full de-correlation. The medium correlation scenario represents a more likely channel situation, while the high correlation scenario describes a situation in which the paths are almost fully correlated—the point at which spatial multiplexing becomes ineffective. It can be seen from Table 6.6-5 that only the low (no) correlation case has been specified; thus the requirement can be mapped to a high SNR radio environment with a nearly perfect uncorrelated channel. The intention of specifying a performance requirement in such conditions is not meant to be indicative of typical spatial multiplexing performance in the real world. Rather the intention is to determine if the UE is capable of reaching the maximum performance in ideal conditions.

To further characterize the UE's throughput performance it is necessary to supplement this scenario with additional minimum requirements covering QPSK, 64QAM, other propagation conditions, and the medium and high correlation matrices.

6.6.3.3 CSI Performance Testing

Channel state information consists of reports sent by the UE to the eNB to provide real-time updates of the downlink channel conditions. CSI is introduced in Section 3.4.6 and includes the PMI, CQI, and rank indication (RI).

PMI reporting is used for closed-loop spatial multiplexing (transmission mode 4). In this mode the PMI report is used to modify the phase and amplitude of the downlink signal in order to optimize decoding performance for channels supporting rank > 1. PMI reporting is also used in transmission mode 6 (UE-specific beamforming) in which the precoding is used to steer the downlink for single rank transmissions.

The CQI reports are provided by the UE to enable the eNB to select the appropriate MCS and RB allocation for each UE, in order to optimize the channel conditions and therefore maximize throughput and use of valuable eNB resources. Incorrect reporting of CQI will generally result in reduced throughput. Both PMI and CQI can be wideband, representing the average across the entire channel bandwidth, or subband, representing the average over a subset of adjacent RBs. Rank indication is reported by the UE to the eNB and indicates the preferred number of layers to be used for transmission modes supporting spatial multiplexing.

The high degree of flexibility in the definition of CSI parameters has led to the definition of several different PUCCH reporting formats. These formats are described in Section 3.4.6 and the use of these formats for various tests in shown in Table 6.6-8. It would be impractical if not impossible to test every combination of CQI, PMI, RI, wideband, and subband, so 36.101 instead defines a limited group of tests to characterize CSI performance. These are summarized in Table 6.6-8.

In Table 6.6-8 specific channel conditions are used for most of the CQI tests, as defined in 36.101[3] Annex B.2.4. For the CQI tests, the following additional multipath profile is used:

$$h(t,\tau) = \delta(\tau) + a\exp(-i2\pi f_D t)\delta(\tau - \tau_d)$$

in continuous time (t, τ) representation, with delay (τ_d), constant (a), and Doppler frequency (f_D).

CQI testing is performed both with and without fading, while all of the tests for PMI and RI requirements include the use of various fading profiles. With only three exceptions all the CSI tests are applicable to both FDD and TDD. The majority of CSI tests are comparative and performed in several stages, and many include repetition for various test conditions and different SNR levels. The RI tests require repetition at different channel correlations.

The most notable real-world omission from these tests is AMC, in which the downlink modulation and coding varies constantly as well as the RB allocation and transport block size (TBS). A possible test might use a fully closed-loop condition in which the CSI performance is inferred from the throughput performance of the system acting upon the real-time CSI feedback. This approach should give a better indication of real-world performance; however, it requires the definition of additional AMC and scheduling behavior in the eNB emulator, a definition that so far has been considered both very difficult and outside the scope of the specifications. For further discussion of testing with AMC, see Section 7.4.5.

The minimum CSI performance requirements are scoped narrowly. Nevertheless, evaluating UE CSI performance under closed-loop conditions with AMC and true dynamic scheduling is a valuable exercise, particularly during R&D, even though there are no formal requirements against which to assess the results. The differences in closed-loop CSI performance among different UE designs has been an issue for UMTS and will be even more so with LTE as a result of the greater flexibility permitted in the definition of the CSI reports. Because there are so many CSI requirements, the remainder of this section will examine representative examples of the various tests.

Table 6.6-8. Summary of CSI test cases

Test title, 36.521 test reference	Channel	SNR options, test count	PUCCH format	Description
CQI reporting under AWGN, 9.2.1	AWGN (1x2)	2, 2	1-0	Compares BLER using CQI median +/−1 values
CQI reporting under AWGN, 9.2.2	AWGN (2x2)	2, 2	1-1	Compares BLER for each codeword using CQImedian +/−1 values
CQI frequency-selective scheduling, 9.3.1	Clause B.2.4 with $\tau_d = 0.45$ µs, $a = 1$, $f_D = 5$ Hz, full correlation	2, 2	3-0	Measures throughput with random eNB subband allocation, then re-measures with UE reported subband allocation. The test checks for a minimum throughput gain using UE-reported subband allocation.
CQI frequency non-selective scheduling, 9.3.2	EPA5, high	2, 2	1-0 to avoid CQI and ACK collisions	Comparison of throughput using UE-reported CQI to throughput using fixed CQI median
CQI frequency-selective interference, 9.3.3 Subband size 6 RB	Clause B.2.4 with $\tau_d = 0.45$ µs, $a = 1$, $f_D = 5$ Hz	2, 1	3-0	Same as 9.3.1
CQI UE-selected subband, 9.3.4.1 Subband size 3 RB	Clause B.2.4 with $\tau_d = 0.45$ µs, $a = 1$, $f_D = 5$ Hz	2, 2	2-0	Same as 9.3.1
CQI UE-selected subband, 9.3.4.2 Subband size 6 RB	Clause B.2.4 with $\tau_d = 0.45$ µs, $a = 1$, $f_D = 5$ Hz	2, 2	PUSCH 2-0 to avoid CQI and ACK collisions	Same as 9.3.1
Single PMI reporting, 9.4.1.1	EVA5, low 2x2	1, 1	PUSCH 3-1	Compares random precoding-matrix-reported TP to UE-reported precoding matrix TP
Single PMI reporting, 9.4.1.2	EVA5, low 4x2	1, 1	PUCCH 2-1 FDD only	Same as 9.4.1.1
Multiple PMI reporting 9.4.2.1	EPA5, low 2x2	1, 1	PUSCH 1-2	Same as 9.4.1.1
Multiple PMI reporting, 9.4.2.2	EVA5, low 4x2	1, 1	PUSCH 2-2	Same as 9.4.1.1
Rank indication reporting, 9.5.1.1	EPA5, low and high 2x2	1, 3	PUCCH 1-1 FDD only	Compares TP with fixed rank to TP with reported rank for three separate channel and rank conditions
Rank indication reporting 9.5.1.2	EPA5, low and high 2x2	1, 3	PUCCH 3-1 TDD only	Same as 9.6.1.2

Wideband CQI Test With AWGN (PUCCH Format 1.0)

CQI is an essential component of the LTE scheduling and link adaptation procedure. Constant feedback of the channel conditions allows the eNB to change the downlink allocation and modulation coding scheme to suit the prevailing channel conditions, thereby maximizing throughput and making best use of the available network resources. The UE must accurately gauge the channel conditions and report in a timely manner.

The wideband CQI requirements specified in 36.101 [3] Subclause 9 follow the precedent set by UMTS. The wideband CQI test with AWGN is a multi-stage requirement based on the conditions shown in Table 6.6-9.

Table 6.6-9. PUCCH 1-0 static test (36.101 [3] Table 9.2.1.1-1)

Parameter		Unit	Test 1		Test 2	
Bandwidth		MHz	10			
PDSCH transmission mode			1			
Downlink power allocation	ρ_A	dB	0			
	ρ_B	dB	0			
Propagation condition and antenna configuration			AWGN (1x2)			
SNR (Note 2)		dB	0	1	6	7
$\hat{I}_{or}^{(j)}$		dB [mW/15kHz]	−98	−97	−92	−91
$N_{oc}^{(j)}$		dB [mW/15kHz]	−98		−98	
Max number of HARQ transmissions			1			
Physical channel for CQI reporting			PUCCH format 2			
PUCCH report type			4			
Reporting periodicity		ms	NP = 5			
cqi-pmi-ConfigurationIndex			6			

Note 1: Reference measurement channel according to Table A.4-1 with one sided dynamic OCNG Pattern OP.1 FDD as described in Annex A.5.1.1.

Note 2: For each test, the minimum requirements shall be fulfilled for at least one of the two SNR(s) and the respective wanted signal input level.

This requirement represents the simplest case for CQI, being wideband (the full channel bandwidth) and having a static propagation channel with the adjacent cell interference represented by AWGN and the internal cell interference represented by the use of the orthogonal channel noise generator (OCNG). The wanted-signal and noise-signal powers per subcarrier are represented by $\hat{I}_{or}^{(j)}$ and $N_{oc}^{(j)}$ with the SNR for the two test conditions of 0 or 1 dB and 6 or 7 dB derived from these two parameters.

The test is performed in multiple stages. The first stage is used to determine the median value of the reported CQI and to establish whether at least 90% of all CQI reports fall within ±1 of the median. The number of CQI reports that have to be captured to obtain the required statistical significance is 2000. The second stage of each test checks the accuracy of the CQI reports around the 10% block error ratio (BLER) target.

The median CQI-1 value derived in the first stage of the test is used by the eNB emulator to measure BLER, which must be less than 10%. The median CQI +1 value is then used by the eNB emulator and the BLER must be greater than 10%. If the UE fails this test using the first SNR value (0 dB), then the test sequence can be repeated using the second value (1 dB). The UE must pass at least one of these two tests. The test is then repeated for the SNR of 6 dB, and if necessary 7 dB.

Frequency Selective Scheduling CQI Test With Fading

This test measures the increase in throughput achieved under fading conditions when the subband with the highest reported differential CQI is used rather than a transmission with fixed frequency allocation. This is a multi-stage test using parameters described in Table 6.6-10. When the channel bandwidth is divided up into subbands, one of the subbands may be a different size than the others. During this test and other frequency selective scheduling tests, if the subdividing of the channel results in a subband that is smaller is size then the other subbands, then this subband is not used, as it would skew the throughput test values.

Table 6.6-10. Subband test for single antenna transmission (FDD) (36.101 [3] Table 9.3.1.1.1-1)

Parameter	Unit	Test 1		Test 2	
Bandwidth	MHz	10 MHz			
Transmission mode		1 (port 0)			
SNR (Note 3)	dB	9	10	14	15
$\hat{I}_{or}^{(j)}$	dB [mW/15kHz]	−89	−88	−84	−83
$N_{oc}^{(j)}$	dB [mW/15kHz]	−98		−98	
Propagation channel		Clause B.2.4 with $\tau_d = 0.45$ μs, $a = 1$, $f_D = 5$ Hz			
Correlation		Full			
Reporting interval	ms	5			
CQI delay	ms	8			
Reporting mode		PUSCH 3-0			
Max number of HARQ transmissions		1			

Note 1: If the UE reports in an available uplink reporting instance at subframe SF#n based on CQI estimation at a downlink subframe not later than SF#(n−4), this reported subband or wideband CQI cannot be applied at the eNB downlink before SF#(n+4)

Note 2: Reference measurement channel according to Table A.4-4 with one/two sided dynamic OCNG Pattern OP.1/2 FDD as described in Annex A.5.1.1/2

Note 3: For each test, the minimum requirements shall be fulfilled for at least one of the two SNR(s) and the respective wanted signal input level.

In the first stage of this test the eNB emulator sends data to the UE under faded conditions and the UE reports wideband and subband CQI until 2000 reports have been gathered. The test requires that a subband differential CQI offset level of 0 be reported at least α % of the time but less than β % for each subband. The α and β values for this test are defined in Table 6.6-11, along with the throughput pass/fail limit γ.

Table 6.6-11. Subband test for single antenna transmission minimum requirement (FDD) (36.101 [3] Table 9.3.1.1.1-2)

Parameter	Test 1	Test 2
α [%]	2	2
β [%]	55	55
γ	1.1	1.1
UE category	1–5	1–5

The median wideband CQI value (CQI_median) is then calculated and used for the throughput stages of this procedure. The eNB emulator takes the CQI median value and transmits in random subband of equal sizes with equal probability. UE subband reports are ignored during this stage of the test. The throughput is measured and referenced as t (median).

The eNB emulator now transmits in the subband with the highest differential CQI, using the CQI value reported for that subband by the UE. The eNB emulator uses the reported CQI value to allocate the same subband to the UE until the UE reports again. The average throughput is computed and is referenced as t (subband). To pass the test, t (subband)/t (median) must be greater than γ in Table 6.6-11 and the BLER must be greater than 5%. If the UE fails the test using the first SNR value (9 dB), then the test sequence can be repeated using the second value (10 dB). The UE must pass at least one of these two tests. The test is then repeated using the SNR value of 14 dB and if the UE does not pass then 15 dB may be used.

PMI Testing

Precoding involves the selection of the most suitable amplitude and phase of the signals transmitted to the UE to suit the downlink channel conditions reported by the UE to the eNB. The UE calculates the most appropriate precoding from a predefined codebook. The use of a codebook reduces feedback accuracy but greatly minimizes the amount of uplink data reported back to the eNB.

Testing for correct PMI reporting will be vitally important for LTE, particularly in environments with varying channel conditions. Incorrect or delayed reporting could result in less optimal signal conditions at the receiver leading to demodulation errors and decreased throughput. There are several PMI reporting tests for both TDD and FDD specified in 36.101 Section 9. Which tests are required depends on the reporting format and whether the PUSCH or PUCCH is being used to report PMI. All the defined tests use transmission mode 6 and a variety of fixed reference channels, which are selected according to the requirements of the specific test. A typical FRC taken from 36.521-1 [27] Table 9.4.1.1.1-1 is shown in Table 6.6-12.

Table 6.6-12. PMI test for single layer (FDD) (36.101 [3] Table 9.4.1.1.1-1)

Parameter		Unit	Test 1
Bandwidth		MHz	10
Transmission mode			6
Propagation channel			EVA5
Precoding granularity		PRB	50
Correlation and antenna configuration			Low 2x2
Downlink power allocation	ρ_A	dB	−3
	ρ_B	dB	−3
$N_{oc}^{(j)}$		dB[mW/15kHz]	−98
Reporting mode			PUSCH 3-1
Reporting interval		ms	1
PMI delay (Note 2)		ms	8
Measurement channel			R.10 FDD
OCNG pattern			OP.1 FDD
Max number of HARQ transmissions			4
Redundancy version coding sequence			{0, 1, 2, 3}

Note 1: For random precoder selection, the precoder shall be updated in each TTI (1 ms granularity).
Note 2: If the UE reports in an available uplink reporting instance at subframe SF#n based on PMI estimation at a downlink SF not later than SF#(n−4), this reported PMI cannot be applied at the eNB downlink before SF#(n+4).

These test conditions relate to the testing of a single PMI using PUSCH format 3-1 for FDD; however, the test conditions are similar for all PMI reporting tests, varying only in the FRC used and results obtained. PMI testing is performed as a three-stage comparative throughput test. The test conditions are stated in 36.521-1 [27].

The first stage of the test is performed to establish the value SNR_{rnd}. This value is the signal to noise ratio that will be used during the second and third stages of the test. 36.101 [3] Annex G.5.2 specifies how to establish SNR_{rnd} by adjusting the SNR until the throughput is settled between 58% and 62% of the calculated maximum throughput t_{rnd}. The second stage of the test is performed using random precoding and the third stage using UE-reported PMI values. Throughput results are obtained with these two stages, and the ratio is expressed as a precoding gain

$$\gamma = \frac{t_{ue}}{t_{rnd}}$$

where t_{ue} represents the throughput using the UE's reported PMI and t_{rnd} represents the throughput value obtained using random PMI values and ignoring the PMI values reported by the UE. The random PMI values used in the second stage change with each subframe. The second and third throughput test stages are both conducted at SNR_{rnd} as established in the first stage of the test.

This test has a single limit for all UE categories, shown in Table 6.6-13. Other defined PMI tests may have different results depending on UE category. A pass is achieved if the ratio γ is exceeded.

Table 6.6-13. Minimum requirement (FDD) (36.101 [3] Table 9.4.1.1.1-2)

Parameter	Test 1
γ	1.1
UE category	1–5

RI Testing

Rank indication is a value computed by the UE and represents the preferred number of layers to be used in the next downlink transmission to the UE. The reported RI represents the maximum number of spatial layers supported by the channel but limited by the UE capabilities. Although LTE provides for up to four layers, the minimum UE requirement is for two layers. RI is computed as follows: for each possible rank, the subbands are ordered by decreasing level of throughput. The best M subbands are selected and the sum throughput is derived. The rank maximizing the sum throughput is selected and reported to the eNB emulator.

A requirement based on 36.101[3] Subclause 9.5.1 verifies that the reported RI accurately represents the channel rank and that using the reported rank as opposed to a fixed rank results in improved throughput. Transmission mode 4 is used for the test sequence. This RI requirement is tested in three separate ways that use varying SNR, restricted codebook subsets, and RI configurations.

Each test has two main test stages. Test stage (a) establishes the value t_{fixed}. Using the restricted codebook subset for single rank, the eNB emulator is configured using the parameters shown in Table 6.6-14. These include the relevant reference channels, propagation conditions, antenna configuration, specific codebook subset restriction bitmap, and SNR. The eNB emulator responds to the UE reported CQI, RI, and PMI with uplink grants to the UE.

Test stage (b) is the stage in which the UE is told to use the codebook subset for the UE-reported RI case (shown in Table 6.6-14) along with all the other parameters needed to establish $t_{reported}$.

For fixed rank 1 transmission, the RI and PMI reporting is restricted to two single-layer precoders. For fixed rank 2 transmission, the RI and PMI reporting is restricted to one 2-layer precoder. For the RI transmissions used in the second test stage, the RI and PMI reporting is restricted to the union of the rank 1 and rank 2 precoders. Channels with low and high correlation are used to ensure that RI reporting reflects the channel condition.

Table 6.6-14. RI test (FDD) (36.101 [3] Table 9.5.1.1-1)

Parameter		Unit	Test 1	Test 2	Test 3
Bandwidth		MHz	10		
PDSCH transmission mode			4		
Downlink power allocation	ρ_A	dB	−3		
	ρ_B	dB	−3		
CodeBookSubsetRestriction bitmap			000011 for fixed RI = 1 010000 for fixed RI = 2 010011 for UE reported RI		
Propagation condition and antenna configuration			2x2 EPA5		
Antenna correlation			Low	Low	High
RI configuration			Fixed RI=2 and follow RI	Fixed RI=1 and follow RI	Fixed RI=2 and follow RI
SNR		dB	0	20	20
$N_{oc}^{(j)}$		dB[mW/15kHz]	−98	−98	−98
$\hat{I}_{or}^{(j)}$		dB[mW/15kHz]	−98	−78	−78
Maximum number of HARQ transmissions			1		
Reporting mode			PUCCH 1-1 (Note 4)		
Physical channel for CQI/PMI reporting			PUCCH format 2		
PUCCH report type for CQI/PMI			2		
Physical channel for RI reporting			PUSCH (Note 3)		
PUCCH report type for RI			3		
Reporting periodicity		ms	NP = 5		
PMI and CQI delay		ms	8		
cqi-pmi-ConfigurationIndex			6		
ri-ConfigurationInd			1		

Note 1: If the UE reports in an available uplink reporting instance at subframe SF#n based on PMI and CQI estimation at a downlink subframe not later than SF#(n−4), this reported PMI and wideband CQI cannot be applied at the eNB downlink before SF#(n+4).

Note 2: Reference measurement channel according to Table A.4-1 with one sided dynamic OCNG Pattern OP.1 FDD as described in Annex A.5.1.1.

Note 3: To avoid collisions between RI reports and HARQ-ACK it is necessary to report both on PUSCH instead of PUCCH. PDCCH DCI format 0 shall be transmitted in downlink SF#4 and #9 to allow periodic RI to multiplex with the HARQ-ACK on PUSCH in uplink subframe SF#8 and #3.

Note 4: The bit field for precoding information in DCI format 2 shall be mapped as:
 For reported RI = 1 and PMI = 0 >> precoding information bit field index = 1
 For reported RI = 1 and PMI = 1 >> precoding information bit field index = 2
 For reported RI = 2 and PMI = 0 >> precoding information bit field index = 0

The ratio of the two throughput values $\gamma = t_{reported} / t_{fix}$ obtained from the two test stages should satisfy the requirements shown in Table 6.6-15.

The second and third tests in the RI requirement repeat the two stages above but with different codebook subset restrictions, varying RI configurations, and different SNR values based on those in Table 6.6-14. Correspondingly the throughput ratio requirements shown in Table 6.6-15 also vary.

Table 6.6-15. RI minimum requirements (FDD) (36.101 [3] Table 9.5.1.1-2)

Parameter	Test 1	Test 2	Test 3
γ_1	N/A	1.05	N/A
γ_2	1	N/A	1.1
UE category	2–5	2–5	2–5

In the first test the throughput values obtained at each of the two test stages should be similar. Due to the low SNR value little or no improvement can be expected as a result of codebook variation. The second test should show a modest throughput improvement but the improvement will be limited because rank 1 is used for both test stages. Test three should show the greatest improvement because this test uses the highest SNR as well as fixed rank 2 for the first stage of the test.

6.6.4 eNB Receiver Performance Characteristics

This section looks into the eNB receiver performance test requirements defined in 36.141 [25] Section 8.

6.6.4.1 Power Definitions for Receiver Performance Requirements

The receiver performance requirements in 36.141 [25] Section 8 define the radio conditions in terms of two main parameters: the power spectral density of an AWGN signal and the required SNR. The latter is defined as the ratio of the wanted signal to the AWGN power. In most test systems, the signal conditions will be controlled by setting the absolute power of the AWGN and the absolute power of the wanted signal, which is derived from the defined SNR. In straightforward cases in which the occupied bandwidth of the wanted signal is the same as that used to define the AWGN power spectral density, it is a simple matter to calculate the wanted signal power by adding the SNR in dB to the absolute AWGN power. The following example is based on the performance requirements for PUSCH in multipath fading conditions for a 10 MHz channel, as defined in 36.141 [25] Section 8.2.1. The AWGN power spectral density is defined in Table 6.6-16 and the required SNR in Table 6.6-17.

Table 6.6-16. AWGN power level at the base station input (from 36.141 [25] Table 8.2.1.4.2-1)

Channel bandwidth [MHz]	AWGN power level
10	−83.5dBm/9MHz

Table 6.6-17. Test requirements for PUSCH, 10MHz channel bandwidth (from 36.141 [25] Table 8.2.1.5-4)

Number of Rx antennas	Cyclic prefix	Propagation conditions (Annex B)	FRC (Annex A)	Fraction of maximum throughput	SNR [dB]
2	Normal	EPA 5Hz	A3-5	30%	−3.6

The FRC for this requirement is defined in Table 6.6-18.

Table 6.6-18. Resource block allocation for FRC A3-5 (36.141 Table A.3-1)

Reference channel	A3-5
Allocated resource blocks	50

Since the occupied bandwidth for FRC A3-5 is 9 MHz (50 * 180 kHz), which is same as the AWGN bandwidth, no power conversion is needed. The wanted signal power is $(-83.5 - 3.6)$ dBm = -87.1 dBm. The total AWGN power (including the 1.25 factor for the AWGN bandwidth discussed in Section 6.5.2) can be calculated as

$$AWGN_{total} = -83.5 + 10\log_{10}\left(\frac{9 \times 1.25}{9}\right) = -82.53 \ [dBm]$$

and therefore the combined power of the wanted signal and AWGN is

$$P_{total} = 10\log_{10}\left(10^{\frac{-87.1}{10}} + 10^{\frac{-82.53}{10}}\right) = -81.23 \ [dBm]$$

A more complex example can be found in the performance requirement for uplink timing adjustment defined in 36.141[25] Section 8.2.2. This example requires two separate UE signals to emulate a multi-user configuration. Both UEs must have the same power and carrier center frequency, with the lower half of the channel allocated to the "moving" UE and the upper half to the "stationary" UE. This is shown in Figure 6.6-4.

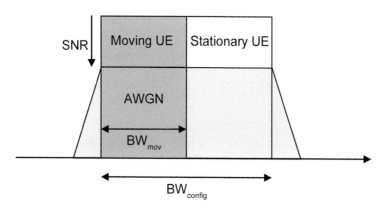

Figure 6.6-4. Moving UE, stationary UE, and AWGN

Tables 6.6-19 to 6.6-21 show the signal definitions for this requirement.

Table 6.6-19. AWGN power level at the base station input for 10 MHz channel bandwidth (36.141 [25] part of Table 8.2.2.4.2-1)

Channel bandwidth [MHz]	AWGN power level
10	−83.5dBm/9MHz

Table 6.6-20. Test requirements for uplink timing adjustment (36.141 [25] part of Table 8.2.2.5-1)

Number of Rx antennas	Cyclic prefix	Channel bandwidth [MHz]	Moving propagation conditions (Annex B)	FRC (Annex A)	SNR [dB]
2	Normal	10	Scenario 1	A7-4	14.4

Table 6.6-21. Resource block allocation for FRC7-4 (36.141 [25], part of Table A7-1)

Reference channel	A7-4
Allocated resource blocks	25

The occupied bandwidth of the moving UE is

$$BW_{mov} = 25 \times 180 kHz = 4.5 MHz.$$

The AWGN power is defined for a 9 MHz bandwidth, so it is necessary to calculate the power of AWGN for BW_{mov} for the SNR calculation as follows:

$$AWGN_{SNR} = -83.5 + 10 \times \log_{10}\left(\frac{4.5}{9}\right) = -86.51 [dBm].$$

The power of the moving UE for an SNR of 13.8 dB is calculated as

$$P_{mov} = -86.51 + 13.8 = -72.71 \ [dBm].$$

Since the stationary UE has the same power as the moving UE, the combined power is

$$P_{UE} = -72.71 + 3.01 = -69.7 [dBm].$$

As before, there is a 9 MHz AWGN flat noise bandwidth, so the total AWGN power is

$$AWGN_{total} = -83.5 + 10 \times \log_{10}\left(\frac{9 \times 1.25}{9}\right) = -82.53 \ [dBm].$$

and the total power is

$$P_{total} = 10 \times \log_{10}\left(10^{\frac{-69.7}{10}} + 10^{\frac{-82.53}{10}}\right) = -69.48 [dBm].$$

A final example involves the PRACH false alarm probability and missed detection test in 36.141 [25] Subclause 8.4.1. The PRACH has a bandwidth that varies according to the PRACH burst format. The AWGN is defined for the 10 MHz channel as before in Table 6.6-22.

Table 6.6-22. AWGN power level at the base station input (36.141 Table 8.4.1.4.2-1)

Channel bandwidth [MHz]	AWGN power level
10	−80.5 dBm/9 MHz

The PRACH requirements are defined in Table 6.6-23.

Table 6.6-23. PRACH missed detection test requirements for normal mode (36.141 Table 8.4.1.5-1)

Number of Rx antennas	Propagation conditions (Annex B)	Frequency offset	SNR [dB]				
			Burst format 0	Burst format 1	Burst format 2	Burst format 3	Burst format 4
2	AWGN	0	−13.9	−13.9	−16.1	−16.2	−6.9
	ETU 70	270 Hz	−7.4	−7.2	−9.4	−9.5	0.5

Depending on the burst format, the PRACH preamble has two different occupied bandwidths based on two different sets of subcarriers and subcarrier spacings as shown in Table 6.6-24.

Table 6.6-24. PRACH burst format and occupied bandwidth

Preamble burst format	Number of subcarriers	Subcarrier spacing	Occupied BW
0–3	839	1250 Hz	1.04875 MHz
4	139	7500 Hz	1.0425 MHz

The correct settings for the PRACH power are calculated in a manner similar to the previous examples.

6.6.4.2 Closed-Loop Test With HARQ Feedback

Several performance tests in 36.141 [25] Section 8 measure receiver performance under fading propagation conditions. Due to the use of fading, these tests are expected to generate cyclic redundancy check (CRC) errors in the eNB. In these cases, the eNB will send retransmission requests to the UE, which will transmit with different redundancy data to help the eNB correctly decode the data. This is the hybrid automatic repeat request (HARQ) process, which is described in Section 3.4. Table 6.6-25 shows some performance tests that use HARQ retransmission.

Table 6.6-25. Performance tests that require closed-loop configuration

Clause	Performance requirements
8.2.1	PUSCH in multipath fading propagation conditions
8.2.2	Uplink timing adjustment
8.2.4	High speed train conditions

To simplify the test system and avoid the need for downlink decoding, a test method is defined whereby a HARQ retransmission request generated by the eNB is fed directly to a base station tester emulating the UE. Figure 6.6-5 shows the test configuration for the uplink timing adjustment requirement, which is one of the test cases that assumes HARQ retransmission.

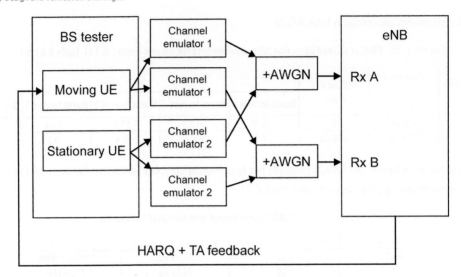

Figure 6.6-5. Uplink timing adjustment test configuration

The HARQ retransmission conditions are defined in Table 6.6-26:

Table 6.6-26. Uplink timing adjustment test parameters for HARQ

Parameter	Value
Maximum number of HARQ transmissions	4
RV sequence	0, 2, 3, 1, 0, 2, 3, 1

When the eNB sends a negative acknowledgement (NACK) message to the base station tester, the redundancy version (RV) index is incremented according to the RV sequence specified in Table 6.6-26. Note that after the fourth NACK request a new packet is sent and the RV index is reset to 0.

The test conditions for the UE timing adjustment test are shown in Table 6.6-27.

Table 6.6-27. Test requirements for uplink timing adjustment (part of 36.141 [25] Table 8.2.2.5-1)

Number of RX antennas	Cyclic prefix	Channel bandwidth [MHz]	Moving propagation conditions (Annex B)	FRC (Annex A)	SNR [dB]
2	Normal	10	Scenario 1	A7-4	14.4
			Scenario 2	A8-4	−1.5

Since the moving UE has moving propagation conditions applied, the probability of correct reception is lower than would be expected under normal conditions. When HARQ retransmission occurs, the time to transmit data packets successfully gets longer and this affects the throughput performance, which is required to be 70% of the maximum throughput with up to four HARQ transmissions. The Agilent N5106A PXB baseband emulator can be configured as a base station tester with the ability to accept HARQ and timing advance (TA) feedback via an auxiliary I/O port and to update the RV index dynamically to simulate the HARQ process. See Figure 6.6-6.

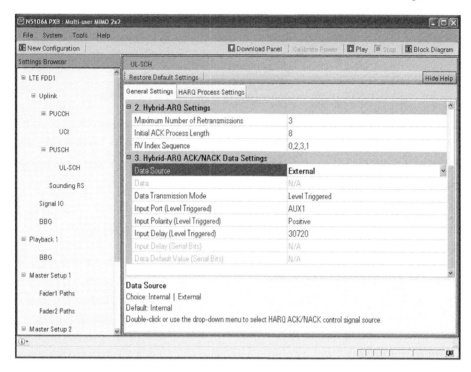

Figure 6.6-6. Hybrid-ARQ ACK/NACK data setting of Agilent N5106A PXB baseband emulator

Timing Adjustment Feedback in Closed-loop Test

The uplink timing adjustment (TA) test in 36.141 [25] Section 8.2.2 requires TA feedback from the eNB in addition to HARQ feedback. The uplink timing adjustment test assumes two UE, one of which has moving propagation conditions applied while the other is stationary. Due to the applied fading, the moving UE will have timing fluctuations too large to be absorbed by the cyclic prefix. In this situation, the eNB signals a TA command to the moving UE to adjust its frame timing relative to the stationary UE. There are two scenarios defined in Table 6.6-28.

Table 6.6-28. Parameters for uplink timing adjustment (36.141 [25] Annex B4)

Parameter	Scenario 1	Scenario 2
Channel model	Stationary UE: AWGN Moving UE: ETU200	AWGN
UE speed	120 km/hr	350 km/h
CP length	Normal	Normal
A	10 µs	10 µs
$\Delta\omega$	0.04 s	0.13 s

The timing difference between the moving UE and the stationary UE is defined as follows:

$$\Delta\tau = \frac{A}{2}\Delta\sin\left(\Delta\omega\Delta t\right).$$

This means $\Delta\tau$ will fluctuate between -5 μs to $+5$ μs over time, which in this case is longer than the normal CP length of 4.69 μs. The TA command is used to compensate timing delay due to $\Delta\tau$ as shown below:

$$\Delta\tau - \left(T_A - 31\right) \times 16T_s$$

where T_s is the symbol length $(1/(15000*2048) = 32.55$ ns) and TA is the timing advance index.

The TA command is 6 bits in length, with a range from 0 to 63. The TA command can adjust the UE timing from -16.1 μs (TA = 0) to 16.7 μs (TA = 63). Because TA = 31, no timing adjustment is applied in this case. The UE is given six subframes to decode the TA commands and apply the new frame timing. The PXB baseband emulator accepts TA commands through an auxiliary I/O port and can adjust subframe timing accordingly.

6.7 Testing Open- and Closed-Loop Behaviors of the Physical Layer

LTE uses a complex suite of feedback mechanisms called control loops to optimize the performance of the UE-to-eNB radio link. The needs of individual users in terms of data throughput, error rate, and delay have to be weighed against those of other users. In addition, channel coding and modulation rates have to be adjusted to match the time-varying quality of the radio channel. A necessary element for achieving optimal system level performance is to verify this dynamic behavior of the UE in the radio link.

This section discusses the messages used to manage the radio link control loops and methods for testing the control loop behavior. Different approaches are described for evaluating control loop operation during system development as well as conformance testing.

Control loops are used to manage an extensive list of radio parameters:
- Transmit power for control and shared channels in the downlink
- Transmit power of each signal type in the uplink, PRACH, PUCCH, PUSCH, and SRS
- Uplink transmission timing
- Modulation and coding rate adaptation in both the downlink and uplink
- Spectrum and time resource allocations (scheduling)
- Choice of the aggregation (repetition) level for the control channels
- Asynchronous downlink hybrid automatic repeat request (HARQ) retransmissions
- Synchronous uplink HARQ retransmissions
- Transmission mode and rank control.

Additional control loops are used in the higher layers to optimize application layer performance based on the radio link performance that is available. For example, radio link control (RLC) second-level ARQ addresses the reality that the HARQ retransmission process in the medium access control (MAC) layer is not immune to receiving false acknowledgements. Discontinuous reception (DRX), a mechanism used to enable inter-cell measurements and reduce UE battery power consumption, is also mentioned briefly in this section and is discussed in more detail in Section 3.6.3.1.

Modeling and testing control system operation as a whole requires very sophisticated evaluation techniques. In practice, radio link performance is assessed during UE conformance testing by testing whether the UE responds appropriately in both static and faded radio conditions. This kind of testing can be carried out using a base station emulator and a channel emulator as shown in Figure 6.7-1. The setup in this example includes an Agilent PXT base station emulator and an RF-to-RF channel emulator constructed from an Agilent PXB baseband channel emulator with RF down- and up-conversion provided by Agilent MXA spectrum analyzers and MXG signal generators.

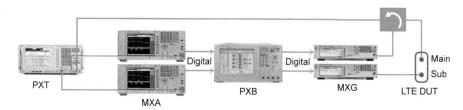

Figure 6.7-1. UE conformance test configuration using base station emulator and channel fading emulator

During UE development or system-level troubleshooting, other kinds of analysis can be used to examine the UE and eNB interactions at the radio link. For instance, it is possible to verify that the UE is correctly averaging the many measurements configured by the eNB that have layer 3 filtering requirements. This information is important because correct measurement averaging directly affects the system's ability to respond to changing radio conditions. Further details of feedback mechanisms are described elsewhere in this book. Examples are found in Sections 6.5.3.3 (HARQ) and 6.6.3.3 (CQI, PMI, and RI reporting).

6.7.1 Overview of the LTE Protocol Layers and Associated Control Messages

Messages used to control the radio link operation are spread across several layers of the protocol stack. Figure 6.7-2 provides a simplified view of all the downlink protocol layers, from the physical layer through layer 3 radio resource control (RRC). The highlighted boxes under the control plane show how the system information block (SIB) and RRC messages travel from the RRC layer down through the lower layers. Note that these messages are not encrypted, which makes decoding them a relatively simple task.

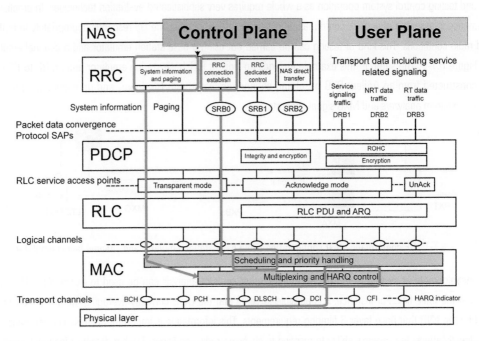

Figure 6.7-2. Downlink signaling protocol layers showing channels and the routing of some key radio link control messages (based on 36.300 [28] and 36.331 [29])

The boxes highlighted at the bottom of Figure 6.7-2 are examples of entities that are located close to the RF physical layer signals. A general rule is that the speed of the control loop response is related to the position of the control message in the protocol stack. As later examples show, HARQ retransmission processes are faster than higher layer processes since they are located closest in the protocol stack to the dynamic conditions of the radio channel.

The specific mechanisms used to operate the radio link control loops include the following:

- Downlink control information (DCI), which provides the parameters that configure the downlink signals; for example, to control the UE control channel and shared channel and the level of coding protection given to control signals themselves, including the uplink physical HARQ indicator channel (PHICH) (Section 4.1.3)
- MAC, examples of which include initial timing advance in the random access response message and uplink timing adjustment sent as a MAC control element (Section 4.1.4)
- RLC (Section 4.1.5), in the form of a second level of ARQ that is used during acknowledged mode (AM) transmission
- RRC, which has a very diverse role, including transmission of information for configuring the format and frequency of channel reports
- Uplink control information (UCI) such as HARQ, PMI, and CQI
- Uplink reports embedded in the PUSCH, such as the power headroom report.

6.7.2 Verification of Configuration and Control Messages

Messages are sent in the downlink to configure and control the uplink. They can be described as either dynamic—e.g., with new information every frame or even within a frame—or semi-static, changing less frequently. This grouping into dynamic and semi-static is important because the rate of change determines the signal triggering needed or the duration of a signal that has to be analyzed to verify the operation of interest. The information in the semi-static and dynamic messages is used to configure the analysis and decoding.

Table 6.7-1 shows downlink signals and messages. These dynamic messages are associated with the DCI signal type and can typically be verified within an individual frame. The timing of SIB messages varies, and higher layer messages may occur only when specific transactions between the eNB and the UE—for example, cell attach—are taking place.

Table 6.7-1. Distribution of dynamic and semi-static signals sent in the downlink to control the uplink

Signal type	PRACH	SRS	PUCCH	PUSCH
P-SS, S-SS		Cell ID		
DCI	Preamble index when DCI format 1A is used for configuration of random access process		RNTI	RNTI, RB allocations, TPC, MCS, frequency hopping bits and flag, CQI requests, DMRS cyclic shift
MIB	PHY signal configuration BW, frame number (for periodic allocations)			
SIB1	Cell frequency band			
SIB2	Detailed configuration, root sequence index	Cell-specific BW, subframe configuration, ACK/NACK–SRS combination. DMRS group and sequence hopping	Allocation starting positions, frequency hopping mode, offset, number subbands, DMRS group and sequence hopping, cyclic shift, etc.	
			nPUCCH(1) logical resource	
Higher layer	Frequency offset, configuration index, high speed flag, zero correlation zone configuration	UE-specific allocations (from RRC messages) PUSCH: HARQ-ACK, RI, CQI offset index PUCCH: nPUCCH(2) CQI/PMI/RI report logical resource allocation SRS: UE-specific BW, frequency position, hopping BW, transmit combining, cyclic shift		

The downlink provides the main source of configuration information for the link, using the DCI, system information, and RRC messages. The downlink also provides control information such as redundancy version, new data indicator, and the PHICH. Thus, even if the downlink is the only signal available for monitoring, it allows a considerable portion of the uplink activity to be inferred.

Elsewhere in this book extensive use is made of vector signal analysis for RF performance measurements. The Agilent 89600 VSA and 89600 WLA software can be used in control loop testing to demodulate the DCI, PHICH, and PUSCH symbol data and then apply further decoding to extract the underlying information. This is shown in Figure 6.7-3, where the 89600 WLA software provides a record of messages over time and uses abstract syntax notation one (ASN.1) decoding to recover higher layer messages, including uplink allocations either sent as DCI 0 commands or extracted from higher layers messages. SIB-2 and RRC messages are particularly significant because they provide the information needed to decode the uplink signal. The center window in Figure 6.7-3 shows the nPUCCH resource index value 36 that the UE should use when sending schedule requests.

Figure 6.7-3. Agilent 89600 WLA software displays showing a list of messages sent in the downlink during connection establishment

The sequence shown in Figure 6.7-3 is extracted from a connection establishment process. The RRC connection setup and system information block, SIB-2, are highlighted because they provide much of the configuration required for reporting on the uplink. Other messages significant for control operation include the acknowledged mode (AM) reports and bearer release commands, which are RRC dedicated commands coming through the signaling radio bearer 1 (SRB 1) shown in Figure 6.7-2. The AM reports and bearer release commands will be relevant to UE battery power consumption tests. See Section 6.13.

Vector signal analysis can provide synchronous downlink and uplink capture and analysis. The left hand side of Figure 6.7-4 shows the downlink RF analysis of a test signal measured using the 89600 VSA software. The decoded data is at the bottom left. This decoding allows verification of DCI messages. Uplink decoding, located at the bottom right, shows HARQ reports and CQI information being transmitted to the eNB.

Not shown, but part of this trace, are the control channel element offset (CCEOffset) value and the aggregation level (L) used for the downlink control channel (PDCCH). It is important to check that the aggregation level is set appropriately for the channel. If the setting is not correct, the UE may not be able to decode important control information and fail to recover the downlink signal. In that case the UE will be unable to recover its allocated data.

Test tools with synchronous capture and display are able to trace faults that are caused by the combined behavior of the downlink and uplink. The results in Figure 6.7-4 come from a recording of the signaling between an eNB and a UE. Several problems were apparent when the recording was reviewed; these included a series of unexpected breaks in the uplink transmissions and uplink allocations. The review showed that resources had been allocated to the UE in DCI 0 transmissions and the resulting UE transmission was acknowledged in the PHICH, even though the UL transmission was not present. In Figure 6.7-4, the center of the display shows an increase in downlink error vector magnitude (EVM) corresponding to the presence of an injected noise burst. The associated dynamic system behavior is explored in a later example.

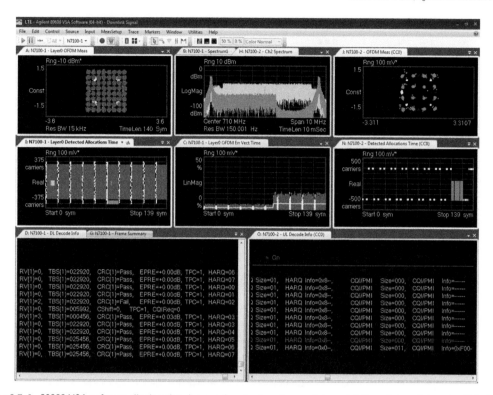

Figure 6.7-4. 89600 VSA software display showing synchronized capture and demodulation of downlink and uplink signals

Scheduling is one of the most fundamental but complex control loop operations carried out by the eNB. The LTE specifications provide mechanisms that enable highly flexible scheduling but the detailed design is left to the E-UTRAN vendors as an implementation challenge and opportunity for product differentiation.

Although the LTE specifications do not define scheduling algorithms (these are considered proprietary), a vector signal analyzer such as the 89600 VSA software can be used to display the downlink and uplink resource allocations made for any given radio link. Figure 6.7-5 shows an example of a TDD signal using aggregated carriers. This kind of display is useful for troubleshooting because it allows confirmation of what is happening down to the subframe level. Individual user allocations are color-coded.

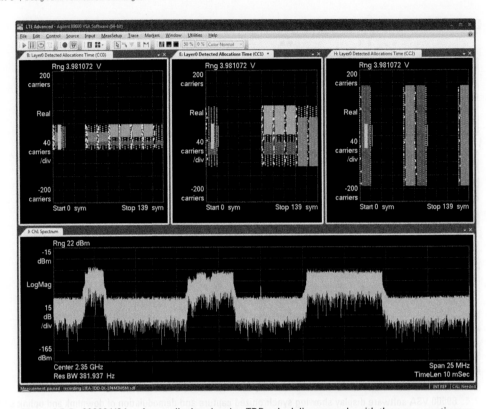

Figure 6.7-5. 89600 VSA software display showing TDD scheduling example with three non-contiguous aggregated carriers across two frames

A standalone uplink demodulator such as the 89600 VSA software does not need or use the downlink signal for timing information. Therefore the location of the uplink signal within the frame structure is not known. Demodulation is synchronized using the channel parameters found in the downlink, while parameters from the RRC connection setup message are used to decode the contents of the uplink control messages. Figure 6.7-6 shows three examples of these parameters.

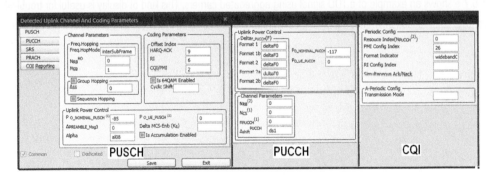

Figure 6.7-6. Examples of uplink coding parameters sent in the RRC connection setup message and used to configure the uplink shared channel, control channel, and CQI reporting

Items such as the HARQ-ACK, RI, and CQI/PMI offset indices are used to populate the demodulation configuration to get the correct reports for the items. The periodic CQI reporting interval and location are defined by the CQI/PMI configuration index value, with a value of 26 shown in Figure 6.7-6. The CQI/PMI data should be sent every 20 subframes on subframe (26-17) = 9. See 36.213 [30] Section 7.2.2, Tables 7.2.2-1A FDD and 7.2.2-1C TDD. Having extracted the reporting information from the connection setup messages, the uplink can be decoded.

Acknowledgements and reports can be delivered on either the PUCCH or PUSCH. The uplink provides feedback control information in the form of reports and transmission requests that influence the behavior of the eNB and therefore the downlink. Higher layer messages are essential components in the signaling transactions in the eNB and the network beyond it. During UE development, messages are typically verified from the transport channels upward depending on the point the design has reached in the development lifecycle. Simulators are often used to test isolated layers with defined boundaries.

6.7.3 Radio Link Control Loop Testing

Vendors are required to implement proprietary control algorithms in the eNB and these are known to be the source of occasional system interoperability problems. However, for the very specific and often open-loop conditions defined in the UE conformance tests, no such degrees of freedom exist in the eNB emulator used for testing, and UEs undoubtedly are being designed to pass these narrowly defined test scenarios.

In contrast, the following test configurations enable many additional test scenarios for gaining deeper insight into the underlying behavior of the UE. This knowledge provides a better understanding of the much broader if not limitless range of conditions a UE will experience in real operation. Figure 6.7-7 shows a generalized test system configuration that is used in the following examples. Most of the analysis capability is provided by a multi-channel signal analysis system, which captures the RF signals for processing by the 89600 VSA and 89600 WLA software. A vector signal generator in the setup is used to add noise bursts to the wanted signal at regular intervals to stimulate responses in the control loops and enable further analysis. The following sections show how this test system can be used to analyze a variety of UE control loop behaviors.

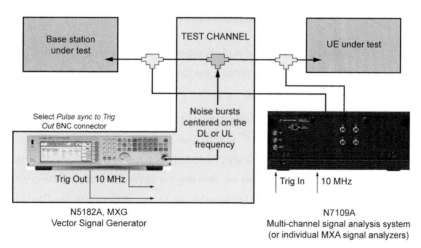

Figure 6.7-7. A practical test configuration for signal capture and control loop response tests

6.7.3.1 Downlink HARQ Retransmission Process Evaluation

As described in Section 3.4, the downlink HARQ processes operate asynchronously, allowing the eNB some flexibility in the timing of rescheduled transmissions. Retransmissions reuse the modulation and coding scheme (MCS) setting of the initial transmission. The following example illustrates the impact of the HARQ retransmission process on performance when the channel capacity is deliberately reduced by lowering the signal to interference plus noise ratio (SINR).

For tests in a channel with a high SINR, where maximum throughput is expected, a graph of redundancy version (RV) can provide a fast and effective indication of correct operation. The graph can be checked to see whether the RV displays the regular cycling expected for the SIBs, identified by the RNTI <FFFF>, or whether any unexpected retransmissions are limiting the link capacity. The operation of RV is described in Section 3.4.1.2. Figure 6.7-8 reveals one important aspect of LTE operation—that signals using very high coding rates can be subject to low rates of retransmission because not every bit of data gets encoded and as a result incorrect decisions are made.

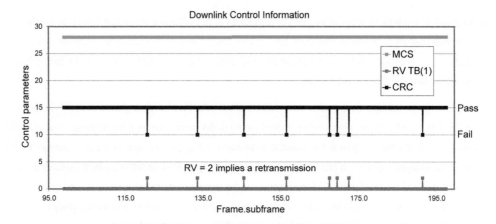

Figure 6.7-8. Occasional downlink retransmissions in a high SINR channel due to the high coding rate associated with the MCS=28

In Figure 6.7-8 the MCS trace is a straight line with a value of 28, shown on the left hand scale. The CRC has a constant value of 1 (pass) shown on the left hand scale. The point of interest is the sporadically spaced jumps in RV, from 0 to 2, shown as a sequence of vertical lines referred to the right hand scale.

In order to test the dynamic behavior of the HARQ processes, the test configuration of Figure 6.7-7 adds bursts of Gaussian noise to the channel to stimulate a response in the eNB and UE. Many parameters may affect the overall response, which can be measured both as throughput and latency. The latency measured in this scenario is represented by timing delays in the delivery of the data from individual HARQ processes. The test uses only the information present in the downlink signal. Figure 6.7-9 provides a guide to the characteristics that may be observed in this type of measurement. The figure shows two of eight possible HARQ processes for the RNTI value 000C.

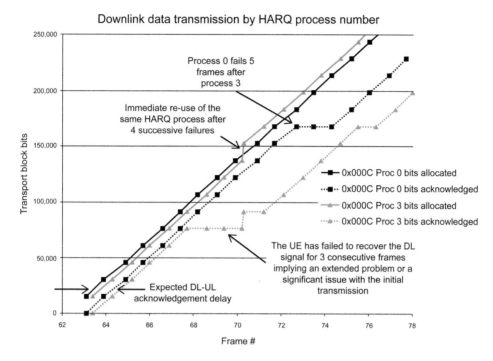

Figure 6.7-9. Graph of allocated vs. acknowledged data bits as a function of occasional retransmission

Figure 6.7-9 shows the variety of ways individual HARQ processes respond to data retransmission. A horizontal line on the acknowledged bits traces indicates where retransmissions are delaying the data content. The slope of the acknowledged bits traces represents frame-by-frame throughput (bits/time). For the majority of the capture, the slope of the allocated bits is constant. Around frame number 70, process 3 takes a step upwards. This occurs immediately after the fourth consecutive failure in reception, as shown by the horizontal line in the process 3 acknowledged bits. After the fourth attempt, the eNB must restart the transmission of the MAC packet data unit (PDU). If the resource allocation is available, the transmission can be restarted in the next available timeslot, causing the vertical jump in allocated transport block bits. This jump represents a realignment of the HARQ process numbering.

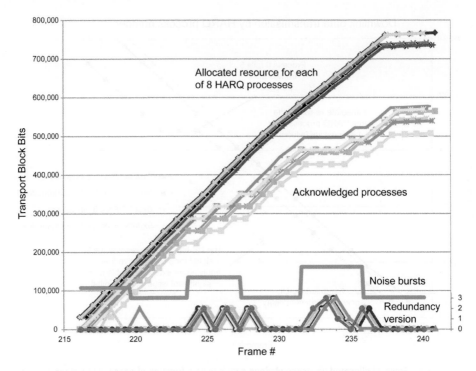

Figure 6.7-10. Impact of added noise on redundancy version and acknowledged data bits

Figure 6.7-10 shows the results of a test using noise bursts that increase in level by 2 dB per step, indicated by the blue trace. Initially the SINR is a good match for the MCS and only two individual retransmissions are seen (bottom left). The retransmissions are indicated by the traces that show RV version per HARQ process. The RV sequences through the values 0, 2, 3, and 1 for successive retransmissions. The graph shows two instances in which the RV for two processes changes from 0 to 2, indicating one retransmission per affected process. As the SINR worsens, the middle noise burst causes three sequences in which RV=2 is applied to all four HARQ processes. The impact on the delay in the acknowledged bits is clear. Since only one retransmission was required, at this SINR the type II recombination in the UE receiver is sufficient to recover the data. The third and highest noise burst causes the RV to run through the complete sequence, 0-2-3-1.

6.7.3.2 Uplink Synchronous HARQ Retransmission Process Evaluation

The uplink HARQ processes run synchronously, making it simpler to track the progress of individual processes. The operation is described in Section 3.4.1.4. Verification can involve a combination of making sure the uplink transmissions occur according to the allocations given by the eNB, and by using the PHICH in the downlink signal to verify if the eNB was able to successfully recover the shared channel data transmitted from the UE. The location of the acknowledgements on the PHICH are a function of the starting point of the uplink allocation to which they apply, so it is possible to use the acknowledgement locations to verify if the mapping is correct.

Graphing techniques similar to those used as examples for the downlink provide similar information about the operational state of the uplink, as shown by the example in Figure 6.7-11.

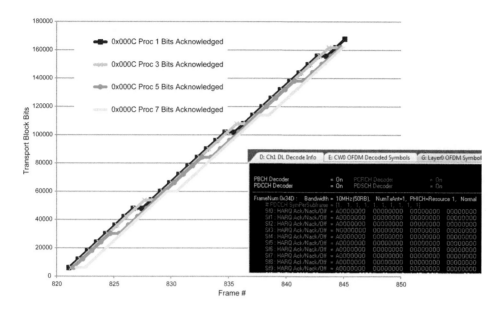

Figure 6.7-11. Example of FDD uplink HARQ processes, tracked through the use of the PHICH

The example in Figure 6.7-11 had a high SINR so there should have been no retransmissions. After further investigation, a baseband coding error in odd numbered processes was discovered. This coding error caused the retransmissions on alternate frames.

6.7.3.3 Downlink Adaptation of MCS Using CQI Reports From the UE

The eNB has to estimate how fast the channel is changing and set the CQI reporting interval accordingly for each UE. The eNB then needs to combine channel quality information with other factors, such as the quality of service (QoS) required for the connection and the available RB allocations. For the control loops to operate effectively, it is important that the UE reports provide an accurate and timely measure of the downlink channel conditions.

Figure 6.7-14 shows a result from using the Figure 6.7-7 test configuration to stimulate the MCS control loop response. A noise burst creates SINR too low for successful demodulation of the current MCS of 28. This causes retransmissions after a seven subframe delay while the other previously programmed processes are transmitted. The retransmissions are shown by the changes in redundancy version at the bottom of Figure 6.7-12. The RV remains at 0 unless there is a retransmission. In this example the packet is lost since the RV is seen to go through the full sequence of 0, 2, 3, 1 before the data has to be resent by the next higher protocol layer.

During the same time, the UE is instructed to report CQI every second frame, which is quite slow. These results are plotted as the boxes in the center of the graph. In control loop terms, there is a fast attack and slow delay to the CQI reporting. The noise exists for only three frames, but it can be seen that the UE is applying some form of averaging function.

It takes three more reporting periods before the CQI report returns to the original value prior to the noise burst. This delay can be investigated further to ensure that any averaging performed by the UE is in accordance with the layer 3 filtering configured by the eNB. There is a five subframe delay from the start of the noise burst to the MCS being reduced in an attempt to match the lower SINR conditions. In combination with the averaging of the CQI by the UE, it takes eight frames before the impact of the three frame noise bursts has settled. Setting a shorter CQI reporting period and less layer 3 filtering would reduce this delay and make the link more responsive to this type of noise burst. The correct settings for the reporting period and filtering are highly dependent on the nature of the propagation channel and interference, however, and these settings are not defined in the specifications but left as a design choice to the eNB vendors and network operators.

The MCS values used by the eNB may be directly mapped from the CQI value reported by the UE. See Section 3.4.1 in this book and 36.213 [30] Table 7.2.3-1.

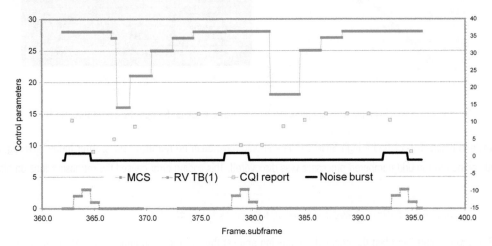

Figure 6.7-12. Noise bursts in the channel, which cause changes in the UE CQI reports and corresponding changes in MCS value used by the eNB for DLSCH transmissions

Another example using the test configuration of Figure 6.7-7 is shown in Figure 6.7-13. This example provides a more comprehensive view in which a number of expected responses are seen along with several aspects of unexpected behavior. This example shows the value of plotting the low level control parameters. These parameters not only allow the control loop response operation to be verified, they also highlight aberrant behavior. In the top trace of the graph the MCS value is used for a single RNTI; the one being examined. Figure 6.7-13 shows the same kind of step response seen in Figure 6.7-12, but other transitions are superimposed. Underneath the square data points showing the reported CQI value and the pulsed trace identifying the noise bursts, there is an additional trace showing the ACK/NACK information sent in the uplink signal.

Point A shows the expected reduction in MCS due to the noise burst. This example also highlights two unexplained events such as point B, where the UE stopped transmitting ACKs or NACKs temporarily, and point C, where the eNB retransmitted signals even though the UE sent ACKs. The interval between noise bursts has been reduced from 150 ms to 100 ms. The noise level is the same as is used for the test in Figure 6.7-12.

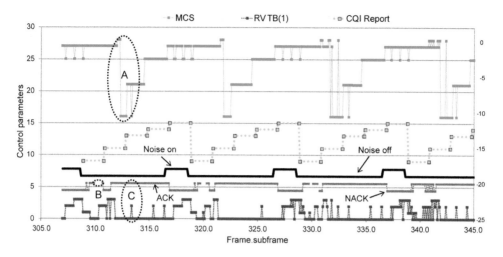

Figure 6.7-13. A more detailed analysis showing different downlink HARQ processes using different MCS values

Toward the top of Figure 6.7-13, the MCS changes apparently more erratically than it does in Figure 6.7-12. In this test configuration, HARQ retransmissions use the same MCS as the first transmission. This reuse allows data recombining in the UE but, as the figure shows, delays the selection of a new MCS more appropriate to the changed SINR conditions.

Note that Figure 6.7-13 shows significantly more, and unexpected, retransmission activity than Figure 6.7-12 shows even though the noise burst duration and resulting SINR are the same in both examples. At the left of Figure 6.7-13, the first noise burst causes a complete 0-2-3-1 RV sequence to be used for all HARQ processes. Then, around frame 310, a second set of retransmissions appears. Point B indicates that the UE has stopped transmitting ACKs or NACKs; in fact, there are no uplink transmissions at all. Since the eNB schedules downlink traffic, it treats the lack of an uplink response as a transmission failure and repeats the transmissions. The analysis shows a problem. The UE log will have to be inspected to determine why it has stopped transmitting.

Point A shows that the effect of the second set of retransmissions is to delay the application of the lower MCS value indicated by the CQI value sent by the UE nearly four frames earlier. A further complication is seen around frame 312, where at point C the first of three apparently isolated retransmissions occurs. This behavior is unexpected because it is clear from the ACK/NACK trace above point C that the UE has not sent any NACKs to trigger a downlink retransmission. Further investigation reveals the presence of uplink allocations and transmissions coincident with the retransmission at point C, but this fact alone does not explain the unexpected retransmissions.

As with the uplink example in Figure 6.7-13, the value of plotting the retransmission and control loop activity is not only to verify the responses and their timing, but also to reveal unexpected behavior that indicates nascent issues in the UE design. An end-to-end performance test might pick up these issues but might also miss them. On the other hand, a low-level analysis of the underlying physical layer procedures in a controlled environment improves the likelihood that underlying problems can be identified.

6.7.4 Conclusions

One of the ways LTE distinguishes itself from other technologies is in the flexibility of transmission formats and spectrum allocations available to match the instantaneous characteristics of the radio channels and thus optimize transmission efficiency. To ensure that this inherent flexibility delivers the expected performance, individual control mechanisms must be thoroughly evaluated at different levels, from simulation all the way through interoperability testing with implementations from different vendors. Running tests with channels that provide deterministic and pseudo-random impairments makes it more straightforward to isolate different effects and reproduce specific test environments. This section has described the major control mechanisms and methods used, together with examples of test configurations and tools to enable these low-level evaluations. These examples have shown both some of the expected responses and some unexpected behavior that would trigger more detailed investigation of the UE design.

6.8 Design and Verification Challenges of MIMO

6.8.1 Introduction

This section describes the design and verification challenges associated with spatial multiplexing and diversity-enabled radios for LTE. The impact of this technology on transmitter and receiver hardware and software will be considered as well as the effects of different antenna arrangements.

It has been said that MIMO is a baseband (digital coding) challenge. There is some truth to this belief, but not all situations are so straightforward. When designs become more compact, the isolation and performance of RF and analog circuits are not ideal. RF currents are notorious for leaking onto non-RF circuits, and the effects can be difficult to filter out. As a result, MIMO performance degrades. The transmit and receive antennas are key components in the signal chain. Their performance is not easy to model, and in the most complex beamforming cases it is necessary to use real-time calibration mechanisms to maintain performance. The receivers in the UE have a particularly demanding task, since they have to recover the MIMO signal using algorithms that balance performance with space in the baseband IC (BBIC). Also, the UE has to report accurate real-time channel state information to the eNB. There are other important factors that impact MIMO performance indirectly, such as battery power consumption in the UE. These factors are dealt with in Section 6.13 on battery drain testing.

To work in practice, spatial multiplexing needs better channel conditions and higher signal-to-noise ratio (SNR) than does equivalent single input single output (SISO) operation. The better conditions required for successful MIMO operation are achieved through a combination of diversity-enabled or beamforming cells, and the use of progressively smaller or isolated cells.

The issues in the downlink paths are considered here first, starting with the eNB transmitters in Section 6.8.2. After noting potential problems such as unwanted signal coupling, this section is broken down to describe a variety of single and multiple input measurement techniques that are useful for isolating problems during development. These include a description of how to use a single input analyzer for MIMO measurements, the use of channel condition number, and insights into in-phase quadrature (IQ) impairments, which give effects such as the "constellation of constellations." Section 6.8.3 provides an example of how a four-input digital oscilloscope can be used to troubleshoot a 4X4 MIMO design.

Section 6.8.4 describes the UE module and eNB receiver implementation issues and test methods. Section 6.8.5 is a description of the additional parameters needed for MIMO performance verification, including the use of fading channels and the use of channel emulation with configurable path correlation. The influence of variables such as antenna spacing, polarization, and path correlation is described. This section also addresses the addition of noise to the signals used for receiver testing in a correlated channel, which relates both to performance assessment and to the measurement of the MIMO channel state information [consisting of channel quality indicator (CQI), precoding matrix indicator (PMI), and rank indication (RI)] that the UE has to provide to the eNB for dynamic control of the link.

The final Section 6.8.6 deals with the question of phase coherence and its impact on design and measurement for the transmitter and receiver when precoding is applied. As a general rule, when a MIMO system is tested, the test equipment should have as many input ports as there are spatial layers, although this is not always the case as will be seen in some of the examples that follow. It is also beneficial and sometimes essential that the measurement channels be phase-coherent and sufficiently time-aligned within a fraction of the cyclic prefix to maintain proper orthogonality between the reference signals (RS). This introduces another testing challenge, that of measuring time skew differences between multiple measurement channels when RS timing issues are debugged.

Table 6.8-1 lists some of the options available for testing LTE MIMO systems. The ideal choice depends on the signal format and the position of the signal within the system block diagram.

Table 6.8-1. Multi-channel test equipment

Agilent product	Signal input format	Multi-channel capability		
		Spatial multiplexing	Beamforming	Bandwidth
SystemVue simulation	Data files, MatLAB, links to test equipment hardware	8 channel	8 channel	Unlimited
RDX cross-domain analyzer	DigRF v3, DigRF v4	Not applicable	Not applicable	Not applicable
M9252A DigRF Host Adapter	DigRF v4	Not applicable	Not applicable	Not applicable
Logic analyzer	Digital baseband	Not applicable	Not applicable	Not applicable
PXB baseband generator and channel emulator	Digital baseband, IQ baseband Use MXA signal analyzer for RF input and MXG signal generator for RF output	Any combination of up to 12 paths (4x2, 4x3)	Any combination of up to 12 paths (12x1)	RF hardware dependent
Infiniium oscilloscope	IQ baseband, RF	4 channel RF	Phase coherent	>1 GHz phase coherent
Dual-channel PXI vector signal analyzer	RF	2		250 MHz/channel
Wideband MIMO PXI vector signal analyzer	RF	2–8 channel		800 MHz/channel
X-Series analyzers	RF	2 channel		25 MHz and upgrades
N7109 multi-channel signal analyzer	IQ baseband, RF	8 channel	8 channel	40 MHz phase coherent
89600 VSA software	IQ baseband, RF	8 channel	8 channel	Hardware dependent

6.8.2 Base Station (eNB) Transmitter MIMO Challenges

Figure 6.8-1 shows a simplified block diagram of an eNB showing separation of RF and digital processing circuits. The development tasks are often carried out separately by different teams or different companies, which requires that subsystem performance testing be done before the modules are brought together. The techniques for using test equipment to simulate SISO RF or baseband signals can be adapted for analyzing individual signal paths. For digitized signal interfaces, it is normal to begin by using repetitive arbitrary waveforms and batch-mode, post-processed measurements.

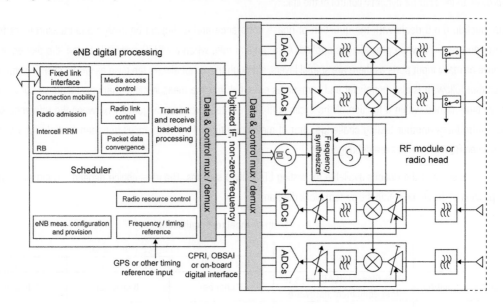

Figure 6.8-1. Typical layout of a MIMO OFDMA eNB

The signal processing path is shared by all signals up to the point at which the scheduler chooses to apply transmit diversity, spatial multiplexing, or a mixture of the two. The MIMO challenge begins with baseband coding. The number and complexity of the LTE signal formats requires step-by-step verification, starting during design simulation and continuing as the design migrates to operation on the target hardware.

Another challenge for multi-channel operation is that of interconnection, whether between modules within the eNB or between the eNB and the test equipment. Digital interfaces such as common public radio interface (CPRI) and open base station architecture interface (OBSAI) are not sufficiently standardized to make it straightforward to plug and play with generic test equipment. However, when vendor-specific options are available, analyzing the signal as digitized IQ is a helpful step in early signal verification, particularly when the specifications are not final.

6.8.2.1 Baseband Coding Assessment Using Spectrogram Pattern Recognition

Software-based analysis using a vector signal analyzer such as the Agilent 89600 VSA software is valuable for fault-finding. To demonstrate how this works, Figure 6.8-2 shows the spectrograms of a four transmitter LTE signal. (Spectrograms are introduced in more detail in Section 6.4.5.3.) Each spectrogram shows time in the vertical axis, and spectral power is identified with white the highest power and black the lowest power. There are patterns clearly visible, and components in the LTE signal structure become readily recognizable after the traces have been studied for a short time. In this example, the different RS structures used for transmitters 2 and 3 on the top left and top right of Figure 6.8-2 are clearly visible and look as expected. However, what is not expected is the truncation of the data that starts in the second subframe. The transmission should have continued to the edge of the channel.

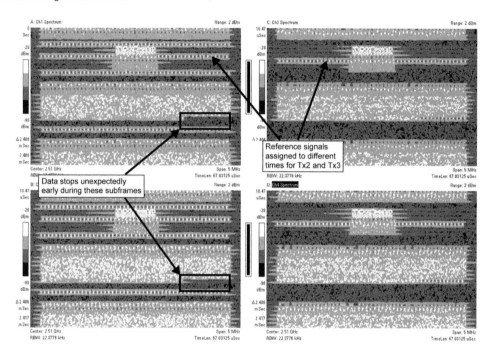

Figure 6.8-2. Using spectrograms for quick diagnosis of basic coding problems of a 4x4 MIMO signal

It can also be seen that the power on Tx1 (bottom left) is slightly higher than the power of the other transmitters. (This higher power is shown as lighter shades in the display, but in the software the spectrogram is displayed in color, giving more clarity to the patterns.) The spectrum gating period is matched to the OFDM symbol length of 66.7 μs, and a Hanning window is used for improved resolution in the frequency domain.

When RF connections are made between the DUT and signal analyzer, the number of available measurement ports determines how the signals can be analyzed. Figure 6.8-3 shows the connection options, starting with direct cable connections on the left and going through to a real or emulated channel on the right. For some measurements, power combiners can be used when only one analyzer input is available, as long as the signals from the transmitters contain some orthogonality in frequency or time.

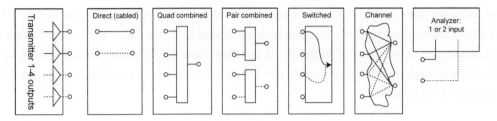

Figure 6.8-3. Connection options between an eNB and signal analyzer

Table 6.8-2 shows which physical signals and physical channels are coupled between the transmitters of the base station. The presence of coupling has an impact on some of the measurement techniques described here.

Table 6.8-2. Application of diversity and spatial multiplexing for different downlink signal types

Physical signal or physical channel	Transmit diversity	Spatial multiplexing	Cyclic delay diversity
Reference signal	Neither diversity nor spatial multiplexing is applied, but the assignment of the sync signals to the transmit antennas may vary (refer to 36.211 [2] subclause 6.11.1.2)		
Primary synchronization signal			
Secondary synchronization signal			
Physical broadcast channel	Used for space frequency block coding (SFBC)	Not used	
Physical downlink control channel			
Physical hybrid ARQ indicator channel			
Physical control format indicator channel			
Physical multicast channel		Used	Not used
Physical downlink shared channel			Used (long)

6.8.2.2 Single Input and Multiple Input Measurement Techniques

If there is little or no coupling between channels, many measurements can be made on a multi-antenna signal using a single input analyzer. In development, it should be straightforward to achieve this signal configuration through the use of codebook index 0. As defined by the specifications, the reference signals for each transmitter are always orthogonal in frequency and time.

There are many possibilities for errors in signal generation in the eNB. Finding errors requires flexible signal analysis. The process of signal analysis for troubleshooting follows similar steps to those used for demodulation in a real receiver. Figure 6.8-4 shows a simplified block diagram of the analysis steps that relate to the signal analyzer's user interface settings. It is normal to alternate between analyzer settings to explore what impairments are affecting the signal. An example of how switching between analysis blocks can be used is to distinguish distortion of a single transmitter (e.g., in a power amplifier) from that on a single layer (e.g., due to a mathematical processing error). Table 6.8-3 shows the range of MIMO measurements possible depending on the number of analyzer inputs.

Table 6.8-3. Measurement possibilities based on number of analyzer inputs

Measurement objective	Number of measurement inputs required		
	1	2	> 2
SISO and MISO errors due to phase noise, timing errors and amplitude clipping	Yes		
Spectrum mask, harmonics and spurious	Yes		
RF phase and baseband timing alignment, using RS-based measurements	Using a power combiner	No combiner needed but errors from the second analyzer input will contribute to result.	
Cross channel isolation using RS-based measurement	Yes	Similar measurements to single input. Can connect to two transmitters at the same time.	
Interference, grounding, transient settling	Yes		
Transmit diversity space time coding (control channel and PDSCH)	Yes		
MIMO spatial multiplexing (with unwanted coupling) and coding verification	Individual (direct mapped) layers	Yes	If > 2 layers

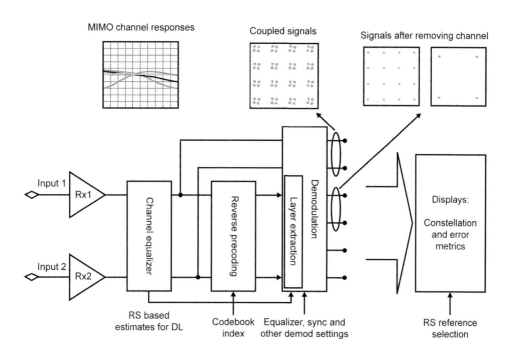

Figure 6.8-4. MIMO signal analysis processing

Figure 6.8-4 is drawn with two measurement inputs but could be extended to four. When cabled connections are used instead of a real or emulated radio channel, the reverse precoding block is required. This block removes the precoding that would have been removed by the radio channel.

6.8.2.3 An LTE MIMO Troubleshooting Process

When two or more layers are in operation, there are additional ways the signal can be impaired. As with a SISO signal, a systematic approach is needed to understand what errors are present. Table 6.8-4 summarizes such a method:

Table 6.8-4. A basic structure for diagnosing multi-transmitter signal impairments

Device configuration and analyzer connection	Analyzer configuration	Measurement steps
Single input measurement; connect to each transmitter output separately. TX Diversity or spatial multiplexing ON	Use the same analysis configuration steps as for SISO signal. Start with analyzer demodulator OFF. Use Hanning window with gate time = 1 symbol (66.7 µs).	Measure power vs. time and gated spectrum to ensure that each channel has the expected power level and structure. Record signal and use spectrogram.
Codebook index = 0	Turn demodulator ON. Display RS, primary and secondary synchronization signals (which may not be configured for all transmitters).	Synchronize to primary synchronization signal or RS Check constellation and EVM of uncoupled signal elements.
	Shared channel and control channel precoding ON.	Check constellation and EVM of diversity and Spatial Multiplexing (SM, MIMO) signals (Table 2.4-1).
Combine signals using a power coupler. Codebook index = 0	Two or four transmitter ports active. Allows precise measurement of relative power, timing and phase.	Check all RS-based measurements.
Measure signals using a two input analyzer for all codebook values.	Two inputs needed to remove the effect of coupling between the transmitters (e.g., precoding). Allows measurement of the residual error that will be seen by the UE receiver.	Check shared channel constellations and EVM with all codebook values.

6.8.2.4 Unwanted Layer Coupling Creating the Constellation of Constellations

Coupling between channels results in each signal having a portion of the other superimposed upon it. This has a distinct effect on the IQ constellation, seen in Figure 6.8-5, informally known as the constellation of constellations. The reason this effect occurs is that the coupled signals are highly correlated in time. Demodulation using the correct code index will remove the expected coupling but any unwanted coupling will remain. The way in which transmitter impairments result in channel coupling varies depending on the precoding.

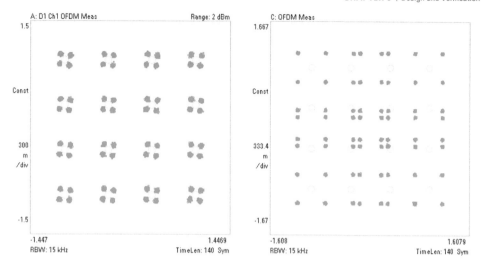

Figure 6.8-5. Constellation of constellations. Left: Unwanted QPSK has coupled into the 16QAM layer. Right: The wrong codebook index has been used.

To determine whether coupling is intentional (due to precoding) or unwanted, and to further understand the likely coupling mechanism, it is necessary to have as many measurement inputs to the signal analyzer as there are layers in the signal. Figure 6.8-4 shows the processing blocks used during analysis and indicates the paths used to produce results when different analysis functions, such as channel estimation and MIMO decoding, are applied.

6.8.2.5 Channel Training Signal Verification Using Equalizer Condition Number

As discussed in Section 2.4.2, the condition number of the matrix representing the channel can be used to show the improvement in SNR required to recover a spatially multiplexed signal. The additional SNR versus condition number is shown in Figure 6.8-6. The measurement and display of channel condition number can also provide a number of useful troubleshooting insights, even when the channel is just a simple cabled connection.

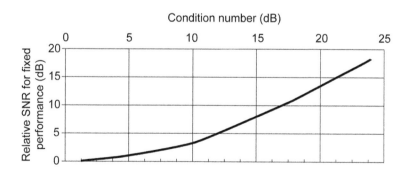

Figure 6.8-6. Additional SNR required versus channel condition number to maintain, constant demodulation performance

Assuming that the signal has been correctly demodulated, the condition number trace provides verification of reference signal coding and the relative power of the transmitters. If precoding is applied creating cross-coupling between the layers, it is not possible to learn from a broadband power measurement whether the relative power assigned to each signal component is correct. The condition number provides a better indication that the relative power is correct.

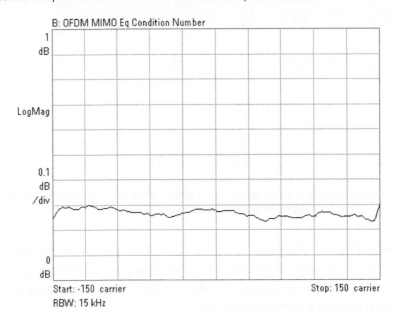

Figure 6.8-7. Matrix condition number trace for a well-behaved two transmitter eNB

The perfect condition number is 0 dB, which occurs when the diagonal values of the matrix representing the received signal are perfectly balanced. To achieve the near perfect performance of the trace in Figure 6.8-7, not only does the precoding have to be correct, but the power of each transmitter needs to be well matched and there has to be little unwanted coupling. The condition number display reflects the equalizer frequency response, since it uses data before normalization takes place.

6.8.2.6 Identifying Physical and Baseband Distortions in a Single Analog Path Using Layer and Channel EVM

If precoding is applied to a two or more layer system, the layers share common analog paths. Distortion in any single path will cause errors in all the layers passing through it. Even without precoding this is generally true at the receiver since any real channel will invariably result in path coupling. From the analysis block diagram in Figure 6.8-4, it can be seen that by turning off the layer extraction block, it is possible to demodulate the signal and measure the error vector magnitude (EVM) of either the received signal or a specific layer encoded onto the signals.

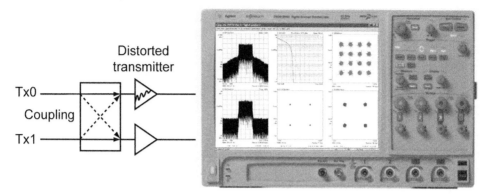

Figure 6.8-8. Distortion of a single transmitter causing impairment of both layers

In Figure 6.8-8, the upper amplifier has been deliberately set to clip, and precoding is applied so that each layer appears on both paths. The bottom left trace of the 89600 VSA software running on the oscilloscope shows the undistorted spectrum of the lower amplifier. The top left trace shows the spectral regrowth caused by the distortion in the upper amplifier. The top center trace shows a complementary cumulative distribution function (CCDF), which indicates that the upper amplifier is in compression. The constellation trace in the bottom center shows what the unclipped signal looks like. The top right and bottom right constellations show distortion in both the original layers after decoding.

It is worth remembering that the level of distortion will depend strongly on the drive level, biasing, and predistortion applied in the power amplifier relative to the signal power. In LTE, the signal power varies rapidly within each symbol at a rate determined by the number of subcarriers. As a result, any distortion results in state spreading as discussed in Section 6.4.1.6. In addition the channel measurement made by the UE is also vulnerable to RF impairments and this in turn can create incorrect cross-channel coupling due to selection of the wrong precoding matrix.

6.8.2.7 Identifying Unexpected Power Peaks Using Cross-Channel Correlation

When two equal, power-correlated signals are combined, the power can increase by 6 dB rather than by 3 dB as occurs when the signals are uncorrelated. Combining correlated signals therefore has serious consequences for all parts of the processing chain, from clipping at baseband to unexpected distortion by the power amplifiers.

Cross channel correlation measurements can be used to investigate whether the transmitter signals are orthogonal (uncorrelated). As noted in Table 6.8-2, how the RS and sync signals are transmitted on each antenna in terms of timing is undefined. Figure 6.8-9 is the result of a cross channel measurement to determine the correlation between the sync signals.

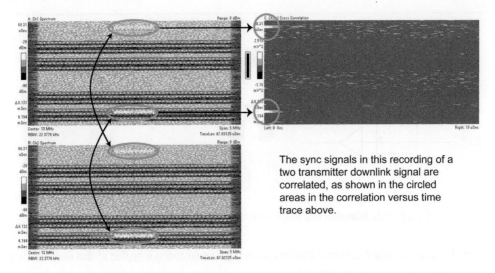

Figure 6.8-9. Using cross correlation to determine sync signal timing between different transmitters. The sync signals and correlation peaks are highlighted.

On the left of Figure 6.8-9 are spectrograms of each signal. The vertical axis is time and the period shown is slightly more than one 10 ms frame. The horizontal axis is frequency and the entire 5 MHz channel is shown. The top right trace shows the time correlation between the two signals. The vertical axis is time and it is aligned with the spectrograms. The horizontal axis is also time, but only 10 μs, a fraction of a symbol. The circled light-colored spots on the vertical axis of the top right trace show where correlation occurs between the signals. Since the correlation occurs for only a short period of time in the vertical axis, only that part of the signal is common. By reference to the spectrogram it can be concluded that the synchronization signals are the common element. Since the correlation occurs on the Y-axis at 0 μs, it can also be concluded that the sync signals were transmitted with no time offset.

6.8.2.8 Measurement of Antenna Signal Relative Timing and Phase

The LTE requirement for alignment between the antenna signals is ±65 ns, which is approximately 0.1% of the symbol period. The cross correlation measurement described in Section 6.8.2.7 does not have sufficient resolution to make the alignment measurement. However, by demodulating the RS it is possible to make relative timing measurements with sub-ns accuracy. Furthermore, since the RS are orthogonal between the antennas, a power combiner can be used to measure the timing relationship using a single input rather than a dual input signal analyzer.

Figure 6.8-10 shows the result from such a measurement. Tx1 is 619 ps behind the reference on Tx0, with a −3.8 degree phase difference.

The same measurement technique involving the combining of signals into one analyzer input can be used to verify the calibration of signals used for phased array beamforming and to calibrate the timing of signals for receiver testing from multiple signal generators with associated RF cabling. The right hand part of Figure 6.8-10 shows the Agilent MXG signal generator user interface for adjusting RF phase and baseband timing.

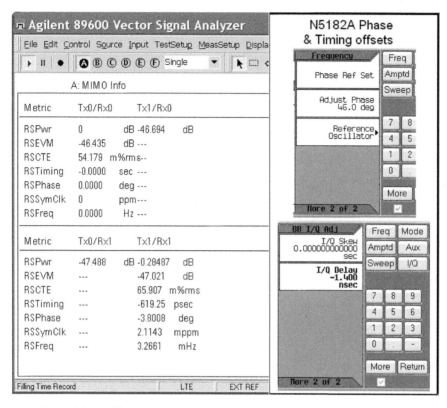

Figure 6.8-10. RS-based timing and phase measurements of a composite downlink
2x2 MIMO signal made using a single input analyzer

6.8.3 Debugging MIMO Designs Using a Digital Oscilloscope

Digital oscilloscopes offer time domain analysis capability, which can be used to troubleshoot the timing issues that affect MIMO transmitter performance, in particular the RF MIMO EVM. This section explains this use of a multi-channel, high performance digital oscilloscope to perform MIMO EVM demodulation measurements on a 4X4 downlink system. The 89600 VSA software is used with the oscilloscope to add capability for RF MIMO modulation domain analysis. The combination of time domain and modulation domain analysis in an oscilloscope creates a powerful tool for troubleshooting MIMO transmitter designs.

As discussed in Section 6.2.6, mixed-signal issues can present system integration risks in a SISO implementation. This risk can be further compounded when the implementation is a complex 2x2 MIMO system and even more so if the implementation is a 4x4 MIMO design. As stated earlier in this section, MIMO presents a primarily baseband (digital coding) challenge, but RF effects occurring in the transmit/receive signal chain can also degrade MIMO performance due to the compact nature of the RF design. Potential issues with the complex hardware field programmable gate array (FPGA) algorithms needed to implement 2x2 or 4x4 MIMO, combined with the complex RF interactions between multiple RF transmitters, receivers, and antennas, can introduce significant system integration risk and be quite difficult to diagnose and debug.

Using a digital oscilloscope with four phase-coherent inputs enables debugging of issues for 2x2 MIMO, 4x4 MIMO, or beamforming applications. Beamforming will be covered in Section 6.9.

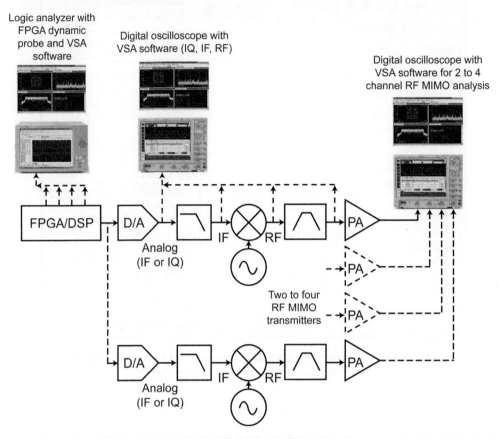

Figure 6.8-11. Example of a mixed signal MIMO test system using a logic analyzer and digital oscilloscope with Agilent 89600 VSA software

An example of a mixed signal test system for MIMO testing is shown in Figure 6.8-11. In this test system, the 32 GHz Agilent 90000 X-Series digital oscilloscope offers four phase coherent inputs with very low noise and jitter, providing an advantage for RF MIMO testing. In combination with the 89600 VSA software, the test system can perform both two channel and four channel phase-coherent MIMO analysis and debugging operations.

Careful consideration should be given to setting the sampling rate of the digital oscilloscope for MIMO applications. The digital oscilloscope in this test system, for example, has a maximum sample rate of 80 Gsa/s for up to two channel measurements or 40 Gsa/s for up to four channel measurements. These maximum sample rates will yield the lowest EVM noise floor for 2x2 MIMO EVM and 4X4 MIMO EVM, respectively, for the various oscilloscope sample rate settings, but there is a tradeoff between measurement time and sampling rate. The sampling rate can be reduced to 20 Gsa/s, 10 Gsa/s, 5 Gsa/s, or other sampling rates to reduce measurement time; however, the EVM noise floor may increase. As an initial starting point, the sampling rate can be set above the Nyquist rate using the 89600 VSA software's minimize mode to quickly see an EVM measurement. Better EVM noise floor performance (residual EVM) may then be achieved by increasing the oscilloscope's sampling rate to user rate or full rate, as illustrated in Figure 6.8-12 with the test setup shown in Figure 6.8-13.

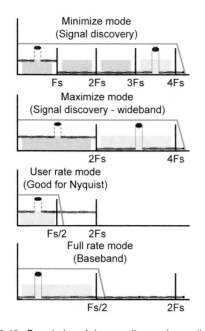

Scope sample mode	Measurement speed	Signal acquire length	EVM (noise floor)
Minimize (Max undersample)	★★★★	★★★★	
Maximize (Least undersample)	★★★★	★★★	★★
User rate (Any rate, good for Nyquist)	★★	★★	★★★ (Recommended)
Full rate (No aliases)	★	★	★★★★

Figure 6.8-12. Description of the sampling modes available with the 89600 VSA software and supported Agilent oscilloscopes

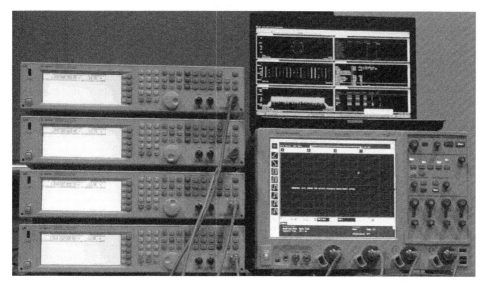

Figure 6.8-13. Four channel LTE MIMO measurement test setup with a 90000 X-Series digital oscilloscope and 89600 VSA software

Figure 6.8-14 shows the 89600 VSA software used with the 90000 X-Series oscilloscope to measure a four channel MIMO test signal, which includes impairments from local oscillator (LO) phase noise; power amplifier gain, phase compression, and distortion; and filter impairments.

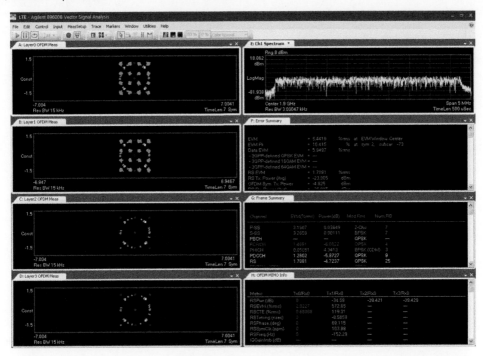

Figure 6.8-14. Four channel LTE MIMO measurement with a 90000 X-Series digital oscilloscope and 89600 VSA software

The four constellation measurements on the left half of the 89600 VSA software display in Figure 6.8-14 (traces A to D) show transmit antenna 0 (labeled Layer 0, top left) through transmit antenna 3 (labeled Layer 3, bottom left), corresponding to the four transmitter channel outputs on a frequency division duplex (FDD) spatially multiplexed, four channel MIMO test signal. The upper right measurement (trace E) shows the channel 1 spectrum centered at 1.9 GHz and configured for a 5 MHz bandwidth. The error summary table (trace F) is showing a composite EVM of 5.4 %. The frame summary (trace G) shows the individual EVMs, relative power levels (dB), modulation formats (Zadoff Chu, BPSK, QPSK, 16QAM PDSCH), and the number of resource blocks (RBs) for each data and control channel. The MIMO information table (trace H) is discussed in greater detail in Figure 6.8-15.

The MIMO information table shown in Figure 6.8-15 includes a number of important MIMO measurements and merits additional discussion. Specifically, the table shows:

- Row 1: Tx1/Rx0, Tx2Rx0, and T3/Rx0—being the crosstalk from transmit antennas 1 to 3 on receive antenna 0
- Row 2: Crosstalk from transmit antennas 0, 2, and 3 on receive antenna 1
- Row 3: Crosstalk from transmit antennas 0, 1, and 3 on receive antenna 2
- Row 4: Crosstalk from transmit antennas 0, 2 on receive antenna 3

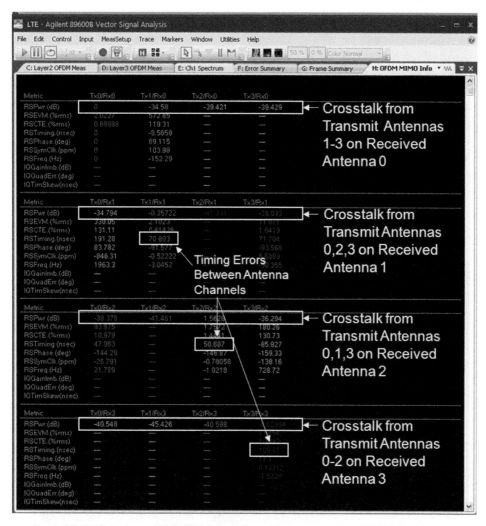

Figure 6.8-15. Four channel LTE MIMO information table measured with a 90000 X-Series digital oscilloscope and 89600 VSA software

Reference signal measurements are also displayed by the MIMO information table. As discussed in Section 2.4.2 and shown in Figure 2.4-3, RS orthogonality in both time and frequency is important in achieving proper MIMO system performance. Consider the effects of timing error on antenna crosstalk measurement results. The symbols between different transmit antennas remain orthogonal as long as the delay between the channels is less than the cyclic prefix duration. However, if this condition is not met, energy from one symbol can leak into the next, which can be perceived incorrectly as crosstalk between the channels.

In Figure 6.8-16 the individual RS EVM values are relatively low (approximately 1.5% to 2.1%) even though there is crosstalk (−34.58 dB to −45.42 dB) between the channels. Since the RS are orthogonal in both time and frequency, RS EVM is typically not affected by antenna crosstalk, unlike the composite EVM (5.4%), which is. However, impairments such as LO phase noise, amplifier gain and phase compression and distortion, and filter impairments do impact RS EVM. Thus both composite EVM and RS EVM can be evaluated, with insight into what types of error mechanisms are contributing to the system performance.

Reference signal timing errors that approach or exceed the LTE cyclic prefix duration (4.69 μs) can cause a loss of RS orthogonality, which affects measurement results. Figure 6.8-15 shows timing errors between the antenna channels ranging from approximately 50.6 ns to 109.9 ns (Tx1/Rx1, Tx2/Rx2, and Tx3/Rx3), which is well within a few percent of the cyclic prefix duration. Thus RS orthogonality is achieved in this example.

In contrast, however, Figure 6.8-16 shows the same four channel MIMO test signal, but with significant timing errors (approximately 2.39 μs to 3.1 μs). In this case the timing errors are significant enough to affect the orthogonality, which in turn affects the crosstalk measurement results. Antenna crosstalk is now measured at approximately −24.95 dB to −39.5 dB. The measurement accuracy has been compromised because there is no longer sufficient orthogonality between the data streams for the 89600 VSA software to make the antenna crosstalk measurement accurately.

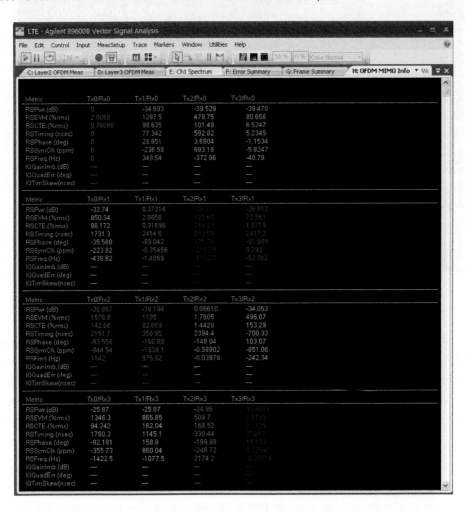

Figure 6.8-16. Four channel LTE MIMO test signal with large RS timing errors

The timing skew between channels can be directly observed and measured using the time domain analysis capability on the high performance digital oscilloscope as shown in Figure 6.8-17. Timing skew is measured at approximately 2.35 µs using the digital oscilloscope's time domain analysis, which correlates to the MIMO information table RS timing measurements, which are made using the 89600 VSA software modulation domain analysis on the oscilloscope.

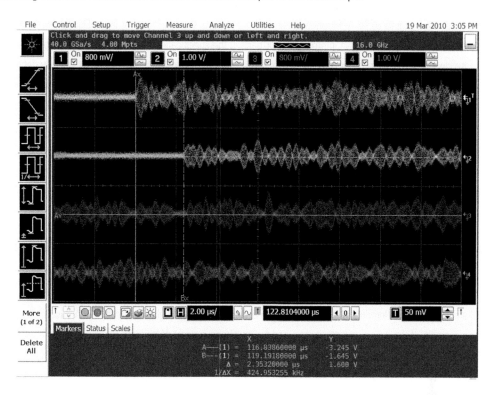

Figure 6.8-17. Timing skew error between MIMO data streams measured with digital oscilloscope

In summary, the baseband/RF mixed-signal performance of a transmitter design can affect 2x2 MIMO and 4x4 MIMO RF EVM performance. Baseband RS timing errors between channels can degrade orthogonality between the antenna channels, which in turn affects antenna cross-talk measurements if the timing errors become a significant percentage of the LTE cyclic prefix duration. In addition to these baseband impairments, RF impairments such as LO phase noise, amplifier gain and phase compression and distortion, and filter impairments can reduce the performance of both RS EVM and composite EVM. Antenna crosstalk may affect composite EVM but should not affect RS EVM if sufficient orthogonality is achieved. The time domain analysis and modulation domain analysis provided by a high performance digital oscilloscope running 89600 VSA software can help solve MIMO EVM performance issues by allowing testing and debugging across multiple domains.

6.8.4 MIMO Receiver Design and Verification Challenges

The challenges for MIMO receiver design are similar to those for the transmitter, ranging from the need to ensure adequate analog circuit performance to verifying baseband coding implementation. Problems can be more difficult to isolate unless the system development environment is available to provide visibility to internal processing steps. The most obvious approach is to start with a SISO test signal to verify synchronization and then introduce a second receiver input. During the early phases of a new standard such as LTE, differences in interpretation or gaps in the specifications can lead to interoperability problems. It is essential to be clear which version of the specification is being used and to document any area not yet fully defined where an implementation choice was made. The analysis tools designed for the eNB transmitter can help isolate differences in interpretation. If the receiver designer has created a test harness with proprietary signals, these should be analyzed with the same independent tool.

This section contains short descriptions of the configurations for testing UE and eNB MIMO receivers. Also covered are the additional complexity of channel fading with a MIMO signal and the performance testing options that address the complexity. It is assumed that the individual input signal paths are already verified using normal SISO techniques as discussed in Section 6.5.

6.8.4.1 UE Module Receiver Verification Challenges

The addition of a second receiving antenna in the UE brings a mixture of RF, power consumption, and digital processing challenges. A successful MIMO implementation not only has to interoperate with other devices through a complex faded channel, but do so while keeping the current drain and complexity of digital processing to a minimum.

Figure 6.8-18 shows the major elements in the block diagram of a two-receiver, single-transmitter UE. An additional transmitter is shown as an example of the further complexity required for multiband support. Within the RF circuitry, maintaining isolation between signals is one of the biggest challenges. Careful filtering of all the power supply and control lines and appropriate levels of isolation in the frequency down-conversion mixers is needed. The down-converted signals can be measured in isolation with the eNB analysis equipment described in previous sections.

If the RFIC has analog IQ outputs, these can be analyzed using an oscilloscope or MXA signal analyzer. If the RFIC interface uses DigRF v3 or v4, the signal can be captured with the Agilent Radio Digital Cross Domain (RDX) tester as described in Section 6.3. These digital or analog IQ signals can then be analyzed with the 89600 VSA software. For the baseband developer, hardware probes are available for analog IQ and DigRF interfaces.

The 89600 VSA software provides numerical EVM performance measurements for verification as well as more detailed graphical information useful during product development to isolate the source of signal impairments. If Gaussian noise is added to the signal as the impairment, a relationship can be drawn between EVM and the raw bit error ratio. Alternatively, the captured IQ signals can be fed into a simulated receiver such as Agilent design software for LTE. The precise design of a receiver determines the performance, so the results will vary depending on the vendor's implementation.

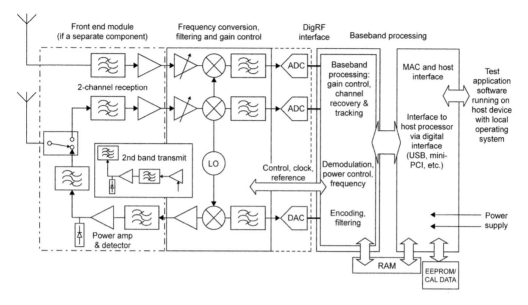

Figure 6.8-18. Simplified diagram of an LTE UE radio

As shown on the right side of Figure 6.8-18, during early development when the radio is in the form of an RFIC or baseband IQ (BBIQ) module, a host environment and test application software are needed for configuration and result display. At this stage open-loop testing without signaling can be used to ensure that design implementation margins meet expectations. It should also be possible to verify the accuracy of channel state information reports that will be transmitted back to the eNB.

It is not always possible or necessary to have a separate signal source to test each receiver input. Table 6.8-5 summarizes what can be achieved for single and dual signal cases. As a corollary to the phase and timing measurement of Section 6.8.2.8, a single source split between inputs removes the possibility of phase variation, although in a receiver test this means that the phase between the inputs is not adjustable.

Table 6.8-5. Tests and fault finding using one or more signal sources

Verification task	Single source	Dual source
Input sensitivity (BER or BLER) due to noise floor, phase noise, RF signal Interference	Yes	
Signal path response matching Amplifier characterization	Yes	
Interference, grounding, transient settling, dynamic performance (e.g., AGC operation)	Yes	Yes
Cross channel isolation	Yes*	Yes
MIMO operation and interoperability Full channel model testing		Yes

*Requires that the receiver be able to distinguish between signals from different inputs

411

In dual source testing, the standard RF phase and baseband timing alignment that can be achieved using a common frequency reference and frame synchronization signal is sufficient for most purposes unless the signals need to be precoded to match the channel or unless beamsteering operation is being evaluated. In these cases special measures are required to provide the necessary phase alignment. See Section 6.8.6 for more information.

6.8.4.2 Base Station (eNB) Receiver Verification

The eNB receiver faces many of the same MIMO challenges that the UE receiver faces, but in addition has to simultaneously receive multiple users. From the point of view of multi-user MIMO (MU-MIMO), each signal comes from a separate UE, so each signal therefore has a completely independent channel, somewhat different power levels, and different timing. These characteristics can be emulated using the Agilent N5106A MIMO receiver tester (PXB) in conjunction with RF signal generators.

The receiver test configuration for the eNB is different from the configuration for the UE. The UE normally sends packet error reports on the uplink back to the test system, whereas the eNB hardware is more likely to make a suitable demodulated signal output available.

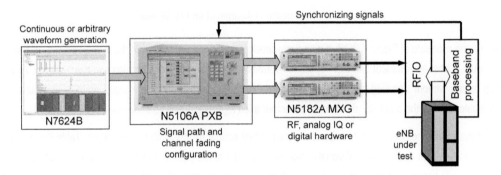

Figure 6.8-19. Continuous faded path receiver test configuration for RF, analog or digital interfaces using the N5106A (PXB) MIMO receiver tester

Figure 6.8-19 shows a typical configuration for eNB receiver test. The synchronization signals are required because it is the eNB that establishes frame timing in the system, and the eNB receiver relies on the UE adapting to the downlink reference.

Verification of UE coding can be achieved using the uplink signal routing configuration shown in Figure 6.8-20. The downlink can be configured as a separate cabled connection, using isolators from one of the eNB transmitters and UE receiver.

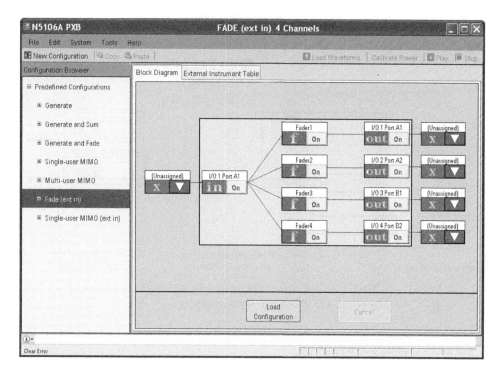

Figure 6.8-20. N5106A PXB user interface showing configuration for single RF input and four RF outputs

6.8.4.3 Verifying Interoperability Using Recorded Multi-Channel RF Signals

One of the most troublesome aspects of receiver design is trying to determine what another designer has done to overcome a design constraint. A troubleshooting technique to facilitate understanding of how a third party signal interacts with a receiver is to capture the signal at RF and then replay it from a signal generator. This enables troubleshooting to take place remotely from the location of the equipment, as might be done by geographically separated teams. The recorded signal can also play a role in regression testing.

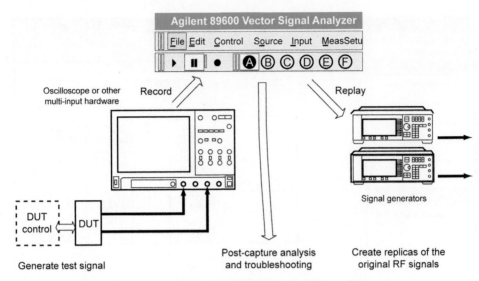

Figure 6.8-21. Dual input signal capture and replay using 89601A VSA, oscilloscope, and MXG or ESG signal generators

Figure 6.8-21 shows one of the possible signal capture methods using an oscilloscope as part of the test setup. The oversampling rate may be reduced to increase the capture time when the carrier frequency is chosen to avoid sampling aliasing products.

6.8.5 MIMO Receiver Performance Testing Using Static and Faded Channels

Some general observations can be made about the antenna systems appropriate to MIMO systems for use in macro-cellular environments. Cross-polarized antennas at the eNB outperform uniformly polarized antennas with equivalent characteristics. The normally wide angle of arrival at the UE antennas helps offset the close physical proximity, but in the lower frequency bands the wavelengths involved make it progressively more difficult to avoid interaction between the antennas.

To support the observations from real systems, many parameters are incorporated into the models used to emulate the physical transmission channel. The following sections assume some understanding of these parameters as they apply to a SISO channel. Described here are the additional considerations needed for MIMO channels.

MIMO channel recovery involves the separation of multiple signal components in the presence of noise and interference. When the signals are transmitted they are orthogonal, but by the time the signals reach the multiple receivers the coupling in the radiated path can reduce the difference between the signals.

Section 6.8.5.1 describes a method of using a fixed tap channel for MIMO receiver evaluation. The remainder of Section 6.8.5 addresses the parameters needed to more realistically represent the channel when a continuous fading test is used. The intention is to provide a test environment for design troubleshooting rather than for performance and conformance testing, which are described in Sections 6.6 and 7.2.

6.8.5.1 Baseband Performance Test Using Fixed Tap Fading Channels

In normal operation the receiver has to deal with a complex and continuously changing channel, but testing using such a fading channel misses the opportunity to ensure that the basic baseband operation is correct. A fading channel built from simple phase and timing differences between paths provides a deterministic signal that can be designed to verify the receiver's performance limits. Adding noise to such a channel can readily create a test signal in which some subcarriers are more difficult to demodulate than others.

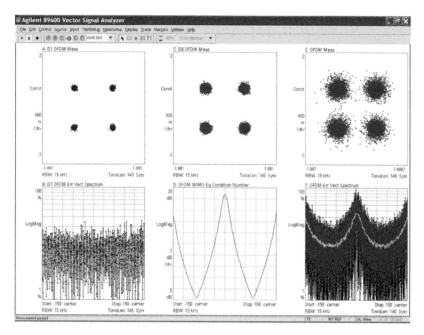

Figure 6.8-22. Example of cross-coupled, fixed tap fading channel for spatial multiplexing test

Figure 6.8-22 shows the results of a fixed tap fading channel. Trace A (top left) and trace B (bottom left) show the SISO RS constellation and EVM spectrum, which are not affected by coupling through the channel. Trace C (top middle) is the uncoupled MIMO constellation prior to passing through the channel. Trace E (top right) shows one of the layers after passing through the channel, with the corresponding EVM spectrum shown in trace F (bottom right). It can be seen how the EVM tracks the channel condition number shown in trace D (bottom middle), which is the result of the 400 ns delay and −2 dB coupling parameters chosen for this example.

6.8.5.2 Spatial Correlation

MIMO systems require a rich multipath environment for optimum operation, and the spatial positions of the multiple transmit antennas, relative to each other and relative to their placement in the surrounding environment, will influence the fading correlation between the different MIMO channels. The same is true for the antenna positions at the receiver. It will be shown in this section that inadequate antenna spacing leads to spatial correlation, which limits MIMO performance gains.

The spatial correlation coefficient ρ_{12}, between two antenna elements, is a function of their spacing and the spatial power distribution of radiated or received signals; i.e., the power azimuth spectrum (PAS) and the gain pattern of the individual elements. Here it is assumed that the antenna elements are identical with the same gain pattern. The correlation coefficient can be calculated using the following equation.

$$\rho_{12} = \frac{\int_{-\pi}^{\pi} e^{-j2\pi\frac{d}{\lambda}\sin(\theta)} PAS(\theta)G(\theta)\,d\theta}{\int_{-\pi}^{\pi} PAS(\theta)G(\theta)\,d\theta} \tag{1}$$

$PAS(\theta)$ and $G(\theta)$ are calculated using one of the equations found in Agilent's application note 5989-8973EN [18]. $PAS(\theta)$ is dependent on the selection of the appropriate distribution; i.e., Laplacian, Gaussian, or uniform, according to the physical environment. The parameter d is the distance between antenna elements. The gain pattern, $G(\theta)$, assumes that the far field assumption holds and that the two antennas have exactly the same radiation pattern and boresight direction.

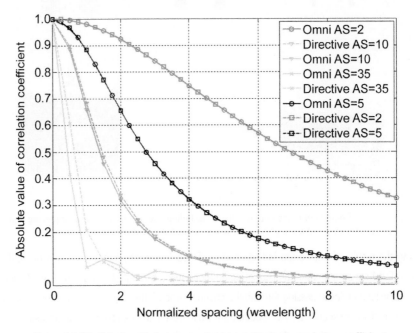

Figure 6.8-23. Relationship between antenna spacing and correlation coefficient using a Laplacian PAS with AoA=200 degrees and $\Delta\theta$ =180 degrees

Figure 6.8-23 shows the absolute value of the correlation coefficient as a function of antenna spacing for several examples of antenna type and azimuth spread (AS). The antenna type was varied between omni-directional and directive using a 3-sector antenna. Each curve represents a different value for AS covering 2, 5, 10, and 35 degrees. These curves assumed a single-modal Laplacian PAS with mean angle of arrival (AoA) of 200 degrees and $\Delta\theta$ of 180 degrees. The correlation coefficient decreases for increasing normalized spacing and for increasing AS. It is worth noting also that for a given antenna spacing and large AS=10 or 35, the directive antennas tend to be slightly more correlated than the omni-directional ones.

The spatial correlation matrix for the complete system can be calculated using equation (1) above and forming the individual spatial correlation matrices at the eNB and the UE. For example, given a 2x2 MIMO system, assume the factors α and β represent the correlation coefficients, calculated using (1), for the eNB and UE antenna pairs, respectively. The correlation matrices for eNB and the UE are represented as

$$R_{BS} = \begin{pmatrix} 1 & \alpha \\ \alpha^* & 1 \end{pmatrix}, \tag{2}$$

$$R_{MS} = \begin{pmatrix} 1 & \beta \\ \beta^* & 1 \end{pmatrix}. \tag{3}$$

The system spatial correlation matrix for the downlink channel can be calculated using the Kronecker product

$$R_S = R_{BS} \otimes R_{MS}, \tag{4}$$

$$R_S = \begin{pmatrix} 1 & \beta & \alpha & \alpha\beta \\ \beta^* & 1 & \alpha\beta^* & \alpha \\ \alpha^* & \alpha^*\beta & 1 & \beta \\ \alpha^*\beta^* & \alpha^* & \beta^* & 1 \end{pmatrix}. \tag{5}$$

These expressions are needed to determine the parameters for the user interface of a fading emulator.

6.8.5.3 Antenna Polarization Correlation

In the previous section it was shown that systems operating with a narrow range of angular spread may require antennas placed physically far apart in order to achieve low spatial correlation. Unfortunately some wireless devices tend to be physically small, thus limiting the antenna separation to less than one wavelength depending on the frequency of operation. In a practical case of a cell site, the issue is often equally prosaic: the cell tower cannot physically support more antennas, or the cost and regulatory issues are significant impediments.

An alternative solution is required to achieve the low channel-to-channel correlation required for successful MIMO operation. One technique to reduce the spatial correlation between two antennas is to cross-polarize the antennas—in other words, position the antenna polarizations in orthogonal or near-orthogonal orientations. As shown in Figure 6.8-24, two closely spaced, vertically polarized (0/0) dipole antennas have a high spatial correlation while two orthogonally polarized (0/90) antennas, one vertical and one horizontal, have a much lower correlation coefficient.

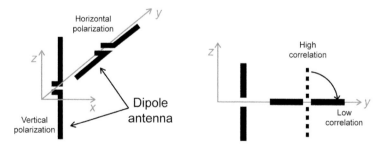

Figure 6.8-24. Diagram showing the effects of relative antenna polarization on correlation characteristics

The use of antennas with different polarizations at the transmitter or receiver may lead to power and correlation imbalances between the various MIMO channels. This implies that one layer of the transmission will have a better average performance than the others.

It is a common practice to derive the correlation matrices using the antenna configuration and angular information of the paths. The definition of channel correlation for LTE is in 36.101 [3] Subclause B.2.3. These correlation matrices are defined based on corner case conditions representing low, medium, and high correlation, independent of the transmitting and receiving antenna configurations necessary to create such correlation. Developing performance requirements for MIMO operation at the corner case correlation conditions helps ensure that the underlying system behaves as expected. The next step, calculating nominal correlation matrices from actual antenna designs at specific frequencies, is outside the scope of the LTE specifications as defined by 3GPP.

Although LTE does not use the physical antenna characteristics to derive the correlation matrices, it is useful to see how this can be done to evaluate the nominal correlation for a particular design. The following example, which shows how to calculate a correlation matrix using nominal antenna configurations, is taken from the conformance test definition for WiMAX [31].

The WiMAX antenna polarization matrix at either the transmitter or receiver is given by

$$S = \begin{bmatrix} S_{vv} & S_{vh} \\ S_{hv} & S_{hh} \end{bmatrix}, \tag{6}$$

where the index v represents vertical polarization and h horizontal polarization. The first index denotes the polarization at the transmitter and the second denotes the polarization at the receiver. Correlation between polarized antennas can be quantified using the cross-polarization ratio (XPR). The XPR is the power ratio between a pair of cross-polarized antennas (v-h or h-v) to that of a co-polarized (v-v or h-h) case. Assume that the XPR = −8 dB, then

$$\frac{|s_{vh}|^2}{|s_{vv}|^2} = \frac{|s_{vh}|^2}{|s_{hh}|^2} = -8\,\mathrm{dB}\,(= 0.1585). \tag{7}$$

For example, consider a 2x2 MIMO system. The eNB polarization matrix with polarization angles α_1, α_2 is

$$P_{BS} = \begin{pmatrix} \cos(\alpha_1) & \cos(\alpha_2) \\ \sin(\alpha_1) & \sin(\alpha_2) \end{pmatrix}. \tag{8}$$

The UE polarization matrix with polarization angles β_1, β_2 is

$$P_{MS} = \begin{pmatrix} \cos(\beta_1) & \cos(\beta_2) \\ \sin(\beta_1) & \sin(\beta_2) \end{pmatrix}. \tag{9}$$

For the downlink case, the total channel polarization matrix is the matrix product of the eNB polarization, channel polarization and UE polarization:

$$Q = P_{MS}{}^T S P_{BS}. \tag{10}$$

Lastly, the polarization correlation matrix is defined as

$$\Gamma = E\left\{ \text{vec}(Q) \cdot \text{vec}(Q)^H \right\}.$$

(11)

For specified polarization angles, such as +45/−45, 0/90, and 0/0, the diagonal elements of Γ have the same value, which means there is no power imbalance between different channels. For arbitrary polarization angles, the diagonal elements of Γ are not equal, which means that polarization leads to an undesired power imbalance between the channels.

Normalization of Γ is required so that the diagonal elements reflect the channel power. In this case the correlation matrix then becomes

$$\Gamma^R = \frac{K}{\displaystyle\sum_{i=1}^{K} \Gamma_{i,j}} \Gamma.$$

(12)

This power normalization process is based on the assumption that the overall channel power is $K = N_r N_t$ for a MIMO system with N_t transmit antennas and N_r receive antennas. The correlation matrix Γ^R will properly reflect the channel imbalance due to polarization, and the diagonal elements in Γ^R relate to the relative power in each channel.

Using a 2x2 MIMO channel as an example, assume that the XPR = −8 dB and the system uses cross-polarized UE antennas (0/90) and slant-polarized eNB antennas (+45/−45). The resulting polarization correlation matrix is

$$\Gamma^R = \begin{pmatrix} 1 & 0 & 0.7264 & 0 \\ 0 & 1 & 0 & -0.7264 \\ 0.7264 & 0 & 1 & 0 \\ 0 & -0.7264 & 0 & 1 \end{pmatrix}.$$

(13)

The diagonal elements of this polarization correlation matrix are all ones, which show that the selected polarization angles do not result in a power imbalance among the different MIMO channels. The other elements in the matrix relate to the correlation between different channels. For example, in the first row, this matrix shows that channel 1 is only correlated to channel 3 with a coefficient of 0.7264. The second row shows that channel 2 is only correlated with channel 4. It can be shown that the use of antennas with differing polarizations at the transmitter and receiver leads to polarization diversity, giving worthwhile performance improvements and enabling antenna designs that do not require the spatial separation otherwise expected.

As another example, consider a case in which the correlation matrix results in a power imbalance. Here, assume that the antenna polarization angles are −10/80 at UE antenna and +30/−60 at eNB. The resulting polarization correlation matrix is

$$\Gamma^R = \begin{pmatrix} 1.3413 & 0.1242 & 0.5911 & 0.2151 \\ 0.1242 & 0.6587 & 0.2151 & -0.5911 \\ 0.5911 & 0.2151 & 0.6587 & -0.1242 \\ 0.2151 & -0.5911 & -0.1242 & 1.3413 \end{pmatrix}.$$

(14)

For this matrix, the diagonal elements are not equal and therefore demonstrate that using this combination of polarization angles leads to a power imbalance among the different channels.

6.8.5.4 Combined Spatial and Antenna Polarization Correlation

Spatial and polarization correlation effects in compound antenna systems are independent and multiplicative. In this case, the corresponding spatial and polarization correlation matrices can be derived separately and combined by an element-wise matrix product. The combined spatial-polarization correlation matrix is then defined as

$$R = R_S \cdot \Gamma^R,\qquad(15)$$

where R_S is the system spatial correlation matrix using equation (5) and Γ^R is the polarization correlation matrix calculated using equation (12).

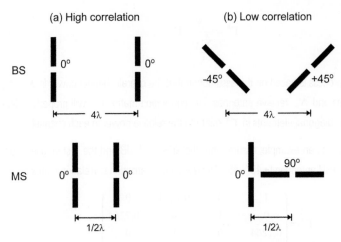

Figure 6.8-25. eNB and UE antenna configurations for (a) high and (b) low channel correlations

Orthogonal antenna positions (0/90) provide the lowest spatial correlation but may not always be required or practical under all conditions. For example, the diagrams in Figure 6.8-25 show two possible 2x2 MIMO configurations for the eNB and UE antenna positioning. In one case, as shown in Figure 6.8-25 (a), all the antennas are vertically polarized resulting in a potentially high level of spatial correlation. To overcome this problem, the eNB antennas are spaced at 4λ to improve the correlation for this typically narrow angle spread condition. The spacing at the UE is fixed at λ/2. This first case is the reference antenna configuration for the high correlation channel model as specified in the WiMAX standard. For this high correlation condition, all the antennas have the same polarization angle, which does not provide any polarization diversity. Therefore the polarization matrix is defined as

$$\Gamma^R = \begin{pmatrix} 1 & 1 & 1 & 1 \\ 1 & 1 & 1 & 1 \\ 1 & 1 & 1 & 1 \\ 1 & 1 & 1 & 1 \end{pmatrix}.\qquad(16)$$

By applying equation (15) to this high correlation antenna configuration, the combined spatial-polarization correlation matrix is the same as the spatial correlation matrix previously defined in equation (5):

$$R = R_S \cdot \Gamma^R = \begin{pmatrix} 1 & \beta & \alpha & \alpha\beta \\ \beta^* & 1 & \alpha\beta^* & \alpha \\ \alpha^* & \alpha^*\beta & 1 & \beta \\ \alpha^*\beta^* & \alpha^* & \beta^* & 1 \end{pmatrix}.$$ (17)

The second case, as shown in Figure 6.8-25 (b), has the antennas at the eNB polarized at ±45 degree orientations while the UE uses orthogonal polarization (0/90). This second combination can greatly reduce channel-to-channel correlation, thus enabling better operation of the MIMO system. This combination is the reference antenna configuration for the low correlation channel model for WiMAX conformance testing. For this configuration, the polarization matrix was defined in equation (13) and the final correlation matrix is defined as

$$R = \begin{pmatrix} 1 & 0 & \gamma\alpha & 0 \\ 0 & 1 & 0 & -\gamma\alpha \\ \gamma\alpha^* & 0 & 1 & 0 \\ 0 & -\gamma\alpha^* & 0 & 1 \end{pmatrix},$$ (18)

where $\gamma = 0.7264$.

Comparing the two correlation matrices found in equations (17) and (18), it can be concluded that introducing different polarization angles at the eNB and UE will lower the channel-to-channel correlations.

6.8.5.5 Configuring the Receiver Tester for Spatial Correlation

It is possible to improve the process of entering the correlation matrices into a wireless channel emulator while minimizing the mathematical complexity and still be capable of modeling realistic wireless channels. The PXB MIMO receiver tester greatly improves the process of MIMO channel emulation by eliminating the need to calculate complex correlation matrices. Users can enter the physical antenna characteristics directly in the instrument. For example, Figure 6.8-26 shows the PXB user interface for entering the receive channel spatial parameters including antenna type, spacing and polarization. A similar table is used to enter the transmit antenna parameters. PXB uses this spatial information along with the AoA and angle of departure (AoD) entries in the fading paths table to automatically calculate the spatial-polarization correlation matrix. This simple entry table eliminates the burden of calculating the correlation matrices and manually entering the coefficients into the emulator.

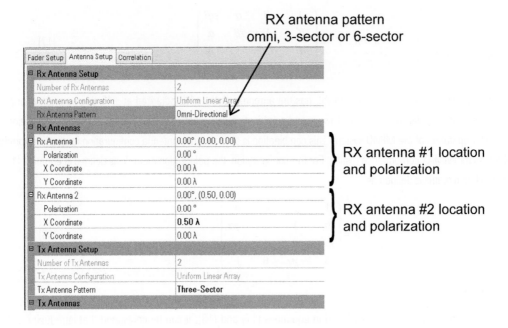

Figure 6.8-26. Agilent N5106A PXB antenna parameter setup screen

6.8.5.6 Per-Path vs. Per-Channel Correlation

When test conditions for emulating MIMO channels are standardized, the correlation properties can be "per-path" or "per-channel." Per-path correlation means that each tap uses a different correlation matrix, while per-channel correlation means that all the taps use same correlation matrix. As explained above, the spatial correlation coefficient between two antenna elements is a function of antenna spacing, the PAS, and the radiation pattern of the antenna elements. The PAS is a function of path AoA/AoD and path AS. In real-world conditions not all paths have the same AoA/AoD and AS values; therefore, different paths could have different correlation coefficients. The use of per-path correlation may improve the fidelity of the channel emulation process. In order to emulate this real-world scenario, the MIMO channel models used for Mobile WiMAX and the WLAN 802.11n standards use the per-path correlation based on different AoA/AoD for each path. While per-path spatial correlation can closely model a real wireless channel, it has a high level of computational complexity. The PXB has predefined channel models to automatically configure the instrument's path correlations.

The LTE requirements, in an effort to reduce MIMO complexity, recommend the per-channel correlation model without considering the path AoA/AoD information. When a MIMO system is tested against these specifications, the PXB can also provide per-channel correlations using predefined models, or, as shown in Figure 6.8-27, provide a simple table entry for custom per-channel emulation models.

When wireless channels are emulated, it is important to understand the test requirements in relation to spatial correlations. The PXB tester has flexibility to support both per-path correlation and per-channel correlation, either individually or at the same time.

		Channel 1	Channel 2	Channel 3	Channel 4
	Channel 1	1.00	-0.74 + 0.67i	0.15 - 0.22i	0.04 + 0.27i
▶	Channel 2	-0.74 - 0.67i	1.00	-0.26 + 0.06i	0.15 - 0.22i
	Channel 3	0.15 + 0.22i	-0.26 - 0.06i	1.00	-0.74 + 0.67i
	Channel 4	0.04 - 0.27i	0.15 + 0.22i	-0.74 - 0.67i	1.00

Fading : Master Setup 1

Restore Default Settings | Correlation:

Fader Setup | Antenna Setup | Correlation

Figure 6.8-27. Agilent N5106A "per-channel" correlation setup screen used for LTE

6.8.5.7 Theoretical MIMO Channel Capacity

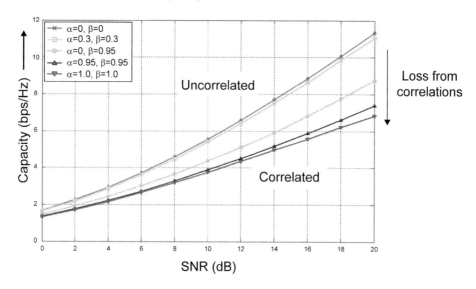

Figure 6.8-28. Ergodic (statistically representative) capacity for a 2x2 MIMO system with different transmit and receive correlation coefficients

To give a more intuitive impression of the channel capacity loss caused by fading correlation, Figure 6.8-28 compares the capacity as a function of SNR for a 2x2 MIMO system with different correlation coefficients at the transmitter (α) and the receiver (β). Compared with the completely independent MIMO channel $(\alpha = \beta = 0)$, the figure shows that there is only a small capacity loss for the low correlated channels $(\alpha = \beta = 0.3)$. For highly correlated channels $(\alpha = \beta = 0.95)$ at high SNR, the capacity decreases by 3.9 bps/Hz compared to the ideal uncorrelated case. For completely correlated channels $(\alpha = \beta = 1.0)$, the capacity decreases by 4.4 bps/Hz at high SNR. Note that even when 1 is the correlation coefficient, there is still an increase in capacity relative to SISO from the increase in number of antenna pairs, though the improvements are small. The largest improvements are achieved when the channels are independent. In this case the MIMO capacity is improved by approximately the SISO capacity multiplied by minimum (N_t, N_r). See "Capacity scaling in MIMO wireless systems under correlated fading" [32]. Note that the median SNR in typical loaded cells is around 5 dB.

6.8.5.8 Using Condition Number to Understand the Performance Limitations of a MIMO Channel

An alternative way to view the impact of the MIMO channel is to use the channel condition number calculation. It provides an understanding of what takes place frame by frame, rather than using the statistical methods described earlier.

In Figure 6.8-29, two samples have been taken from a Pedestrian A channel. Although not primarily intended for faded channel measurements, the 89600 VSA software is able to demodulate this relatively benign channel and display the results. The noise in the channel was set by adjusting the input level of the measurement system.

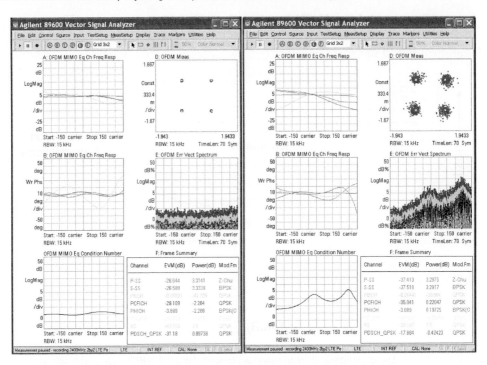

Figure 6.8-29. Measurements from a Pedestrian A faded channel showing channel frequency responses, condition number, and effect on EVM versus subcarrier

The matrix condition number is calculated for each subcarrier frequency and plotted below the corresponding channel amplitude and phase responses. As expected, in the 2x2 system shown, four traces are used to represent each component of the channel. In the left half of Figure 6.8-29, the condition number response, shown in the bottom left trace, is approximately 10 dB. This value is typical of a good MIMO channel. The resultant EVM is about 3%. It is plotted in the center right trace on a logarithmic scale to allow a direct comparison with the condition number.

On the right half of Figure 6.8-29, a second sample shows a very different response. This is not unusual for a faded channel. The condition number trace shows several peaks, and a gradual increase in condition number from left to right, which indicates that the upper subcarriers have a poorer MIMO channel than the lower subcarriers. The EVM plot in the center right trace shows how the degradation in condition number points to worsening demodulation performance. In practice, this would imply using closed-loop operation with frequency-selective scheduling to avoid transmitting the MIMO signal at the upper part of the channel. There is another aspect of the MIMO channel that is not immediately obvious: in any particular channel, the performance for different layers may not be the same. This topic is discussed further in Section 6.8.5.12.

6.8.5.9 Configuring the Channel Emulator to Achieve the Desired Correlation

Under ideal channel conditions, MIMO systems can provide dramatic channel capacity gain through increased spatial diversity. However, the capacity gain is reduced if the fading characteristic among various channels is correlated. Many wireless standards, such as WiMAX and LTE, recommend test scenarios that use correlated channel matrices. One approach that is widely accepted for defining the correlation properties of a MIMO channel relies on the "λ-parameter." In this case, λ provides an indication of the correlation as it relates to capacity. The capacity, operating under a specified SNR, is defined as a linear interpolation of the capacity for a completely correlated MIMO channel to that of an uncorrelated MIMO channel. Using the λ-parameter, the resulting capacity is defined as

$$C_\lambda = (1 - \lambda)C_0 + \lambda C_1,$$ (19)

where C_0 is the channel capacity without correlation and C_1 is the channel capacity when the channels are completely correlated. With this approach, it is simple to specify the expected correlation degree through the λ parameter. For example, in the LTE channel model for 2x2 antenna configurations, the medium and high correlation matrices are defined using values of $\lambda \approx 0.5$ and $\lambda \approx 0.9$, respectively. For a system operating with a target value for λ and under a specified SNR, the correlation matrix can be tuned to achieve the desired correlation.

The correlation matrix that can guarantee the expected capacity, C_λ, is not unique and different correlation matrices can be chosen to satisfy this capacity requirement. One very flexible method of achieving the desired correlation matrix is by adjusting the antenna configuration, such as element spacing and polarization. For example, the eNB antenna spacing is adjusted using a 2x2 MIMO configuration with vertically polarized antennas, as was shown in Figure 6.6-25 (a), until the desired correlation is achieved. The antenna parameters for the UE are fixed with values shown in Table 6.8-6. The receiver correlation coefficient, β, is calculated using equation (1).

Table 6.8-6. UE (receiver) antenna configuration

	Antenna spacing in wavelength	Antenna type	AS (degrees)	AoA (degrees)	Correlation coefficient (β)
UE	0.5	Omni	35	67.5	−0.6905 + 0.3419i

For this example, the eNB uses a 3-sector antenna configuration with AS = 2 degrees and AoD = 50 degrees. The eNB correlation coefficient, α, changes when the spacing between the eNB antenna elements is adjusted. With this configuration, the combined spatial-polarization correlation matrix can be calculated using equation (17). If the calculated channel capacity is lower than the desired channel capacity, the antenna spacing is increased to reduce the correlation and thus increase the channel capacity. By iteratively adjusting the antenna spacing, the desired λ can be achieved. Table 6.8-7 shows correlation index λ as a function of eNB antenna spacing for this 2x2 MIMO example.

Table 6.8-7. Relationship between eNB antenna spacing, correlation coefficient and channel capacity under specified SNR values

Antenna spacing d	correlation coefficient α	λ SNR=10 dB	λ SNR=20 dB
0	1.0000	0.9060	0.9445
0.5	−0.7390 + 0.6700i	0.9004	0.9270
1.0	0.0969 − 0.9854i	0.8921	0.8806
1.5	0.5827 + 0.7857i	0.8543	0.8189
2.0	−0.9433 − 0.1881i	0.8252	0.7598
3.0	−0.2687 + 0.8779i	0.7591	0.6542
4.0	0.7955 + 0.3350i	0.6958	0.5636
5.0	0.3854 − 0.7028i	0.6246	0.4951
6.0	−0.6061 − 0.4196i	0.5704	0.4389
7.0	−0.4388 + 0.5106i	0.5232	0.3971

6.8.5.10 Adding Noise to a Multiple Input Receiver Test

Adding noise to SISO signals is necessary to allow the baseband developer to understand the performance limitations of the demodulator. When a multiple input receiver is verified, the correlation of the noise between channels becomes an important factor to address.

6.8.5.11 SNR for SISO and Uncorrelated MIMO Channels

A convenient place for setting the channel's SNR is typically at the receiver. The signal power can be accurately measured with a power meter and the channel emulator can generate the required noise according to the desired SNR. This technique is valid for SISO systems and for MIMO systems that have uncorrelated channels. When the MIMO channels are correlated, an alternate approach to measuring the signal power and generating noise is required.

For SISO systems, the received signal, Y, is defined as

$$Y = HX + N, \qquad\qquad (20)$$

where X is the transmitted data, H is the channel coefficient, and N is the noise. For a specified SNR, the signal power S, is first measured at the output of the channel emulator in the absence of noise. The covariance of the noise, σ^2, being a random Gaussian process, can be calculated and added by the channel emulator to simulate the effect of applying noise to the SISO channel. As shown in Agilent application note 5989-8973EN Appendix B [26], this technique is also valid for uncorrelated MIMO channels.

n this case, the signal at the receiver can also be defined using equation (20) where X is now a vector of M_t transmitted signals, H is the channel coefficient matrix with M_r rows and M_t columns, and Y is a vector of M_r received signals. In the MIMO case, N is an M_r row of random Gaussian processes. It is also shown in the same reference that the signal power can be measured at either the receiver or the transmitter for the uncorrelated MIMO system.

6.8.5.12 SNR for Correlated MIMO Channels

When the MIMO channels are correlated, the measured signal power at the receiver side can be dependent on the correlation of the channels. This correlation dependency prevents a channel emulator from accurately configuring the MIMO system for a desired SNR using power measurements at the receiver. To overcome this difficulty, the channel emulator can use measurements of the signal power at the transmitter to set the required SNR appropriately. The following derivation shows a simplified example using a 2x1 MISO system to demonstrate an appropriate measurement technique for configuring the SNR in a channel emulator when the channels are correlated. The MISO precoding matrix is defined as

$$\frac{1}{\sqrt{2}}\begin{bmatrix} 1 \\ e^{j\theta} \end{bmatrix}. \tag{21}$$

The signal transmitted from antenna 1 is $X/\sqrt{2}$, the signal transmitted from antenna 2 is $Xe^{j\theta}/\sqrt{2}$, and the transmitted signal power from each antenna is S. The channel between transmit antenna 1 and the receive antenna is H_1. The channel between transmit antenna 2 and the receive antenna is H_2. Using equation (20), the received signal becomes

$$Y = \frac{X}{\sqrt{2}}H_1 + \frac{Xe^{j\theta}}{\sqrt{2}}H_2. \tag{22}$$

If H_1 is independent with H_2, the received signal power is

$$E(YY^*) = (\overline{H}_1 + \overline{H}_2)S, \tag{23}$$

where $\overline{H}_1$ and $\overline{H}_2$ represent average channel gains of H_1 and H_2 respectively. When $\overline{H}_1 = \overline{H}_2 = \overline{H}$ then $E(YY^*) = 2\overline{H}S$.

If $\overline{H}_1$ is completely correlated with $\overline{H}_2$, meaning that $\overline{H}_1 = \overline{H}_2 = \overline{H}$, then the received signal becomes

$$Y = (1 + e^{j\theta})H\frac{X}{\sqrt{2}}, \tag{24}$$

and the received signal power is

$$E(YY^*) = 2(1 + \cos(\theta))\overline{H}S. \tag{25}$$

When $\theta = \pi/4$, the received signal power becomes $2(1 + \sqrt{2}/2)\overline{H}S$, which is different from the case with independent channel conditions. Therefore, if the measured signal power at the receiver is used to calculate the noise power required for a specific SNR, then the added noise power will vary according to the fading correlation property. Continually adjusting the noise power as a function of correlation property introduces unnecessary complexity into the measurement and may result in reduced accuracy when the required SNR is configured.

To overcome this difficulty, the PXB tester defines the SNR relative to the transmitted signal power and uses the following SNR definition:

$$SNR = \frac{S_1 \overline{H}_1 + S_2 \overline{H}_2}{\sigma^2},$$ (26)

where S_1 and S_2 are the signal powers from each transmitter. With this definition, the PXB measures the signal power at the transmitters prior to fading and then adds the appropriate noise power to achieve the desired SNR. In this technique the noise contribution can be determined without considering the fading correlation property of the channel.

6.8.6 Requirements for Phase Coherence

This section describes the impact of phase coherence and potential solutions. For directly mapped, open-loop MIMO testing, the phase relationship between the test signals does not affect the performance of the receiver because orthogonal signals have to be coupled twice for vector addition to take place. In closed-loop systems, the phase between test signals needs to be constant during the period when the channel is sampled, allowing any coupling coefficients to be calculated and applied. This may require the system to be stable rather than phase-locked.

It was shown earlier that the channel condition number can be used to determine the SNR needed to achieve a specific performance at the demodulator. The condition number gives a measure of the composite channel performance. Each layer of the MIMO signal may actually have a different performance. The plots in Figure 6.8-30 show the demodulated signals from a single frame of an LTE signal. The channel was "flat-faded" (no frequency selectivity).

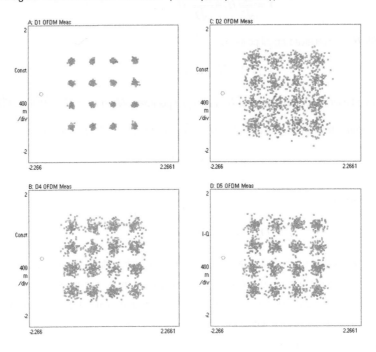

Figure 6.8-30. Spatial multiplexed signals without precoding (top) and with precoding (bottom) to match the channel

The two constellations at the top of the figure show the two layers of the MIMO signal. It is clear that the constellation on the left is tighter, which results in a lower BER in a real receiver. If the channel characteristics are known—e.g., by the UE sending channel state information to the eNB—the mismatch in performance can be dealt with in either of two ways. The layer with better performance can be given a larger payload using higher order modulation or less channel coding protection, or precoding can be applied to equalize the performance of the two layers.

In LTE, the codebook index method is used to facilitate channel precoding, with a small number of codes used to minimize the system overhead in signaling. This means that the codebook index provides an approximation to the channel, implying some level of residual error. Figure 6.8-31 shows that once a codebook is chosen to equalize the EVM, the actual EVM still depends on the phase match between the transmitters.

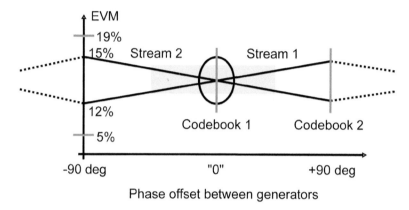

Figure 6.8-31. Impact of phase errors on precoding effectiveness, with example performance values

The rectangular block at the center of the figure represents the region in which codebook 1 would be chosen as the best fit. The diagonal lines show how the EVM of each stream varies with phase error. EVM is used as a performance metric, but BER could also be used. In the best circumstance, the codebook exactly fits the channel state and the performance of both layers is made the same. This is the case for the constellations in the lower half of Figure 6.8-30. As the phase between the transmitters varies—indicating that a mismatch in the codebook choice or variation in the channel occurred after the channel station information was provided—the layer performance separates. At extremes, the performance of the layers can be swapped.

For receiver measurements, the significance of precoding errors can be seen in the need for a fixed RF phase relationship at the output of the signal generators being used for a test. The term phase coherence is used to signify that the RF phase at the outputs of two or more generators is being controlled for a specified frequency.

When it is necessary to guarantee that phase will not change versus frequency, a test configuration such as that shown in Figure 6.8-32 can be used. Note that when two generators are present, the external LO is not required.

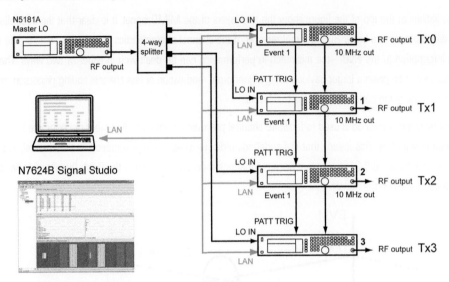

Figure 6.8-32. Configuring multiple signal generators for timing and phase synchronization

6.8.7 Conclusions

Multi-antenna techniques offer significant performance benefits when they are applied correctly to suit the channel conditions. This MIMO section has examined some of the challenges of implementing MIMO techniques in the transmitters and receivers of LTE eNBs and UEs. A number of measurement techniques have been identified that allow the impact of cross-coupling, timing errors, and distortion to be analyzed in the transmitter.

The issues for the receiver include the need to test performance with a wide variety of impaired signals. Impairments in this case include noise, interference, and channel fading. The extensive features available in the N5106A PXB MIMO receiver tester provide the flexibility needed to make these tests.

6.9 Beamforming

This section briefly recaps multi-antenna techniques before introducing the concept of beamforming, its advantages, and its use within modern wireless communications systems such as LTE.

Multi-antenna beamforming measurement challenges are discussed from an eNB test perspective, and the importance of calibration is highlighted when it comes to testing the performance of a beamforming transmission system.

6.9.1 Multi-Antenna Techniques Summary

Various multi-antenna techniques employed by LTE are introduced in Section 2.4. These radio access techniques are illustrated in Figure 2.4-1, which captures the concept of single input single output (SISO), single input multiple output (SIMO), multiple input single output (MISO), and multiple input multiple output (MIMO).

SISO is the most basic radio channel access mode and sets the baseline for minimum transmission performance. However, SISO does not provide any diversity protection against channel fading.

SIMO provides additional receive antenna redundancy compared to the SISO baseline, which allows the use of receive diversity techniques such as maximum ratio combining in the receiver. This flexibility improves signal to interference noise ratio (SINR) observed at the device receiver and can help improve robustness under channel fading conditions.

MISO provides additional transmit antenna redundancy, allowing the use of transmit diversity techniques such as Alamouti symbol coding, or space frequency block coding (SFBC) as is the case for LTE. Similar to SIMO, MISO also provides an improvement in the observed SINR at the device receiver which helps protect against channel fading.

In contrast MIMO provides both additional transmit and receive antenna redundancy. This redundancy can be used to improve the SINR at the device receiver using transmit and receive diversity techniques. Alternatively some or all of the potential SINR performance improvement can instead be traded off to obtain an improved spectral efficiency by employing spatial multiplexing transmission techniques. This improved spectral efficiency can be realized either in the form of increased data rate throughput for a single user device using single-user MIMO (SU-MIMO) techniques, or alternatively in the form of increased system cell capacity using multi-user MIMO (MU-MIMO) techniques. In addition to the diversity and spatial multiplexing techniques summarized above, it is possible to use the multi-antenna path redundancy to support transmit and receive beamforming techniques to improve system performance, as will be introduced next.

6.9.2 Introduction to Beamforming

The basic principles of transmit diversity, spatial multiplexing, and beamforming are compared in the context of a multi-antenna transmitter. Figure 6.9-1 illustrates the basic concepts.

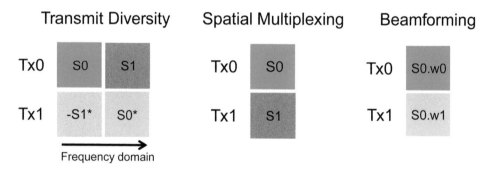

Figure 6.9-1. Comparison of transmit diversity, spatial multiplexing and beamforming

In the case of transmit diversity, orthogonally modified redundant copies of an information symbol pair, S0 and S1, are transmitted simultaneously in time across the multiple antenna elements. The example at the far left of Figure 6.9-1 shows SFBC transmit diversity for a subcarrier pair. The symbols S0 and S1 are transmitted on different subcarriers on Tx0 and the complex subjugate of the symbols (denoted by *) is transmitted on Tx1, with S1 also being negated. The benefit is improved SINR, observed at the device receiver, plus improved robustness to channel fading.

In the case of spatial multiplexing, separate and unique information symbols, S0 and S1, are transmitted simultaneously in time across the multiple antenna elements, as shown in the center example of Figure 6.9-1. The benefit is improved spectral efficiency observed as either increased individual user throughput or increased cell capacity.

In the case of beamforming, weighted copies of an information symbol, S0, are transmitted simultaneously in time across the multiple antenna elements, as shown in the example at the far right of Figure 6.9-1. In this example w0 and w1 represent the applied complex per antenna weightings. The benefits are improved SINR observed at the receiver of the primary target device, resulting from a coherent signal gain, plus the ability to minimize interference to other devices within the system.

6.9.2.1 Beamforming Selectivity

Beamforming techniques are used within many different technologies such as radar, sonar, seismology, radio astronomy, acoustics, and wireless communications.

In the general case, transmit beamforming works by exploiting the interference patterns observed whenever the same signal is transmitted from two or more spatially separated transmission points. A similar principle applies whenever the same signal is received from two or more spatially separated reception points, which is exploited by receive beamforming techniques.

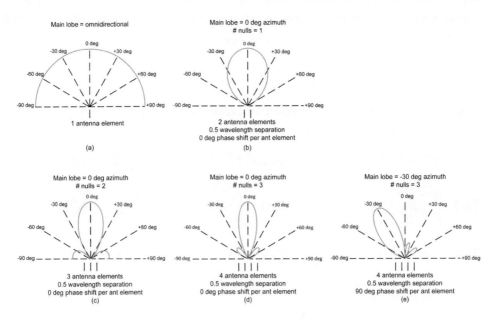

Figure 6.9-2. ULA beamforming examples

A simple example is the case of an RF wireless signal transmitted from a single omnidirectional antenna. The resulting signal relative field strength is shown in Figure 6.9-2 (a) represented as a solid blue line.

To enable transmit beamforming a second identical omnidirectional antenna element is added, separated from the first element by half the RF carrier wavelength, as shown in Figure 6.9-2 (b). In this example both antenna elements carry identical copies of the signal information symbol to be transmitted. However, in the azimuth directions around 0 degrees azimuth, where constructive (or in phase) interference occurs, the combined field strength increases, producing an effective coherent signal power gain in those directions. In contrast the azimuth directions around ±90 degrees, where destructive (or out-of-phase) interference occurs, result in a decreased or attenuated combined field strength in those directions.

Adding a third antenna element, separated along the same axis as the first two elements by half the RF carrier wavelength, improves the spatial selectivity of the combined relative field strength as shown in Figure 6.9-2 (c). In this example the array elements are co-polarized, correlated, and uniformly separated along a single antenna element axis, creating a uniform linear array (ULA) antenna system. The formation of a single main lobe in the azimuth direction of 0 degrees relative to the ULA broadside can clearly be seen. This main lobe region is where maximum constructive (or in phase) interference occurs, producing a power gain maximum within the combined field strength beam pattern. The formation of two distinct power attenuation nulls, one either side of the main lobe located at ±42 degrees azimuth, can now be observed. These two power minimum locations represent the azimuth directions in which maximum destructive (or out-of-phase) interference occurs within the combined field strength beam pattern.

Finally, adding a fourth antenna element to the ULA further improves the main lobe selectivity as shown in Figure 6.9-2 (d). The number of power nulls has also increased from two to three. Two nulls are now located at ±30 degrees azimuth, with the third located on the ULA antenna axis line. Two distinct power side lobes are now clearly observed, located at ±50 degrees azimuth. Both side lobes appear at reduced power levels relative to the main lobe. The resultant beam pattern is determined not only by the ULA physical geometry and element separation but also by the effects of the relative magnitude and phase weightings applied to each information symbol copy transmitted on each antenna element.

This sensitivity of the lobe to power and phase changes can be demonstrated by now introducing a +90 degrees relative phase shift weighting across each of the four antenna elements. The result is a shift of the main beam location from 0 degrees azimuth to -30 degrees azimuth as shown in Figure 6.9-2 (e). Note that the null and side lobe locations have also been affected by the new weighting values.

With careful design of the beamforming antenna array geometry and accurate control of the relative magnitude and phase weightings applied to each of the antenna elements, it is possible to control not only the selectivity shape and azimuth direction of main lobe power transmissions but also the power null azimuth locations and side lobe levels.

6.9.2.2 Beamforming Gain

At this stage it is important to note that the combined beam pattern in Figure 6.9-2 focuses on the spatial selectivity improvements, and as such the main lobe peak powers of plots (b) through (e) are shown normalized to the single antenna plot (a) case. The next example considers how adding more antenna elements affects the effective power gain of the resultant beam pattern observed at a target device receiver.

Figure 6.9-2 plot (b) shows the addition of a second antenna element, which transmits an exact symbol copy of what is being transmitted on the first antenna element. In this case, the constructive in-phase signal summation results in a 6 dB coherent power gain improvement, observed by a target device receiver positioned at the 0 degrees azimuth main beam location. If plot normalization had not been applied, the main lobe maximum of the plot (b) two antenna case would in theory be twice the main lobe maximum of the plot (a) single antenna case.

433

The 6 dB coherent gain improvement is considered to be the beamforming gain improvement observed at the target device receiver and is the result of using two spatially separated antenna elements relative to a single antenna transmission. In practice the symbol power levels transmitted on each of the two antenna elements may be reduced by 3 dB to half the original single antenna symbol power level, maintaining the same total transmitter power as the single antenna case. Even so, the result will still be a 3 dB beamforming gain observed at the target device receiver relative to a single antenna transmission.

6.9.2.3 Beamforming Advantages

The use of multi-antenna beamforming transmission is very attractive in modern wireless communication systems because it offers the advantages of beamforming selectivity, interference management, and coherent signal gain. Some important concepts and terminology used to describe beamforming transmissions are summarized below and illustrated in Figure 6.9-3.

- Main lobe: the primary maximum transmission power lobe, usually directed at the target device or a transmission path that will reach the target device by reflections in the radio propagation channel.
- Side lobes: the secondary power transmission lobes which can produce unwanted interference that affects other user devices within the serving or adjacent cells.
- Power null: locations of minimum power within the transmission beam pattern which the system may choose to exploit and control in order to mitigate interference to other devices within the serving or adjacent cells.
- Main beam width (Φ): selectivity of the main lobe transmission measured as the degree azimuth spread across the 3 dB points of the main lobe.
- Main lobe to side lobe levels: the selectivity power difference of the desired main lobe transmission power relative to the unwanted side lobe transmission power.

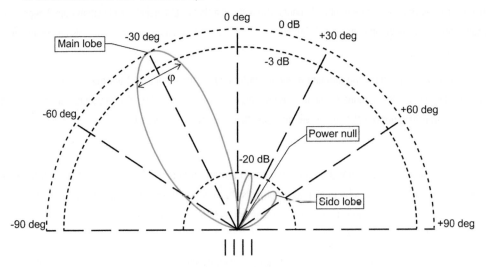

4 antenna elements
Correlated co-polarized 0.5 wavelength separation
with 90 degree phase shift per antenna element

Figure 6.9-3. Beamforming terminology

Figure 6.9-4 shows two practical scenarios, both of which exploit the advantages of beamforming to improve performance within a modern cellular wireless communication system.

Figure 6.9-4 (a) depicts two adjacent cells, each communicating with a UE located at the boundary between the two cells. The illustration shows that eNB1 is communicating with target device UE1. The eNB1 transmission is using beamforming to maximize the signal power in the azimuth direction of UE1. At the same time eNB1 is attempting to minimize interference to UE2 by steering the power null location in the direction of UE2. Similarly eNB2 is using beamforming to maximize reception of its own transmission in the direction of UE2 while minimizing interference to UE1. In this scenario, it is clear that the use of beamforming can provide considerable performance improvements, particularly for cell edge users. The beamforming gain can also be used to increase the cell coverage where required.

Figure 6.9-4 (b) depicts a single cell (eNB3) communicating simultaneously with two spatially separated devices (UE3 and UE4). Since different beamforming weightings can be applied independently to each of the spatial multiplexing transmission layers, it is possible to use space division multiple access (SDMA) in combination with MU-MIMO transmissions to deliver an improved cell capacity.

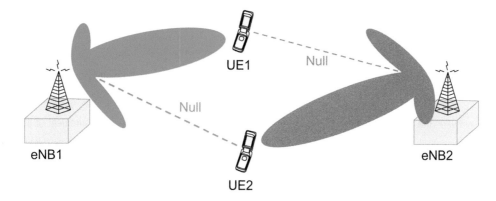

Figure 6.9-4 (a). Beamforming for cell-edge performance improvement

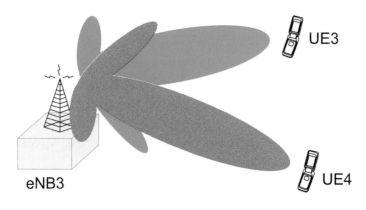

Figure 6.9-4 (b). Beamforming for cell capacity improvement using MU-MIMO

6.9.2.4 Beamforming Implementation Techniques

Two different beamforming implementation techniques are illustrated in Figure 6.9-5.

Figure 6.9-5 (a) shows an example of a fixed conventional switched beamformer consisting of an eight-port Butler matrix beamforming network. This network implementation consists of a matrix of different selectable fixed time or phase delay paths, implemented using a combination of 90 degree hybrid couplers and phase shifters.

The number of fixed transmission beams produced is equal to the number of antenna elements N used to form the Butler matrix network. (The example shown uses eight antennas, producing eight selectable beams.) This is sometimes also referred to as a "grid of beams" beamforming network, and it supports selection of any individual or combination of the N fixed transmission beams in order to maximize the SINR at the device receiver.

In a wireless network, optimal eNB downlink transmission beam selection would be driven primarily by some knowledge of the UE position within the cell. This knowledge can be directly obtained through measurement of the uplink signal angle of arrival (AoA) across the eNB receive antenna array, or indirectly derived from uplink control channel quality feedback information.

In contrast, Figure 6.9-5 (b) shows an example of an adaptive beamformer. As the name suggests, an adaptive beamformer has the ability to continually adapt and recalculate the optimal applied transmission beamforming complex weighting values in order to best match the channel conditions.

Because the adaptive beamformer weightings are not fixed, they can both optimize the received SINR at the target UE and also better adapt the selectivity and power null positioning to minimize interference to other users.

In a wireless network, the eNB would typically estimate the optimal weightings through direct measurement of the received uplink reference signals observed across the eNB receiver array. This information can be then be used to calculate the uplink AoA as well as decompose the channel characteristic matrix.

For the case of a frequency division duplex (FDD) system, in which both the downlink and the uplink use different RF carrier frequencies, the applied beamforming transmission complex weightings will be driven primarily by measured AoA information derived for both the target UE, as well as any other UEs within the cell. In addition, weighting estimation can be aided by channel feedback information reported by the UE on the uplink.

For the case of a time division duplex (TDD) system, since the downlink and uplink share the same RF carrier frequency, channel reciprocity may be assumed. The applied beamforming transmission complex weightings may therefore be chosen to best match the decomposed channel characteristic matrix eigenvectors, as derived from the eNB received signal. These channel-matched beamforming weightings can help optimize the SINR observed at the target UE receiver. For this reason beamforming in a TDD system can outperform what is possible in an FDD system. Note that for the TDD case, the eNB is not reliant on channel feedback information supplied by the user device on the uplink, although in practice channel feedback may still be used in the eNB beamforming weighting estimation process.

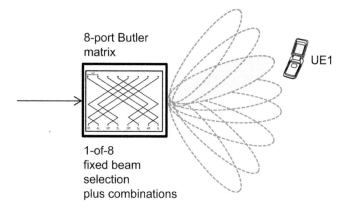

Figure 6.9-5 (a). Fixed conventional switched beamformer

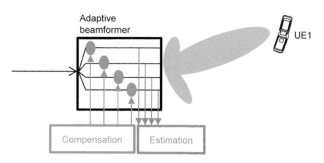

Figure 6.9-5 (b). Adaptive beamformer

6.9.3 Beamforming in LTE

One of the biggest challenges in any modern wireless cellular communications system is the performance at the cell edge. This is a major reason why beamforming technology has a key role to play in delivering LTE services.

6.9.3.1 LTE Downlink Transmission Mode Support for Beamforming

LTE defines many downlink transmission modes, which are listed in Table 2.4-1. Of particular interest from a beamforming point of view are transmission modes 7, 8, and 9. Release 8 introduced TM7, which supports single layer beamforming on antenna port 5. Release 9 added TM8, which supports dual layer beamforming on antenna ports 7 and 8. Finally, Release 10 added TM9, which supports up to eight layer transmission on antenna ports 7 to 14.

It should be noted that the ports mentioned above are all virtual antenna ports representing particular configurations of reference signals. The physical geometry and number of antenna elements are not defined in the LTE specifications. In practice each virtual port's physical realization may comprise four or more spatially separated physical antenna elements.

The following examples focus on TM7 and TM8, which are the focus of development for initial TD-LTE market deployments.

6.9.3.2 Signal Processing for TM7 and TM8

A summary of the defined downlink signal processing flow for TM7 and TM8 is shown in Figure 6.9-6. As with other transmission modes, the PDSCH data transport block information is channel-encoded and has rate matching applied, producing either one or two code words, which are then mapped onto layers.

It's worth noting that for TM7 and TM8, the precoding block is not codebook based. It is left up to the eNB to determine the optimal beamforming precoding to apply. This coding can be derived by the eNB from direct measurement of the received uplink sounding reference signal, and can include the use of any configured UE channel feedback (CQI/PMI/RI) information. Also worth noting is that the beamforming precoding can be dynamic and vary on a per subframe and resource block basis to adapt to changing channel conditions.

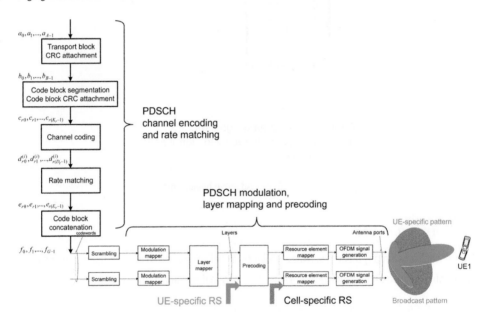

Figure 6.9-6. Signal processing flow for TM7 and TM8

For demodulation purposes, TM7 and TM8 include the mapping of UE-specific reference signals (UE-specific RS), also known as demodulation reference signals (DMRS) in each PDSCH resource block. The UE-specific RS undergo the same beamforming precoding as the associated PDSCH. This concept is shown in Figure 6.9-6 where the UE-specific RS feed into the precoding block. The beamforming precoding is calculated primarily to maximize the SINR observed by the target UE, but the precoding will also attempt to minimize interference to other UE within the serving or adjacent cells.

n addition to producing user-specific beam patterns, the base station has the ability to choose a different sector-wide broadcast beam pattern for common control channel content, which is received by all UE within the cell. This beamforming of the control channels is possible when the number of beamforming antenna elements is greater than the number of configured cell RS ports, as shown in Figure 6.9-6.

6.9.3.3 LTE UE-specific Reference Signals Structure

To support beamforming for TM7, TM8, and TM9, UE-specific RS are defined for port 5 and ports 7 through 14. The physical structure of the UE-specific RS is shown in Figure 6.9-7 for the TM7 and TM8 cases.

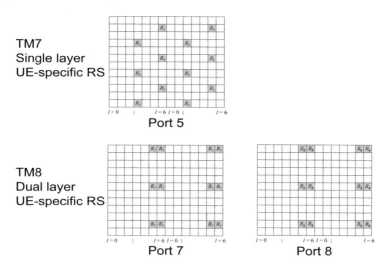

Figure 6.9-7. UE-specific RS structure for TM7 and TM8

Transmission mode 7 supports only single layer beamforming transmissions. For this purpose port 5 UE-specific RS resource element mappings are defined in time and frequency within each scheduled PDSCH resource block assignment as shown in Figure 6.9-7. Since the UE-specific RS undergo the same beamforming weight precoding as the associated PDSCH data, it is possible for the target UE to directly demodulate the precoded PDSCH using the similarly precoded UE-specific RS as the reference.

Transmission mode 8 extends beamforming to dual layer spatial multiplexing. For this purpose ports 7 and 8 UE-specific RS resource element mappings are defined. Each port corresponds to a different spatially multiplexed MIMO transmission layer. It is worth noting that the same physical resource elements are used by both port 7 and port 8. In order that the UE can correctly separate these simultaneously transmitted UE-specific RS, orthogonal UE-specific RS sequences are used.

The orthogonality of UE-specific RS resource mappings for ports 7 and 8 are further extended using a combination of frequency division multiplexing (FDM) and code division multiplexing (CDM) resources in order to support ports 9 through 14 as required for TM9. From a test point of view, it is essential that the UE-specific RS content for TM7, TM8, and TM9 be verified for baseband correctness as well as for relative magnitude and phase weighting accuracy observed at the calibrated RF output of the antenna element array.

6.9.4 TD-LTE MIMO Beamforming Test Setup

This section introduces a typical TD-LTE eNB antenna configuration and system test setup used to verify the performance of downlink beamforming and spatial multiplexing signals used for TM7 and TM8.

Examples are included that demonstrate how a phase-coherent multi-channel signal analyzer along with appropriate measurement software can be used in the beamforming signal verification process, enabling visualization of the beamforming signal at the RF antenna array. The importance of calibration when it comes to verifying the performance of a beamforming transmission system is also given special attention.

6.9.4.1 TD-LTE eNB Antenna Configuration

Figure 6.9-8 shows a typical eNB RF antenna configuration used in TD-LTE cellular networks that support TM7, TM8, and TM9 MIMO beamforming signals.

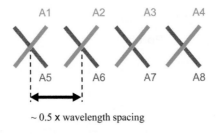

~ 0.5 x wavelength spacing

+45 degree polarization = A1, A2, A3, A4
-45 degree polarization = A5, A6, A7, A8

Figure 6.9-8. Typical eight antenna configuration for TD-LTE TM7, TM8, and TM9

The example is an eight-element physical antenna, configured with two groups of antenna elements. Each group is orthogonally cross-polarized at 90 degrees to the other group. Antenna group 0 consists of antenna elements 1 through 4, polarized at plus +45 degrees. Antenna group 1 consists of antenna elements 5 through 8, polarized at −45 degrees.

Each of the elements within a given group are spatially separated by approximately half the RF carrier wavelength. This provides a high degree of antenna element correlation within the antenna group, which is good for coherent beamforming. Since each of the two groups are cross-polarized relative to each other, there is a low correlation between each of the two antenna groups, which is good for spatial multiplexing. Thus a typical TD-LTE eNB RF antenna physical configuration attempts to satisfy the desirable but conflicting correlation requirements for MIMO spatial multiplexing and coherent beamforming.

6.9.4.2 TD-LTE eNB Test System Configuration

A typical configuration for testing a TD-LTE MIMO beamforming TM7 and TM8 eNB is shown in Figure 6.9-9. Starting at the left, the two main eNB blocks are shown: the baseband (BB) and the remote radio head (RRH). The RRH provides eight antenna feeds, which are connected for test purposes to an RF antenna calibration coupler unit. Note that calibration of the RF antenna elements is achieved using a dedicated calibration port, located between the RRH and the RF antenna calibration coupler.

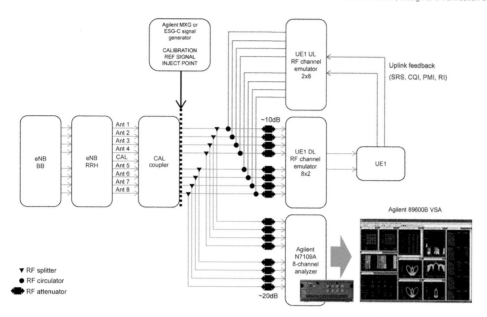

Figure 6.9-9. Typical TD-LTE beamforming test system configuration

The RRH is capable of generating a known calibration signal to be used as a common magnitude and phase reference. This calibration signal is periodically injected via the calibration port into the RF antenna network. The eNB is then able to measure the RF-coupled calibration signal observed on each of the eight receiver ports of the RRH. This allows the eNB to monitor and correct for per antenna magnitude and phase variations, which are inherent in the system due to the antenna feed cabling and coupler variations. The periodicity of eNB in-service calibration measurements can vary and will depend on environmental operating conditions. Verification of the eNB calibration performance is an important aspect of beamforming test.

The calibration coupler output is typically fed into an RF downlink channel emulator, shown here in an 8x2 configuration, to emulate the downlink channel characteristics. The two RF outputs of the channel emulator are connected to the UE. In this example, the UE transmits the uplink signal on two output ports, which can be connected to an uplink RF channel emulator in a 2x8 configuration, to allow emulation of the uplink channel characteristics.

Finally, to complete the UE feedback loop, the eight RF outputs from the uplink channel emulator are coupled back into the eNB's eight receive antenna ports using RF circulators.

6.9.4.3 Beamforming Measurement Challenges

One of the main test challenges for beamforming is the need to verify and visualize the beamforming signal performance at the physical RF antenna array, in order to validate the following:

- eNB RF antenna calibration accuracy
- Baseband encoded beamforming weighting algorithm correctness
- MIMO single and dual layer EVM at the RF antenna.

This test challenge can be met using the Agilent N7109A Multi-Channel Signal Analyzer and the Agilent 89600 VSA software installed with TD-LTE measurements. The multi-channel signal analyzer can support eight phase-coherent RF measurement channels and, along with the appropriate RF splitters and attenuators, can easily be integrated into a typical TD-LTE base station test setup, as shown in Figure 6.9-9.

In this example, the 89600 VSA software provides a correction wizard for the signal analyzer that is used along with an Agilent MXG or ESG-C signal generator and an appropriate high quality calibration two-way RF power splitter to correct for all the RF cabling and connectors used within the test system. Note that the quality of the corrections will be determined by the quality of the power splitter and any connectors required between the power splitter and the measurement cables. The signal generator is used to create a broadband calibration reference signal output that is connected to the input of the two-way power splitter. The desired beamforming measurement verification point in Figure 6.9-9 is indicated by a dotted line at the output of the RF antenna calibration coupler. It is essential to compensate for any magnitude and phase mismatch inherent in the measurement cables, connectors, splitters, and attenuators used between the RF antenna calibration coupler's eight output ports and the signal analyzer's eight input channels.

The correction wizard guides the calibration process, prompting the user to connect the signal analyzer channel 1 measurement cable to the first output port of the two-way calibration splitter at the injection point represented by a dotted line in Figure 6.9-9. Note that all cross-channel characterization measurements will be made referenced to channel 1. The user is then prompted to connect each of the remaining channels 2 through 8 measurement cables (located on dotted line) one at a time to the second output port of the two-way calibration splitter. In this way the correction wizard is able to characterize the cross-channel corrections required to compensate the signal analyzer beamforming measurements for all mismatch effects inherent in the measurement cables, connectors, splitters, and attenuators. As a result direct, corrected measurements of the antenna beamforming performance can be observed at the RF antenna output.

The importance of test system calibration of magnitude and phase variations due to RF cabling and connectors cannot be overstated. Calibration is covered in more detail in Section 6.9.4.5.

6.9.4.4 Verification and Visualization of MIMO Beamforming Signals

This section discusses some of the useful verification measurements that can be made using the test system shown in Figure 6.9-9.

The 89600 VSA software and multi-channel signal analyzer are first used to display the time-synchronized RF signal capture from all eight antenna elements as shown in Figure 6.9-10. Any fundamental RF power or timing performance impairments can be identified quickly, before the more advanced demodulation measurements are attempted.

The 89600 VSA software spectrogram feature provides useful insight into the frequency resource activity as shown in Figure 6.9-11. Spectrograms allow a quick picture to be built up of RF activity on a per subframe basis for user-specific resource block scheduling, and on a per symbol basis for common control channels and signals. This feature does not require demodulation and so is a very simple and useful debugging tool for investigating unexpected RF or scheduling related issues, especially when those issues might prevent measurements that rely on demodulation of the signal.

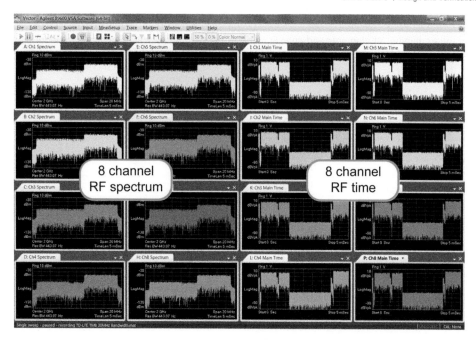

Figure 6.9-10. Time-synchronized capture of eight-antenna transmission

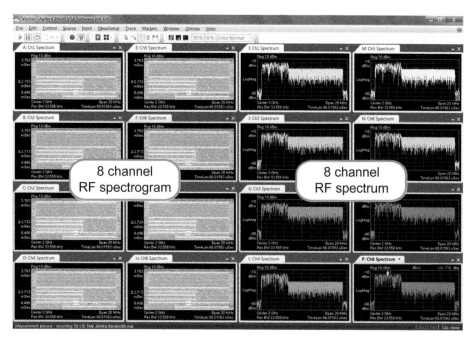

Figure 6.9-11. Spectrogram of eight-antenna transmission

Prior to demodulating the TD-LTE signal it is important to properly configure the 89600 VSA software antenna group parameter with the appropriate number of elements and spacing used to match the physical RF antenna configuration shown in Figure 6.9-12.

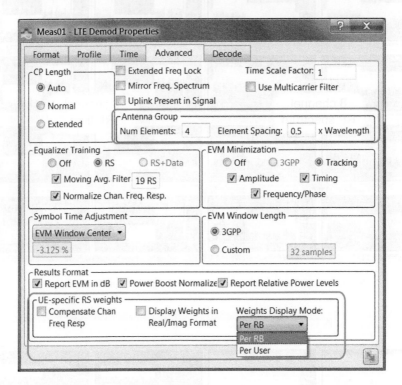

Figure 6.9-12. Configuration of Agilent 89600 VSA software for TD-LTE beamforming verification

As mentioned earlier, the beamforming weightings on each resource block may be changing; therefore, the UE-specific weighting results may be viewed either per resource block or per user allocation.

The 89600 VSA software TD-LTE measurement application provides a rich set of demodulation results for verifying downlink MIMO beamforming signals. These include IQ constellations, EVM result metrics, detected resource allocations, UE-specific RS weights, cell-specific RS weights and impairments, and UE-specific and common broadcast antenna beam patterns.

The demodulated IQ constellations are displayed per spatial multiplexing layer, as shown in Figure 6.9-13 traces A and L, and provide a quick visual indication of the signal's modulation quality correctness.

The frame summary shown in Figure 6.9-13 trace D provides access to individual EVM and power metrics associated with each channel and signal type. It also provides a color key (not shown here) for all channel type results, which is reused throughout the 89600 VSA software traces.

The detected allocations displayed in Figure 6.9-13 trace B shows the resource block allocations for each user-specific transmission, plus resource allocations used by common control channels.

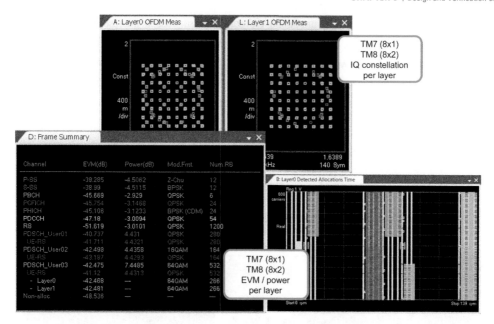

Figure 6.9-13. Constellation diagrams, frame summary, and detected resource allocations

Measured UE-specific RS weights are presented in table format for each of the eight antenna elements, as shown in Figure 6.9-14 trace E. Weightings can be evaluated in both magnitude and phase down to the individual resource block allocations associated with each user transmission. Separate UE-specific RS weights traces are available for each spatial multiplexing layer.

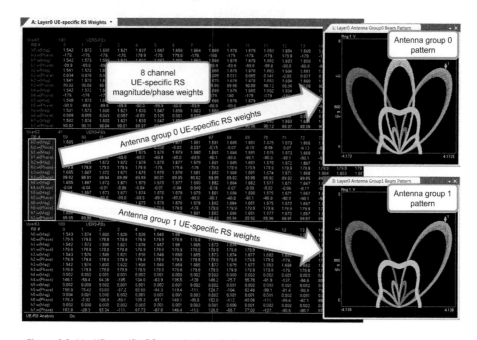

Figure 6.9-14. UE-specific RS magnitude and phase weightings and corresponding beam patterns

To give a picture of beamforming performance, the 89600 VSA software also presents the resulting combined beam pattern trace associated with each antenna group. Measured UE-specific RS weights from the first four channels are used to compute the antenna group 0 beam pattern as shown in Figure 6.9-14 trace L. This process is repeated using the second four channels, producing the antenna group 1 beam pattern results shown in Figure 6.9-14 trace B. Note that a separate beam pattern trace can be shown for each resource block associated with each user device.

Similar to the way the IQ constellation provides a quick visual check of modulation quality, the antenna group beam pattern trace provides a quick visual check of beamforming baseband encoding and RF calibration quality. Any identified anomalies can be investigated in detail using the UE-specific RS metrics.

Channel frequency, magnitude, and phase response traces can be viewed simultaneously for all eight antenna elements, along with the 89600 VSA software common tracking error trace shown in Figure 6.9-15 traces A, L, and B, respectively.

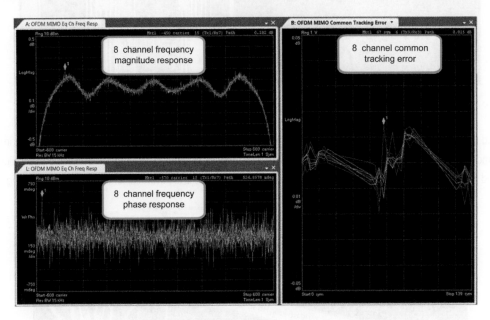

Figure 6.9-15. Channel frequency, magnitude, and phase responses

The 89600 VSA software MIMO Info traces A and B shown in Figure 6.9-16 reports cell RS (CRS) metrics and impairments measured for all eight antenna elements. The reported metrics include CRS power, EVM, timing, phase, symbol clock, and frequency error, and these metrics make it possible to verify the common broadcast beam pattern weightings associated with each antenna element.

The 89600 VSA software also extracts these relative antenna weightings in order to produce the CRS-derived, sector wide broadcast beam pattern results. The broadcast beam pattern results are shown in Figure 6.9-16 traces L and D, associated with antenna groups 0 and 1, respectively.

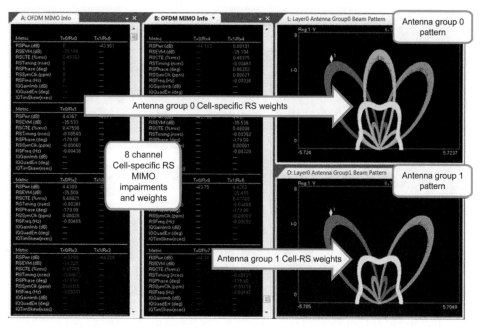

Figure 6.9-16. MIMO info display and cell-specific RS broadcast beam pattern

The user-specific and common broadcast beam pattern results can be viewed in either IQ polar format or log magnitude (dB) format as shown next in Figure 6.9-17 traces A and L. Both formats support markers for easy tracking of the main lobe peak levels and azimuth locations during live measurement updates. Markers can also be used to read out various beam pattern characteristics such as null depth, azimuth locations, and main lobe to side lobe levels.

For the signal in these examples, the channel frequency magnitude response derived from the CRS content varies by as much as 0.6 dB across the 20 MHz transmission bandwidth, as can be observed in Figure 6.9-15 trace A. The magnitude response variation is caused by the transmission filtering used, and so it applies equally to all channel types including UE-specific RS weights and associated PDSCH content. This magnitude response variation is also observed in the user-specific and common broadcast beam pattern results of Figure 6.9-17 trace A. Each beam pattern corresponds to a different resource block allocation, and just as the channel frequency magnitude response is attenuated at the transmission bandwidth lower and upper edges, so the per resource block beam pattern magnitudes are also attenuated at the transmission bandwidth lower and upper edges.

A key metric to be verified in TD-LTE beamforming transmissions is beamforming gain. The 89600 VSA software has a beamforming gain results trace for this purpose that is shown in Figure 6.9-17 trace B. This trace reports the dB difference between each UE-specific beam pattern and the common CRS broadcast beam pattern, producing a beamforming gain trace result for each user allocation. The beamforming gain results can be viewed for each individual resource block associated with each user's allocation.

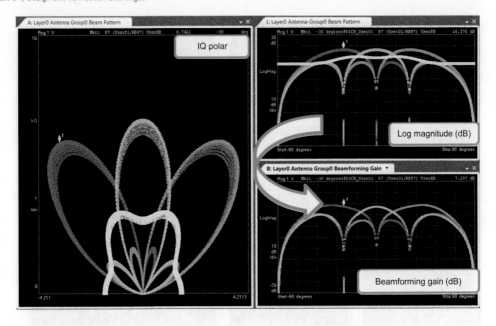

Figure 6.9-17. User-specific and cell broadcast beam patterns in polar and log magnitude format plus beamforming gain

6.9.4.5 Calibration of the Beamforming Test System

The importance of correcting measured beamforming results for all RF cabling and connectors used in the test setup cannot be overstated. This section investigates the reasons why, using the simple calibration setup illustrated in Figure 6.9-18.

The calibration test system in this example consists of an Agilent MXG RF signal generator connected via a 1-to-4-way RF splitter network to the four RF inputs of an N7109A signal analyzer, which is controlled by the 89600 VSA software.

As mentioned in Section 6.9.4.3, the 89600 VSA software provides a correction wizard utility program for the signal analyzer that can be used to correct for magnitude and phase delay variations observed between each of the analyzer's RF input channels. These magnitude and phase variations are in part due to slight performance variations within each of the analyzer's multi-channel phase coherent receiver paths. Perhaps more importantly these variations may be the result of magnitude and phase delay variations associated with the RF cabling, splitters, and adaptors used between the RF antenna calibration reference point and the signal analyzer RF input ports.

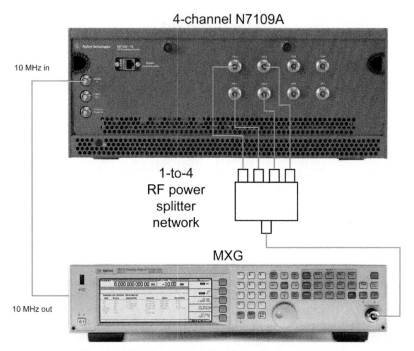

Figure 6.9-18. Example beamforming calibration system

The correction wizard can directly control the MXG to generate an appropriate wideband modulated calibration signal to exercise the frequency band of interest. The wizard also directly controls the 89600 VSA software to measure the cross-channel frequency response trace results for all RF input channels relative to channel 1, which is used as the reference channel. In this way the correction wizard can measure and characterize both the magnitude and phase response variations between each of the signal analyzer RF input channels (including measurement cabling, splitters, and adaptors) at the RF carrier measurement frequency and band of interest. By loading the per channel characterized magnitude and phase responses into the 89600 VSA software fixed equalization feature, it is possible to compensate and correct all 89600 VSA software results for the relative channel-by-channel magnitude and phase response offset variations.

Figure 6.9-19 shows the uncorrected magnitude and phase response plots for measurement channels 2, 3, and 4, all displayed relative to the reference measurement channel 1. In this example the uncorrected relative channel magnitude responses can vary as much as −0.95 dB, as is the case for channel 3 relative to channel 1, shown in trace C. The uncorrected relative phase variation between channels is expected to fall anywhere between ±180 degree range, as indicated by the spread of uncorrected relative phase results in Traces F, G, and H.

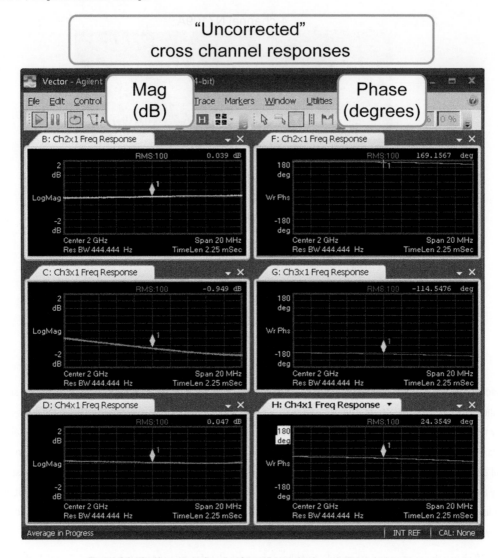

Figure 6.9-19. Uncorrected cross-channel magnitude and phase responses

Figure 6.9-20 shows the equivalent corrected magnitude and phase response plots for measurement channels 2, 3, and 4 relative to reference channel 1. As shown the corrected relative channel magnitude and phase responses now appear as flat responses in contrast to the uncorrected results. The maximum magnitude response variation has dropped to around −0.02 dB, and the maximum phase response variation has dropped to <0.1 degree after applying the 89600 VSA software cross-channel corrections.

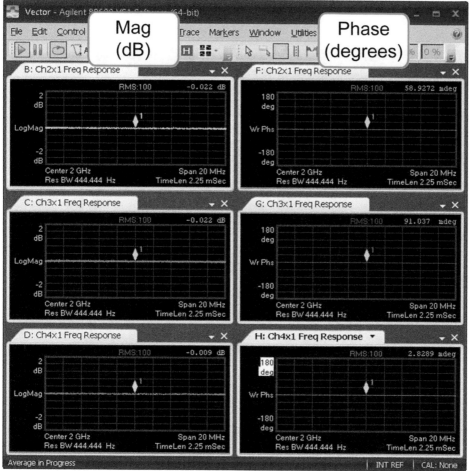

Figure 6.9-20. Corrected cross-channel magnitude and phase responses

Using Agilent N7625B Signal Studio for 3GPP LTE TDD application software, it is possible to generate and download a 20 MHz TD-LTE downlink signal configured for a single user scheduled with TM7 (port 5) PDSCH plus associated UE-specific RS resource allocations in subframes 0 and 5. Figures 6.9-21 and 6.9-22 show the resulting uncorrected and corrected 89600 VSA software demodulation results for comparison purposes.

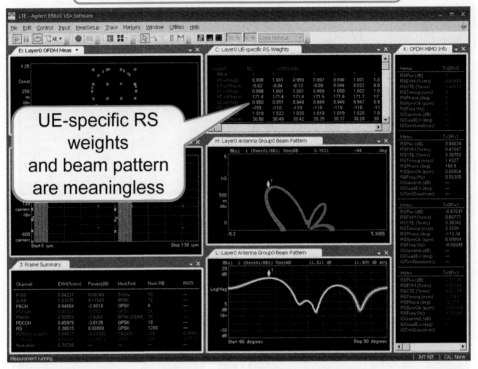

Figure 6.9-21. Uncorrected beam patterns

Figure 6.9-21 shows the uncorrected results. The antenna group 0 beam pattern results in trace H are not what might be expected given the simple test setup in which the MXG output RF source is split directly to each of the four signal analyzer input channels. Theoretically a main lobe should appear at 0 degrees azimuth, as was the case in Figure 6.9-2 plot (d). Also, the power null depth is around 25 dB down from main lobe level. The reality is that the uncorrected variations in relative per channel magnitude and phase responses of the RF splitters, cabling, adaptors, and analyzer receiver paths all affect the reported UE-specific RS magnitude and phase results shown in trace C. These per channel response variations also affect the derived antenna group 0 beam pattern results shown in traces H and L.

The 89600 VSA software per-channel fixed equalization feature produces the corrected demodulation results shown in Figure 6.9-22. The results now align very well with the theoretical results of Figure 6.9-2 plot (d), and it is now possible to identify a distinct power main lobe at 0 degree azimuth. Also, the power null depths are much improved at > 40 dB down from main lobe level.

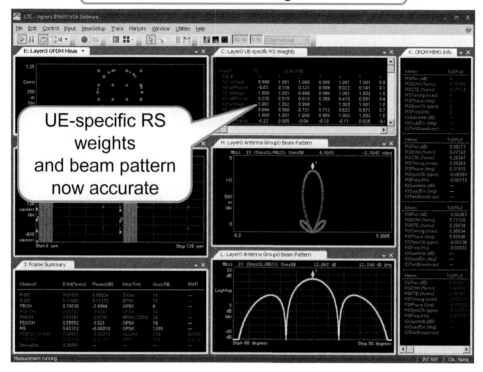

Figure 6.9-22. Corrected beam patterns

Beamforming measurement results are sensitive to any uncalibrated changes to the test setup. To illustrate this fact, two physical test setup changes are made to the original test setup in Figure 6.9-18 that has just been calibrated.

First a short N-type male-to-female extension adaptor is added to the signal analyzer input channel 3 N-type cable connection. The effect of this change is to introduce a −85 degree phase shift impairment to the channel 3 UE-specific RS results shown in Figure 6.9-23 trace C.

The result of this phase shift impairment is seen in the antenna group 0 beam pattern results of traces H and L. The main lobe power has been reduced from 12.04 dB to 10.34 dB as reported by the trace L marker readouts. Also the left side lobe power level is greatly increased and the power null depths observed in trace L are degraded.

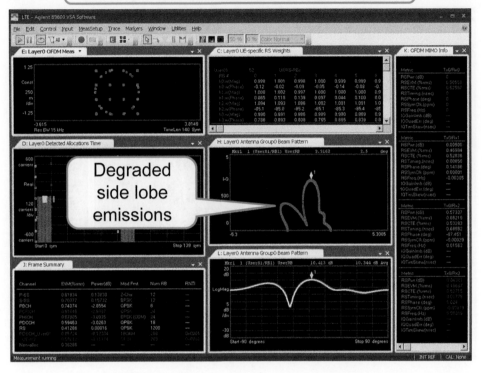

Figure 6.9-23. Effect of adding uncalibrated RF adapter

Next, the original calibrated 1 meter N-type RF cable for signal analyzer channel 3 is replaced with an uncalibrated 4 meter cable. This introduces a 165 degree phase shift impairment to the channel 3 UE-specific RS results as shown in Figure 6.9-24 trace C. Again the effect of this phase shift impairment is seen clearly in the antenna group 0 beam pattern results of traces H and L. However, in this case there also appears to be a distinct azimuth spreading of the reported beam pattern trace results. To explain this, it is first important to understand that trace H and L actually contain a separate beam pattern trace result plot for each of the UE device PDSCH resource block allocations. The UE-specific RS results for each separate resource block are reported within trace C as a separate column entry. It can be observed from trace C that the input channel 3 phase error is actually changing slightly for each measured resource block. The 165 degree error reported for RB0 is reduced to 160 degrees for RB5. This changing phase error can be explained by the fact that introducing a longer RF cable to channel 3 in effect introduced a fixed time delay impairment to the channel 3 reported results. This fact is confirmed by the "RSTiming" metric, reported in MIMO Info trace K as 17 ns for input channel 3. Since each resource block occupies a different carrier frequency region, the fixed time delay impairment results in a different phase shift for each resource block. The effect on the combined antenna beam pattern trace is a distinct spreading of the observed beam pattern results corresponding to each frequency resource as seen in Figure 6.9-24 trace H.

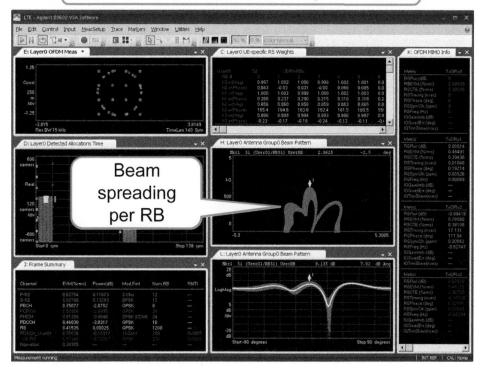

Figure 6.9-24. Effect of adding uncalibrated RF cable

The key point here is that in order to make accurate measurements of eNB MIMO beamforming signal performance and eNB beamforming calibration accuracy, it is essential that all the physical cabling, adaptors, splitters, and attenuators used in the measurement test setup be included within the calibration correction procedure. Also the calibration should be repeated whenever the physical configuration of the measurement setup is changed.

6.10 SISO and MIMO Over-the-Air Testing

6.10.1 Introduction

Traditionally, most testing of mobile devices has been done through directly cabled (galvanic) connections to the device's temporary antenna connectors, which are the ports used for conformance testing. This method is appropriate for the vast majority of performance tests since it is convenient and is not susceptible to radiated noise or interference in the test environment. The downside of this type of conducted testing, however, is that it ignores the performance of the device's antennas, and any fault or performance problem related to the antenna design or manufacture goes unnoticed.

Earlier generation mobile phones, operating perhaps in only one band with a traditional pull-out "whip" quarter wavelength antenna, had intrinsically good antenna performance such that conducted measurements of parameters such as maximum output power and reference sensitivity were good indicators of how the device performed "over the air" (OTA). But with the introduction of multiband devices and the continued desire to reduce device size, and more recently with the introduction of MIMO technology, the pressure on antenna design has significantly increased to the point where it is no longer safe to assume that conducted measurements bypassing the antennas will be a good indicator of radiated OTA performance.

The first OTA tests were standardized for single input single output (SISO) devices by CTIA in October 2001 [33]. Work to define OTA tests for multiple input multiple output (MIMO) devices started around 2007. At the time of this writing, MIMO OTA standardization activities have not yet completed and so this section will provide an interim summary of the status of this important work.

6.10.2 SISO OTA Overview

The first SISO OTA test specifications were published in "CTIA ERP Test Plan for Mobile Station Over the Air Performance" [33] in October 2001 by CTIA. These tests defined two metrics for the device; total radiated power (TRP) and total isotropic sensitivity (TIS). TRP is defined as the integral of the power transmitted in different directions over the entire radiation sphere. Total radiated sensitivity (TRS) is a similar measure, but it represents the reference sensitivity of the DUT receiver averaged over the same sphere. The first CTIA specification defined a test procedure for measuring TRP and TIS inside an anechoic chamber. Most of the work in defining the test procedure was related to the calibration of the test system, which was calculated— by means of a very detailed error model of some 20 terms—to be around ±2 dB. This uncertainty figure was subsequently confirmed by a substantial measurement campaign, at which time reference devices were circulated among many laboratories. It is important to note that the measurement uncertainty obtained by the new test procedure is not as good as can be obtained by conducted test methods, but the advantage of including the antenna in the overall DUT performance far outweighs this slight loss in accuracy. It is worth noting that measurement uncertainty is defined only for a small test volume within the chamber known as the "quiet zone." The size of the quiet zone is correlated to the size of the anechoic chamber and inversely with the frequency of test, with most chambers aiming to keep a distance of three meters between the DUT and the probe antenna.

Once standard metrics and a test procedure with bounded uncertainty had been established, it was possible for device vendors and network operators to independently measure legacy and new devices in order to compare radiated performance. In the early days of testing, significant differences in device performance were uncovered that were attributed to the free space antenna performance. In addition, tests were defined that included the loading effects of a head "phantom," intended to emulate the electrical properties of a human head These tests uncovered further performance differences between devices which were attributable to antenna design. CTIA did not set specific performance requirements but did enable the industry to comparably measure SISO OTA performance. The results were used subsequently by operators who could then set their own requirements as part of device acceptance testing (see Section 7.5).

In June 2006 3GPP published their first SISO OTA test specifications in TR 25.914 [34]. Most of the procedural aspects were similar to CTIA but 3GPP further defined performance requirements as part of the conformance test regime. This step took considerable time because, although the TRP and TIS metrics were conceptually simple; many implementation factors complicated the performance definition. These factors included the frequency band in question, the presence of other frequency bands, the primary operating band of the device, and the device "mechanical" mode—e.g., whether the device was open or shut, whether the antenna was extended, or how the device was being held. Within the 3GPP OTA specifications, the CTIA TIS metric was renamed TRS and the two terms can now be considered synonymous. The actual conformance tests based on the requirements in 25.914 are specified in 34.114 [35].

Figure 6.10-1 shows two different ways that the probe antenna in the anechoic chamber can be positioned in elevation relative to the DUT.

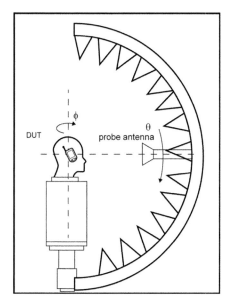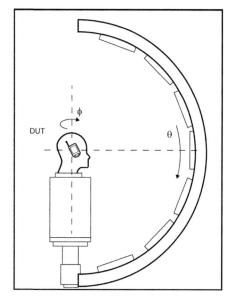

Figure 6.10-1. Example of a spherical positioner system with a moving probe antenna (left), and with multiple probe antennas (right) (25.914 Figure 7.1)

In the left hand figure the probe antenna can move in an arc to vary the elevation angle θ by arbitrary increments, whereas in the right hand figure multiple antennas are mounted at discrete angles and the elevation is altered by switching between antenna elements. Other methods can be used including hybrid systems that mix antenna switching with physical movement. The device itself is shown on a rotating table where the azimuth angle φ can be varied. A sampling grid of 15 degrees in both azimuth and elevation is considered sufficient to maintain the required accuracy. Since TRP measurements are quick to make, this sampling frequency is not a problem. However, TRS measurements require a search to be made to find the reference sensitivity for each angle of arrival (AoA) and therefore a coarser sampling grid of 30 degrees can be used with some loss off accuracy but considerably reduced test time. Given the number of test permutations and taking into account frequency bands, mechanical modes, and phantom loading, SISO OTA testing can take many days of expensive anechoic chamber time per device. Since the time and cost of SISO OTA test is significant, 3GPP further specified in 25.914 an alternative test method based on a stirred-mode reverberation chamber as shown in Figure 10.6-2.

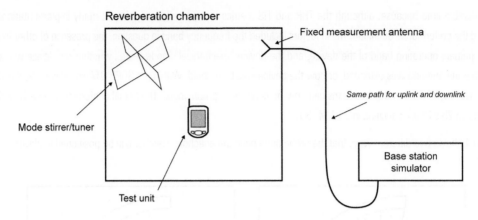

Figure 6.10-2. Schematic picture of the reverberation chamber measurement setup (25.914 Figure E.1)

The reverberation chamber is a metal box in which the test signal is launched from a fixed measurement antenna. The signal then reflects from the internal surfaces to create a standing wave pattern. A mode stirrer consisting of a rotating angular reflector is then rotated slowly over time to create a long term uniform field. Due to the nature of the reverberation chamber it is not possible to predict the exact field at any one time. But it has been shown empirically, after careful chamber calibration, that TRP and TRS results obtained using the reverberation chamber method are of a similar accuracy to those performed by averaging many single AoA measurements performed in an anechoic chamber. The primary advantage of the reverberation chamber is its much smaller size and lower cost compared to an anechoic chamber.

6.10.3 MIMO OTA Overview

Although the introduction of SISO OTA test methods and requirements has made a significant contribution to the design of better antennas, the TRP and TRS metrics do not scale up to SIMO and MIMO conditions. In the simple case of testing a DUT with receive diversity, the anechoic chamber test method using a single AoA and single polarization presents the DUT with an unrealistic signal that has been shown to interact with DUT smart antenna algorithms, thus producing unstable or incorrect results. The solution to this problem is defined by CTIA and involves testing each receive antenna individually. Such an approach is expedient but does not fully characterize how the DUT would perform in a real environment containing multiple simultaneous angles of arrival and polarizations.

Although SISO methods can be extended to SIMO, albeit with some loss of efficacy, the same cannot be said for MIMO devices. The performance of a MIMO device is a complex real-time interaction between the correlation, gain, and polarization properties of the transmitting antennas, the radio channel, and the receiving antennas. The performance is further affected by the signal quality, which may be degraded by noise and interference. Figure 6.10-3 shows the Shannon-bound capacity of a 2x2 MIMO channel as a function of correlation within the channel and of signal to interference plus noise ratio (SINR).

The lowest curve in Figure 6.10-3 represents the SISO channel capacity as a function of SINR. The highest curve represents the theoretical performance of rank 2 spatial multiplexing (MIMO) as a function of SINR.

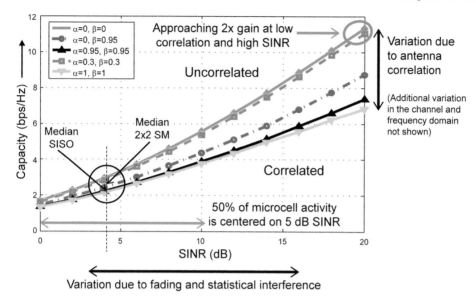

Figure 6.10-3. 2x2 MIMO channel capacity as a function of correlation and SINR

Note that at high SINR, the rank 2 performance approaches double what is possible with a single rank (SISO) channel. The curves in between these rank 1 and rank 2 curves represent different correlation values obtained by varying the correlation properties of the transmitting antennas (α) and the receive antennas (β). It can be seen that the fully correlated case where α and β equal 1 equates to the rank 1 SISO condition since with fully correlated antenna no spatial multiplexing gain is possible. The upper curve, which is perfect rank 2 performance, is achieved when both α and β equal zero; i.e., the transmit and receive antennas are fully decorrelated. The actual performance possible at any given time depends on the instantaneous SINR and the additional effects of the radio propagation performance of the channel (which for simplicity is not represented in the figure). A typical loaded microcell will experience a median SINR of around 5 dB. From this it is possible to see that the spatial multiplexing gain, assuming the antennas and radio channel produce a correlation of 0.5, is perhaps around 20%.

Two significant consequences for MIMO OTA testing can be deduced from Figure 6.10-3: first, the expected performance is highly affected by the propagation conditions, all antennas, and the SINR; and second, under median conditions the spatial multiplexing gain is quite low. This last point is significant because uncertainty in the test system will affect the efficacy of any test that attempts to measure the small difference between SISO and MIMO throughput.

6.10.4 The Importance of the Channel Model and Interference Conditions

Since the channel model and any additional interference conditions have such a large influence on expected performance, much of the MIMO standardization activities are focused around this complex topic. For SISO OTA, the DUT performance was only a function of the DUT and its antennas and could be measured in a simple anechoic chamber with a single antenna. But for MIMO OTA, the complete opposite is true: to fully characterize the DUT it is necessary to create an arbitrarily complex radio environment to emulate what might be the experience in a real environment.

This challenge is new to the test industry and largely explains why significant work remains before the problems of DUT characterization are solved at an acceptable cost and accuracy.

The channel model and interference challenge can be summarized as follows. In any wireless system, a signal propagating through a terrestrial channel can arrive at the destination along a number of different paths, referred to as multipath. Each path can be associated with a time delay and a set of spatial angles, one angle at the transmitter and one at the receiver. For example, Figure 6.10-4 shows a simplified diagram of a 2x2 MIMO system operating in a multipath environment. Between the Tx1 and Rx0 antenna pair, the figure shows a direct line-of-sight (LOS) path and several other non-LOS (NLOS) paths. Each transmit and receive antenna pair would ideally have an uncorrelated set of multipath characteristics depending on the antenna spacing and polarizations. The multipath characteristics arise from scattering, reflection, and diffraction of the radiated energy by objects in the surrounding environment. The various propagation mechanisms influence the channel's spatial characteristics, signal fading, Doppler, and path loss. For a more detailed explanation of channel propagation see Sections 6.6.2 and 6.8.5.

At each receive antenna, multipath propagation results in both a time spreading, referred to as delay spread, and spatial spreading, as transmitted signals follow unique transmission paths from the eNB to the UE antennas, arriving at different angles with different path losses. At the receive antenna, each path can be associated with an AoA as measured relative to the array normal. These signal paths are also associated with an angle of departure (AoD) as the transmitted signals leave the eNB antenna and enter the channel. The AoDs are measured relative to the normal of the transmit antenna array. The spatial characteristics of the wireless channel are modeled using a spatial distribution referred to as power angle spectrum (PAS). Because spatial correlation introduced by the PAS and by the antenna characteristics strongly influence MIMO performance, it is the initial focus of MIMO OTA standardization activities to measure MIMO devices using an OTA test system that can accurately emulate realistic channels and include the effects of angular spread, direction of arrival, antenna gain, antenna spacing, and antenna polarization. The difference between antenna patterns as a function of angle may help to de-correlate the signals arriving at each receive antenna; therefore, it is important to include the antenna characteristics in any MIMO channel model and associated OTA test system in order to get an accurate measurement of the MIMO UE performance. The spatial and temporal properties of any additional interference also need to be considered.

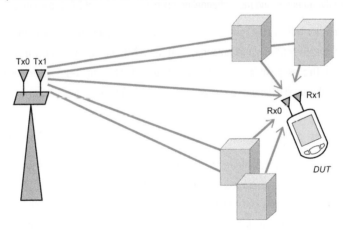

Figure 6.10-4. Simplified model of a 2x2 multipath environment

6.10.5 MIMO OTA Standardization Activities

CTIA started working on MIMO OTA testing in 2007 and 3GPP followed in 2008. The study item within 3GPP recently concluded with the publication of 37.976 [36]. Due to the complexity of the problem and its possible solutions, the study did not conclude on a single test method but instead recommended the next steps to be carried out in the work item phase, which will be documented in 37.977 [37]. Unlike SISO OTA which initially had only one main test method, there are currently seven candidate test methods being considered for MIMO OTA by 3GPP. These fall into two groups: five based on anechoic chambers and two based on reverberation chambers.

Anechoic chamber candidate methods are the following:

1. Multi-probe method (arbitrary number and position)
2. Ring of probes method (symmetrical)
3. Two-stage method
4. Two-channel (decomposition) method
5. Spatial fading emulator method.

Reverberation chamber candidate methods are the following:

1. Basic or cascaded reverberation chamber
2. Reverberation chamber with channel emulator.

The anechoic and reverberation methods take fundamentally different approaches toward achieving the same end goal, which is the creation of a spatially diverse radio channel. In the case of the anechoic chamber, multiple probes are used to launch signals at the DUT to create known AoAs, which map onto the required channel spatial model. This is a powerful approach, although to achieve arbitrary channel model flexibility, large numbers of probes are required, which is costly and challenging to calibrate due to issues such as backscatter. In the reverberation chamber method, the spatial richness is provided in 3D by relying on the natural reflections within the chamber, which are further randomized by use of mode stirrers that oscillate to provide a spatial field that approaches a uniform field over long periods of time. However, the instantaneous spatial field is not uniform, which means that the reverberation chamber can be used to measure spatial multiplexing gain in decorrelated antennas. Each method will now be briefly introduced. For further details, refer to 37.976 [36] and 37.977 [37].

6.10.5.1 Multi-Probe Method

The goal of the multi-probe method shown in Figure 6.10-5 is to create the desired channel model by positioning an arbitrary number of probe antennas in arbitrary positions within the anechoic chamber. The antennas are all located equidistant from the DUT, and each antenna is faded by a channel emulator to provide the desired temporal component. By careful choice of the number and position of the probe antennas, it is possible to construct an arbitrarily complex radio propagation environment. The method is conceptually simple since there is a direct relationship between the required angular spread of the channel and the physical location of the probes.

A simple, single cluster channel model with a narrow angular spread can be emulated using four antennas in a relatively small anechoic chamber. The DUT is placed at one end of the anechoic chamber and the probes at the other end. More complex multi-cluster conditions can be generated by increasing the number of probes and the size of the anechoic chamber, in which case the DUT is placed in the center and the probes on the perimeter. The simplest configurations locate the probes in the same azimuth plane to create a 2D environment. More complex 3D fields are created by locating antennas on different planes. The direct relationship between the probe antenna positions and the emulated channel model means that to test the DUT from all angles, it must be mounted on a rotating and tilting platform.

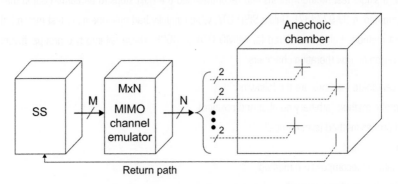

Figure 6.10-5. Multi-probe configuration (37.977 [37] Figure 6.3.1.1-2)

6.10.5.2 Ring of Probes Method

The ring of probes method is based on a symmetric ring of probe antennas equidistant around the DUT, which is placed at the center of the anechoic chamber as shown in Figure 6.10-6. As with the multi-probe method, each probe is fed by a channel emulator to generate the temporal characteristics of the desired channel model. Where the symmetrical ring of probes method differs from the basic multi-probe method is that there is no longer a fixed relationship between the probe antenna positions and the angle of departure. Instead, the spatial components of the channel model are mapped onto the equally spaced probe antennas in such a way that an arbitrary number of clusters with associated angular spreads can be generated. This more flexible approach allows any 2D spatial channel model without having to reposition (and recalibrate) the probe antennas.

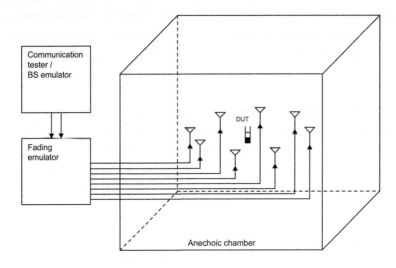

Figure 6.10-6. Example of MIMO/multi-antenna OTA test setup (37.977 [37] Figure 6.3.1.2-1)

The number of antennas in the ring affects the accuracy with which the spatial dimension of the channel model can be implemented. A typical configuration is a 22.5 degree raster with 90 degree cross polarization at each location, giving a total of 32 probes each independently driven by a channel emulator.

6.10.5.3 Two-Stage Method

The two-stage method takes a fundamentally different approach to creating the necessary conditions to test MIMO performance. It is illustrated in Figure 6.10-7. The first stage involves the measurement of the 3D antenna pattern of the DUT using an anechoic chamber of the size and type used for existing SISO tests. To measure the antenna pattern non-intrusively (i.e., without modification of the device or the attachment of cables), a special test function is required in the DUT to report the received power per antenna and relative phase between antennas for a given received signal.

The second stage takes the measured antenna pattern and convolves it with the desired channel model using a channel emulator. The output of the channel emulator then represents the faded downlink signal modified by the spatial properties of the DUT's antenna. This signal is then connected to the DUT's temporary antenna connectors as used for traditional conducted testing. The second stage does not require the use of an expensive anechoic chamber.

The accuracy with which the DUT can measure the received power per antenna and the relative phase between the antennas is not critical since both the absolute accuracy and, if necessary, the linearity of the DUT measurements can be calibrated by the test system. The only important criteria is that the measurements be repeatable in the same test conditions.

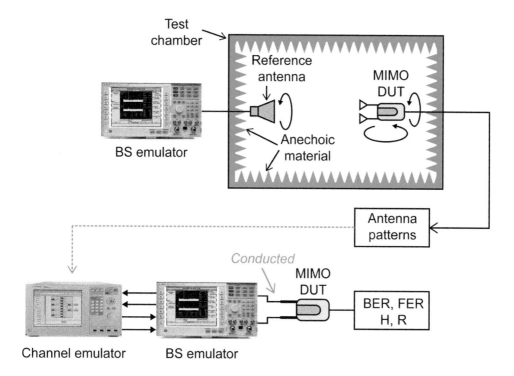

Figure 6.10-7. Proposed two-stage test methodology for MIMO OTA test (37.977 [37] Figure 6.3.1.3.1-2)

Since the 3D antenna pattern can be measured easily, the two-stage method can be used to emulate any arbitrary 3D channel propagation condition. The rotation of the DUT relative to the channel model is accomplished by synthesis within the channel emulator. In the basic form of the two-stage method, wherein the antenna pattern is measured at a power well above reference sensitivity, the impact of self interference is not captured.

Since spatial multiplexing requires relatively good SINR to provide gain, the spatial multiplexing performance at low signal levels at which self-interference occurs is unlikely to be of significance. Standards exist for measuring SISO OTA at reference sensitivity and it is the intent to extend these simpler tests for SIMO operation. However, since self interference is included in the other MIMO candidate methodologies, work is underway to extend the two-stage method to include the evaluation of self interference. It has been shown that reference signal received quality (RSRQ) measurements made by the UE are usable for calculating the received noise due to radiated effects. This calculated noise can be added during the second stage throughput measurements to emulate the impact of radiated noise.

6.10.5.4 Two-Channel (Decomposition) Method

The two-channel method shown in Figure 6.10-8 is a special case of the multi-probe method and uses just two probes with no channel emulator. The angle of departure of the two downlink test signals can be configured for any elevation or polarization. The principle of the method is to evaluate the impact of the direction and angular separation of the two signals on the DUT performance. By carrying out a large number of tests using different combinations of angles, statistical analysis can be used to derive figures of merit for the DUT. Direct comparison with results achieved using more complex spatial signals with temporal variations are not possible but results show this method to provide similar DUT ranking. the addition of channel emulation is being studied.

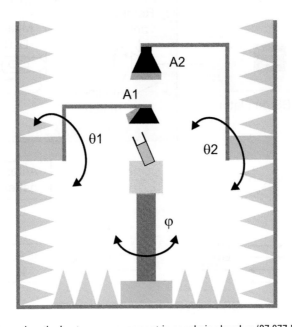

Figure 6.10-8. Two-channel method antenna arrangement in anechoic chamber (37.977 [37] Figure 6.3.1.4.1-1)

6.10.5.5 Spatial Channel Emulation Method

The last of the anechoic chamber methods is a variation of the ring of probes method in which the channel emulation function is provided by a much simpler programmable attenuator and phase shifter per antenna. This method is shown in Figure 6.10-9.

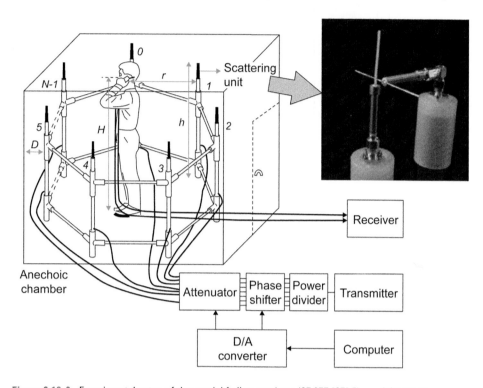

Figure 6.10-9. Experimental setup of the spatial fading emulator (37.977 [37] Figure 6.3.1.5.1-1)

By controlling the amplitude and phase in real time, a Rayleigh distribution or other relevant multipath distribution can be obtained.

6.10.5.6 Reverberation Chamber Method

The first of the reverberation-based methods uses the intrinsic reflective properties of the mode-stirred reverberation chamber to transform the downlink test signal into a rich 3D multipath signal. This is shown in Figure 6.10-10. The spatial characteristics of the signal are random and over time can be shown to be uniform, but when observed over the time period of a demodulated data symbol, they are known to be highly directional. This non-uniformity provides the DUT with diverse signals on each antenna thus enabling spatial multiplexing gain.

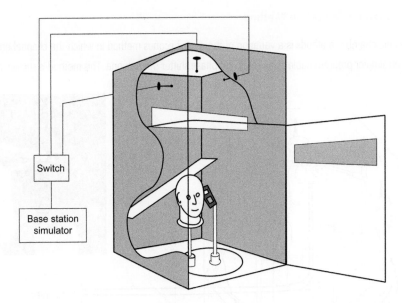

Figure 6.10-10. Reverberation chamber setup for devices testing with single cavity
(37.977 [37] Figure 6.3.2.1-1)

The natural time domain response of the chamber can be modified by the use of small amounts of RF absorptive material. The basic reverberation chamber is limited to a single power delay profile and a relatively slow Doppler spectrum determined by the speed of the mode stirrer. Further control of the power delay profile and spatial aspects can be obtained by cascading two or more reverberation chambers as shown in Figure 6.10-11, and there has also been research using nested chambers.

In addition to the conventional Rayleigh 3D isotropic fading scenario emulated by single-cavity reverberation chambers, multi-cavity multi-source mode-stirred reverberation chambers employ de-embedding algorithms for enhanced repeatability and have added capabilities to emulate different K-factors for Rician fading, different non-isotropic scenarios including single and multiple-cluster with partial door opening, and standardized or arbitrary amplitude power delay profiles (e.g. 802.11n, Nakagami-m, on-body and user-defined) using sample selection techniques.

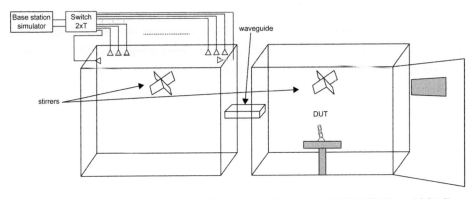

Figure 6.10-11. Reverberation chambers with multiple cavities (part of 37.977 [37] Figure 6.3.2.1-2)

6.10.5.7 Reverberation Chamber and Channel Emulator Method

The final method shown in Figure 6.10-12 addresses the limitation of the basic or cascaded reverberation chamber by adding a channel emulator to the downlink prior to launching the signals into the chamber. This method allows the temporal aspects of the desired channel model to be fully controlled, although the underlying natural and very short decay time of the chamber will slightly spread the power delay profile.

With the use of a channel emulator capable of negative time delay (inverse injection), multiple cavity mode-stirred reverberation chambers can accurately emulate the power delay profiles of 3GPP SCME channel models.

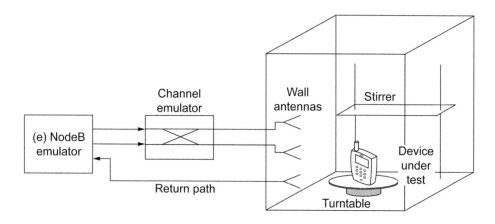

Figure 6.10-12. Test bench configuration for test using channel emulator and reverberation chamber for a 2x2 MIMO configuration 37.977 [37] Figure 6.3.2.2.1-1)

6.10.5.8 Comparison of Methods

All seven methods have unique attributes, some of which are desirable and others less so. Section 9.1 of 37.976 provides an extensive list of these attributes. A simplified summary of the key points is given in Table 6.10-1. This includes an assessment of the key technical areas still under study.

Table 6.10-1. Comparison of candidate methodologies

Method	Pros	Cons	Future work
A1 Multi-probe	Conceptually simple	Limited flexibility, cost per probe	Calibration and validation
A2 Ring of probes	Arbitrarily flexible	Cost, 3D very costly	Calibration and validation
A3 Two-stage	Low cost including 3D	Requires DUT test mode	Self interference solution
A4 Two-channel	Very low cost	No temporal and limited spatial control	Correlation with other methods, addition of channel emulation
A5 Spatial emulator	Low cost	Limitations in channel models	Calibration and correlation with other methods
R1 Reverb	Very low cost	Limited temporal and no spatial control	Calibration and evaluation of spatial aspects
R2 Reverb plus fader	Low cost	No spatial control	Calibration and evaluation of spatial aspects

6.10.6 Results from Two-Stage Method Tests

6.10.6.1 Stage 1: Antenna Pattern Measurements

Figure 6.10-13 shows typical antenna pattern measurements of two different devices. Two dimensional cuts from the overall 3D measured pattern are shown. Device 1 has a highly directional first antenna and a more omnidirectional second antenna. Device 2 appears to have a significant problem with the gain of the sub-antenna. This device would not likely show good spatial multiplexing gain since the antennas have a high gain imbalance.

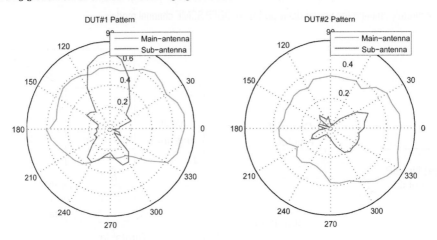

Figure 6.10-13. 2D antenna patterns that can be generated from antenna gain measurements performed in stage 1 of the two-stage method

A full 3D antenna pattern is shown in Figure 6.10-14.

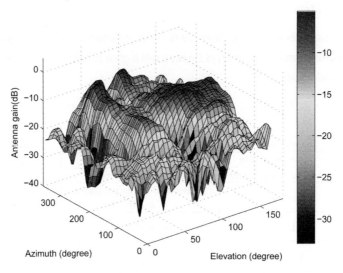

Figure 6.10-14. 3D antenna patterns that can be generated from antenna gain measurements performed in stage 1 of the two-stage method

An advantage of the two-stage method over other methods that provide throughput results only is that the first stage provides valuable design information about the antenna performance. It is not yet clear if antenna metrics such as correlation or gain imbalance will be specified, but such measures are highly valuable to the antenna designer. The view of the antenna as seen through the DUT receiver when a special antenna test function is used cannot be derived from end-to-end throughput measurements.

6.10.6.2 Stage 2: Throughput Measurements

Once the 3D antenna pattern is measured in the anechoic chamber, all further measurements of throughput are done in conducted mode, without the need of an anechoic chamber, through the DUT's temporary antenna ports using a channel emulator such as the Agilent PXB, which convolves the measured antenna pattern with the desired channel model. A received signal is created at the DUT's temporary antenna port that matches what the device would have seen had it been placed in the actual radiated field.

Figure 6.10-15 shows an example of throughput as a function of azimuth angle for different downlink power levels. The test conditions were fixed channel coding and forced rank 2 as specified in 37.977 [37]. In the two-stage method the azimuth angle of the DUT relative to the channel model is simulated inside the channel emulator. This type of testing can identify angles at which the DUT does not perform well.

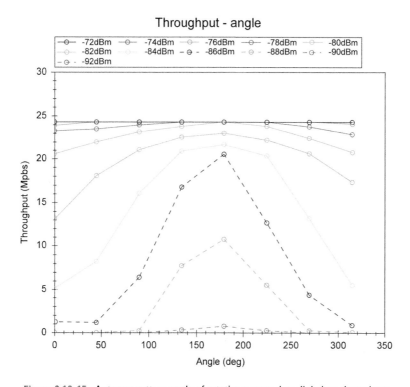

Figure 6.10-15. Antenna pattern angle of rotation versus downlink data throughput

Figure 6.10-16 plots the downlink power level versus throughput, but this time each data point represents the averaging of the throughput values for the eight angles of rotation tested at each downlink power.

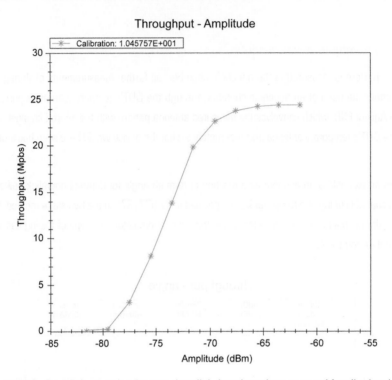

Figure 6.10-16. Downlink power level versus downlink data throughput averaged for all azimuth angles

An advantage of the two-stage method here is that it is not necessary to re-measure the DUT in the anechoic chamber for each desired channel model since the antenna pattern is only a function of the DUT and not a function of the channel model. Once the antenna pattern is known, it is possible to measure the MIMO performance with any channel model. Figure 6.10-17 shows the results.

Another advantage of the two-stage method is that it is possible to substitute the measured antenna patterns used for the second stage throughput measurements with simulated patterns in order to accurately assess the overall DUT performance, without having to physically design and implement alternative antennas. This can save significant time during the design phase. It is even possible to measure the antenna pattern of one device then apply it to the receiver of a different device.

An example of antenna pattern simulation using Agilent's EMPro antenna design software is shown in Figure 6.10-18. The figure shows that the loaded antenna pattern is significantly distorted from the ideal free space design. Using the two-stage method it would be possible to evaluate a real receiver's performance using either of the simulated antenna patterns without having to actually build a prototype.

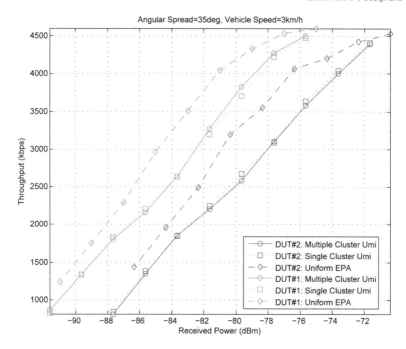

Figure 6.10-17. Second stage throughput results showing differences in performance between channel models

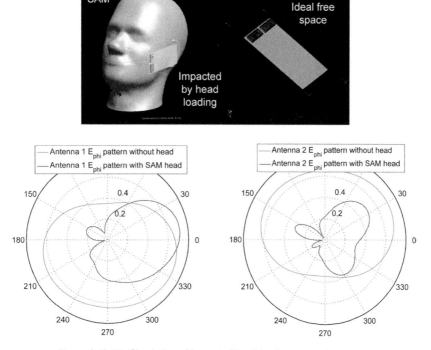

Figure 6.10-18. Simulation of impact of head loading on patch antenna

6.10.6 Conclusion

OTA testing determines how antenna design and multipath radio environments influence a device's radiated performance. Testing using the two-stage method yields additional data that can be used by antenna designers and device manufacturers to demonstrate that products meet critical performance criteria.

6.11 Signaling Protocol Development and Testing

6.11.1 Introduction

Given the complexity of the LTE UE, successful products rely on a program of testing that goes beyond the requirements of the standards. As part of that program, the stability of the software in the UE needs to be the subject of relentless and systematic test. That includes the signaling software, which is a major contributor to the overall stability of the UE.

This section discusses the challenges associated with developing the UE signaling protocols and describes a test environment in which engineers can write and execute test applications.

6.11.2 Challenges of UE Signaling Protocol Development

6.11.2.1 Development Complexity

By necessity the LTE UE supports many signaling protocols to enable communication (1) with the base station (eNB) for the air interface connection and (2) with the evolved packet core network. In many cases protocols are replicated to accommodate multiple radio access technologies (RATs) such as GERAN, UTRAN, cdma2000, and LTE E-UTRAN. In addition, protocols must be developed to interact with the core network and the application entities that make the UE's location known to the network, and to manage the packet-switched and circuit-switched connections along with the associated services and subscriber applications. The complexities faced by protocol stack developers are evident. When a new technology feature is added to the UE, whether specific to a RAT or spanning multiple RATs, the new feature must be tested for each protocol and for its ability to operate correctly both in isolation and in interactions with legacy features. The legacy features must also be tested to ensure that their operation has not been compromised.

The work of UE developers who are creating or enhancing protocol stacks can therefore be divided into two categories, each with its own test requirements:

- Developing and testing new features and their interactions with legacy features. This work requires the development of new test applications.
- Regression testing of legacy features using existing test applications along with test automation functions.

6.11.2.2 Test Objectives and Strategies

Testing new features is one of the most important aspects of software development. When protocol developers create test strategies, they keep several goals in mind.

Considerations for protocol testing:

- Test all the requirements
- Be prepared for the unexpected
- Aim for early detection of problems
- Improve the process whenever possible.

Testing based on these considerations is performed at many stages in the UE development cycle. For example, initial testing may include some inspection of system design documents and models. Software test harnesses can be developed that are collections of software and test data configured into an automated virtual environment for testing and logging performance under varying conditions. These are used to verify many of the signaling features prior to testing on the actual UE processors. At the UE system level, testing is performed by many different engineering teams: development teams, UE verification teams, pre-conformance tests teams, independent conformance test labs, and operator acceptance test teams.

Signaling development often proceeds with daily, weekly, monthly, and even quarterly cycles. At each stage of integration and testing, plans are developed that define the test coverage for each signaling software release. Testing varies from limited-in-scope automated regression testing performed overnight for the benefit of internal development teams through much more extensive and less frequent testing performed by independent test teams for major software releases. LTE signaling protocols must also be tested in accordance with 3GPP requirements, which involves developing tests and test processes to verify the proper implementation of each signaling feature in the LTE UE.

Formal signaling conformance tests, such as those required by the Global Certification Forum (GCF) and PCS Type Certification Review Board (PTCRB), are written to verify expected UE signaling protocol operation under normal conditions and under a range of anticipated abnormal conditions. The GCF and PTCRB LTE signaling conformance test cases are written in testing and test control notation (TTCN-3) [38] and are traceable to the 3GPP-defined protocol conformance specification 36.523-1 [39]. The signaling conformance tests for LTE are further discussed in Chapter 7.3.

Network operators require that user equipment meet GCF and PTCRB test requirements. In addition, operators often define acceptance tests of their own prior to promoting the use of a UE on their network. (Note that failure to pass operator acceptance tests cannot be used by that operator to refuse service to any UE that has successfully passed the narrower type-approval tests that form the basis of national regulation).

While operator acceptance tests sometimes involve modest extensions to existing GCF and PTCRB signaling and RF conformance test plans, they more frequently focus on other types of testing that measure the UE performance most likely to be noticed by the end user—for example, data throughput, system-selection, voice quality, inter-RAT handover and other multi-RAT interactions, location-based services, messaging, Internet and other application performance, operating system and application-based network loading, and battery drain. These aspects of testing are covered later in this chapter. For UE signaling protocol developers, testing based on operator acceptance test plans often exposes large numbers of defects not revealed by the GCF and PTCRB conformance tests.

Along with the test requirements imposed by certification groups and network operators, UE developers must also meet in-house test requirements. For example, system architects generate test requirements to verify the proper implementation of the UE system design. Tests are defined to verify expected operation of the UE under both normal and error conditions. Tests are also defined to verify the proper operation of subsystems and interfaces at the system level and at the level of individual subsystems.

The combined tests for GCF and PTCRB conformance, network operator acceptance, and in-house system design form a library of requirement-based tests. These tests typically consist of a well-defined set of pass/fail conditions that follow a well-considered sequence of test steps starting from a defined initial condition. The UEs are often reset between tests to achieve known starting conditions and ensure consistent results. While these formal signaling tests can detect many thousands of potential defects, many more may exist that are not covered by the required tests. UE developers therefore need to expect the unexpected and build test strategies that will detect additional defects. Examples include:

- Repeated running of GCF/PTCRB signaling conformance tests and operator acceptance tests without cycling power or fully resetting the UE, looking for UE crashes
- Stress tests with high data rates and multiple applications running in parallel, looking for degradations in performance due to processor capacity or memory shortage
- Simulated subscriber use patterns extending for weeks of continuous operation
- Randomized test sequences
- Application of nonstandard message lengths and sequences to force pointer over-runs
- Extended simulated network testing with handovers and faded channel conditions.

Statistical techniques are often used to track the number of software defects detected per test hour. Requirement-based testing is deterministic and repeatable. As defects are found and resolved, it is reasonable to expect the find rate for those particular kinds of defects to trend to zero as tests are rerun and defect resolution is verified. In contrast, testing using the broader techniques listed above may produce find rates that are reduced only slowly or remain flat even as defects are resolved. Statistics in this case may reveal that modern and complex software systems contain a defect count so large that it can be modeled as essentially infinite. This huge count may be the consequence of creating new defects during the resolution of other defects.

6.11.2.3 Case for Early Detection

Latent defects in software design or implementation that are not detected and corrected during testing can have serious consequences once the product is released in large numbers to the field. Users may become disillusioned with unreliable devices and possibly with the service provider, too. The brands of both the manufacturer and the service provider could suffer.

Ensuring software quality and stability requires tens of thousands of test hours with many test systems running in parallel to exercise the UEs in a large variety of test scenarios. Such tests need to be automated and run 24 hours per day. As part of the process, each new software release will go through regression testing and statistics will be collected with an expectation that the mean time between failures has improved over earlier revisions.

Comprehensive testing of UE software may appear to be time-consuming and expensive, but the cost of resolving defects will only get higher as the development cycle progresses. Defects identified by software developers before integration can be resolved at almost no cost. Defects found after commercial launch may require user software updates and product recalls. The case for early detection is therefore strong.

The test process itself can be continuously improved to reduce cost and effort. For example, software defects that have gone undetected in earlier test phases are often identified during interoperability testing, which is performed using real infrastructure in lab or field environments. This learning can be incorporated into earlier test phases to enable early detection. Based on this experience, new test cases can be defined and added to the library of tests performed during the software release process. As the test library expands, recurrences of defects or regressions can be prevented. Similar principles can be used at all test stages to move defect detection upstream as much as possible.

Process improvement projects can analyze defects found at each phase of the development and test cycle. Statistics can be used to categorize defects by their root cause, which may be related to requirements, design, standards interpretation, coding, and so forth. Pareto analysis can be used to highlight the most productive areas for process improvement, and steps taken to improve both development and test processes to catch the various classes of defect prior to interoperability testing and commercial deployment.

6.11.2.4 Test Solution Requirements

Test equipment has been developed to enable UE developers to implement efficient and comprehensive test programs. Important characteristics of this test equipment include the following:

- Execution of GCF/PTCRB signaling conformance test cases and additional signaling protocol scenarios in RF conformance tests
- Execution of user-developed TTCN-3 test cases
- Functional test plan coverage including representative operator test plans
- High performance, real time protocol implementation supporting high end-to-end application data rates
- Inter-RAT support
- Integration of FTP, HTTP, SMS, IMS, VoIP, and other application layer servers
- Integrated RF and battery drain measurements
- Efficient user-development environment for functional test suites
- Test sequencing, UE control, and automation
- Logging and diagnostics capability with the ability to pass logs from site to site within a geographically distributed organization
- Worldwide support.

6.11.3 Test System Overview

To illustrate some of the test concepts just described, the remainder of this section will refer to the signaling protocol test system based on the Agilent E6621A PXT wireless communications test set, which is shown in Figure 6.11-1 alongside the E6615A 8960 2G/3G wireless communications test set.

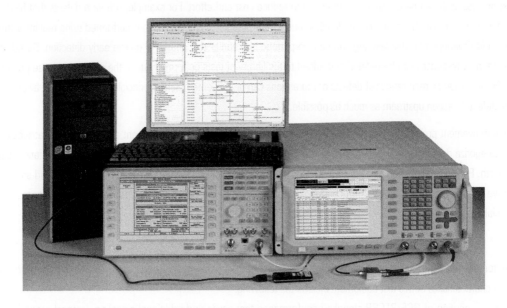

Figure 6.11-1. Agilent E6615A 8960 and E6621A PXT wireless communications test sets

6.11.3.1 Base Station Emulation and Logging

The PXT test set is a versatile test platform for functional and conformance-based protocol testing. It has many essential features for LTE signaling protocol testing, including the following:

- Up to 6 GHz MIMO RF in/out to cover current and anticipated LTE bands, along with support for multiple cells
- Built-in base-station emulation with high performance processing to support high end-to-end data rates, message customization, and logging
- Automation interfaces for TTCN-3 control and functional test automation and as part of user-developed custom systems.

The PXT is able to operate with self-contained base-station protocol emulation for high performance signaling operation. When running TTCN-3 test cases, the PXT suspends the operation of the self-contained base-station protocol emulation. Signaling functionality is provided in this mode by the TTCN-3 software running in an external PC.

The PXT is used in conjunction with the 8960 test set to emulate non-LTE multi-RAT environments. The 8960 is able to support a large variety of GERAN, UTRAN, and cdma2000 cell configurations. The PXT and 8960 work with external Agilent software to capture and replay multi-layer protocol logs.

Figure 6.11-2 shows the Agilent N6061A LTE Protocol Logging and Analysis application. Logs can be saved for later replay and analysis, and they can be emailed to remote teams who are investigating UE protocol issues. The logging tool provides a detailed and independent record of message exchanges to support operation with the PXT's self-contained base station protocol emulation and to provide a more detailed supplement to TTCN-3 logging.

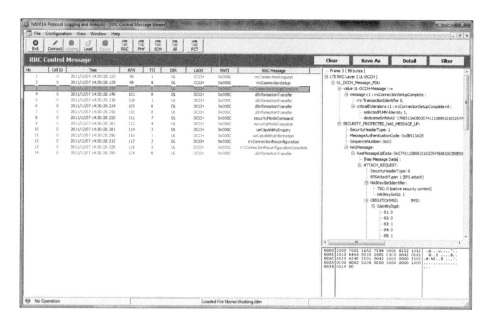

Figure 6.11-2. Agilent N6061A LTE Protocol Logging and Analysis

The Agilent N6062A LTE Message Editor application enables users to modify detailed protocol settings. Used in conjunction with the PXT's self-contained base station protocol emulation, the Message Editor allows signaling scenarios to be customized and downloaded to run real-time inside the PXT. The Message Editor is shown in Figure 6.11-3.

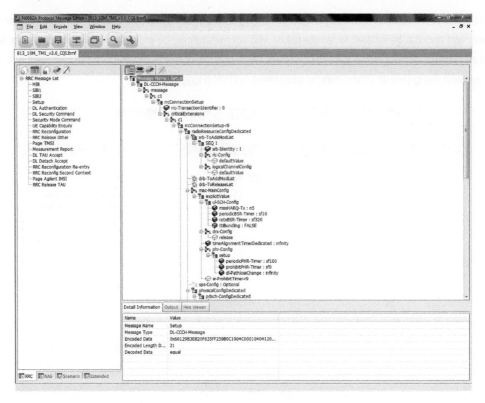

Figure 6.11-3. Agilent N6062A LTE Message Editor

6.11.3.2 TTCN-3 Environment

For comprehensive signaling test, a system is required that has the ability to execute standards-based tests such as the GCF- and PTCRB-validated test cases. The Agilent N6070A Series Signaling Conformance Test and Development System combines a powerful TTCN-3 development and execution environment with the PXT LTE test set. The system is illustrated in Figure 6.11-4. System software integrates the TTCN-3 test cases defined in 3GPP TS 36.523-1 [39] with a TTCN-3 test adaptor. This TTCN-3 test adaptor incorporates the LTE message codec and the system and platform adaptors, allowing the implementation-independent TTCN-3 test cases to be executed on the PXT. The TTCN-3 messages are converted into ASN.1-encoded messages and routed to and from the UE under test while other aspects of the test setup such as the RF parameters and the MIB and SIB configuration are controlled. Users can modify TTCN-3 source code to add or modify test cases.

A TTCN-3 execution environment such as this enables users to run tests individually or run entire test campaigns. Campaign loader files (CLF) can be edited to select the tests to be executed together with the number of runs, retries, and actions on fail. CLF files also contain user-editable parameters such as protocol implementation conformance statements (PICS) [40] and protocol implementation extra information for testing (PIXIT) to allow simple customization of test settings. PICS settings define the capabilities of the UE and must be declared by the UE manufacturer when a device is submitted for certification. PIXIT settings define additional parameters such as network, USIM, IP version, band, and bandwidth to provide some flexibility in the test system configuration. Although PICS settings remain constant in a particular UE implementation, PIXITs can be changed to alter the test conditions for verifying UE behavior under the many different operating conditions that may be encountered in deployed networks. Meta-campaigns can be created to load and execute sequences of multi-test campaigns, each with a unique list of tests and parameters.

As tests are executed, message flowcharts are displayed that show each message being sent to and from the UE, between the protocol layers, or between other key components in the system. Users can click on message arrows or TTCN-3 message template matches and mismatches in the flowchart to display detailed message contents.

Log files can be saved, emailed, and later reloaded. The logs retain a message flowchart with a fully interactive user interface for analyzing and diagnosing the messages. The system also enables reports to be saved in a variety of formats including PDF, HTML, and XML.

In addition to TTCN-3 based logging, this system supports the parallel capture of independent logs with protocol logging and analysis software. The software displays, decodes, and saves multi-layer protocol logs captured directly from the PXT test set with time-coordinated TTCN-3 event references.

UE control is provided using a number of user-defined interfaces. Standard attention (AT) commands [41] can be combined with proprietary UE control messages to enable unattended testing. Automation can be managed from the TTCN-3 execution environment graphical user interface (GUI) or via a command line interface (CLI) under the control of an external test executive.

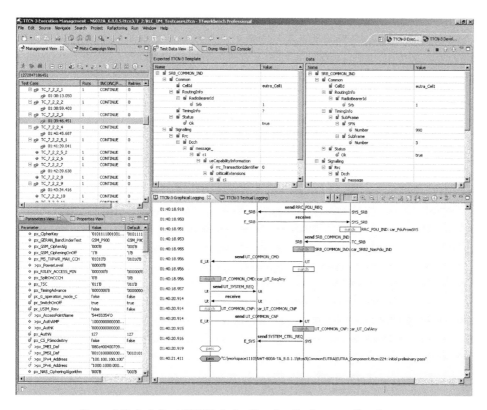

Figure 6.11-4. Agilent N6070A Series Signaling Conformance Test System

6.11.4 Summary

As UEs become increasingly complex and subscribers become more demanding, achieving steadily improving levels of UE stability has become a way for UE developers to differentiate their products. This can be achieved only with effective test strategies and tools. By employing the tools described in this section, UE reference designers and UE developers are able to implement test programs that ensure UE conformance, quality, and stability. Testing can be performed on each UE software revision and at several stages during the software release and integration cycle. Tests can be automated to enable continuous and parallel testing of many test scenarios with integration and test engineers supervising many test systems.

Test systems can be deployed by conformance test labs and network operators to certify devices ready for deployment. With experience, test suites can be expanded to prevent reoccurrences of defects discovered during interoperability testing and other downstream activities, enabling the defect density of each release to improve.

6.12 UE Functional Testing

6.12.1 Introduction

The earlier sections of this chapter have presented detailed descriptions of a variety of testing challenges and solutions that will benefit R&D engineers engaged in different parts of a UE design and development project. Chapter 7 focuses on testing the completed product addressing the conformance testing that is typically required as part of the operator device acceptance process for any new product. This section of Chapter 6 also concerns itself with testing the complete design, with a focus not on conformance aspects, but rather the end-to-end (E2E) functional performance and end user experience.

Functional testing is carried out for multiple reasons by different parties within the design lifecycle; for example, by

- R&D engineers looking to validate and benchmark key performance indicators (KPIs) for the overall performance of their design
- Validation engineers doing regression testing in order to check for errors following a change to the software
- Operators seeking to prove out the user experience prior to accepting a new device on their network (see Section 7.5).

Each of these different engineering groups is likely to have its own specific requirements. Therefore, while functional testing follows similar themes across the different users, it is not something that traditionally has been defined and managed by standards organizations or industry groups. The situation is changing, however, with the Global Certification Forum (GCF) now looking at benchmarking application layer throughput performance. This development will be described in Chapter 7.4.5.

The functional test space covers a broad range of tests which can be categorized into the following general areas:

- Signaling protocol functional test
- E2E throughput testing
- Handover testing
- Testing of services
- Characterization of battery life performance (see Sections 6.13 and 7.5).

In many cases the functional test engineer will perform testing using a real-time eNB emulator (as opposed to an emulator using scripted signaling protocols, as is the norm in conformance testing). The eNB emulator gives greater test coverage in terms of being able to stress the UE with many different variables at the same time. The key element of this testing is that it is defined in terms of the end users' activity with the goal to test the performance of the device at the limits of its specification. As a result, testing seeks to uncover issues that relate to the overall design and which may not be observed when the individual components are tested. For example, the RF/modem design may work well at the highest data rate using randomly generated test traffic, but when tests are done with the full protocol stack and real IP data, the user may uncover processing or memory issues which lead to a degradation in performance.

The following sections describe each of these areas by drawing on examples relating to the challenges posed by LTE.

6.12.2 Signaling Protocol Functional Test

Figure 6.12-1 shows a test setup for a typical configuration used for functional test of signaling protocols. At the core of the setup is an eNB emulator. The eNB emulator is a piece of test equipment that seeks to emulate the behavior of a real base station by sending protocol messages such that the UE believes it is connected to a real, live network. In addition the eNB emulator will typically have a programmable interface, which allows it to be used as part of an automated test setup, as well as diagnostic tools for debug.

The test setup in Figure 6.12-1 shows the eNB emulator feeding an external PC, which is capturing a log of the signaling protocol exchange between the DUT and the eNB emulator. This tool can provide invaluable information when a test is post-processed and anomalies are sought in the protocol exchanges. Figure 6.12-1 also shows how the eNB emulator provides an IP interface that allows the UE to make a connection and then access servers on an IP network. This capability is essential for testing the E2E performance of a given application.

The requirements for functional test of signaling protocols typically become more complex with the development of any given technology. For the validation engineer this process requires constantly adding new tests as well as maintaining a database of legacy tests in order that a comprehensive set of tests can be used to do regression testing for each new software build. Often these tests will be categorized into groups depending on functionality. This gives the validation manager an opportunity to select what tests are felt to be necessary based on the changes relevant to that particular software build.

In the early days of a new technology the tests tend to be focused on signaling protocol features that are required to make a connection to the network—for example, registration, authentication, and security. More advanced features such as support for IPv6, idle mode, or testing of handover procedures come later. Given the flexibility of the standards, a comprehensive set of signaling protocol tests can be an invaluable tool for checking new functionality and for regression testing of old features on new releases.

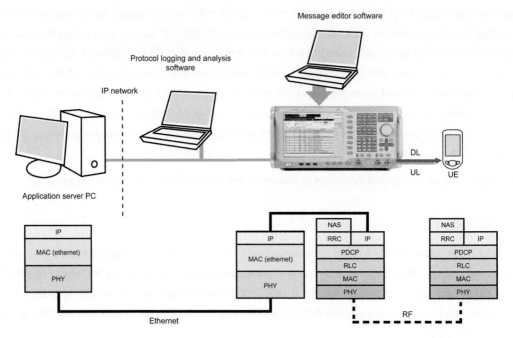

Figure 6.12-1. Test configuration for signaling protocol functional test

6.12.3 E2E Throughput Testing

In addition to telephony services UEs provide access to data services and the Internet. Therefore a key parameter for assessing UE performance is the peak E2E data throughput that the UE can reliably sustain. Testing peak throughput represents a stress condition that is particularly useful as a short "one-off" test that demands high performance within many areas of the UE. During the test, the central processing unit (CPU) in the UE supports the end-user application, the IP stack, and the data flows at the higher layers. In addition a lower level processing unit in the UE provides the RF modem functionality required to support the high speed wireless link. These two parts of the UE will be operating at their limits and must also interwork correctly. Under real network conditions the theoretical maximum throughput will seldom be available due to radio conditions and network loading; however, the peak configuration is still an essential test in the laboratory since it stresses the phone at the limits of its design. The test is done using ideal RF conditions to ensure that the throughput is not reduced by retransmissions.

6.12.3.1 Application Layer Throughput Test System

Testing the throughput of an IP-based packet-switched data connection typically requires both a data source and a data sink. The requirements on the data source are usually straightforward and can often be summarized as "send data as fast as the air interface will allow." Most test environments use a generic PC or server connected to the LAN of an eNB emulator. The data sink is the UE itself or another PC, often referred to as the "client PC" connected to the UE and using the UE as a modem. Figure 6.12-2 shows a typical test system for evaluating application layer throughput.

Two common approaches to throughput testing involve sending either user datagram protocol (UDP) or transmission control protocol (TCP) data as fast as the air interface will allow.

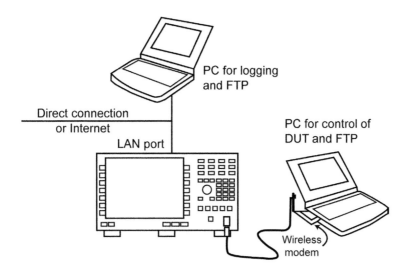

Figure 6.12-2. Test system for evaluating application layer throughput

UDP does not impose any throughput restrictions of its own and is a "connectionless" protocol; that is, it does not require any acknowledgements of the received data. UDP is a good mechanism for testing the raw downlink data throughput capability of the UE to the theoretical limits. However, because UDP is a transparent data service with no retransmission, UDP throughput testing does not require higher layer flow control or error correction and so it is not completely representative of the end user experience for many data services.

TCP is used by common protocols such as hypertext transfer protocol (HTTP) and file transfer protocol (FTP)and so is a good choice for measuring throughput closer to what a real user will experience. However, E2E throughput can be restricted by TCP mechanisms in the server PC (such as receive window size, flow control algorithms, and congestion control) that are not related to over-the-air capacity. TCP is a connection-orientated protocol and makes use of retransmission to ensure the integrity of the at the expense of raw data throughput.

6.12.3.2 Testing Dynamic E2E Throughput in Real Life Conditions

The conditions for application layer throughput testing and the results can vary considerably:

- By using static conditions with no noise, fading, or interference, it is possible to measure the baseline throughput performance using UDP data throughput testing. As previously described this test provides an opportunity to compare the baseline UE performance with an expected theoretical performance.
- By using user data from a real application, it is possible to establish the capability of the processing and memory management functions that cannot be fully assessed with data generated for test purposes.
- Fading, noise, or interference can be added to verify the UE's data retransmission processes, which include UE-HARQ operation and requesting or generating retransmissions, as well as testing the CQI reporting function.
- Different transmission modes can be used to test spatial multiplexing performance to test the precoding matrix indicator (PMI) and rank indication (RI) reporting aspects.

Once a physical layer baseline throughput has been established using static conditions, the UE should be stressed under adverse conditions. A highly variable channel for testing can be implemented by adding noise, interferers, or a fading profile, and it may be useful to verify the UE's response to known conditions, which are more easily replicated. For example, by deliberately sending NACKs on the physical hybrid indicator channel (PHICH) even when the uplink data has been received correctly, a designer can assess the UE's capability to provide retransmissions in a timely manner. Similarly, if some known portion of the downlink transmission is deliberately corrupted, it is possible to test whether the UE correctly requests retransmission of the corrupted data blocks. Such a test can also be useful for repeatable testing of PMI and RI with a spatially multiplexed channel.

To assess how a UE will perform in the real world, data must be transmitted all the way up through the protocol stack, stressing all components and layers at the same time. This is commonly achieved on real networks; however, an eNB emulator can do such testing on the bench, which is much easier, more convenient, and more repeatable. Using TCP as the transport layer protocol allows data to be sent all the way up the protocol stack to the application layer and thus all layers, as well as the interactions between each layer, can be tested.

Note that for all these UE tests it is usual only for the downlink side of the channel to be tested. While the uplink performance in terms of ACK/NACK reception by the eNB will clearly be relevant to the downlink throughput, this behavior is associated with the receiver performance of the eNB and is covered by separate eNB tests (see Section 6.6). In a typical test configuration, fading is applied only on the downlink, and the uplink conditions are set so that the eNB emulator is able to reliably recover the ACK and NACK information. This configuration separates the eNB emulator receiver performance from the goal of the testing, which is the UE receiver.

6.12.4 Handover Testing

LTE is attracting a lot of interest in the industry with many network operators making significant investments and rolling out networks. However, for most geographies it will take a sustained period of investment before the LTE coverage can reach a level that is close to what operators can provide with their legacy networks. It is therefore necessary for operators to support strong interworking between LTE and legacy networks to ensure that their customers with the latest LTE devices have a good user experience when they roam around the network. For this reason handover testing is very important.

Table 6.12-1. Types of handover testing

Testing type	Purpose	Reference	Solution
Signaling protocol conformance test	Test for signaling protocol behavior by sending known sequences of commands	3GPP 36.523	TTCN scripts driving multi-cell system simulator with limited RF flexibility
Radio resource management (RRM)	Test for correction functional behavior and performance by adjusting radio conditions and measuring delays in state transitions	3GPP 36.521-3	Multi-cell system simulator with greater RF flexibility
Application layer	Test functional behavior and E2E application performance	Vendor or operator proprietary test plans	Similar to RRM but testing E2E performance with application servers

Handover testing takes on different forms throughout the design lifecycle. Table 6.12-1 shows the three main types of handover testing. Signaling protocol conformance test and RRM testing concern themselves primarily with the radio and lower layers of the protocol stack and not with the functional behavior of E2E services. The signaling conformance test regime for LTE includes a series of tests for handover. Signaling conformance testing uses a scripted protocol messaging approach to guide the UE through a very specific and repeatable sequence of predetermined signaling protocol messages. The UE is required to respond in strict accordance with the specifications and reach a predetermined end state. This testing is done during the device design phase and as part of conformance testing.

RRM testing is also part of conformance testing. For RRM the system simulator also measures time transitions between states in order to ensure the UE is performing handovers within the necessary time limits.

Application layer handover testing is not part of conformance testing since it uses servers to test the E2E handover behavior of the UE and get closer to the real user experience of an application during the handover process. A real service (for example, video streaming or FTP download) is used and the test checks that IP data continuity is maintained and the service continues to function after the handover event. Application layer testing is carried out first by the UE design team and then by network operators, often as part of acceptance testing. Figure 6.12-3 shows a typical configuration of emulation-based application layer handover testing. This configuration is similar to Figure 6.12-2 but now includes two eNB emulators in order to support handover scenarios.

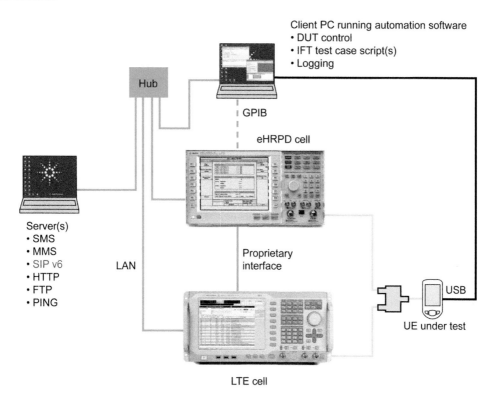

Figure 6.12-3. Test configuration for handover testing

With the ongoing development and investment in new physical layer technologies, the radio and signaling protocol complexity associated with handovers continues to grow. Figure 6.12-4 is based on 36.331 [29] Figure 4.2.1-1 and shows the E-UTRA states and inter RAT mobility procedures for 3GPP technologies (GSM, UMTS, and LTE). Figure 6.12-4 further shows how at any given time an individual UE will either be in an "idle" or "connected" state. If the UE is in idle mode then it is not providing any services to the user, although it is contactable by the network. In this state the UE takes responsibility for cell change events, which are termed cell reselections. If the UE is in "connected" state then it will have an active session (which could be a circuit-switched voice call or a packet-switched data connection). In connected state the network tells the UE to change cells by issuing instructions to perform a handover assisted by radio environment measurement reports from the UE.

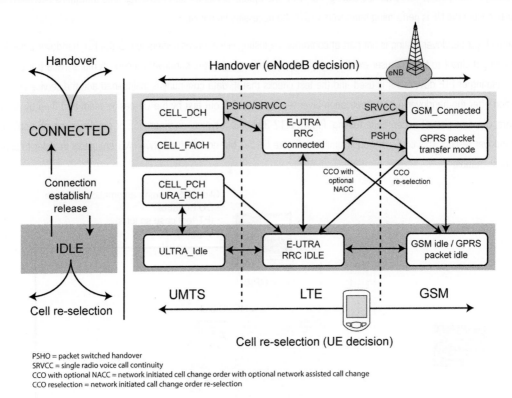

PSHO = packet switched handover
SRVCC = single radio voice call continuity
CCO with optional NACC = network initiated cell change order with optional network assisted call change
CCO reselection = network initiated call change order re-selection

Figure 6.12-4. State transitions for 3GPP inter RAT handover (based on 36.331 [29] Figure 4.2.1-1)

Figure 6.12-5 is a simplified representation of Figure 6.12-4 highlighting the cell reselection process (labeled 1) and two different handover mechanisms. The simpler of these handover mechanisms is called an RRC_release handover (labeled 2) and is a network-initiated cell reselection in which the UE is advised to drop from connected mode to idle mode and then is directed to join a target eNB. This handover mechanism minimizes the signaling needed while the UE is connected by requiring the UE to first release the connection and go through the attach process from idle in order to join the target cell. The second mechanism is the E-UTRA connected mode handover (labeled 3), which is more complex and requires additional network coordination prior to the handover event. The serving eNB pre-negotiates registration and resources from the target cell and then communicates directly with the UE so that it can move directly to the target cell and be given the resources that it needs. While this handover as more complexity on the network side, it is able to provide a faster cell transition, which is important for some real-time services such as voice. The E2E functional test system allows applications to be run concurrently with the handover tests in order to check that IP connectivity is maintained and observe the impact of different handovers on a particular application.

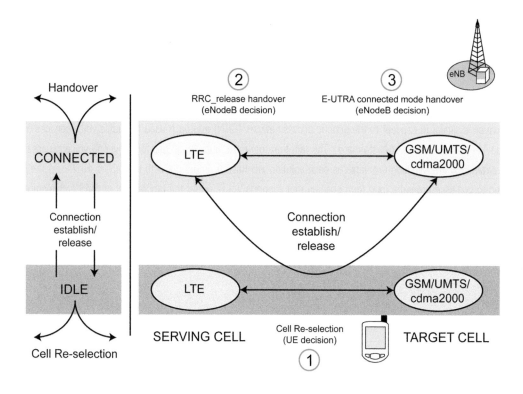

Figure 6.12-5. Simplified state transition diagram

6.12.5 Testing of Voice Services

This section looks in more detail at how voice services will be provided in networks employing LTE technology. Today's mobile devices offer a rich variety of voice and data services including messaging, location services, and applications that access the internet for their data. Having the ability to test these services within an automated environment is a useful tool for engineers looking to perform functional testing on a complete device. In many cases it is possible to provide this testing using the test configuration described in Figure 6.12-2 along with a readily available IP-based server. In some cases it is also necessary to provide a server that is not IP based; for example, a server for SMS/MMS testing.

While a wide range of services are now available, the importance of voice remains paramount. For GSM and W-CDMA, the air interface was developed to support voice telephony using a circuit-switched connection, with packet-switched connections used solely for data services. The evolved packet core (EPC) network that supports LTE has no support for circuit-switched services and therefore forces voice into the packet-switched domain.

The long term plan for voice over LTE (VoLTE) is to implement the IP multimedia system (IMS) in the EPC. IMS enables support for multiple services on an all IP core network and provides the essential interworking with legacy circuit-switched services. This includes a voice telephony service using session initiation protocol (SIP) to establish peer-to-peer links between UEs. Prior to the full availability of voice using an IMS architecture, network operators need interim solutions for voice support alongside their LTE networks.

There are two primary choices for how vendors can add voice using their existing infrastructure; namely, circuit-switched fall back (CSFB) to legacy networks and use of a dual-radio approach to offer voice through legacy networks. Other voice solutions are also possible, including "over the top" solutions such as Skype that use a packet data connection. Although expedient, such solutions do not provide a seamless user experience since they offer completely different voice client, numbering, and billing models to the legacy circuit-switched services. Another approach that has been proposed is voice over LTE generic access (VoLGA), which is similar in concept to the generic access network (GAN) approach used to enable voice services delivered over the packet-based wireless LAN standard. The fact that there are multiple options for delivering voice services for LTE devices creates new interworking problems as each solution will have to interwork with the others. However, the industry is beginning to coalesce around a long term position that supports VoLTE under the umbrella of an IMS architecture. In the shorter term, voice delivery options for network operators are largely determined by any existing legacy infrastructure.

Figure 6.12-6 shows the basic sequence of events for CSFB. A more in-depth explanation of the CSFB handover process can be found in 23.272 [42], "Circuit Switched Fallback in Evolved Packet System."

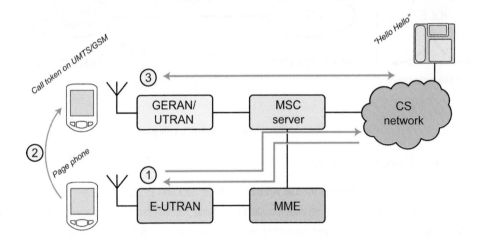

①   UE service request or network page (sent over LTE)

②   Handover or cell re-selection to GSM or UMTS

③   CS domain service established over GSM or UMTS

Figure 6.12-6. Overview of CSFB procedure

When the network needs to page a UE to initiate a voice call, it signals the UE to hand over the call to a legacy network capable of handling circuit-switched services. With LTE being a 3GPP technology, CSFB from LTE to GSM or UMTS is relatively straightforward, but for CSFB to cdma2000 it is necessary to provide pre-registration with the cdma2000 network using tunneling over the S101 interface, which allows messages to be sent via the LTE air interface before being "tunneled" over to the cdma2000 network. The need for interworking between two separate core networks adds more signaling protocol complexity, although this has not deterred implementation in the industry.

With CSFB implemented over UMTS, simultaneous voice and data can still be supported on a single radio by establishing bearers on both the CS domain and PS domain. This is also possible in GSM networks for devices that support dual transfer mode; however, support for this feature is not universal.

In order to reduce the complexity and time delays involved with CSFB, the alternative dual radio approach can be used. With this technique, shown in Figure 6.12-7, the UE contains two separate radios with no interworking requirements. This example shows how a device could use a cdma2000 connection for voice and also have a simultaneous LTE connection for data. The added complexity of managing different air interfaces for voice and data is something that has been solved previously in cdma2000 networks since the radio bearer that supports data services, 1xEV-DO, does not support voice. The downside of this approach is that the UE now has two complete radios, which adds more circuitry, cost, and power consumption to the UE.

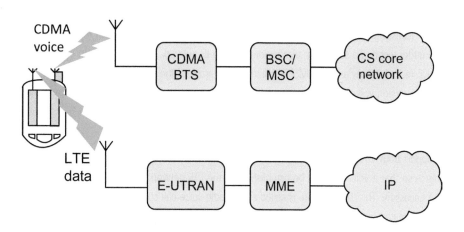

Figure 6.12-7. Dual radio voice solution

Initially LTE smartphones will use technologies such as CSFB and the dual radio approach to support voice. The ability to perform functional testing of these voice technologies is clearly an important part of testing the expected user experience with early-to- market LTE UE. However, CSFB and the dual radio approach are interim solutions and IMS is considered by most operators to be the long term technology for providing VoLTE.

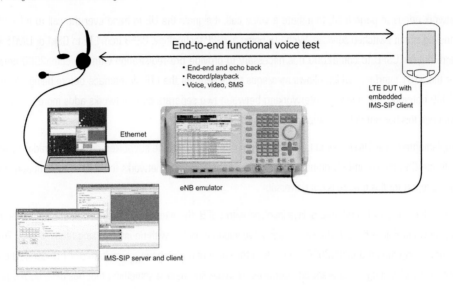

Figure 6.12-8. Functional test system for VoLTE

Figure 6.12-8 shows a functional test configuration that can be used for VoLTE testing. In this solution the LTE eNB emulator supports a packet-switched data connection with the UE and backhaul connected to an IP network. The UE has an embedded IMS-SIP client that is accessing the IMS-SIP server on the IP network. The IMS-SIP server in turn is supporting a voice call between the device under test and a soft client, also resident on the IP network. This configuration can be used for functional testing of both the UE radio features and the IMS-SIP client. For example if the radio has been developed to support VoLTE, then it is likely that the design will include features such as semi-persistent scheduling (SPS), robust header compression (RoHC), and transmission time interval (TTI) bundling. These features are all designed to support VoLTE. For the IMS client itself different codecs can be tested as well as the signaling compression protocol (SIGCOMP).

When operators support both VoLTE and IMS, they will have the ability to either support LTE islands with IMS or provide handover mechanisms so that a voice call can be established on the LTE network and then handed over to a legacy circuit-switched network if necessary. This latter feature is called single radio voice call continuity (SRVCC) and is illustrated in Figure 6.12-9. This handover is quite complicated and requires careful testing.

In addition to the functional testing described, parametric voice quality testing can be performed using a setup such as the one shown in Figure 6.12-10. For this measurement a head and torso simulator is used to simulate a human body. An audio signal is played out of a speaker inside the torso of the simulator. This audio signal is recorded by a microphone, placed close to the mouth, as well as being sent over the LTE link using the device under test. An audio quality analyzer then makes voice quality measurements based on a comparison of the audio waveform captured by the microphone and the signal sent over the LTE link. A similar measurement can also be made for the voice traffic sent over the downlink. Historically, voice quality testing was done statistically by analyzing the perceived voice quality as experienced by a number of users. This testing can now be done using algorithms that mathematically derive an approximation to the perceived voice quality. Perceptual evaluation of speech quality (PESQ) is such an algorithm that was developed to model subjective tests and is now a worldwide industry standard used for speech quality testing.

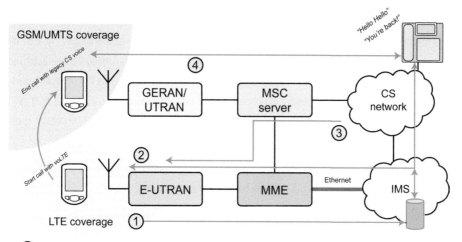

① UE registers with IMS server

② UE establishes voice call using IMS and VoLTE

③ Network instructs UE to do a connected mode handover between LTE and GSM or UMTS

④ CS domain service established over GSM or UMTS and the IMS call is torn down

Figure 6.12-9. Functional test system for SRVCC

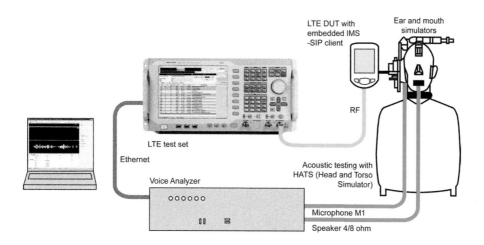

Figure 6.12-10. Parametric voice quality test system

6.12.6 Functional Test Automation Environment

This section of the book has described a variety of different types of functional test (for example battery life, E2E, and handover) that are used by both validation engineers for regression testing and operators for benchmarking the characteristics of phones. In many cases the testing is quite complex, incorporating multiple functional blocks, and hence it is helpful if testing takes place within an automated environment. Some test engineers choose to add test equipment and servers into their own automation environment, while others prefer to use vendor-supported tools that are capable of complex functional testing within an automated environment.

The Agilent N5970A/71A Interactive Functional Test (IFT) System, shown in Figure 6.12-11, is one example of such a solution. It provides a highly productive way for users to run complex functional test scenarios.

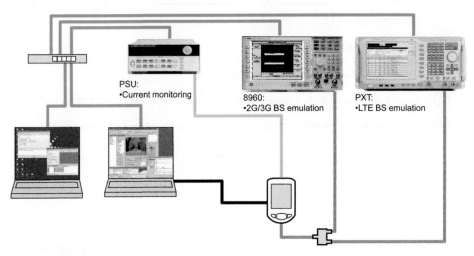

Figure 6.12-11. Agilent N5971A Interactive Functional Test System

The IFT System includes a GUI shown in Figure 6.12-12 for manual test control, a test executive for automation, drag-and-drop scripting, and a Visual Basic environment for more complex test development. UE control and automation is implemented using modem attention (AT) commands together with other proprietary and customizable interfaces. The system includes FTP, HTTP, UDP, SMS, IMS, VoIP, video, and many other servers to enable complex functional test scenarios to be quickly established.

Power supplies integrated into the system enable real-time current monitoring for battery drain analysis. Multiple test sets can be controlled for handover and InterRAT handover testing. Test suites are available that implement selected network operator test plans.

Users can guide test scenarios manually using the IFT GUI to initiate sequences of parallel events, or automatically under Visual Basic control. Example tests can include FTP transfer rates, checking e-mail, SMS, IMS, web surfing, handovers, and voice and video calling. Tests can be designed to stress the UE to measure stability or can be used to assess UE battery drain and data rate performance with new UE software releases.

LTE messages can be customized using Agilent LTE Message Editor software. The IFT system also operates with Agilent LTE Protocol Logging and Analysis software to allow logs to be saved for later analysis. Logs can be automatically inspected by the IFT system so that the detailed message contents can be used as part of test pass/fail criteria. The IFT system also includes an interface to Microsoft Excel ® that allows the generation of user-defined test reports.

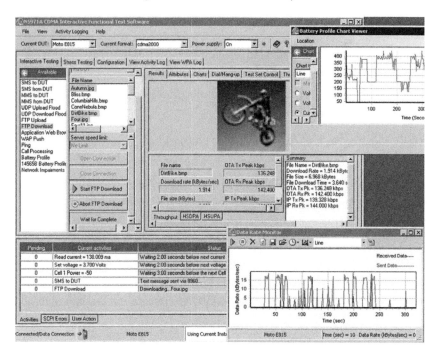

Figure 6.12-12. Interactive Functional Test system graphical user interface

6.13 Battery Drain Testing

6.13.1 Introduction

The focus of this chapter thus far has been the design and measurement challenges associated with the radio and network aspects of LTE/SAE. However, one topic that is crucial to bringing successful products to market is the humble battery. Indeed, it was primarily concern with battery performance that led 3GPP to define SC-FDMA for the LTE uplink rather than the more power-hungry OFDMA used for the downlink.

Although advances in battery capacity continue to be made, they are being outpaced by the demands of modern UE. It is not just the primary radio that requires power; power is required by multiband multi-RAT support, receive diversity, MIMO, interference cancellation, ever-higher data rates, Wi-Fi, *Bluetooth*®, FM Radio, MP3, MP4, GPS, larger brighter displays and, in the not so distant future, integrated video projection. Installing a larger battery is usually not an option; consequently, an increasing amount of R&D effort has to be directed towards designing, measuring, optimizing, and verifying UE current consumption in an ever wider set of use cases.

The need to measure current drain exists throughout the design lifecycle.

- During product development: evaluating and analyzing current drain to identify anomalies and the root causes leading to changes in design to optimize run time.
- During design validation: checking performance against benchmarks for varying operating conditions such as standby time, talk time, web browsing, performance in low battery conditions, extreme environmental conditions, etc.
- During software development: evaluating and validating software changes and the impact on current drain though a suite of established regression tests.
- During product acceptance: validating performance against industry standards and in operator-specific acceptance tests.

The task of current drain analysis can be made substantially easier through the use of advanced tools. For example, the Agilent 66319D/66321D and N6781A are DC source/measurement units were designed specifically for wireless device current drain testing. The DC sources can be used as battery emulators or in a special zero voltage configuration to measure the performance of the mobile device battery, commonly called battery run down testing. The 66319D/66321D DC sources are used in conjunction with the Agilent 14565B battery drain analysis software, enabling the designer to carry out advanced current drain analysis either manually or with full automation at all stages of the product design lifecycle. The N6781A is used in the N6705B DC Power Analyzer mainframe with advanced current drain analysis, which can be further enhanced using the optional 14585A software. Three basic measurement modes are supported:

- Waveform mode provides an oscilloscope-like capability for capturing and analyzing current drain signals from tenths of milliseconds to seconds in duration.
- Data logging mode provides extended current drain measurement and analysis for up to 1,000 hours of testing.
- Complementary cumulative distribution function (CCDF) mode provides statistical profiling to display current drain performance for up to 1,000 hours operation.

The following sections give examples of the measurement process and examples of waveform and CCDF current drain measurements and how they are used to carry out essential battery performance measurements.

6.13.2 Measurement Process

A typical battery run down test and measurement setup is shown in Figure 6.13-1.

The DC source can be used just as a battery emulator, but in the configuration of Figure 6.13-1 it is connected in series with the real device under test (DUT) battery in what is known as a zero-burden active shunt mode. In this mode the DC source acts as a logging ammeter utilizing the built-in basic digital volt meter (DVM) input to measure the voltage. Figure 6.13-2 shows a conceptual schematic of the DC source operating as a zero-burden shunt.

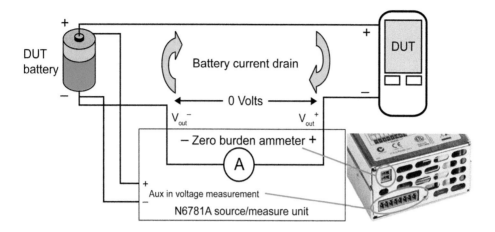

Figure 6.13-1. Typical DUT battery drain measurement setup

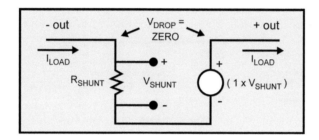

Figure 6.13-2. Schematic for zero-burden active shunt concept

A zero-burden shunt in combination with an active opposing voltage source cancels the voltage drop across the shunt resistor. The benefit of doing this is that the DUT sees the full voltage from the battery or source powering the DUT. In addition, since the voltage drop across the shunt resistor is no longer an issue, a larger value of shunt resistor can be used to improve the dynamic range and resolution of measurements. The current and voltage of the battery can then be measured and logged for periods of up to 1,000 hours. Underlying sampling is 64 KSa/s for the 66319D/66321D and 195 KSa/s for the N6781A.

6.13.3 Waveform Measurements

Figure 6.13-3 shows a typical voltage and current waveform profile created as part of a test to analyze the run down performance of the battery.

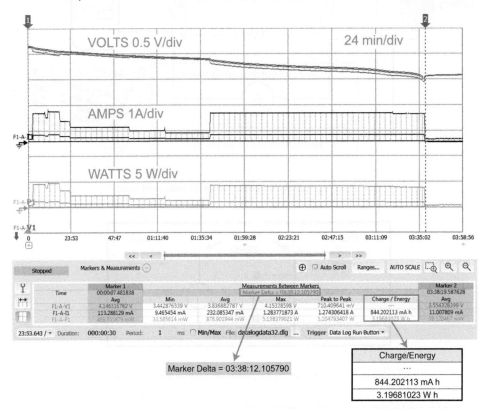

Figure 6.13-3. Run down performance of battery showing peak, average, and minimum profiles of voltage, current, and power

The test was carried out over a period of 4 hours, during which time the DUT was executing a dynamic sequence: changing active output power levels, then switching through other operating modes (standby, off, active) as part of a benchmark profile. Numerical values displayed include minimum, maximum, and average current, voltage, and power. Cumulative values include run time, amp-hours, and watt-hours actually delivered to the DUT. It can be seen from the figure that the device switched off after approximately 3 hours and 40 minutes. This type of measurement is essential to ensure correct operation of the circuits and software responsible for shutting down the UE at the appropriate time, before the radio performance begins to degrade. In the trace, vertical markers have been placed at the start point and at the end point where the phone shut off, indicated by the current dropping to zero and the battery voltage recovering due to removal of the load. The numeric values are recalculated based on the markers and show that the UE ran for 3 hours and 38 minutes and consumed 844 mA-hours and 3.197 watt-hours.

This integrating feature of the software enables the large quantity of results to be reduced to a meaningful set of critical figures that enable a fuller understanding of the battery and DUT performance. Zoom and marker controls allow closer analysis of subsections of the data log.

In addition to all the peripheral features on modern phones that have relatively simple current demands from the battery, the underlying primary radio function has a complex relationship with current drain. The talk time (or data connect time) as well as the standby time of the radio is a function of the radio conditions and network parameters. To control all the variables that have an impact on current drain, a base station emulator such as the Agilent E5515C (8960) or E6620A (PXT) is required. These emulators can provide a controlled radio and network environment in which the designer can experiment with the many parameters that impact performance. Testing in such environments can be challenging and so the Agilent N5970A Interactive Functional Test Software has been created to automate current drain testing using the base station emulator. See application note 5989-9153EN [43] for more details.

The standby time of the same UE in different networks can be vastly different, and it is therefore essential that the designer understand the underlying capabilities of the product independent of any specific network implementation. To set guidelines for the process of battery drain performance testing, industry standards have been developed such as the "Battery Life Measurement Technique" produced by the European Conference on Technology-Enhanced Learning (ECTEL) and the GSM Association [44], which defines the essential parameters that affect current drain so that products can be fairly compared.

The waveform example in Figure 6.13-3 demonstrates what can be done over long periods of time, but the sampler used in the 66319D/66321D and N6781A runs at 64 KSa/s and 195 KSa/s respectively and is capable of measuring fast current profiles over very short durations. The RF designer can use this level of resolution to monitor the profile of current drain over periods shorter than a single SC-FDMA symbol. An example in which this technique may be useful is in the profiling of the uplink sounding reference signal (SRS), which is a pulse lasting just over 70 μs used to train the eNB receiver. This signal needs to have an accurate power profile, be flat in the frequency domain, and have low error vector magnitude (EVM). Problems with supplying sufficient current from the off state to full power in perhaps 20 μs could contribute to signal impairments..

6.13.4 Statistical Measurements

Figure 6.13-4 shows an example of how the underlying data logged by the DC source can be displayed in the form of a CCDF profile. This is a powerful technique that enables the designer to quickly and easily see the probability of a particular current drain and then link to particular activity factors such as talk time, standby, etc. The task otherwise requires working directly from time-based log plots and is time-consuming and difficult.

The figure shows three different user profiles: idle time, talk time, and typical PC data use. The CCDF is using a double log scale with probability of occurrence on the vertical scale and current on the horizontal scale. Any point on the graph represents the probability that the measured current over the duration of the measurement period will exceed the value shown.

The idle time trace shows the lowest current drain. The near flat line at 100% extending out to the inflection at 45 mA and 20% indicates that for 80% of the time the UE is drawing almost no current. This figure should align with the design goals of the UE and will reflect the operating parameters in the network. From the inflection at 45 mA and 20% there is a near constant decline to 200 mA at 3% before the curve decreases rapidly to around 350 mA at 0%, which is the maximum current observed during idle mode.

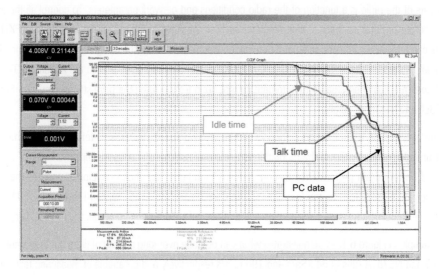

Figure 6.13-4. CCDF current profile for three different use cases

The talk time curve shows that for 50% of the time the current is below 3 mA. This is to be expected and indicates that the voice activity factor for speech is less than 50% transmit. The large increase out to 170 mA at 20% represents the bulk of the current drain. The current continues to rise out to 900 mA at 0.6 % before reaching a maximum of around 1.2 A. The reason for the peak current being over five times the average during the transmit periods is not obvious. It may be due to other background factors such as backlighting or to peaks in the audio currents driving the speaker. Correlating the peak current with specific activity on the UE would lead to an understanding of whether this is an expected or unexpected result.

The PC data curve is a hybrid of the two previous curves. The curve starts out like the idle time curve, showing little current drain for 80% of the time, as would be expected during the periods when no data is being transmitted or received. The first inflection occurs at 200 mA and 70%, with a slow decline to 300 mA at 60% and then a sharp decline to 400 mA at 1.5% indicating the effective peak current. The trace trails off to a maximum current of 600 mA. It is interesting that the peaks of data current are much lower than voice peaks. This may be due to the audio circuits not being active.

For further reading see application note 5989-6016EN, "Battery Drain Analysis Improves Mobile-Device Operating Time" [45] and application note 1427, "Evaluating Battery Run-down Performance" [46].

6.14 Drive Testing

6.14.1 Do We Need Drive Test in LTE?

In LTE as with other cellular technologies, drive testing is a part of the network deployment and management lifecycle from on the early stages of deployment. Despite the goals of minimization of drive test (MDT, see Section 8.3.4.2), drive testing provides an accurate real-world capture of the RF environment under a particular set of network and environmental conditions. The main benefit of drive testing is that it measures the actual network coverage and performance that would be experienced by a user on the route of the drive. It is argued that in modern networks with modern simulations, network engineers can mathematically model how a network will perform. While this is true to a certain extent, it is also essential to do drive testing as network parameter settings alter how the UEs interact and deal with the network environment. Such interactions cannot be wholly predicted through mathematical modeling. Figure 6.14-1 shows how drive test is used in the network planning lifecycle.

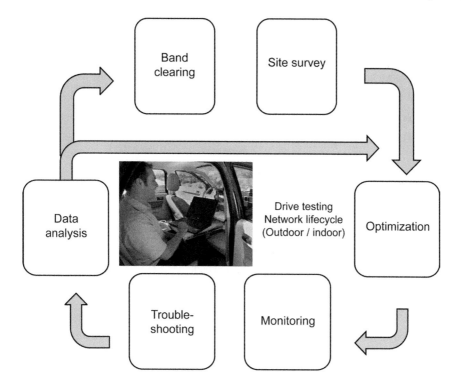

Figure 6.14-1. Network planning cycle

6.14.2 Components of a Drive Test System

Drive test systems are generally built around two measurement components: instrumented mobile phones (test engineering phones) and measurement receivers. Each has its own characteristics with associated benefits and drawbacks. Phone-based systems can respond to problems within network-controlled constraints. Receiver-based systems give a complete overview of RF activity but cannot duplicate network-related problems. These measurement devices are controlled with data logging software on a laptop PC together with a GPS receiver to provide geo-location of the collected data. The collected data can be analyzed using software that allows the results to be plotted on digital maps, which allows visualization of the RF environment.

The measurements carried out during drive testing have evolved over time, from the early days when the focus was purely on parametric RF measurements, to the wide variety of application-based data performance measurements integrated into modern drive test systems. Many of these changes in test methodology have been driven by the growing complexity in services that LTE intends to enable. Network operators have shifted their focus from purely measuring RF performance to measuring customer experience, and this has driven the integration of many data application tests—such as video streaming and VoIP—into drive test systems so that engineers can correlate end-user application performance with detailed RF measurements. With LTE, many of the measurements themselves have had to take into account the much higher data rates that LTE provides.

Another evolution is the move from single-band, single-RAT networks to multiband, multi-RAT networks. LTE is not going to exist in isolation but will be overlaid and integrated with the existing UMTS/HSPA and cdma2000 1xRTT/HRPD (1xEV-DO) networks. Drive test tools need to embrace this multi-RAT, multiband environment. An area of key interest to cellular operators will be the interaction of LTE with their existing infrastructure and in particular the crossover points where handovers take place.

6.14.3 Importance of Drive Test for LTE

Many different strategies and methods to monitor network performance have been defined by network equipment manufacturers and test equipment vendors. Network "probing" in which the signaling traffic is monitored at control points and then centrally analyzed can provide valuable insights into the overall network health. This strategy works well where, as in GSM and UMTS networks, much of the control traffic is consolidated through RNCs or base station controllers (BSCs) so that by monitoring relatively few major interfaces, a concise view of a variety of base station end-points can be obtained.

As the industry has moved ahead with HSPA and now LTE technologies, more of the network intelligence has moved out from the core to the edge of the network and into the eNBs, taking the traffic management also towards the network edge. This is shown in Figure 6.14-2. The consequence is that much of the control and decision making is now deployed within the eNBs, and interaction between the UE and base stations can no longer be observed using network monitoring tools but has to be monitored by instrumented phones involved in the actual transactions. This move of the decision point in traffic management is needed to realize the reduced latency requirements for LTE performance—signaling control traffic no longer has to traverse multiple network nodes when a change is made for a UE.

While the term "drive test" comes from the fact that measurement equipment is generally driven around in a vehicle, the importance of measuring indoor coverage is increasing. For indoor performance measurements, there is no "view of the sky" meaning GPS cannot be used for positional information.

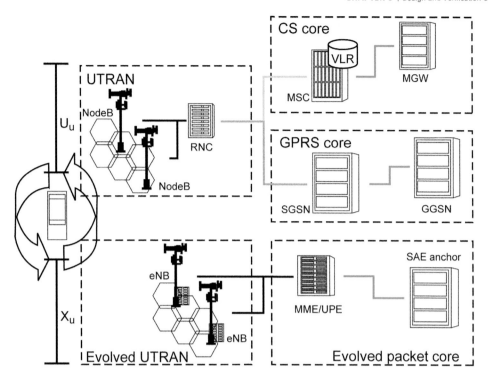

Figure 6.14-2. Control functions of the RNC have moved into the eNB

Instead, test engineers typically enter their measurement location on a scanned floor plan of a building using point-and-click navigation. Typical logging software enables true geo-location of the corners of the floor plan so that the indoor measurement results can be combined with outdoor results in a final overall analysis. The massive increase in processing power available on smartphones has allowed the creation of handheld drive test devices that can be used when the need for instrument portability takes priority over depth of functionality. Handheld devices lend themselves well to indoor measurement environments in which discretion during the data collection process may also be a factor. (Historically, data collection systems required bulky equipment to be carried on a backpack in the target location—for example, a shopping mall—which may draw the wrong kind of attention in today's security conscious environment. Newer instrumented smartphones have alleviated this issue.) The range of measurements on a handheld device is as extensive as that on a full drive test system and data may be analyzed and amalgamated in post-processing.

6.14.4 Phone-Based Drive Test for LTE

Phone-based drive test systems are useful for evaluating basic network performance and are essential for characterizing how the end user experiences the network. Phone-based systems address the need to verify network settings such as cell selection and reselection boundaries and to measure voice and data application performance in the live network. Most modern UE chipsets have engineering measurement capabilities built into them, which were used during the UE's design process. These same parameters are exploited in drive test software to provide new value to the RF engineers deploying the final network.

Table 6.14-1. Top 14 phone-based measurements for LTE drive testing

Measurement	Description	Use
Protocol log	RRC and NAS protocol log	Capture of cell broadcast and system information messages. Used to perform calculation of timing of events such as handover latency and cell re-selections. Verification of network management procedures and troubleshooting failures.
Cell-ID/E-ARFCN	Physical cell identity/E-ARFCN	Identifies the cell sector and base station in use. Confirmation of the specific carrier in use.
RSRP	Reference signal receive power	3GPP-defined measure of the downlink power received by the UE
RSRQ	Reference signal receive quality	3GPP-defined measure of the downlink signal quality received by the UE
SINR/CINR/SNR	Signal quality measures	Supplemental to RSRQ. Mobile-manufacturer-specific implementations (not defined by 3GPP)
DL-SCH/UL-SCH throughput	DL/UL shared channel throughput	Measure of throughput that the mobile is achieving on shared DL or UL channels
DL-SCH/UL-SCH BLER	DL/UL shared channel BLER	Measure of error ratio that the mobile is experiencing on the DL or UL shared channels
#DL RB/UL RB	Number of DL/UL resource blocks allocated	Measure of how much of the shared channels are being allocated to the mobile by the eNB scheduler
#DL MCS/UL MCS	Index number of DL/UL modulation and coding scheme allocated	Indication of the capacity of the transport channel under current radio conditions. Incorporates the modulation depth, coding rate and transport block size.
Wideband/narrowband CQI	Wideband/narrowband channel quality indicator	3GPP-defined UE measure of the channel quality across defined resource blocks. Used for link adaption.
Rank	Number of transmission layers in use	Determines when spatial multiplexing is preferred or active
UE TX Power	Mobile transmit power	UE report of current uplink power. Used for link adaption.
Application throughput	Throughput performance of end user application	Measure of the end user experience. Both UDP and TCP measures are made.
Ping	Round trip time to reach a specific IP address	Measure of network latency

With radio resource management taking place in the eNB, suitably instrumented UEs can be used to monitor the performance of the physical layer including modulation schemes, access procedures, synchronization, and power control. Some of the most common measurements are shown in Table 6.14-1.

The same types of parameters are measured for LTE as for other cellular technologies. Beyond the essential protocol log, which provides visibility of the fundamental interaction of the UE with the network, the initial focus of the testing campaign is on RF coverage and quality. In LTE these equate to reference signal receive power (RSRP) and reference signal receive quality (RSRQ), which are measures of the reference signal's strength and quality.

The results of these measurements become the major components of network-based decisions about whether to keep a UE on its current cell or hand it over to an adjacent cell. Additional measurements are used to assess the link quality and include channel quality indicator (CQI) and block error ratio (BLER).

Although RSRQ is a 3GPP-defined measure of signal quality that all UEs must implement, this metric is slightly influenced by network loading since it is measured on all RS subcarriers, including symbols that may carry cell traffic. Therefore many LTE UEs also implement custom carrier-to-noise ratio measurements, which are independent of loading and can be used to further assess the channel quality. These additional carrier-to-noise ratio measurements are not reported back to the network, but they are available within the drive test logs and can be used by the RF engineering teams to get extra insight as to how the mobiles are perceiving the RF environment.

Instrumented UEs can also report the measured channel state information (CQI, PMI, RI) and hybrid ARQ (HARQ) statistics. The number of resource blocks assigned to a device at a particular time, together with the modulation and coding scheme applied, can be used to evaluate the eNB scheduler performance. These types of tests are of particular interest at the early stages of deployment of a new network but also need to be monitored as the network loading increases and true end-user traffic patterns are established.

One aspect of the LTE network of great interest to RF optimization engineers are the effects of MIMO with spatial multiplexing and antenna diversity on the end-user experience. Drive-test-enabled devices can log the current rank and number of transmit and receive paths in active use, together with the reported availability of antennas. They can also report the signal strength and quality of each of the UE's antennas individually. This information can be correlated with the measured data application performance to establish the effects of MIMO on network performance. The full implementation of MIMO requires real time feedback between the UE and the network, so an instrumented UE that is part of the real time connection is the only way to evaluate completely the impact of MIMO on the system.

As LTE networks are being deployed alongside existing cellular networks, the efficient use of each network resource and the transition points between the network technologies are of particular interest to cellular operators. Drive testing is used extensively to monitor the handover points between LTE and the legacy technologies. The signal strength, quality, cell-ID, and neighbor information both before and after a handover are analyzed and optimized. The length of time it takes to complete an initiated handover, the success rates, and the end-user data-interruption time (during the actual transition between technologies) are all key performance indicators and are closely monitored.

End-user data-throughput performance and latency are the two key measures of a network's optimization. If the network is not achieving the expected data performance, it is important to be able to analyze the signaling performance and settings at each signaling layer including the RRC, RLC, and MAC. Monitoring the resources allocated to a UE together with the measured network conditions, available neighbor cells, and power levels will allow troubleshooting and optimization of network settings.

As LTE evolves, it is not only data services that are of interest. Voice is still the key service that many end users rely on. For LTE, in the initial stages of deployment, circuit-switched fallback (CSFB) is the initial voice solution being deployed ahead of full IP voice over LTE (VoLTE). See Section 6.12.5. Phone-based drive test systems are being used to ensure that these voice services operate as designed. The length of time required to set up voice calls is monitored through the recorded protocol signaling.

Also monitored are the effects on user data activity that may be in process at the time a voice call is made and the behavior of the data service afterwards.

As network technologies advance and move towards IP service delivery mechanisms, drive test systems are being increasingly used in combination with other test instruments. For signaling analysis, this includes protocol testers, which monitor the signaling between the core and the eNB. The drive test system provides the last link in the chain to the end user, so combining the measurements made with the drive test system and core network protocol analyzers can give a full end-to-end picture of network performance. This is illustrated in Figure 6.14-3.

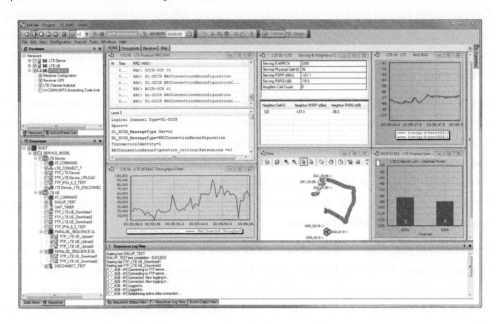

Figure 6.14-3. Example of a drive test system capture showing a variety of logged parameters including location information

6.14.5 Receiver-Based Drive Test for LTE

Receiver-based systems are used to obtain a "raw" view of the RF environment. They can measure the entire spectrum and are not constrained by network operator settings. These systems are useful for activities such as band clearing and general coverage estimation but they cannot give a true measure of end user experience as they do not physically interact with the network under test. With new network technologies, receivers have the benefit that they are generally available before fully fledged mobile devices, which means that propagation models can be validated at a very early stage in the network life cycle.

In many regions LTE is being deployed in re-farmed spectrum; e.g., the "digital dividend" analog television spectrum around 800 MHz in Europe. This has led to a renewed interest in spectrum clearing activities to ensure that there are no rogue interfering signals left from other technologies prior to deploying LTE. Drive test is used extensively in this activity where the entire band can be monitored using a receiver which can be set to trigger if a signal is detected above a user-defined threshold.

Figure 6.14-4 shows an example of a spectrogram or "waterfall" display of the time-varying nature of RF signals. Frequency is shown on the X axis with time on the Y axis. The intensity of the RF signal is indicated by the color. This form of signal analysis provides a powerful tool to simultaneously analyze signal activity across a wide range of frequencies.

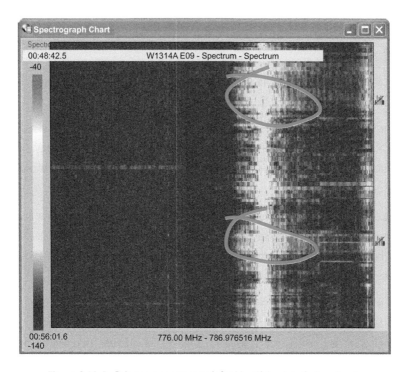

Figure 6.14-4. Drive test spectrograph for identifying interfering signals

Modern drive test receivers include specialist measurements targeted at specific RATs. To be useful, it is necessary to identify which signals are contributing positively to the mobile environment and which are negative influences. For LTE, identification of the reference signals which the UEs use to lock onto the network are key measurements. The RSRP, RSRQ, and physical cell identity (PCI) are the key measurements required. Some receivers also report custom carrier-to-interference ratio measurements, similar to what is implemented in some UEs described above. The ability to extract the LTE PCI allows the next level of identification of the cell in the same way that the scrambling code would be used to identify a UMTS cell, or how a base station identity code (BSIC) would identify a GSM cell. Receivers also have a role to play in assessing MIMO deployment in the network. A receiver must be able to report the RF characteristics received from each of the transmission antennae of the eNB.

As has been mentioned, LTE is being deployed in conjunction with existing RATs and in multiple bands both indoors and outdoors. For this reason, it is essential that a receiver used for drive test is capable of measuring LTE and any legacy RATs that are co-deployed. Indoor drive test receivers are typically battery operated and this requires low power consumption and careful hardware cooling design ensure that they can be utilized in the indoor environment where battery power is the norm and equipment is usually enclosed in a backpack.

The fact that a receiver is independent of the network has the advantage that it can be used to monitor both the network of interest and also other, perhaps competing networks, without any direct interaction with either. For this reason, receivers are often employed in RF coverage benchmarking to plot the coverage of one network/RAT combination versus another.

6.14.6 Benefits of Combining Phone- and Receiver-Based Test

While both phone-based and receiver-based systems have their place, the real advantages come when the measurements from both systems are combined to allow troubleshooting that cannot be done or is difficult to do when only one type of system is used.

In all RATs, identifying missing neighbor cells is a classic problem that can be addressed by the use of a combined phone- and receiver-based system. The phone reports and measures how the network has set the neighbor list while the receiver reports the actual impartial neighbor list. Combining the results of these two independent measurements allows optimization of the network settings for the current RF environment.

In LTE, which integrates a new RAT with existing RATs, optimizing traffic loading and handover points is another area where receivers can be used to provide insight into the complete RF environment. A multiband, multi-RAT receiver is able to simultaneously monitor both the LTE network and the legacy network. With the growth in network deployment, it is necessary to simultaneously monitor multiple carrier frequencies.

Identifying interference is also facilitated by the use of a combined receiver- and phone-based test solution. The phone part of the system is able to establish network connection and report application performance, and the receiver part of the system is able to monitor any external RF sources that may be adversely affecting the radio link. Thus network-dependent application performance and independent RF measurements are combined to provide information to solve end-user application problems.

6.14.7 Verifying LTE Application Performance Using Drive Test

Delivering user data rates of 100 Mbps and above is one of the key challenges for LTE. Therefore, in addition to RF measurements, data application testing is a critical activity during network deployment. LTE moves wireless communications to an all-IP network. Bridging the gap between RF performance and end-user IP services such as VoIP, video telephony, and video streaming is a challenge for network operators who need to ensure that these new offerings can be added to their infrastructure without affecting the quality of existing services.

Drive test solutions need to include a broad portfolio of instrumented data test applications for services such as video streaming, video telephony, HTTP, FTP, e-mail, SMS, MMS, and WAP. These test applications are essential to allow network engineering departments to measure application performance in conjunction with the RF environment. Many such test applications have been used in deploying previous technologies but changes have been made to accommodate the higher data rates associated with LTE. An example is the simple FTP test, which has long been a favorite of RF engineering departments to measure the end-user TCP throughput. However, as a result of LTE's massive increase in available data bandwidth, a single FTP connection is no longer sufficient to fully exercise the channel capacity. For this reason, multi-segment FTP testing has been introduced, wherein multiple simultaneous socket connections are established between the drive test UE and the network based FTP server.

As shown in Figure 6.14-5, when the number of simultaneous connections increases, so does the network-delivered throughput. If just a single FTP session were tested, the result might appear to show that the network is delivering only a fraction of what it is capable of.

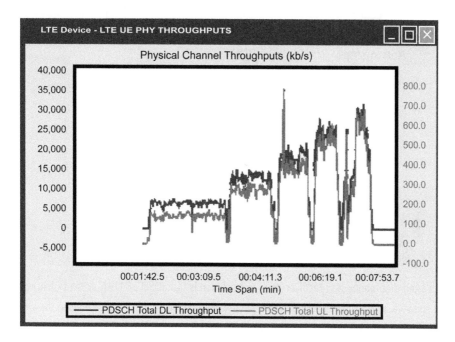

Figure 6.14-5. Multi-segment FTP test showing increasing throughput as segments increase

Note that download time duration decreases and throughput increases with more FTP segments.

A new test introduced to the cellular testing environment with LTE is the iPERF test. iPERF has traditionally been used in wireline testing and is a client/server application that is capable of measuring network bandwidth and throughput. The main attraction of iPERF for LTE deployments is that it can measure both UDP as well as TCP throughput and the user can specify the test bandwidth. If a service is not performing as expected, the RF performance and network configuration information is available alongside application performance data. GPS geo-location data provides precise geographic location to allow problem identification and troubleshooting.

Figure 6.14-6 shows a typical network optimization and troubleshooting cycle involving analysis of data.

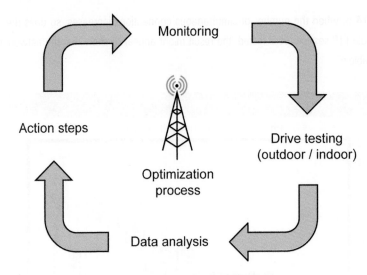

Figure 6.14-6. RF network optimization and troubleshooting cycle

6.14.8 Drive Test Solutions for LTE

An example of an integrated phone and receiver drive test solution is the JDSU E6474A Wireless Network Optimization Drive Test system software, which creates links between the test engineering phones and receivers in the system so that the receivers can dynamically track and measure the channels being used by the UE. The E6474A system address the challenges of optimizing LTE network performance by quickly and accurately identifying problems. The E6474A system shown in Figure 6.14-7 can be used in spectrum clearing, site evaluation, cluster and initial optimization, system acceptance, and ongoing optimization and troubleshooting.

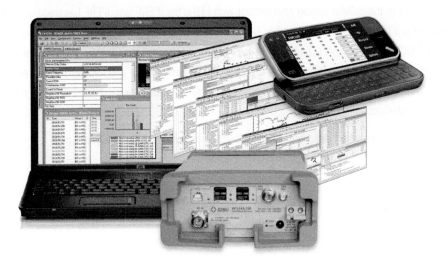

Figure 6.14-7. JDSU E6474A drive test system for network planning, deployment, and maintenance

The drive test system can be scaled from a single phone-based solution to a multi-phone, multi-RAT receiver and phone combined system covering all the major RATs that are deployed in wireless networks worldwide: LTE, HSPA, UMTS, GSM, GPRS, EDGE, 1xEV DO, cdma2000, TD-SCDMA, iDEN, and WiMAX. The system also encompasses a wide range of application tests including TCP/UDP, messaging, analog and VoIP testing, web-oriented and video streaming components, and the system is designed for both indoor and outdoor use.

With the flexibility to combine multiple test engineering phones and the JDSU multiband, multi-RAT receiver, this drive test system allows multiple technologies to be measured simultaneously with a single computer. The unique graphical data test sequencer facilitates construction of a range of end-user scenarios to simplify network optimization and determine quality of service from the end-user perspective.

To simplify configuration and reduce costs the JDSU W1314A measurement receiver used in the E6474A drive test system can measure up to eight frequency bands in a single drive, allowing multiband, multi-RAT coverage to be verified using a single piece of measurement hardware. Powerful DSP technology allows for software upgrades to keep the measurements current as the technology evolves.

When extreme portability and ease of use are most important, JDSU has smartphone-based drive test solutions that seamlessly integrate with the full laptop solution for in-depth analysis.

6.15 UE Manufacturing Test

This section looks at trends in wireless device manufacturing with an emphasis on the non-signaling test techniques that are being incorporated into the production of LTE UEs.

From the early days of analog cellular telephones through the proliferation of 2G and 3G digital cellular handsets, production test was implemented using standardized Layer 3 signaling messages that allowed the device to be controlled by a base station emulator. This base station emulator usually took the form of a test set that also performed RF parametric measurements to check the RF performance of the device. Using such a test set greatly simplified the control aspect of the testing process and had the additional benefit of checking the call-establishment, handover, and call-release functionality that is used during standard network operation. Additionally, the test set could be leveraged easily between R&D and manufacturing test use. However, as wireless devices became more complex, a separation grew between the test needs of R&D and those of manufacturing, requiring new test equipment and functionality in each case. The increase in device complexity in the form of multiple radio formats and operating frequency bands significantly increased the cost of manufacturing test. With pricing pressures dominating the highly competitive consumer wireless device market, there was continual pressure to develop more efficient test processes to offset the increasing test burden and keep manufacturing costs down.

6.15.1 Trends in Manufacturing Test

As each new generation of cellular technology is rolled out, user equipment has to maintain compatibility with legacy networks until the technology transition is complete. As a result, UEs today generally must support multiple radio formats for the different technologies as well as the associated frequency bands, which can vary by geography. For manufacturing test, that means test equipment must be flexible enough to support different format combinations and the equipment must be able to calibrate and verify all the required frequency bands.

Most LTE smartphones are expected to support some combination of W-CDMA, cdma2000, GSM, and their enhanced variants. Testing each of these formats and bands multiplies manufacturing test times and correspondingly increases the test costs—see Figure 6.15-1. The situation is further exacerbated by the use of MIMO techniques, which effectively multiplies the cost-of-test per device either by (1) increasing the number of test equipment "ports" required to test the multiple antennas simultaneously or (2) increasing the overall test time if the multiple antennas are tested sequentially.

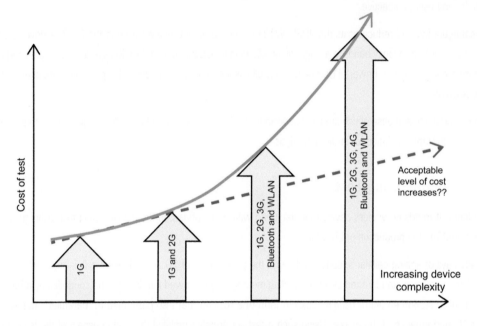

Figure 6.15-1. Devices include more formats and bands, increasing the cost of traditional signaling-based test at an unsustainable rate

Non-signaling test offers a solution to the growing test burden. "Non-signaling" describes a new generation of industry-developed proprietary manufacturing test techniques that eliminate Layer 3 signaling to decrease test time overhead and test equipment cost during UE manufacturing, thereby helping reduce the cost of test per device. Figure 6.15-2 shows the test setup for traditional signaling-based testing and for the non-signaling-based alternative. Whereas traditional Layer 3 signaling has served the manufacturing community well for over 20 years, the robustness and hence overhead built into this signaling was designed to handle the unpredictable nature of real network operation. The radio environment seen in manufacturing, however, is far more benign and thus allows shortcuts to be made to get the device into the appropriate state for test.

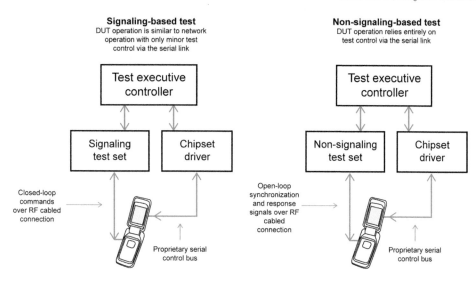

Figure 6.15-2. Comparison of traditional signaling and non-signaling test setups

User equipment manufacturers have been aware over the years that mature phone designs rarely fail functional tests unless there is an underlying software or physical defect, the latter most likely to be discovered during basic parametric testing of the radio early in the manufacturing process. Ultimately, the extent to which individual vendor chipsets have been designed with non-signaling test modes determines the amount by which the test time can be reduced. Initially non-signaling test was focused on the earlier calibration part of the manufacturing process, in which predefined transmissions using test modes were widely employed for a number of years. Increasingly, however, chipsets have been designed to also incorporate test modes that enable non-signaling verification. By adopting non-signaling verification, the entire UE manufacturing process now can be carried out with simpler and faster non-signaling test equipment leading to lower test equipment costs and faster test times.

Non-signaling test has been around for many years but is becoming increasingly sophisticated, increasing the gap between the way the UE is operated on a real network and how it is controlled during manufacturing. Chipset design engineers are now implementing fast-sequenced test modes in UE chipsets, stringing together multiple non-signaling test sequences as single tests in order to greatly reduce communication overheads, which in turn reduces the overall test time and, by extension, the cost of test. It is critical, therefore, that manufacturing test engineers understand non-signaling test mode requirements so that they can implement test plans that exploit the non-signaling capabilities of the next generation of non-signaling test equipment.

Another factor that influences manufacturing test is the desire of chipset vendors to continue the pursuit of higher levels of integration. Until recently, the primary goal was to merge RF and baseband integrated circuits or, when that was not possible, to integrate the ICs into a system-in-package (SiP). However, the development of application processors has emerged as the new frontier, addressing the insatiable demand for more processing power in the UE to cope with the flood of mobile applications that consumers are downloading for gaming, messaging, graphics, web browsing, etc.

Although the push to integrate baseband ICs and application processor units does not yet directly impact UE manufacturing processes, as silicon devices become more tightly integrated, accessing the UE's underlying radio functionality becomes more difficult. It is worth noting here that gaining access to the special chipset control methods required for non-signaling testing is inherently at odds with the trend toward increased integration.

A final, overarching concern that is common to all UE manufacturers is the need to balance the speed and cost of test with the resulting device quality, as these are often opposing requirements. A UE must be calibrated accurately during manufacturing to achieve proper, specified performance when the product is in the consumer's hands, and to this extent the UE needs to be verified—but ideally never more than absolutely necessary (and, in a perfect world, not at all). Poor or defective devices will damage a company's reputation and cost valuable market share. But if UE testing is too thorough or the manufacturing processes are too expensive, the company may have trouble competing with other handset manufacturers on price. This balance between cost and quality is a recurring theme in wireless device manufacturing.

6.15.2 Rollout of Non-Signaling Techniques

All of the trends just described are gaining acceptance in the wireless manufacturing industry. The remainder of this section focuses on one trend in particular, the use of non-signaling test techniques.

6.15.2.1 Test Modes

Non-signaling test techniques are possible only with appropriate chipset control over the device under test (DUT). This control is implemented through what are referred to here as proprietary manufacturing test modes, accessed usually through a direct, cabled connection such as a universal serial bus (USB). A non-signaling test mode can be thought of as a proprietary engineering mode within a device, designed specifically to fulfill the requirements for testing. The test mode is enabled by commands sent from the test executive via the chipset driver and a set of application programming interfaces (APIs). Once the UE is under test mode control it can be commanded into a known state, ready to receive known downlink test signals or to transmit a predefined RF signal, usually a sequence across a power and frequency range. This signal is then measured by the test equipment.

Because both the chipset driver and the test equipment are programmed and automated by a test script, the device and test equipment control need to be developed together. This is the downside of non-signaling test: traditional methods relied on standardized Layer 3 protocols implemented by all devices whereas the protocols required for non-signaling test fall outside of the scope of the 3GPP specifications and every chipset vendor will take a different approach to optimizing device control.

The 3GPP specifications do define standardized UE conformance test functions for LTE in 36.509 [47] but it should be noted that the proprietary manufacturing test modes described here are distinct from these.

6.15.2.2 Using Non-Signaling Techniques With Test Modes

Making the transition from signaling-based test to non-signaling requires the test engineer to take responsibility for the configuration of any required downlink signals, the configuration of the DUT to prepare it for reception and transmission, and the control of subsequent measurements made by the test system. In a signaling environment much of this overhead is automated but for reasons described earlier, the test is not particularly fast. The additional effort required in non-signaling to manually configure the test system is rewarded by a faster test process. The following sections will explain the basic principles behind non-signaling test as it applies to receiver and transmitter testing.

Receiver Testing

Receiver testing in the manufacturing of digital phones has traditionally been based around bit error ratio (BER) or block error ratio (BLER) measurements made at reference sensitivity in a non-faded radio environment. These measurements are in effect a measure of the noise figure of the receiver expressed as an error ratio. Receiver measurements are often a bottleneck in manufacturing and have long been a target for test process improvement. Since the first digital phones were introduced around the early 1990s there have been five distinct methods for measuring receiver quality, each offering speed benefits over its predecessors:

1. Signaling-based loopback using fully coded downlink and defined Layer 3 conformance test modes
2. Non-signaling-based loopback using fully coded downlink
3. Single-ended BER (SE-BER) using predefined (open-loop) fully coded downlink
4. SE-BER using spectrally correct downlink modulated with random data
5. DUT self-noise estimation techniques.

The first method is the one used for conformance testing and it relies on the special conformance test functions defined for each radio standard. In the case of LTE the special test functions are defined in 36.509 [47]. The principle behind this loopback method is that the test system sets up a call in the traditional manner using Layer 3 signaling and then provides a fully coded downlink signal with a data sequence unknown to the DUT. The DUT loops this data back to the test system where it is compared with the original data, and the test system can then calculate the BER or BLER (or throughput, which is a derived quantity) without the DUT having had any part in the computation of the result. This is considered the most robust way of measuring the DUT's receiver quality and closest to the actual user's experience.

In the early days of GSM manufacturing a second method was developed that dispensed with the traditional Layer 3 control and instead directly configured the same fully coded downlink signal and then manually configured the DUT to synchronize to it and loop back the data on the DUT uplink for subsequent measurement. This method speeded up the configuration of the BER measurement by dispensing with the need to set up a call. Otherwise the end result was identical to the first method.

The third method, SE-BER, marked the first major departure from what had come before by allowing the DUT to calculate its own BER from a known signal. There are two main methods for doing this. The first requires the UE to synchronize to a known pseudo random binary sequence (PRBS) pattern in the encoded data from which the BER can be calculated. The second method is simpler and uses cyclic redundancy check (CRC) calculations on the correctly encoded data frames to calculate the BLER, which provides an indication of the underlying BER. Due to the channel-coding gain, CRC-based BLER measurements are not as sensitive as BER measurements in identifying how close the DUT is to its noise floor; however, they are easier to make. For the DUT to measure BER, it would need to synchronize to a known PRBS pattern in the encoded data. The main difference between SE-BER and loopback BER is that the responsibility for the measurement has shifted from the test system to the DUT. Single-ended BER may not be considered an appropriate option for conformance testing because of the reliance on the UE to make the pass/fail judgment, but for the purposes of manufacturing it can be assumed that the ability of the DUT to correctly report BER or BLER has been fully verified by the design verification process. It is reasonable to exploit the considerable measurement capabilities of the DUT in a manufacturing context when the results can be shown to correlate with independent loopback measurements.

The fourth method described here is a further evolution of SE-BER, which differentiates itself from the third method by dispensing with the need for a fully coded downlink emulating a real base station. The SE-BER described in the third method relies on the existing receiver algorithms developed for operation in a real network, hence the need for the fully coded downlink signal. With the fourth method, the algorithm in the DUT is specially developed just to decode a downlink signal that is simply modulated with a known PRBS. Once the DUT has synchronized to the downlink, it can directly calculate the BER.

The fifth and final receiver test method exploits the nascent noise estimation capabilities built into all devices as part of the radio resource management (RRM) requirements. An essential part of normal network operation requires that the UE be able to accurately estimate the downlink signal quality. Each radio system has its own UE measurements for this purpose and in the case of LTE it is the reference signal received quality (RSRQ) measurement that provides the network with this capability. (See Section 3.6.1.2) The requirements around RSRQ are defined for RRM in faded interference-limited radio conditions but the underlying noise estimation algorithms can be adapted for the clean radio environment used in manufacturing test in order to measure the self-noise of the DUT.

Although there is no direct relationship between self-noise estimation and BER or BLER, it is certainly possible to derive a meaningful relationship between the quantities in order to develop figures of merit suitable for verifying that the DUT will meet the desired target BER or BLER value for a given estimate of self-noise. The primary advantage of the noise estimation technique is speed and sensitivity since it is both fast and provides a continuously variable measurement of receiver quality, which enables the detection of small changes in device quality that can be easily compared against design margins. In contrast, BER measurements provide less warning of how far the DUT is from failing, and BLER measurements provide even less warning due to channel coding gains and can only be used as a pass/fail indication.

Figure 6.15-3 shows single-ended BER for non-signaling and fast-sequenced non-signaling. Depending on the DUT's level of sophistication, it may need to resynchronize to the downlink after changing channels before SE-BER can resume on the new channel. This need can be addressed by constructing the downlink arbitrary waveform file to repeat certain data patterns following a channel change.

**Non-signaling
SE-BER receiver test**

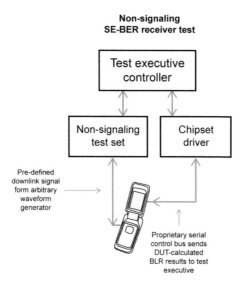

Figure 6.15-3. Non-signaling single-ended BER receiver test

Transmitter Testing

As with receiver testing there has been continual evolution in the sophistication of transmitter testing. Due to the proprietary nature of non-signaling evolution there are no exact descriptions of this evolution but the following are indicative of the major steps:

1. Signaling-based DUT control using fully coded downlink to provide synchronization and layer 3 messages to control the DUT transmitter for measurements

2. Non-signaling-based fully coded downlink to provide synchronization for the DUT with manual control of the DUT transmitter via proprietary serial bus

3. Early implementations of open-loop sequencing of DUT transmitter configurations with synchronization to measurement provided by simplified downlink

4. Fast-sequenced device control.

The traditional signaling-based DUT control was limited in its speed by the minimum time required to set up a call and perform Layer 3 reconfiguration using handover commands of the DUT transmitter that typically took 0.5 seconds per step. In the early days with slow measurements this was not considered a problem but with measurement speeds now much faster, the signaling overhead of handover became the bottleneck. The introduction of non-signaling transmitter test in the second method maintained the need of a fully coded downlink for the DUT to synchronize to but replaced the call setup and handover messages by direct control of the DUT over the proprietary serial bus. This technique was sometimes referred to as the "set and measure" method since the DUT would be manually sequenced through each state with measurements in between.

Figure 6.15-4 shows a typical configuration for non-signaling transmitter test.

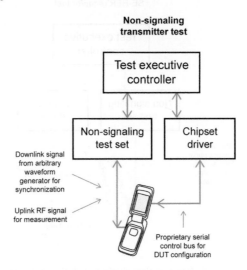

Figure 6.15-4. Transmitter testing using non-signaling methods requires synchronization on the downlink

The third method saw the transition towards basic sequencing of the DUT through a predefined set of states following a trigger. The measurements would also step through the same sequence. The use of a trigger has the potential to simplify the need for a fully coded downlink signal for synchronization. An example of this method is seen in Figure 6.15-5.

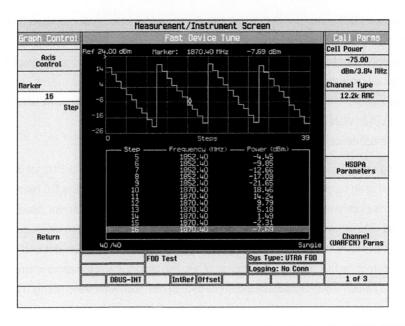

Figure 6.15-5. Example of non-signaling transmitter test using sequencing with the Agilent 8960 (E5515C) test set

The final and most sophisticated transmitter test is called fast-sequenced non-signaling test. This method is not fundamentally different from the third method other than in the speed at which the DUT and the measurement equipment can change state and maintain synchronization through a complex sequence of changes that includes power, frequency, and format. These improvements are facilitated by both the DUT and the tester having a "stored sequence" of "hardware states." Upon sequence initiation, both the DUT and the tester step through these sequences at the same rate without further need for synchronization commands, thus very significantly reducing the communication overhead of commands from the test controller to both the DUT and tester.

The potential for fast sequencing is a result of the continued evolution of chipset test modes to replace earlier non-signaling algorithms that were based on fully coded downlink signals. With fewer demands being placed on downlink synchronization, it has become possible to generate test signals using arbitrary waveform playback rather than the real-time state machines required previously to generate the fully coded signals. Test equipment vendors have taken advantage of this fact and are developing a new generation of non-signaling test equipment that uses fast sequencing to complement the new chipset capabilities.

6.15.2.3 Non-signaling Test and LTE

With the emergence of LTE, there has been an acceleration in the move to non-signaling test modes that has been influenced in particular by manufacturers' experiences with WLAN (IEEE 802.11) and WiMAX (IEEE 802.16) technologies. Specifically, manufacturers have learned that (1) non-signaling does in fact work for volume manufacturing; (2) one technology can leverage test methodology from another, especially when both are OFDM-based systems; and (3) when legacy signaling architectures are not involved, the test mode design for manufacturing can be optimized to minimize test development effort and test execution times, thus creating an overall improvement in the efficiency of manufacturing test.

In general, signaling has not been used for the production test of short range, high data rate wireless systems such as WLAN, and manufacturers instead have relied on "set and measure" parametric tests under test mode control. As WLAN evolved from direct sequence spread spectrum technology into an OFDM–based system, and as WiMAX grew out of the WLAN world to address longer communication ranges, this segment of the wireless industry continued to develop the manufacturing test mode concept. Now with the emergence of LTE, which adopts an evolved version of OFDM, many LTE devices are being offered with non-signaling test modes from the outset.

Additionally, LTE chipset vendors have been able to implement baseband stacks with a minimal need for downlink synchronization information, whereas earlier 2G and 3G designs carried a certain amount of technical "baggage" from the previous generation technology within the baseband and stack architecture of the chipsets. Designers were required to support this legacy architecture, which put limits on the implementation of new innovations such as manufacturing test modes. The introduction of LTE provides a clean slate in many ways, giving designers much greater flexibility in how they implement the baseband architecture with regard to the protocol stack, the manufacturing test mode features, and the interplay of the two (or lack thereof). This flexibility has ultimately helped clear the way for the wider adoption of non-signaling test methodologies.

6.15.3 Typical LTE Manufacturing Test Plan

The starting point for many LTE UE manufacturing test plans is the list of RF conformance tests defined in 36.521-1. However, it has long been understood that for the purpose of manufacturing test, which is to ensure process control, it is not necessary to run a complete set of RF conformance tests on every device. Table 6.15-1 shows a summary of the main RF conformance tests and the shorter list of tests supported by the Agilent E6607B (EXT) typically carried out in manufacturing.

Table 6.15-1. RF Conformance test plan with applicability to typical manufacturing test plans

3GPP TS 36.521-1 subclause	UE transmitter test	Agilent EXT
6.2.2	UE maximum output power (MOP)	Yes
6.2.3	Maximum power reduction (MPR)	Yes
6.2.4	Additional maximum power reduction (A-MPR)	Yes
6.2.5	Configured UE transmitted output power	Yes
6.3.2	Minimum output power	Yes
6.3.3	Transmit OFF power	Yes
6.3.4	ON/OFF mask	Yes
6.3.5	Power control	N/A
6.5.1	Frequency error	Yes
6.5.2.1	Error vector magnitude	Yes
6.5.2.2	IQ component	Yes
6.5.2.3	In-band emissions for non-allocated RB	Yes
6.5.2.4	Spectrum flatness	Yes
6.6.1	Occupied bandwidth	Yes
6.6.2.1	Spectrum emission mask	Yes
6.6.2.2	Additional spectrum emission mask	Yes
6.6.2.3	Adjacent channel leakage power ratio (ACLR)	Yes
6.6.2.4	Additional ACLR	Yes
6.6.3.1	Transmitter spurious emission	No
6.6.3.2	Spurious emission band UE coexistence	No
6.6.3.3	Additional spurious emission	No
6.7	Transmitter intermodulation	Yes

In a mature manufacturing test process, the performance and performance variation of a device are characterized extensively during the design verification test and early, lower volume new product production runs, which are often referred to as "pilot" production. During pilot production the manufacturing engineer will focus testing on the essential areas of device calibration, verification of operation, and identification of any issues related to process stability or areas of marginal performance identified during the product design phase. These essential tests are performed typically on a fully assembled UE, often including the consumer packaging, with access to a USB port for device control. There are antenna connections for cabled RF testing, although sometimes RF measurements are made in an enclosed fixture with calibrated coupling to the antenna.

It is expected that some manufacturers of LTE equipment will continue to make sample or limited batch tests using signaling, but the bulk of production will adopt non-signaling techniques on the assumption that baseband operation will not be subject to variations in production methods and component tolerances.

6.15.3.1 Calibration

The first step in manufacturing is to calibrate any parameters of the UE that are not otherwise correct by design. The most common form of calibration relates to output power, which is not tightly controlled by design but can be easily calibrated. This allows lower cost but otherwise stable components to be used in the design while maintaining sufficient performance. In LTE the calibration stage of manufacturing test proceeds in a manner similar to other CDMA or OFDM-based cellular technologies. The fully assembled device, from digital baseband to antenna port, needs to be "through" calibrated as a system. This calibration must include all the baseband, RF modem, front-end power amplifier, filter, and duplexer elements. The calibration process ensures that the overall LTE radio performance will meet specification, taking into account the parametric spread introduced by each of the key components.

Calibration values for each of several power and frequency compensation settings are typically held in a set of non-volatile registers in the UE. At the start of the calibration process, default values are loaded into these registers. During the calibration process, these values are updated with device-specific values derived from the measurements and specific algorithms chosen during the chipset design process.

A frequency offset measurement is used to establish the correct UE transmission frequency. Additional frequency offset measurements may be used to characterize the linearity of the tuning range. The main part of the calibration process, however, is transmitter calibration in which the gain variations of the components in the transmitter chain are calibrated. For power calibration, a modulated signal, usually with fully allocated resource blocks, is used for the main transmitter power calibration, which begins with the UE in test mode transmitting a "staircase" of successive power levels. See Figure 6.15-7. The test equipment then measures the power in each step sequentially. This procedure is repeated on different channels as necessary in each band of operation. Thus an array of compensation values covering the spread of operating power levels and frequencies can be calculated for the individual device.

Power amplifier calibration may be performed separately over a smaller power range. In this case power is measured for an array of gain settings to profile the combination of the baseband and RF stages, and the frequency response is profiled for all bands and channels that will be used in the UE under test.

In mature volume manufacturing processes, measuring and calibrating compensation for each resource block allocation, supply voltage, and temperature effect is usually unnecessary. However, as received power reporting is key to the operation of the LTE system, the UE receiver reference signal receive power (RSRP) performance needs to be characterized, to compensate for effects such as low noise amplifier gain variation and surface acoustic wave (SAW) filter ripple. A series of fast receiver measurements are made at multiple levels and frequencies, typically using a CW signal. This is simpler and faster than using a fully coded downlink and making proper RSRP measurements on the RS symbols, and the results can be transformed into the necessary calibration of proper RSRP measurements. Both main and diversity receiver paths need to be calibrated. As with the transmitter calibration, compensation factors are calculated and downloaded into the device.

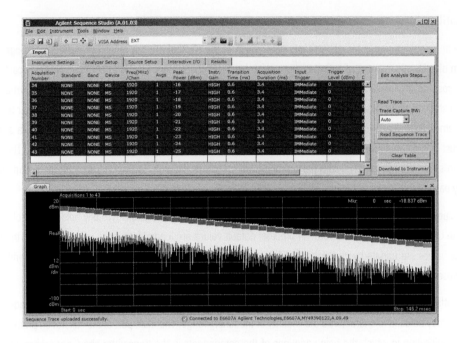

Figure 6.15-7. Screenshot of EXT performing power calibration

6.15.3.2 Verification

Once calibration is complete and the correct calibration values have been downloaded into the UE, LTE operation and performance can be verified across the relevant bands, channels, and power levels. Verification is usually done in accordance with a subset of LTE-specific transmitter and receiver measurements defined in 36.521-1 [28].

Transmitter Verification

Transmitter verification includes the following measurements:

- Power range—Maximum, minimum, and mid-range channel power measurements are made, usually on only one of the bandwidth configurations and with a fully allocated channel.
- Modulation accuracy—Error vector magnitude (EVM) is most often used, in accordance with the 3GPP conformance specifications. In manufacturing, this measurement is made on only a limited number of bandwidth configurations with known resource block allocation, likely fully allocated, and with as few channels per band as the design data indicates will be sufficient to prove operation. This is shown in Figure 6.15-8.
- Spectral flatness—In manufacturing this measurement helps ensure that the carriers at the edge of the channel are not rolling off too quickly and that sub-carrier ripple is within the specification limits. This measurement is particularly important for some of the narrower channel bandwidth configurations and certain bands that require aggressive out-of-band filtering for co-existence with adjacent frequency systems.

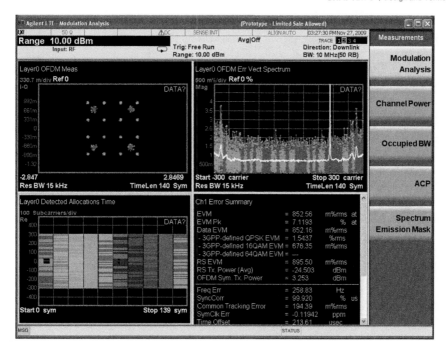

Figure 6.15-8. EXT LTE modulation accuracy measurement

A check of IQ origin offset power is often required on the uplink as this impairment is a common problem in OFDM-type systems. This impairment shows up as a power spike at the center of the channel caused by direct current offsets at the modulator. (See Section 2.1.5.) Because the offset power does not vary with output power, problems can occur at lower UE output powers. In manufacturing, measuring one channel per band at a single power level is usually sufficient.

Adjacent channel leakage ratio (ACLR) is the most commonly used spectral measurement in manufacturing verification because network efficiency requires that UEs not interfere with signals on the adjacent channel. ACLR is normally tested on at least one channel per band, again using full power and fully allocated channels. Some manufacturers may also check the spectral emissions mask, depending on how thoroughly they have characterized their device during design verification testing and how much risk they are willing to assume in exchange for a lower cost of test.

Figure 6.15-9 shows an example of an LTE ACLR measurement.

Receiver Verification

Receiver verification measurements include receiver sensitivity and RSRP, which is normally verified to the conformance specification limits for at least one or two channel bandwidths. For most LTE devices, receiver sensitivity can be verified using SE-BER.

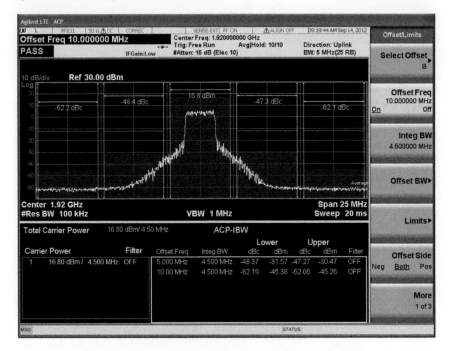

Figure 6.15-9. One-button screenshot of an ACLR measurement

6.15.4 Using Fast Sequencing to Optimize the LTE Test Plan

While device calibration has long been performed using rapid sequences of signal transmission and measurement, verification tests have more typically used the traditional "set and measure" process in which both DUT and instrument commands are executed between each sub-step in a serial order.

However, test equipment, and increasingly chipsets, offer the possibly of using fast-sequencing techniques also for the verification tests. When both the DUT and the test equipment know the complex sequence, it is possible for the test process to execute using no more than a start trigger thus eliminating all further communication overhead. The test plan can be enhanced even further if the DUT and test equipment support multi-standard sequencing, so that a fast sequence can be run not only for a specific standard, but for multiple technologies one after another.

Sequencing alone, however, doesn't provide all the speed enhancements that are possible. Within the test equipment, techniques such as single acquisition multiple measurements (SAMM)—also sometimes referred to as single acquisition combined measurement (SACM)—coupled with powerful and flexible sequencing engines can orchestrate parallel analyzer operations in which multiple different measurements are made on the same sample signal, minimizing test time by eliminating repetitive signal captures. While SAMM/SACM techniques can be used on their own to achieve some additional level of optimization in a non-signaling test plan, they provide much better optimization when coupled with the fast-sequencing technique discussed earlier.

An example of the new generation of flexible non-signaling tests sets developed specifically for the needs of UE manufacturing is the Agilent EXT. Figure 6.15-10 shows the EXT sequence studio interface, which allows the user to position the timing of signal acquisition and measurement periods using a graphical user interface to enable measurement of a fast-sequenced transmission from the UE.

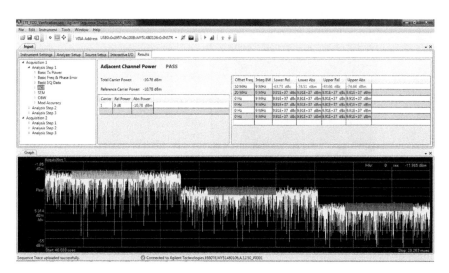

Figure 6.15-10. Agilent EXT Sequence Studio with LTE DUT

6.15.5 Additional Efficiency Enhancements

6.15.5.1 Testing Multiple Devices in Parallel

Connection and start-up of the UE can take significant time, particularly for smartphones that will make up a significant proportion of LTE devices. In manufacturing, one way to optimize test time and test station efficiency is to configure the test station to handle multiple devices, overlapping their start-up times. This is sometimes referred to as a "ping pong" test methodology.

Other elements of the test process may also be performed in parallel. For example, receiver tests in verification can take a high proportion of the overall test time and could be performed for several devices simultaneously by splitting the output signal from the test equipment to several devices. This can be done with a splitter, though such a generic approach has technical limitations and performance drawbacks due to signal attenuation. Test equipment vendors are now making available multiport adapters, which use low-loss switching to enable increased test hardware utilization.

Figure 6.15-11 shows the Agilent E6617A multiport adapter, which allows eight UE to share the same test set. The multiport adapter uses real-time port balancing to enable metrology-grade precision on all I/O ports.

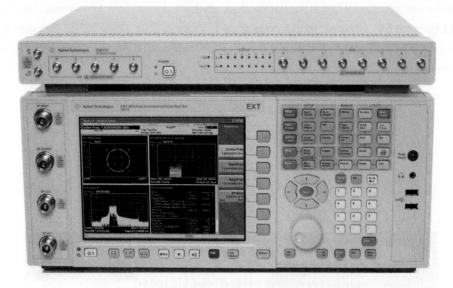

Figure 6.15-11. Agilent EXT with E6617A multiport adapter

6.15.5.2 Evolution of Test Sequences

Many LTE devices support at least one 3G radio and more 2G radios, along with *Bluetooth* and probably WLAN. It is likely that in the future multi-format verification test sequences will be supported directly in chipsets and in test equipment, opening up another avenue for test time reduction. Since non-signaling test modes are inherently chipset-specific (and thus chipset-vendor-specific), each vendor will be working hard to provide new calibration and verification test plans, features, and functions that incorporate multi-format sequences, which will enable further reductions in test time and cost of test. These enhancements are likely to include reduction in the calibration and verification segment sizes and in the time steps between them.

Manufacturers are also looking at ways to reduce the manufacturing test load by performing more thorough characterization in the design and verification processes. Measurements such as SE-BER that are typically the longest tests are being replaced by alternative noise-based measurement techniques that are equally able to verify receiver performance.

Finally, dynamic manufacturing test flows are a frontier that manufacturers may explore. In the majority of testing today, the test flow is exactly the same for each device. In the future, with the ability to run very precise, predefined sequences in conjunction with "smart" test executive software and robust device characterizations, manufacturers may be able to dynamically change the test plan from device to device so that not only is the performance verification maintained to the level required, but the device can shed unnecessary testing and further reduce overall test costs.

6.15.6 Implications of LTE-Advanced on Manufacturing Test

The introduction of LTE-Advanced in Release 10 has significant implications for UE radio design whose impact on UE manufacturing needs to be understood.

6.15.6.1 Multi-Antenna Operation

For the UE receiver, the change making headlines from Release 8 to Release 10 is the increase from four to eight spatial streams. For a UE to implement eight-layer decoding it must have eight discrete receivers. This requirement directly affects test system connectivity and cost as well as test time. However, given the fact that there are no performance requirements through Release 11 for spatial multiplexing beyond two streams, it seems unlikely that four-receiver UEs (let alone eight receiver UEs) will become predominant any time soon. Most of the focus of eight-antenna eNB transmission is for the purposes of beamforming in 4x2 or 8x2 configurations that can be exploited by a UE with only two receivers.

For the UE transmitter, the evolution to Release 10 marks the introduction of uplink spatial multiplexing of up to four streams. This complicates UE design since it mandates the use of two or four transmitters. For the next several years it is unlikely that more than two transmitters will be seen; however, even the addition of a second transmitter affects existing connectivity and will double the test burden for the uplink.

6.15.6.2 Carrier Aggregation

Carrier aggregation (CA) may also have an impact on UE manufacturing test. Although Release 10 specifies CA for up to five non-contiguous channels, all evidence suggests that only two carrier aggregation is likely in the near future. For the receiver one might expect carrier aggregation to increase test time although the opposite may be true. It is already necessary to test a multiband device separately in each band, but with the introduction of CA it may be possible to test both receivers simultaneously.

For the UE transmitter, dual-channel CA has an effect similar to that of dual-stream MIMO in that two transmitters will be required unless the carriers are contiguous and implemented using a wideband amplifier.

6.15.6.3 Combination of Carrier Aggregation and Multi-Antenna Operation

It should not be forgotten that CA and higher order MIMO can be implemented independently and thus a device using two aggregated carriers and two stream MIMO would need to have four transmitters and four receivers. For the reasons already given it is unlikely such a device will be commercially viable for the uplink any time soon. For the downlink, the implementation of inter-band CA may require separate antennas, which, when taken with the Release 8 requirement for receive diversity, means that downlink CA will require a minimum of four UE receivers.

6.15.6.4 Wider Channels

LTE-Advanced allows for up to five aggregated carriers, up to a maximum of 100 MHz aggregated bandwidth. However, even for cases in which the carriers are contiguous, no performance requirements have yet been defined for channels wider than the widest component carrier, which is still 20 MHz through Release 11. Therefore, regardless of the CA combinations, all RF parametric measurements can be carried out using measurement bandwidths of 20 MHz. The ability to capture more than one carrier using a wideband receiver is still attractive, although the actual measurements performed on the sample will still be made per component carrier.

6.16 References

[1]   3GPP TS 36.212 V11.0.0 (2012-09) Multiplexing and Channel Coding

[2]   3GPP TS 36.211 V11.0.0 (2012-09) Physical Channels and Modulation

[3]   3GPP TS 36.101 V11.2.0 (2012-09) UE Radio Transmission and Reception

[4]   3GPP TS 36.104 V11.2.0 (2012-09) Base Station Radio Transmission and Reception

[5]   IST-WINNER II Deliverable 1.1.2 v.1.2, "WINNER II Channel Models", IST-WINNER2, Tech. Rep., 2007. Available from www.ist-winner.corg/deliverables.html.

[6]   ITU-R M.[IMT-EVAL] "Guidelines for evaluation of radio interface technologies for IMT-Advanced" August 2008. Available from www.itu.int/md/R07-SG05-C-0069/en.

[7]   3GPP TR 36.814 V9.0.0 (2010-03) Further advancements for E-UTRA physical layer aspects

[8]   "Theory, Techniques and Validation of Over-the-Air Test Methods for Evaluating the Performance of MIMO User Equipment" Application Note Agilent Technologies, Literature Number 5990-5858, Dec 2010.

[9]   3GPP TR 36.815 V9.1.0 (2010-06) "LTE-Advanced feasibility studies in RAN WG4"

[10]   Chunming Chao et al. "Constrained Clipping for Crest Factor Reduction in Multiple-User OFDM" IEEE Radio and Wireless Symposium, Jan 2007.

[11]   Voelker, Ken, "Vector Modulation Analysis and Troubleshooting for OFDM Systems." Paper presented at 2002 Wireless Systems Design Conference.

[12]   Cutler, Bob, "Effects of Physical Layer Impairments on OFDM Systems." RF Design, May 2002. Available from http://rfdesign.com/images/archive/0502Cutler36.pdf.

[13]   Wright, Tom and Gorin, Joe and Zarlingo, Ben, "Bringing New Power and Precision to Gated Spectrum Measurements," High Frequency Electronics, August 2007, www.highfrequencyelectronics.com/Archives/Aug07/HFE0807_Zarlingo.pdf.

[14]   "Spectrum Analyzer Basics," Application Note 150. Agilent Technologies, Literature Number 5952-0292.

[15]   "Spectrum Analyzer Measurements and Noise," Application Note 1303. Agilent Technologies, Literature Number 5966-4008EN.

[16]   Matsuoka, H., and Zarlingo, B., "Analysis of Baseband IQ Signals." Paper presented at IMS2008. Available from www.agilent.com/find/IMS2008-MicroApps.

[17]   Zarlingo, B., and Gorin, J., "Spectrum Analyzer Detectors and Averaging for Wireless Measurements." Proceedings of the Wireless and Portable by Design Symposium, Spring 2001.

[18]   Rumney, Moray, "3GPP LTE: Introducing Single-Carrier FDMA, "Agilent Measurement Journal, Issue 4 2008, p 18-27, Literature Number 5989-7680EN.

[19]   Rumney, Moray, "Understanding Single Carrier FDMA—The New LTE Uplink." Webcast March 20, 2008, at TechOnline, www.techonline.com.

[20]   Zemede, Martha, "RF Measurements for LTE." Webcast August 7, 2008, at TechOnline, www.techonline.com.

[21]   "N9080A LTE Measurement Application Technical Overview with Self-Guided Demonstration." Agilent Technologies, Literature Number 5989-6537EN.

[22] 3GPP TS 37.104 V.11.2.1 (2012-09) Multi-Standard Radio (MSR) radio transmission and reception

[23] 3GPP TS 37.141 V.11.2.1 (2012-09) Multi-Standard Radio (MSR) Base Station (BS) conformance testing

[24] 3GPP TR 36.942 V11.0.0 (2012-09) Radio Frequency (RF) system scenarios

[25] 3GPP TS 36.141 V11.2.0 (2012-09) Base Station (BS) conformance testing

[26] "MIMO Channel Modeling and Emulation Test Challenges," Application Note. Agilent Technologies, Literature Number 5989-8973EN, October 2008.

[27] 3GPP TS 36.521-1 V10.3.0 (2012-09) User Equipment (UE) conformance specification Radio transmission and reception; Part 1: Conformance Testing

[28] 3GPP TS 36.300 V11.3.0 (2012-09) Evolved Universal Terrestrial Radio Access (E-UTRA) and Evolved Universal Terrestrial Radio Access Network (E-UTRAN); Overall description; Stage 2

[29] 3GPP TS 36.331 V11.1.0 (2012-09) Evolved Universal Terrestrial Radio Access (E-UTRA); Radio Resource Control (RRC); Protocol specification

[30] 33GPP TS 36.213 V11.0.0 (2012-09) Physical Layer Procedures

[31] MIMO channel model for TWG RCT Ad-Hoc Proposal, V16, pp. 6-8.

[32] Chen-Nee Chuah et.al., "Capacity scaling in MIMO wireless systems under correlated fading." IEEE Transactions on Information Theory, Volume 48, Issue 3, Mar 2002, pp 637-650.

[33] CTIA ERP Test Plan for Mobile Station Over the Air Performance v1.0 October 2001.

[34] 3GPP TR 25.914 V11.2.0 (2012-09) Measurements of radio performances for UMTS terminals in speech mode

[35] 3GPP TS 34.114 V11.2.0 (2012-09) User Equipment (UE) / Mobile Station (MS) Over The Air (OTA) antenna performance; Conformance testing

[36] 3GPP TR 37.976 V11.0.0 (2012-03) Measurement of radiated performance for Multiple Input Multiple Output (MIMO) and multi-antenna reception for High Speed Packet Access (HSPA) and LTE terminals

[37] 3GPP TR 37.977 V0.2.0 (2012-09) Verification of radiated multi-antenna reception performance of User Equipment (UE)

[38] ETSI ETS 201 873 Series (http://www.ttcn-3.org/StandardSuite.htm)

[39] 3GPP TS 36.523-1 V11.0.0 (2012-09) Evolved Universal Terrestrial Radio Access (E-UTRA) and Evolved Packet Core (EPC); User Equipment (UE) conformance specification; Part 1: Protocol conformance specification

[40] 3GPP TS 36.521-2 V10.3.0 (2012-09) Radio transmission and reception; Part 2: Implementation Conformance Statement (ICS)

[41] 3GPP 27.007 V11.4.0 (2012-09) AT command set for User Equipment (UE)

[42] 3GPP TS 23.272 V11.2.0 (2012-09) Circuit Switched (CS) fallback in Evolved Packet System (EPS); Stage 2

[43] "Agilent N5970A Interactive Functional Software," Application Note. Agilent Technologies, Literature Number 5989-9153EN, July 2008.

[44] ECTEL/GSM Association, "Battery Life Measurement Technique," October 1998.

[45] Brorein, Edward, "Battery Drain Analysis Improves Mobile-Device Operating Time," Application Note. Agilent Technologies, Literature Number 5989-6016EN, February 2007.

[46]   "Evaluating Battery Run-down Performance," Application Note 1427. Agilent Technologies, Literature Number 5988-8157EN, December 2006.

[47]   TS 36.509 V10.0.0 (2012-09) Evolved Universal Terrestrial Radio Access (E-UTRA) and Evolved Packet Core (EPC); Special conformance testing functions for User Equipment (UE)

Links to all reference documents can be found at www.agilent.com/find/ltebook.

Chapter 7

Conformance and Acceptance Testing

This chapter covers user equipment (UE) and evolved Node B (eNB) conformance testing. This chapter also explains the role of certification bodies such as the Global Certification Forum (GCF) and PCS Type Certification Review Board (PTCRB) in UE certification and in network operator acceptance testing, which—although not formally part of the conformance testing regime—is becoming an increasingly important step for equipment vendors to take before their products are actively marketed by specific operators.

7.1 Introduction to Conformance Testing

The goal of the LTE conformance tests is to ensure a minimum level of performance. Although the list of tests can seem very large, it should be understood that passing all the conformance tests represents only a minimum level of performance and there are many other kinds of testing that equipment vendors will need to carry out, as illustrated in Figure 7.7-1. This includes a more thorough investigation of performance margins, since conformance testing gives a pass/fail result with no indication of how close the product is to a particular limit.

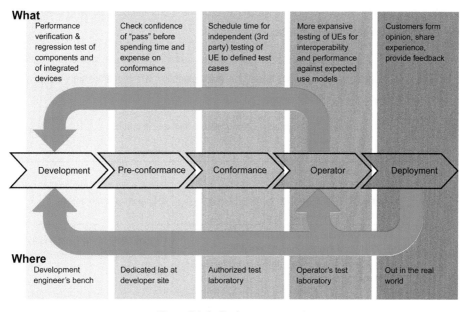

Figure 7.1-1. Equipment test cycle

LTE and the Evolution to 4G Wireless: Design and Measurement Challenges, Second Edition, Edited by Moray Rumney.
Copyright Agilent Technologies, Inc. 2013. Published by John Wiley & Sons, Ltd.

From the end users' perspective there is a need to test applications since the conformance tests are very much aimed at a lower level of capability, ensuring that the underlying transport mechanisms are in place to carry end-user services. Operator acceptance testing is another step in the process of getting a product to market and will include additional user-centric tests not otherwise covered. Thus conformance testing is an important and essential step towards the successful deployment of a new system, but it is by no means the beginning or end of the test process.

For the UE, conformance testing is divided into three parts: radio frequency (RF), radio resource management (RRM), and signaling. For the eNB or base station, only RF conformance tests are defined.

7.1.1 Common UE Conformance Test Specifications

Table 7.1-1 shows two important specifications common to the UE conformance tests. 3GPP Test Specification 36.508 [1] aims to define the typical operating conditions of a real network and is developed based on the input received from network operators. It describes the following features:
- Reference conditions
- The test frequencies to use for each of the defined frequency bands
- Test signals
- The default parameters for the system information blocks and all signaling messages. It also describes the network simulation for intra LTE, LTE-UTRAN, LTE-GERAN and LTE-cdma2000 networks.
- Reference radio bearer configurations used in radio bearer interoperability testing
- Common radio bearer configurations for other test purposes
- Generic setup procedures for use in UE conformance tests. These procedures describe the sequence to bring the UE into a stable state before starting the test sequence.

3GPP Test Specification 36.509 [2] defines any special functions needed for testing that are not otherwise defined in the core specifications. The most important of these functions is the UE loopback capability, which enables signals sent on the downlink to be returned on the uplink for analysis by the test system

Table 7.1-1. Common UE conformance test specifications

Specification	Title	Purpose
36.508 [1]	Evolved Universal Terrestrial Radio Access (E-UTRA) and Evolved Packet Core (EPC); Common Test Environments for User Equipment (UE) Conformance Testing	Provides the call setup and bearer definitions used for testing
36.509 [2]	Evolved Universal Terrestrial Radio Access (E-UTRA) and Evolved Packet Core (EPC); Special Conformance Testing Functions for User Equipment (UE)	Defines test-specific functionality in the UE not otherwise defined in the core specifications

7.2 RF Conformance Testing

This section describes the RF conformance tests as they apply to the UE and eNB. Comparison is made where appropriate to the UMTS conformance test specifications. Because of space limitations, it is not possible to go through all the tests in detail; they can be read directly from the specifications. Rather, the intent is to provide the overall context within which the RF tests exist.

7.2.1 Conformance Test Specifications

The key specifications relating to UE RF conformance test are defined in Table 7.2-1.

Table 7.2-1. UE RF FDD and TDD conformance test specifications

Specification	Title	Purpose
36.124 [3]	Electromagnetic Compatibility (EMC) Requirements for Mobile Terminals and Ancillary Equipment	Tests for EMC emissions and immunity
36.521-1 [4]	User Equipment (UE) Conformance Specification; Radio Transmission and Reception Part 1: Conformance Testing (FDD/TDD)	The RF conformance tests primarily based on 36.101 [5]
36.521-2 [6]	User Equipment (UE) Conformance Specification; Radio Transmission and Reception Part 2: Implementation Conformance Statement (ICS)	Definition of applicability of tests for different UE capabilities
36.521-3 [7]	User Equipment (UE) Conformance Specification; Radio Transmission and Reception Part 3: Radio Resource Management Conformance Testing	The RRM conformance tests based on 36.133 [8]

Base station RF FDD and TDD conformance test documents are listed in Table 7.2-2.

Table 7.2-2. Base Station (eNB) RF FDD and TDD conformance test documents

Specification	Title	Purpose
36.113 [9]	Evolved Universal Terrestrial Radio Access (E-UTRA); Base Station (BS) and Repeater Electromagnetic Compatibility (EMC)	Tests for EMC emissions and immunity
36.141 [10]	Evolved Universal Terrestrial Radio Access (E-UTRA); Base Station (BS) Conformance Testing	The RF conformance tests primarily based on 36.104 [11]
36.143 [12]	Evolved Universal Terrestrial Radio Access (E-UTRA); FDD Repeater Conformance Testing	The repeater conformance tests primarily based on 36.106 [13]

The UE and eNB EMC tests in 36.124 [3] and 36.113 [9] are largely independent of the radio system and are not discussed further here.

7.2.2 Scope of RF Conformance Tests

The complexity of the air interface continues to grow from one cellular system to the next. One of the system design goals for LTE was to reduce this complexity in the UE. From the baseband perspective, it can be argued that OFDM is indeed simpler to implement than CDMA. Nevertheless, complexity is seen from a test perspective and is defined by the number of permutations of the RF configuration that can impact performance. LTE introduces more degrees of freedom than any previous radio standard, and these degrees of freedom have a direct impact on the number of tests that could be carried out. LTE inherits some of the degrees of freedom from UMTS, including the three modulation formats (QPSK, 16QAM, and 64QAM), but many more variables have been added.

When UMTS was first specified, it had only one FDD frequency band, but more were added over the years. Based on the Release 11 specifications LTE is defined for 28 FDD frequency bands and 12 TDD bands. Probably no one product will ever implement all these options, but the wide range of choices adds to the complexity of specifications since many requirements become a function of the band and position within the band. Furthermore, for any one band, UEs have to support up to six different channel bandwidths. The use of OFDMA and SC-FDMA rather than CDMA introduces more variability since the transmission and reception requirements of OFDMA and SC-FDMA are a function of the allocation bandwidth, which can range anywhere from 180 kHz to 18 MHz. The position of the allocation within the channel can also affect performance since allocations towards the edge of the channel are influenced by the roll-off of channel and duplex filters aimed at meeting out-of-channel and out-of-band requirements.

When all the flexibility of the LTE air interface is considered, it quickly becomes apparent that the number of permutations that could be tested is enormous. Moreover, the list of possible tests needs to include those already defined for other radio access technologies (RAT) supported by the UE, as well as any new inter-RAT tests required by the addition of LTE. In selecting the configurations for the LTE RF conformance tests, 3GPP made considerable effort to identify those combinations of parameters that represent the most difficult operating conditions so that when a UE or eNB passes the tests, the design engineer can be reasonably confident that the device will perform satisfactorily in many more combinations than those explicitly tested. Prior to reaching conformance testing, a wider set of testing will have been carried out during product development. Chapter 6 of this book explores such testing in greater detail. The remainder of this chapter focuses on those tests that are mandatory to obtain conformance certification.

7.2.3 Generic Structure of RF Conformance Tests

The structure of a UE RF conformance test follows a set pattern comprising the following steps:
- Test purpose
- Test applicability
- Minimum conformance requirements
- Test description, including initial conditions, test procedure, and message contents
- Test requirements.

The eNB RF conformance tests cover the same list in a slightly different order, and the message contents are not required since the eNB tests are done without signaling. Each step will now be described.

7.2.3.1 Test Purpose

Every test case starts with a description of the test purpose. Although the purpose of a test may seem obvious from its title, in some cases a test may appear to fail for reasons that have nothing to do with the test purpose. Some tests generate intermediate results that could be meaningful to other tests, but it is only the specific items listed in the test purpose that actually determine the pass/fail result. A clearly written test purpose helps clarify what is and what is not important, especially when additional minimum requirements that are not to be tested get copied from the core specifications to the minimum conformance requirements subclause.

7.2.3.2 Test Applicability

This subclause is generally used to identify the specification release to which a test applies, although it may also identify particular UE capabilities that are required for the test; for example, support of FDD or TDD. Because FDD and TDD modes are more closely integrated in LTE than in UMTS, an important goal of LTE conformance test development has been to make as many tests as possible applicable to both FDD and TDD modes. Only when necessary are independent FDD and TDD test cases being drafted. Since UE requirements for Release 8 may differ from those of later releases, test applicability indicates explicitly for which release a test case applies (for example, Release 8 and forward, Release 9 only, and so forth).

7.2.3.3 Minimum Conformance Requirements

The term "minimum requirements" is used in the RF specifications to define the lowest level of performance that the UE should meet. The subclause defining the minimum requirements is typically a direct copy of the minimum requirements listed in one of the so-called "core" specifications. In the case of RF and RRM tests, this document will be either 36.101 [5] or 36.133 [8], respectively. To avoid any doubt should a conflict arise between what is written in the test specification and what is written in the core specification, the core specification always takes precedence. Usually the subclause on minimum conformance requirements ends with a statement such as "The normative reference for this requirement is 36.101 Subclause A.B.C." The use of "normative" is intended to stress that the test subclause contains information that is subservient to the core requirement.

7.2.3.4 Initial Conditions

The initial conditions define the test environment, which will be normal or extreme as defined in 36.508 [1] Subclause 4.1. The initial conditions may specify the temperature range and battery voltages as well as the bands, channel frequencies, and channel bandwidths that need to be tested. When test cases are developed, great care must be taken to ensure that the list of initial conditions is no longer than necessary, as every binary variable added to the list can double the number of times that the test needs to be executed. For example, a quad-band UMTS phone may take up to 1600 hours to execute a complete set of conformance tests.

In addition to defining the starting conditions for the test, the initial conditions subclause also defines the procedure for getting the UE into the correct state to start the test. Connection diagrams are usually included that show how the UE should be connected to the test system. Once the physical connections are made, the test will usually require that the UE be put into a specific state according to generic procedures in 36.508 [1]. For some tests the initial conditions may include reference to test-specific message contents not defined in 36.508.

In UMTS the wording of the initial conditions subclause mandated that a particular setup procedure be followed every time the test was executed. In LTE, however, the emphasis is on reaching the desired initial state rather than the procedure used to get there. Each test is written as a standalone procedure, but in the interest of saving time it is acceptable to bypass those parts of the initial condition procedures that do not change the state of the UE prior to the test. For example, a test system may concatenate two tests to save time, in which case it may not be necessary to switch the UE off at the end of the first test and on again at the start of the second. If a bearer of the correct type is already established, it may not even be necessary to end the call from the previous test.

7.2.3.5 Test Procedure

This subclause contains the main substance of the test case, which includes the collection of measurement results used to determine pass or fail.

7.2.3.6 Message Contents

This subclause defines any additional test-specific message contents not already covered in the initial conditions or in the exceptions to the default message contents defined in 36.508.

7.2.3.7 Test Requirements

The final subclause of the test case provides the limits, known as test requirements, against which the results acquired during the test procedure are compared to determine a pass/fail verdict. The verdict is determined only for those results that are relevant to the purpose of the test. The test requirements are usually a copy of the minimum requirements from the earlier subclause of the same name, but modified by test tolerances to take into account the test system uncertainty. Test tolerances are discussed fully in the next section.

7.2.4 Test System Uncertainty and Test Tolerances

The concept of test tolerances that was first introduced in UMTS has been adopted for LTE also. The principle behind test tolerances is the relaxation of minimum requirements by an allowance based on the expected uncertainty of measurement in the conformance test system. The minimum requirements for UMTS and LTE were developed with an implementation margin to take into account practicalities such as component tolerances in the UE or eNB, but no further allowance was made for uncertainty in the measurement process. Since conformance testing is performed on a single sample of a product, it was decided that the probability of passing a good UE or eNB should not be reduced because of uncertainty in the test system. For that reason the minimum requirements for most parametric tests are relaxed by a test tolerance to produce a test requirement against which any pass/fail criteria are compared.

The shared risk principle defined in ETR 273 [14] applies to UMTS and LTE. The principle comprises three parts:
- An agreed method of calculating measurement uncertainty
- A maximum acceptable value of measurement uncertainty (stated in the standard)
- An agreement to use the numerical value of a measurement as the pass/fail criteria.

The principle states that allowances for measurement uncertainty shall not be used to modify the measurement result since the uncertainty is not correlated with the DUT performance and is as likely to increase the chance of a pass as to decrease it—hence the risk from the verdict is shared between the DUT and the overall system. However, strict interpretation of the shared risk principle states that measurement results shall not be modified by measurement uncertainty. UMTS and LTE both employ a mechanism in which the minimum requirements are modified by a test tolerance to create a test requirement against which unmodified measurements are then directly compared to generate a pass/fail verdict. The practice of modifying the requirement by the measurement uncertainty before making the measurement avoids having to later modify the measurement result by the uncertainty. This practice reflects technical adherence to the shared risk principle. However, the effect is much like having adopted the principle "never fail a good DUT" since it shifts the balance of performance in favor of the DUT at the expense of the system. The consequences on the system are considered minimal and are difficult to assess. The more important point is that measurement uncertainty is minimized and constrained within the specifications.

7.2.4.1 Calculation of Conformance Test System Uncertainty

In many cases, the measurement uncertainty for a test is a direct function of a single parameter of the test system. Examples include power measurement accuracy and EVM accuracy. There are, however, some tests that involve more complex interactions between many variables. The most complex of these are often the tests in support of RRM, which can include many different stimuli, each with its own uncertainty influencing the behavior of the UE. However, if the uncertainty of every variable were stacked end to end to create a worst case uncertainty, the test requirement would then be so relaxed that the test would no longer have any value. Even a bad UE implementation would have no trouble passing the relaxed limit. In fact most uncertainties in the test system are uncorrelated, and therefore to calculate a realistic uncertainty, the individual components are added using a root sum square (RSS) approach to better predict the distribution of likely performance. The application of linear (worst case) addition or RSS addition varies depending on the test case, and each has to be analyzed in its own right to fairly represent a more probable uncertainty than that predicted by worst case analysis.

7.2.4.2 Test System Uncertainty Confidence Levels

The test system uncertainty, once defined, might appear to be a hard limit. In practice the uncertainty of any measurement can be expressed statistically, often as a normal distribution with a standard deviation σ. Within the 3GPP specifications, the norm for defining test system uncertainty is to adopt a 2σ figure, which for a normal distribution represents approximately 95% of the population. This approach allows a tighter limit to be specified for the test system but it does mean that 5% of results are expected to have an accuracy outside of the specified limit. This consequence is accepted in the conformance test regime because the required confidence level for the test result is also 95%. In most cases the results of performance testing fall well inside the test requirements, and in these cases the test equipment confidence level is assumed to have little effect on the confidence level of the results. Only when the DUT is on the very limit of the test requirement would the test system confidence level fully impact the confidence of the pass/fail verdict.

It is important to note that warranted specifications for test equipment are often derived for confidence levels higher than the 95% required for conformance testing. For example, Agilent Technologies usually specifies a 3σ figure representing approximately 99% of the population of results. Regardless of the confidence level, any specification falling outside of a warranted limit would be considered a warranty failure and corrected accordingly. If test equipment warranted specifications based on 3 σ fall within the 3GPP requirements for 2 σ, then there can be no doubt that the test equipment will provide superior performance during conformance testing and reduce the chance of a good DUT being failed or a bad one being passed. Sometimes the specifications for test equipment provide 2 σ uncertainty figures that will look better than 3 σ figures, but which simply represent a smaller population of the test equipment performance. To correctly compare specifications it is important to know the confidence level that each represents.

7.2.4.3 Application of Measurement Uncertainty to Test Tolerances

In some regulatory tests with requirements that come from outside of 3GPP, test tolerances are set to zero despite the non-zero measurement uncertainty involved. Perhaps the most familiar example is the −36 dBm spurious emissions limit from ITU-R SM 329-10 [15]. This limit is non-negotiable and therefore must be used in the test requirement without relaxation. However, not relaxing the test requirement by the measurement uncertainty does not mean that the DUT is then vulnerable to unlimited measurement uncertainty. Even though the test tolerance is set to zero, for such tests there is still a defined limit on the allowed test system uncertainty, which constrains the test system performance and impact on the DUT.

The sequence for determining the limits that apply for a particular test follow these steps. First the uncertainty of the test system is analyzed and limits set. In most cases this limit is then adopted as the test tolerance for that test and the minimum requirements are relaxed accordingly. For a few tests, the test tolerance must be set to zero, meaning no relaxation, but the uncertainty of the test system is still defined and thus constrained.

7.2.4.4 Excess Test System Uncertainty

One further provision is made for the case in which a test system does not meet the requirement for test system uncertainty. In this case it is allowable to use a less accurate system provided the test tolerance is reduced by the amount of the excess uncertainty. This narrows the range over which the DUT can pass the test, but if the DUT does pass the test under these conditions, it clearly would also pass if a more accurate test system were used. The choice of whether or not to use the excess uncertainty provision is entirely up to the test house. If the excess test system uncertainty principle is used, the test system can safely pass a good DUT, but a fail verdict does not prove that the DUT is bad. If the DUT performance is just inside the test requirements, a test system meeting the test system accuracy requirements would correctly pass the DUT.

7.2.5 Statistical Testing

The majority of minimum requirements are expressed as absolute limits that have to be measured with 95% confidence as already discussed. This is straightforward for cases in which the uncertainty can be linked directly to the test system error. Two other scenarios require special treatment. These are (1) tests with requirements that rely on error ratios, such as receiver tests, and (2) tests with requirements that are expected to be met only a percentage of the time, such as most RRM tests.

7.2.5.1 Statistical Receiver Testing

Receiver minimum requirements are expressed in terms of a percentage of a maximum throughput, typically 95%, which can be directly mapped to a block error ratio (BLER). This ratio will by default be measured using a 95% confidence level, which should not be confused with the throughput requirement. The two variables are independent and just happen to have the same value. The 95% confidence level creates a difficulty for the test design for the following reason.

While it is straightforward to deal with the uncertainties associated with setting the signal levels that define performance, the actual BLER measurement involves a binomial statistical phenomenon that does not exhibit the continuous variation associated with analog results. To put this another way, when the test system reports the status of an individual block, there is no uncertainty; the block is either received correctly or it has an error. In order to assess a throughput or BLER, many blocks must be observed before an error ratio can be calculated.

Given that the verdict needs to be assessed with a 95% confidence level, a determination must be made regarding how many blocks to measure to achieve the desired confidence before declaring a pass or fail verdict. The answer is not obvious. For instance, if 100 blocks are measured and 5% of them are in error, does this indicate a pass or a fail with 95% confidence? Or, if only 50 blocks are measured and two are in error, does this indicate a pass? What is the maximum length of time that the test should be run? The mathematics behind this determination are complex but the measurement period has been explicitly defined for each test in the specifications. For the receiver conformance tests, in addition to the 95% confidence level and the minimum requirement (which is >95% throughput of the reference measurement channel), two other factors must be taken into account in calculating how long to run the test. The first is the possibility of early decision. It can be shown that when a receiver test is run, based on continuous analysis of interim results, the test can sometimes be stopped early with a pass or a fail verdict and this verdict will still meet the criteria for 95% confidence. The conditions under which this can happen occur when the DUT produces either very good or very bad results. Consider the example of early failure: if the criterion for a pass is defined as >95% throughput and, after running the test for more than 5% of the defined maximum time, the analysis indicates continued failure, the test can be aborted early with a fail verdict since nothing that happens after that point can change the verdict. The principle of early decision is illustrated graphically in 36.521-1 [4] Annex G. It is a powerful technique for speeding up throughput tests that get repeated hundreds of times; for example, blocking tests.

The second factor that must be taken into account in calculating the minimum test time is known as the "bad DUT" factor. Consider the case in which a DUT is right at the limit of the test requirement. How long should the measurement continue to ensure 95% confidence? The answer is that the test would have to run forever. Obviously this is not possible, so to prevent such an asymptotic situation from occurring, the fail limit is raised by the bad DUT factor such that if the higher limit is reached, the test can be stopped.

7.2.5.2 Statistical Performance Testing

The same concepts used for statistical testing of the receiver are applied to the performance tests as well. An important difference is that performance tests use fading channels (rather than white noise, which is used in the receiver tests), so the statistics of the fading profile must be considered and the test times adjusted accordingly. This impact on the minimum test duration is most noticeable for the slow fading profiles at 3 km/h, which can take tens of seconds to repeat. Another difference is that the performance tests are based on 30% and 70% throughput targets of the reference measurement channel (RMC) maximum throughput. For these reasons the minimum test time calculations and early pass/fail criteria are calculated separately from the receiver tests.

7.2.5.3 Statistical Radio Resource Management Testing

The other major use of statistical testing is for radio resource management (RRM) testing. The RRM requirements generally have to be met more than 90% of the time. An example is the cell reselection delay requirement. System integrity does not rest on every single cell reselection occurring within the target value, and the variables that can influence the reselection time are numerous. For this reason the best approach is to set a tighter limit to be met 90% of the time and not put constraints on the remaining 10%. From a testing perspective this again means that calculations have to be done to determine how many times the test must be run to reach 95% confidence. Since some RRM tests take a long time to run, it is desirable to repeat them as few times as possible.

7.2.6 Typical RF Test System Configuration

The configuration of an RF conformance test system varies depending on the complexity of the tests. Some tests can be performed with just an eNB emulator (sometimes referred to as a system simulator or one box tester). Other tests require additional equipment including an additional eNB emulator to provide multi-cell test capability for the RRM tests, interference sources for blocking tests, and spectrum analyzers for spurious emissions tests. Figure 7.2-1 shows how some of these pieces of equipment might be used in an RF conformance test system.

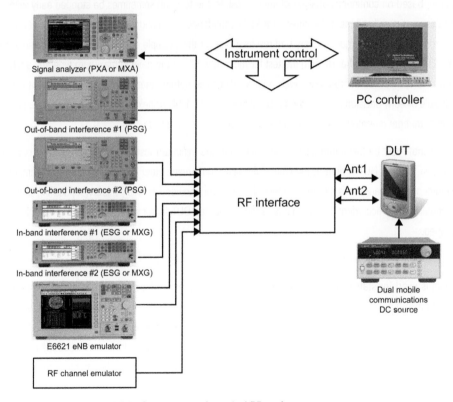

Figure 7.2-1. Components of a typical RF conformance test system

The RF interface module provides the essential connectivity between the test equipment and the UE under test. It contains the switching required to run the tests automatically as well as essential filtering to ensure that the correct RF conditions can be achieved.

The Agilent T4010S CT is a solution for LTE RF conformance testing of LTE UEs and is part of the Agilent T4010S family of automated RF testers. This conformance test system provides a comprehensive set of tools that help the user through the process of entering DUT data into the test system, defining the test plan to be executed, configuring the system to execute the tests according to the specific UE characteristics, analyzing the test results, and producing the associated test reports.

7.2.7 UE RF Conformance Test Case Overview

The UE RF conformance tests defined in 36.521-1 [4] are split into four main sections: RF transmitter characteristics; RF receiver characteristics; RF performance characteristics; and reporting of channel quality indicator (CQI), rank indication (RI), and precoding matrix indicator (PMI). The intention here is to provide an overview of the overall scope with some LTE-specific discussion rather than to go through every test in detail.

Most UE requirements are defined relative to uplink and downlink RMCs. These are defined in 36.101 [5] and are described in Section 2.1.3.

7.2.7.1 UE RF Transmitter Characteristics

Table 7.2-3 lists the UE RF transmitter test cases defined in 36.521-1 [4] with references to further information in this book. The scope of these RF transmitter tests will be familiar from UMTS and are modified only in the details as they pertain to LTE and the SC-FDMA uplink modulation format. One point worth noting is the addition of tests for spectrum emission mask (SEM) and adjacent channel leakage ratio (ACLR). The concept of "additional" tests is new to LTE and comes from the need to control out of band emissions for certain combinations of bands. To achieve the desired performance the concept of network has been introduced to indicate to the UE that dynamic requirements have to be applied. This is discussed more fully in Section 2.1.4. In essence, when the network sends a particular network value, the UE has to meet additional emission requirements for coexistence with adjacent bands.

The transmitter tests are carried out using uplink RMCs. The RMCs fall into three main categories—fully allocated, partially allocated, and single RB—and were defined based on the simulation assumptions used to derive the requirements. The number of different RMC configurations defined for testing is a balance between thoroughness and excessive test time.

Table 7.2-3. UE RF transmitter test cases

36.521-1 [4] subclause	Test case	Sections in this book with further information
6.2.2	UE maximum output power	2.1.4
6.2.2A	UE maximum output power for intra-band contiguous carrier aggregation (CA)	2.1.4
6.2.3	Maximum power reduction (MPR)	2.1.4
6.2.4	Additional maximum power reduction (A-MPR)	2.1.4
6.2.4A	Additional maximum power reduction (A-MPR) for intra-band contiguous CA	2.1.4
6.2.5	Configured UE transmitted output power	2.1.4
6.3.2	Minimum output power	2.1.5.1.1
6.3.4.1	General ON/OFF time mask	2.1.5.1.3
6.3.4.2.1	PRACH time mask	2.1.5.1.3
6.3.4.2.2	SRS time mask	2.1.5.1.3
6.3.5.1	Power control absolute power tolerance	2.1.5.1.4
6.3.5.2	Power control relative power tolerance	2.1.5.1.4
6.3.5.3	Aggregate power control tolerance	2.1.5.1.4
6.5.1	Frequency error	2.1.6.1, 6.4
6.5.2.1	Transmit modulation—error vector magnitude (EVM)	2.1.6.1, 6.4
6.5.2.1A	Transmit modulation—PUSCH-EVM with exclusion period	2.1.6.1, 6.4
6.5.2.2	Transmit modulation—carrier leakage	2.1.6.1, 6.4
6.5.2.3	Transmit modulation—in-band emissions for non-allocated RB	2.1.6.1, 6.4
6.5.2.4	Transmit modulation—EVM equalizer spectrum flatness	2.1.6.1, 6.4
6.6.1	Occupied bandwidth	2.1.7.1, 6.4
6.6.2.1	Out-of-band emission—spectrum emission mask	2.1.7.2, 6.4
6.6.2.2	Out-of-band emission—additional spectrum emission mask	2.1.4, 6.4
6.6.2.3	Out-of-band emission—adjacent channel leakage power ratio (ACLR)	2.1.7.3.1, 6.4
6.6.3.1	Transmitter spurious emissions	2.1.7.4.1, 6.4
6.6.3.2	Spurious emission band UE coexistence	2.1.7.4.1
6.6.3.2_1	Spurious emission band UE coexistence (Release 9 and forward)	2.1.7.4.1
6.6.3.3	Additional spurious emissions	2.1.4
6.7	Transmit intermodulation	2.1.7.5.1

7.2.7.2 UE RF Receiver Characteristics

Table 7.2-4 lists the UE receiver test cases defined in 36.521-1 [4]. Further discussion of receiver testing is provided in Section 6.5.

Table 7.2-4. UE RF receiver test cases

36.521-1 [4] subclause	Test case
7.3	Reference sensitivity level
7.4	Maximum input level
7.5	Adjacent channel selectivity
7.6.1	In-band blocking
7.6.1A	In-band blocking for CA
7.6.2	Out-of-band blocking
7.6.2A	Out-of-band blocking for CA
7.6.3	Narrowband blocking
7.6.3A	Narrowband blocking for CA
7.7	Spurious response
7.7A	Spurious response for CA
7.8.1	Wideband intermodulation
7.9	Spurious emissions

The test cases above are similar to the equivalent UMTS test cases. One difference worth noting is that the receiver minimum requirements for UMTS were typically specified in terms of a bit error ratio (BER) that was distinct from the BLER used in the UMTS performance tests. This difference was due to somewhat arbitrary choices made during the early development of UMTS when simulation work for some requirements was done using BER and for others using BLER. Because a verifiable BER result requires the transmitted data to be looped back to the test system, it is a more difficult measurement than simply counting the UE ACK and NACK reports that are necessary for calculating BLER. That said, BER is better able to pick up small variations in performance compared to BLER, and remains a useful diagnostic during product development.

For LTE the receiver minimum requirements are expressed in terms of a percentage throughput (>95%) of the RMC used in the test. Since BLER can be mapped directly to throughput, the LTE receiver tests are brought in line with the performance tests that have always been based on BLER and throughput.

7.2.7.3 UE RF Performance Requirements

Table 7.2-5 lists the UE performance test cases defined in TS 36.521-1 [4].

Table 7.2-5. UE RF performance test cases

36.521-1 [4] subclause		Test case
FDD	TDD	
8.2.1.1.1	8.2.2.1.1	PDSCH single antenna port performance
8.2.1.1.1_1	8.2.2.1.1_1	PDSCH single antenna port performance (Release 9 and forward)
8.2.1.1.2	8.2.2.1.2	PDSCH single antenna port performance with 1 PRB in presence of MBSFN
8.2.1.2.1	8.2.2.2.1	PDSCH transmit diversity 2x2 performance
8.2.1.2.1_1	8.2.2.2.1_1	PDSCH transmit diversity 2x2 performance (Release 9 and forward)
8.2.1.2.2	8.2.2.2.2	PDSCH transmit diversity 4x2 performance
8.2.1.2.2_1	8.2.2.2.2_1	PDSCH transmit diversity 4x2 performance(Release 9 and forward)
8.2.1.3.1	8.2.2.3.1	PDSCH open-loop spatial multiplexing performance
8.2.1.3.2	8.2.2.3.2	PDSCH open-loop spatial multiplexing 4x2 performance
8.2.1.4.1	8.2.2.4.1	PDSCH closed-loop single-/multi-layer spatial multiplexing performance
8.2.1.4.1_1	8.2.2.4.1_1	PDSCH close- loop single-/multi-layer spatial multiplexing performance (Release 9 and forward)
8.2.1.4.2	8.2.2.4.2	PDSCH closed-loop single-/multi-layer spatial multiplexing 4x2 performance
8.2.1.4.2_1	8.2.2.4.2_1	PDSCH closed-loop single-/multi-layer spatial multiplexing 4x2 performance (Release 9 and forward)
	8.3.2.1.1	PDSCH single-layer spatial multiplexing on antenna port 5 (Release 8 and forward)
	8.3.2.1.1_1	PDSCH single-layer spatial multiplexing on antenna port 5 (Release 9 and forward)
	8.3.2.1.2	PDSCH Single-layer Spatial Multiplexing on antenna port 7 or 8 without a simultaneous transmission (Release 9 and forward)
	8.3.2.1.3	PDSCH single-layer spatial multiplexing on antenna port 7 or 8 with a simultaneous transmission (Release 9 and forward)
	8.3.2.2.1	PDSCH dual-layer spatial multiplexing (Release 9 and forward)
8.4.1.1	8.4.2.1	PCFICH/PDCCH single antenna port performance
8.4.1.2.1	8.4.2.2.1	PCFICH/PDCCH transmit diversity 2x2 performance
8.4.1.2.1_1	8.4.2.2.1_1	PCFICH/PDCCH transmit diversity 2x2 performance (Release 9 and forward)
8.4.1.2.2	8.4.2.2.2	PCFICH/PDCCH transmit diversity 4x2 performance
8.4.1.2.2_1	8.4.2.2.2_1	PCFICH/PDCCH transmit diversity 4x2 performance (Release 9 and forward)
8.5.1.1	8.5.2.1	PHICH single antenna port performance
8.5.1.2.1	8.5.2.2.1	PHICH transmit diversity 2x2
8.5.1.2.1_1	8.5.2.2.1_1	PHICH transmit diversity 2x2 (Release 9 and forward)
8.5.1.2.2	8.5.2.2.2	PHICH transmit diversity 4x2
8.5.1.2.2_1	8.5.2.2.2_1	PHICH transmit diversity 4x2 (Release 9 and forward)
8.7.1.1	8.7.2.1	Sustained data rate performance provided by lower layers (Release 9 and forward)

he performance requirements are written around the baseline UE capability, which has two receivers. It is still open to debate vhether UE implementations meeting the dual receiver requirements with only one receiver will be allowed.

he UE RF performance tests are performed with throughput measurements for the downlink physical channels specified in Section 3.2.5. The tests apply uncorrelated fading conditions and AWGN signals to each receiver antenna connector.

7.2.7.4 UE Reporting of Channel State Information

Table 7.2-6 lists the test cases for reporting of CSI defined in 36.521-1 [4].

Table 7.2-6. Test cases for reporting of channel state information

36.521-1 [4] subclause		Test case
FDD	TDD	
9.2.1.1	9.2.1.2	CQI reporting under AWGN conditions—PUCCH 1-0
9.2.2.1	9.2.2.2	CQI reporting under AWGN conditions—PUCCH 1-1
9.3.1.1.1	9.3.1.1.2	CQI reporting under fading conditions—PUSCH 3-0
9.3.2.1.1	9.3.2.1.2	CQI reporting under fading conditions—PUCCH 1-0
9.3.3.1.1	9.3.3.1.2	CQI reporting under fading conditions and frequency-selective interference—PUSCH 3-0
9.3.4.1.1	9.3.4.1.2	CQI reporting under fading conditions—PUSCH 2-0
9.3.4.2.1	9.3.4.2.2	CQI reporting under fading conditions—PUCCH 2-0
9.4.1.1.1	9.4.1.1.2	PMI reporting—PUSCH 3-1 (single PMI)
9.4.1.2.1	9.4.1.2.2	PMI reporting—PUSCH 2-1 (single PMI)
9.4.2.1.1	9.4.2.1.2	PMI reporting—PUSCH 1-2 (multiple PMI)
9.4.2.1.1_1	9.4.2.1.2_1	PMI reporting—PUSCH 1-2 (multiple PMI) (Release 9 and forward)
9.4.2.1.1_1	9.4.2.1.2_1	PMI reporting—PUSCH 1-2 (multiple PMI) (Release 9 and forward)
9.4.2.2.1	9.4.2.2.2	PMI reporting—PUSCH 2-2 (multiple PMI)
9.5.1.1	9.5.1.2	RI reporting—PUCCH 1-1

The CSI reporting tests are explained in Section 6.6.3.3. The tests apply uncorrelated fading conditions and AWGN signals to each receiver antenna connector. The CSI indicators are CQI, RI, and PMI and are introduced in Section 3.4.6.

7.2.7.5 MBMS Performance

Table 7.2-7 lists the MBMS performance test case defined in 36.521-1 [4]. This test case is applicable for Release 9 and forward UEs.

Table 7.2-7. MBMS performance test case

36.521-1 [4] subclause		Test case
FDD	TDD	
10.1	10.2	MBMS performance (fixed reference channel)

7.2.8 UE RRM Conformance Test Case Overview

The RRM requirements are defined in the core specification 36.133 [8] and the conformance tests are defined in 36.521-3 [7]. The RRM requirements are complex due to the number of variables that can affect performance; therefore, the core specification36.133 [8] includes Annex A, which provides guidance on test case configuration for conformance testing. The RRM conformance tests are based on this annex rather than direct reference to the core requirements as is the case with the RF conformance tests and 36.101 [5]. The RRM requirements from 36.133 [8] are discussed in Section 3.6.3. They are divided into six main areas and the annex follows the same six-part structure. Table 7.2-8 lists the general categories of the RRM test cases defined in 36.521-3 [7] with reference to the test configurations defined in 36.133 [8]. Due to their number, the individual test cases in each category are not listed.

Table 7.2-8. UE RRM test cases

36.133 [8] subclause	36.521-3 [7] subclause	Category of test	Number of test cases
A.4.2	4.2	Cell reselection in E-UTRAN RRC_Idle state	6
A.4.3	4.3	E-UTRAN to UTRAN cell reselection	8
A.4.4	4.4	E-UTRAN to GSM cell reselection	2
A.4.5	4.5	E-UTRAN to HRPD cell reselection	1
A.4.6	4.6	E-UTRAN to cdma2000 1XRTT cell reselection	1
A.5.1	5.1	Handover delay in E-UTRAN RRC_Connected state	6
A.5.2	5.2	E-UTRAN handover to other RATs	10
A.5.3	5.3	E-UTRAN handover to non-3GPP RATs	4
A.6.1	6.1	RRC connection mobility control RRC reestablishment	4
A.6.2	6.2	RRC connection mobility control random access	4
A.7.1	7.1	Timing and characteristics—UE transmit timing	2
A.7.2	7.2	Timing and characteristics—UE timing advance	2
A.7.3	7.3	Timing and characteristics—UE radio link monitoring	8
A.8.1	8.1	E-UTRAN FDD intra-frequency measurements	5
A.8.2	8.2	E-UTRAN TDD intra-frequency measurements	4
A.8.3	8.3	E-UTRAN FDD inter-frequency measurements	5
A.8.4	8.4	E-UTRAN TDD inter-frequency measurements	5
A.8.5	8.5	E-UTRAN FDD – UTRAN FDD Measurements	3
A.8.6	8.6	E-UTRAN TDD – UTRAN FDD Measurements	1
A.8.7	8.7	E-UTRAN TDD – UTRAN TDD Measurements	3
A.8.8	8.8	E-UTRAN FDD – GSM Measurements	2
A.8.9	8.9	E-UTRAN FDD – UTRAN TDD measurements	1
A.8.10	8.10	E-UTRAN TDD – GSM Measurements	2
A.8.11	8.11	Monitoring of multiple layers	6
A.9.1	9.1	RSRP measurements	8
A.9.2	9.2	RSRQ measurements	6
A.9.3	9.3	UTRA FDD CPICH RSCP	2
A.9.4	9.4	UTRAN FDD CPICH Ec/No	2
A.9.6	9.6	GSM carrier RSSI	2

The UE RRM performance requirements defined in 36.133 [8] are considered to be independent from all bands. Therefore, the required performance in the respective test cases (with the exception of inter-band tests) can be verified in any one of the bands supported by the UE. The test case configurations in Subclauses A.9.1 and A.9.2 are considered to be band-dependent and therefore apply in all of the supported bands in the UE.

7.2.9 Base Station RF Conformance Test Overview

Base station (eNB) conformance testing for LTE is similar to UMTS except for those areas of testing affected by the change to using an OFDMA modulation scheme. The eNB RF conformance tests based on the core specification 36.104 [11] are defined in 36.141 [10]. There are three main sections: RF transmitter characteristics, RF receiver characteristics and RF performance characteristics. These tests are listed in the following sections with comments highlighting differences from UMTS.

7.2.9.1 BS RF Transmitter Characteristics

Table 7.2-9 lists the eNB RF transmitter characteristics test cases defined in 36.141 [10].

Table 7.2-9. eNB RF transmitter characteristics tests

36.141 [10] subclause	Test case
6.2	Base station output power
6.2.6	Home BS output power for adjacent UTRA channel protection
6.2.7	Home BS output power for adjacent E-UTRA channel protection
6.3.1	Resource element (RE) power control dynamic range
6.3.2	Total power dynamic range
6.4.1	Transmitter OFF power
6.4.2	Transmitter transient period
6.5.1	Frequency error
6.5.2	Error vector magnitude (EVM)
6.5.3	Time alignment between transmitter branches
6.5.4	Downlink reference signal power
6.6.1	Occupied bandwidth
6.6.2	Adjacent channel leakage power ratio (ACLR)
6.6.3	Operating band unwanted emissions
6.6.4	Transmitter spurious emissions
6.7	Transmitter intermodulation

The eNB transmitter characteristics tests follow very closely the pattern from UMTS with differences mainly due to the use of OFDMA. The test for time alignment between the transmitter branches is particularly important to LTE because of the widespread use of transmit diversity, spatial multiplexing and beamsteering. The requirement is for a time alignment of 65 ns. This requirement is the same as that in UMTS, which was ¼ of a chip (65 ns).

The downlink RS power test is the equivalent of the primary common pilot channel (CPICH) power accuracy test from UMTS.

545

7.2.9.2 BS RF Receiver Characteristics

Table 7.2-10 lists the eNB RF receiver characteristics test cases defined in 36.141 [10].

Table 7.2-10. eNB RF receiver characteristics tests

36.141 [10] subclause	Test case
7.2	Reference sensitivity level
7.3	Dynamic range
7.4	In-channel selectivity
7.5	Adjacent channel selectivity (ACS) and narrowband blocking
7.6	Blocking
7.7	Receiver spurious emissions
7.8	Receiver intermodulation

Of note is the in-channel selectivity test. This is unique to OFDMA and is a test of the receiver's ability to maintain a particular throughput on an allocation on one side of the DC subcarrier when a larger signal is present on the opposite side. This test checks for IQ distortion in the receiver and is the reverse of the UE transmitter IQ image requirement for in-band emissions.

7.2.9.3 BS RF Performance Requirement

Table 7.2-11 lists the eNB RF performance test cases defined in 36.141 [10].

Table 7.2-11. eNB RF performance tests

36.141 [10] subclause	Test case
8.2.1	Performance requirements of PUSCH in multipath fading conditions transmission on single antenna port
8.2.1A	Performance requirements of PUSCH in multipath fading propagation conditions transmission on two antenna ports
8.2.2	Performance requirements for UL timing adjustment
8.2.3	Performance requirements for HARQ-ACK multiplexed on PUSCH
8.2.4	Performance requirements for high speed train conditions
8.3.1	ACK missed detection requirements for PUCCH format 1a transmission on single antenna port
8.3.2	CQI performance requirements for PUCCH format 2 transmission on single antenna port
8.3.3	ACK missed detection for multi-user PUCCH format 1a
8.3.4	ACK missed detection for PUCCH format 1b with channel selection
8.3.5	ACK missed detection for PUCCH format 3
8.3.6	NAK to ACK detection for PUCCH format 3
8.3.7	ACK missed detection for PUCCH format 1a transmission on two antenna ports
8.3.8	CQI performance requirements for PUCCH format 2 transmission on two antenna ports
8.4.1	PRACH false alarm probability and missed detection

7.2.10 Base Station Test Signals

7.2.10.1 Downlink Test Models

The eNB transmitter conformance tests are carried out using downlink configurations known as E-UTRA test models (E-TM). This concept has been inherited from UMTS although any similarity stops there. The highly flexible nature of the downlink OFDMA modulation scheme means that a large number of parameters are required to fully define any signal. An inspection of the definition of the E-TM in 36.141 [10] Subclause 6.1.1 clearly shows how much more complex the signal structure is compared to UMTS. There are three distinct classes of test model defined, known as E-TM1, E-TM2, and E-TM3. The first and third classes have further subclasses. All test models share the following attributes:

- Defined for a single antenna port, single codeword, single layer with no precoding
- Duration of one frame (10 ms)
- Normal cyclic prefix
- Localized virtual resource blocks, no intra-subframe hopping for PDSCH
- Cell-specific reference signals only—no use of UE-specific reference signals.

The data content of the PDSCH is generated from a sequence of zeros scrambled using a length-31 Gold code according to 36.211 [16]. The reference signals and the primary and secondary synchronization signals are also defined according to 36.211 [16]. The physical channels PBCH, PCFICH, PHICH, and PDCCH all have detailed definitions. For each E-TM every physical signal and physical channel is allocated into the channel at a specific power relative to the RS power. There are six different mappings for each E-TM to take account of the six different channel bandwidths. For E-TM employing power boosting or de-boosting of specific RBs, there is an additional table defining which RB this applies to as a function of the channel bandwidth. Each E-TM is defined for specific use according to Table 7.2-12.

Table 7.2-12. Evolved Test Model mapping to test cases

E-TM	Notes	Test case
E-TM1.1	Maximum power tests	Output power, occupied bandwidth, ACLR, operating band unwanted emissions, transmitter spurious emissions, transmitter intermodulation, reference signal absolute accuracy
E-TM1.2	Includes power boosting and de-boosting	ACLR, operating band unwanted emissions
E-TM2	Minimum power tests	Total power dynamic range (lower OFDM symbol power limit at min power, EVM of single 64QAM PRB allocation (at min power), frequency error (at min power)
E-TM3.1		Total power dynamic range (upper OFDM symbol power limit at max power with all 64QAM PRBs allocated), frequency error, EVM for 64QAM (at max power)
E-TM3.2	Includes power boosting and de-boosting	Frequency error, EVM for 16QAM
E-TM3.3	Includes power boosting and de-boosting	Frequency error, EVM for QPSK

An example of a 5 MHz E-TM3.3 configuration using Agilent Signal Studio signal creation software is shown in Figure 7.2-2.

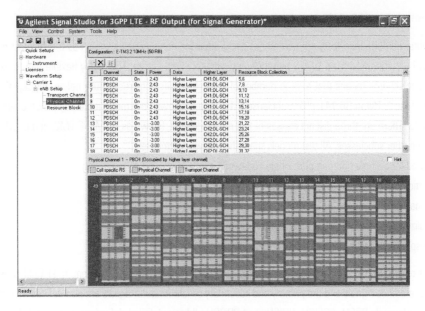

Figure 7.2-2. E-TM3.3 for 5 MHz

This particular signal has had amplitude clipping added to emphasize the impact this type of distortion has on EVM vs. time across the subframe. A measurement of this signal using the Agilent 89600 VSA software can be seen in Figure 7.2-3. The variation in EVM versus time is seen in the top right trace.

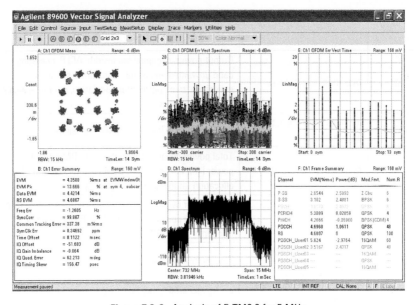

Figure 7.2-3. Analysis of E-TM3.3 for 5 MHz

7.2.10.1 Uplink Fixed Reference Channels

The eNB receiver and performance tests are carried out using uplink fixed reference channels (FRCs) in a similar way to UMTS. The eNB FRC is similar in concept to the RMCs used for UE testing. In most cases these are single-ended signals that can be generated in a signal generator without the need for any real-time feedback.

Table 7.2-13. FRC parameters for performance requirements (64QAM 5/6) (36.141 Table A.5-1 [10])

Reference channel	A5-1	A5-2	A5-3	A5-4	A5-5	A5-6	A5-7
Allocated resource blocks	1	6	15	25	50	75	100
DFT-OFDM symbols per subframe	12	12	12	12	12	12	12
Modulation	64QAM	64QAM	64QAM	64QAM	64QAM	64QAM	64QAM
Code rate	5/6	5/6	5/6	5/6	5/6	5/6	5/6
Payload size (bits)	712	4392	11064	18336	36696	55056	75376
Transport block CRC (bits)	24	24	24	24	24	24	24
Code block CRC size (bits)	0	0	24	24	24	24	24
Number of code blocks–C	1	1	2	3	6	9	13
Coded block size including 12 bits trellis termination (bits)	2220	13260	16716	18444	18444	18444	17484
Total number of bits per subframe	864	5184	12960	21600	43200	64800	86400
Total symbols per subframe	144	864	2160	3600	7200	10800	14400

This example in Table 7.2-13 uses a code rate of 5/6, which is intended for testing the highest throughput requirements. For the 100 RB case of A5-7, there are 86,400 bits per 1 ms subframe indicating a maximum throughput of 86.4 Mbps. The eNB performance requirements measured under fading conditions will be based on reaching a percentage of the maximum throughput under particular conditions. An example from 36.141 [10] Table 8.2.1.5-6 shows that a two channel eNB receiver operating in a pedestrian A channel with 5 Hz Doppler is required to reach 70% of the A5-7 FRC maximum throughput when the SNR is above 19.7 dB.

7.3 UE Signaling Conformance Testing

Section 6.11 discussed the development of signaling protocols and a typical test environment. This section covers the formal UE signaling conformance tests that would be carried out on such systems. Due to the number of signaling conformance tests, it is not possible to go into detail. The intent here is to explain the scope of the testing and the role of each functional area.

7.3.1 Signaling Conformance Test Specifications

Table 7.3-1 shows the specifications that define the signaling conformance tests.

Table 7.3-1. Signaling conformance test specifications

Specification	Title	Purpose
36.523-1 [17]	User quipment (UE) conformance specification; Part 1; Protocol conformance specification	Defines the overall test structure, the test configurations, the conformance requirements and reference to the core specifications, the test purposes and a brief description of the test procedure, the specific test requirements and short message exchange table
36.523-2 [18]	User Equipment (UE) conformance specification; Part 2; Implementation Conformance Statement (ICS) Performance Specification	Based on UE implemented features, provides the ICS proforma in compliance with the relevant EPS requirements and a recommended applicability statement for the 36.523-1 [17] test cases
36.523-3 [19]	User Equipment (UE) conformance specification; Part 3; Abstract Test Suite (ATS)	Provides a detailed and executable description of the test cases written in the test language TTCN-3

7.3.2 Signaling Conformance Test Categories

The signaling conformance test cases for LTE are similar to the UMTS signaling test cases with respect to design and methodology. However the LTE test cases are written in the newer Testing and Test Control Notation (TTCN-3) [20] rather than the TTCN-2 used for UMTS. The test cases are organized based on the layer of the stack being tested. The number of test cases by function is shown in Table 7.3-2.

Table 7.3-2. Signaling conformance test cases by layer

Functional area	Number of tests
Idle mode	73
Layer 2 (MAC/RLC/PDCP)	94
RRC	88
EMM	135
ESM	19
General tests	4
Radio bearer	8
Multi-layer procedures	34
ETWS	2
Non-3GPP	2
Total	**457**

Due to the large number of signaling conformance tests, test development and implementation has been split into four groups based on priorities defined by the network operators. Table 7.3-3 shows the number of test cases in each group. The test cases have been selected to cover the high priority (P1) to low priority (P4) features identified for LTE with some still to be confirmed (TBD). Note that these test cases apply to both FDD and TDD technologies. The GCF manages the priority of FDD test cases in Work Items 81 and 82 and TDD in Work Items 91 and 92.

Table 7.3-3. LTE-only FDD signaling conformance test cases by layer and priority

36.523-1 [17]	Category	Description	GCF work item	P1	P2	P3	P4	TBC	Total
Section 6	Idle mode	Idle mode selection in pure E-UTRAN environment	WI-81/91	2	9	3	7	5	26
Section 7.1	MAC layer	RACH, UL, and DL SCH functionality	WI-81/91	34	4	7	5	0	50
Section 7.2	RLC	UM and AM support	WI-81/91	24	6	0	0	0	30
Section 7.3	PDCP	PDCP functionality— ciphering, deciphering, integrity protection, and handover	WI-81/91	7	7	0	0	0	14
Section 8	RRC	RRC functionality— establishment, release, reestablishment, LTE handover, and measurement reporting	WI-81/91	10	20	11	11	2	54+1
Section 9	EPS mobility management	Attach, detach, tracking area update, paging, authentication, and security procedures	WI-82/92	9	29	51	13	0	102
Section 10	EPS session management	EPS bearer activation/ deactivation, PDN connectivity	WI-82/92	0	8	5	6	0	19
Section 11	General tests	SMS over SGs	WI-82/92	0	4	0	0	0	4
Section 12	E-UTRA radio bearer	Test MIMO configurations	WI-82/92	1	1	4	2	0	8
Section 13	Multi-layer procedures	Multiple procedures	WI-82/92	2	0	3	0	1	5+1
Section 14	ETWS	Earthquake and tsunami warning system	WI-82/92	0	0	0	2	0	2
TOTALS				**89**	**88**	**84**	**46**	**8**	**317**

In addition to the LTE-only test cases, inter-RAT test cases have also been defined as part of 36.523-1 [17]. The GCF has grouped these test cases together and broken them out into several different work items, depending on the technologies required for testing in different bands. Table 7.3-4 shows the test cases defined for UMTS and LTE handovers. Similar groupings of test cases exist for GSM and cdma2000.

Table 7.3-4. UMTS and LTE handover tests

Category ID	Category	Description	GCF work item	P1	P2	Total
Section 6	Idle mode	Idle mode selection in pure E-UTRAN environment	WI-86	15	1	**16**
Section 8	RRC	RRC functionality— establishment, release, reestablishment, LTE handover, and measurement reporting	WI-86	9	2	**11**
Section 9	EPS mobility management	Attach, detach, tracking area update, paging, authentication, and security procedures	WI-86	31	0	**31**
Section 13	Multi-layer procedures	Multiple procedures	WI-86	11	1	**12**
TOTALS				**66**	**4**	**70**

The total number of handover test cases, including the GSM and cdma2000 tests, equals 140, which added to the LTE-only tests gives an overall total of 457 Release 8 test cases.

7.3.3 Signaling Conformance Test Overview

The following sections describe each functional area of the test cases outlined in Table 7.3-2.

7.3.3.1 RRC Idle State

The RRC idle state test cases shown in Table 7.3-5 are split into three subgroups.

Table 7.3-5. 36.523-1 [17] Section 6 RRC idle state tests

Subgroup	Description
Pure E-UTRAN environment	Used to test UE cell reselection and PLMN selection in a pure E-UTRAN environment
Multimode environment	Used to test UE inter-RAT PLMN selection, cell selection/reselection in an environment consisting of ETRAN, UTRAN, and GERAN cells
Closed subscriber group cells	Used to test UE in closed subscriber group cells

7.3.3.2 Layer 2

The layer 2 test cases shown in Table 7.3-6. are split into three subgroups.

Table 7.3-6. 36.523-1 [17] Section 7.1 Layer 2 testss

Subgroup	Description
MAC	Mapping between logical channels and transport channels
	RACH
	DL-SCH data transfer
	UL-SCH data transfer
	MAC reconfiguration
	DRX Operation
	Transport block size support
	Reporting of rank indicator
RLC	Unacknowledged mode
	Acknowledged mode
PDCP	Maintenance of PDCP sequence numbers for radio bearers
	PDCP ciphering/deciphering
	PDCP integrity protection
	PDCP — handover
	PDCP — others

7.3.3.3 RRC Connected State

The RRC connected state test cases shown in Table 7.3-7 are split into five subgroups.

Table 7.3-7. 36.523-1 [17] Section 7.2 RRC connected state tests

Subgroup	Description
RRC connection management procedures	Paging, RRC connection establishment, and RRC connection release procedures
RRC connection reconfiguration	Bearer setup, radio resource reconfiguration, radio bearer release, and intra-LTE cell handover procedures
Measurement configuration control and reporting	Intra-LTE measurements, inter-RAT measurements (GERAN, UTRAN, HPRD, 1XRTT cells), and self optimized networks
Inter-RAT handover	E-UTRA to UTRA
	UTRA to E-UTRA
	E-UTRA to GERAN
	GERAN to E-UTRA
	E-UTRA to HRPD
	HRPD to E-UTRA
	E-UTRA to 1xRTT
Other	Radio link failure, DL direct transfer, UL direct transfer, and UE capability transfer

7.3.3.4 EPS Mobility Management

The EPS mobility management (EMM) test cases shown in Table 7.3-8 are split into five subgroups.

Table 7.3-8. EMM test cases

Subgroup	Description
EMM common procedures	GUTI reallocation procedure, authentication, security mode control procedure, identification, and EMM information procedures
EMM specific procedures	Attach, detach, and tracking area update procedures
EMM connection management procedures	Service request and paging procedures
NAS security	NAS ciphering and integrity protection
Other	Radio link failure, DL direct transfer, UL direct transfer, and UE capability transfer

Both success and failure scenarios are tested for these procedures.

7.3.3.5 EPS Session Management

The EPS session management (ESM) test cases are shown in Table 7.3-9.

Table 7.3-9. 36.523-1 [17] Section 10 ESM test casess

Functional area	Tests
EPS session management	Dedicated EPS bearer context activation
	EPS bearer context modification
	EPS bearer context deactivation
	UE requested PDN connectivity
	UE requested PDN disconnect
	UE requested bearer resource allocation
	UE requested bearer resource modification
	UE routing of uplink packets

Both success and failure scenarios are tested for these procedures.

7.3.3.6 General Tests (36.523-1 [17] Section 11)

The general tests cover SMS over SGs tests.

7.3.3.7 Radio Bearer (36.523-1 [17] Section 12)

The radio bearer tests cover radio bearer interoperability under various conditions with and without MIMO configured. A generic radio bearer test case verifies all the possible combinations of unacknowledged mode (UM) and acknowledged mode (AM) data radio bearers for each UE capability.

This test case is based on the generic radio bearer SRB1 and SRB2 for

DCCH + n * AM DRB + m * UM DRB, where $n = 1..N$ and $m = 0..M$.

7.3.3.8 Multi-Layer Procedures (36.523-1 [17] Section 13)

This section covers intra-system mobility within E-UTRAN and inter-system mobility between E-UTRAN (FDD and TDD) and UTRAN (UMTS), GPRS, cdma2000, and GSM. Table 7.3-10 lists the tests.

Table 7.3-10. Multi-layer procedures

Functional area	Tests
Multi-layer procedures	E-UTRA call setup and activation/deactivation of additional radio bearers
	E-UTRA connection re-establishment after link failure
	Inter-system packet connection re-establishment, E-UTRAN FDD and TDD to UTRAN, GPRS, and cdma2000
	Intra- and inter-frequency mobility, E-UTRA FDD and TDD
	Inter-system mobility for packet data, E-UTRA FDD and TDD to UTRA, GPRS, and cdma2000
	Inter-system mobility for voice, E-UTRA FDD and TDD to UTRA, cdma2000, and GSM CS
	Inter-system session management

7.3.3.9 Earthquake and Tsunami Warning System (36.523-1 [17] Section 14)

The earthquake and tsunami warning system (ETWS) is designed to broadcast emergency warning messages to UE in both idle and connected states.

7.4 UE Certification Process (GCF and PTCRB)

7.4.1 Introduction

The 3GPP core specifications are necessary to design the UE, and the 3GPP conformance tests define how to measure compliance against the core specifications. The last step of the sequence is to deliver certified devices to the market. This step involves the validation of the test systems used to carry out conformance testing and then the execution of the conformance test suites under controlled conditions. For UMTS and LTE, these two final steps are carried out under the management of certification bodies.

There are two main certification bodies involved with LTE. The first certification body is the Global Certification Forum (GCF). This organization was established in 1999 under the umbrella of the GSM Association (GSMA), the industry body representing GSM and UMTS operators worldwide. The GCF was formed in 1999, two years before the European Radio and Telecommunications Terminal Equipment (R&TTE) directive superseded the previous Terminal Directive of 1991. The new directive significantly reduced the scope of the tests required for type approval to just those tests essential to ensure efficient use of the spectrum and no harm to the network. The wider performance requirements that previously formed the bulk of the type approval tests in the 1991 directive were dropped. To fill the vacuum left by this change in European regulations the GCF was formed to promote an industry-backed certification scheme with a remit closer to the scope of the original Terminal Directive. This scheme is non-binding but GSMA members are encouraged to promote only UEs that have achieved GCF certification. In 2008, GCF became a private company.

Although GCF is not responsible for providing type approval certification—that remains the responsibility of accredited test laboratories—GCF is, in effect, managing the entire testing process, which incorporates those tests required for type approval. Type approval tests are the only tests that the UE must pass in order to be operated within the European Union.

The second major certification body was formed in the United States in 1997 at the introduction of the personal communications system (PCS), which was a variation of GSM for the 1900 MHz band. This body is known as the PCS Type Certification Review Board (PTCRB). The PTCRB provides UE certification services for North American operators.

There are other industry certification bodies such as the CDMA Certification Forum (CCF), which provides certification of performance for products designed to the 3GPP2 CDMA standards.

The roles of GCF and PTCRB are largely similar although they focus on those frequency bands and specific test lists appropriate to their markets. Over the years the co-operation between GCF and PTCRB has grown.

Figure 7.4-1 shows the two routes that a UE takes to reach the market. The regulatory certification under the administration of national and regional authorities such as the Federal Communications Commission (FCC) in the US and the CE mark administered by the European Commission are mandatory to get to market. The industry certification provided by GCF, PTCRB, and others is optional but very much encouraged to reach a minimum performance level and enable stable international roaming.

Initially GCF and PTCRB started with certification of products based on the GSM standard. Their scope was expanded to GPRS, EGPRS, W-CDMA, HSDPA, and now HSUPA, with LTE being the next standard covered. More information is available at the following websites: for GCF, http://www.globalcertificationforum.org; for PTCRB, http://www.ptcrb.com.

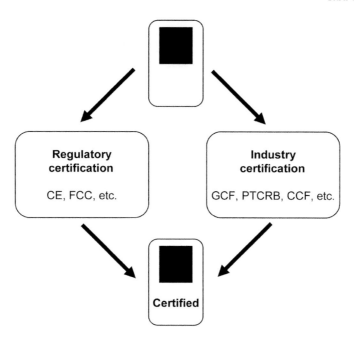

Figure 7.4-1. Regulatory certification and industry certification

7.4.2 Preparation Process for Mobile Handset Certification

Figure 7.4-2 shows the relationship between the standards development body, certification bodies, mobile phone manufacturers, and test industry. Test industry here refers to the test system vendors and the certified test laboratories that validate the test systems and carry out the conformance tests.

The certification bodies do not automatically adopt the entire 3GPP conformance test specifications. Each certification body selects a group of test cases (called a work item) and defines its priority based on operators' deployment plans. In previous systems such as UMTS there were far more conformance tests defined than adopted by GCF, but there is better coordination now between operators, GCF, and 3GPP to ensure that only those tests likely to be implemented and used for certification are developed in detail within 3GPP.

Once a GCF work item is clearly defined, test system vendors develop the test cases required by this work item on their test platforms. Each test system vendor selects a certified test laboratory to validate their test cases on the test platform using reference UEs that support required features for the designated work item.

The test laboratory submits a validation report to the relevant certification body for approval, which then enables the test system to be used to certify any UE. When sufficient tests in a work item have been validated (usually 80%), the certification body declares the work item active and UE certification can start.

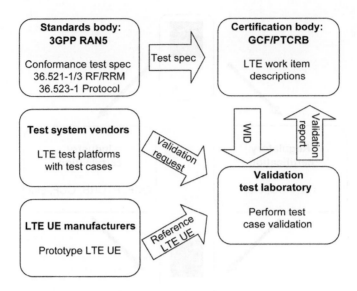

Figure 7.4-2. Development and validation of conformance test cases

7.4.3 Test Case Validation Process

Figure 7.4-3 illustrates in more detail the process of test case validation in accordance with 3GPP conformance test specifications. The test system vendor provides a measurement uncertainty analysis document for the test laboratory to review. This document is then used to validate conformance with 3GPP requirements for test system uncertainty.

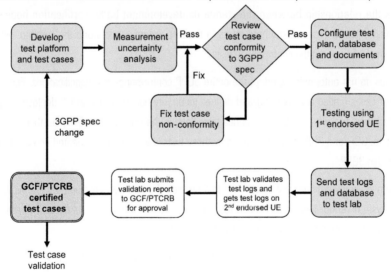

Figure 7.4-3 Test case validation process

Normally test case validation is performed on two reference UEs that use different chipsets in order to ensure that the test cases are usable with various UEs.

7.4.4 LTE UE Certification Plan

The way that GCF has approached LTE UE certification is to divide the work across work items in which FDD and TDD are treated separately, as follows:

- E-UTRA RF Rel-8 FDD
- E-UTRA Protocol Rel-8 FDD
- EPC Protocol Rel-8 FDD
- E-UTRA RF Rel-8 TDD
- E-UTRA Protocol Rel-8 TDD
- EPC Protocol Rel-8 TDD

Each work item (WI) contains a number of the UE conformance tests developed by 3GPP based on predetermined selection criteria that involves all the companies supporting the work items. In this way, GCF reflects market priorities for the first LTE devices. A minimum set of tests needs to be validated from each work item before UE certification can start. This establishes a wide spread of tests that the devices must pass in order to be certified.

Table 7.4-1 describes the current status for each of the work items focused on testing Release 8 E-UTRA functionality. The table shows the speed with which the industry responded to the need for LTE certification testing. The first band, Band XIII, became "active"—that is, ready for certification testing—in Dec 2010, less than two years after the specification was frozen. Additionally, the worldwide adoption of LTE is highlighted by the proliferation of bands being managed by the GCF process. Historically GCF has had a European focus; however, for LTE certification, bands from America, China, Korea, and Japan are now part of the process.

Table 7.4-1 GCF work item status

Work item description	WI number	Bands supported	WI status (August 2012)
E-UTRA RF Rel 8 FDD Tests	WI-80	1, 3, 4, 5, 7, 11, 12, 13, 14, 18, 20, 25	Bands 1, 3, 4, 5, 7, 13, and 20 are active with status frozen. Bands 11, 12, 14, 18, and 25 are active work items.
E-UTRA Protocol Rel 8 FDD Tests	WI-81		
EPC Protocol Rel 8 FDD Tests	WI-82		
E-UTRA RF Rel 8 TDD Tests	WI-90	38, 39, 40, 41	Band 40 is active with status frozen. Bands 38, 39, and 41 are active work items.
E-UTRA Protocol Rel 8 FDD Tests	WI-91		
EPC Protocol Rel 8 FDD Tests	WI-92		

7.4.5 The Performance Agreement Group (PAG)

The work of the GCF is divided into different groups. Historically there have been two key groups focusing on UE acceptance testing: the Field Trial Agreement Group (FTAG) and the Common Agreement Group (CAG).

The GCF originated with a group of operators and, later, device manufacturers who came together with a goal to simplify the requirements for UE acceptance testing on their networks. This work was advanced within GCF by further division into the aforementioned groups focused on field trial testing (FTAG) and conformance testing (CAG). In 2011 GCF created a new group called the Performance Agreement Group (PAG).

This group was formed as a result of two primary drivers:

1. Many operators are performing benchmark testing of key performance indicators (KPIs). Defining some common tests will help unify this testing and improve time to market.

2. The GCF certification regime administered by the CAG uses tests taken from 3GPP RAN WG5, which were developed to test the performance requirements defined in the 3GPP radio specifications. The radio performance requirements are written around functionality defined in the different layers of the signaling and radio protocol stacks. The testing therefore focuses on the individual layers and not on the end-to-end performance of a real application as seen by the end user. For example, current conformance tests are based on fixed reference channels and do not test the closed-loop behavior implemented in networks based on adaptive modulation and coding. Other adaptive algorithms are used further up the signaling protocol stack to optimize data transmission in dynamic conditions. However, to thoroughly test a UE, it is also necessary to test the full end-to-end application performance and not just the performance of individual layers as done during conformance testing.

As mandated by GCF, PAG looks at key performance tests such as end-to-end IP data throughput, battery life, and over the air (OTA) antenna testing (see Section 6.10). This application testing is not intended to form part of the certification regime and no performance requirements will be developed to determine pass or fail limits. The purpose of the work is to define detailed test scenarios so that it will be possible to benchmark device performance in controlled conditions. Any performance expectations will be governed by proprietary agreements between operators and UE vendors.

The GCF does not see its role as that of a standards development organization. As an organization formed of operators and device manufacturers, the GCF makes a contribution by establishing a collective view of how to prioritize tests and then applies those tests to manage a thorough certification process. Historically, GCF has worked with 3GPP RAN WG5 for radio conformance testing and with GSMA for battery life testing. In the new area of end-to-end IP data throughput testing, which is a key area for UE performance, the GCF will work with 3GPP RAN WG5 to define tests that the PAG can adopt. The development of end-to-end performance testing and benchmarking is an area that will see increased focus over the next few years. The first application layer data throughput performance tests are defined in 37.901 [21].

7.5 Operator Acceptance Testing

7.5.1 Introduction

The relatively recent introduction and dramatic growth of so-called "smart devices" such as smartphones, tablets, and increasingly sophisticated machine-to-machine (M2M) modules have resulted in an equally dramatic growth in demand for data and services. LTE and LTE-Advanced will help satiate this demand and enable the new data-centric devices to deliver a plethora of services, entertainment, and business-critical functions.

During this time, network operators have shifted their position from simply supplying "dumb pipes" through which wireless devices communicate with each other to becoming gate-keepers who put demanding performance and quality metrics as entry qualifications to their network. This is done by instituting test methodologies for verifying whether the devices have adequate performance and security functions in place before they are launched on the network. While type approval is the only formal step currently required of all devices before they are allowed to connect to the network, operators frequently do more rigorous conformance and acceptance testing of devices before marketing them to customers. Figure 7.5-1 illustrates the influence that network operators have on the supply chain.

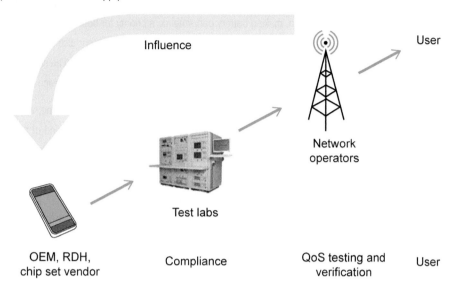

Figure 7.5-1 Network operator influence on the supply chain

Operator acceptance tests consist of performance and functional tests as outlined in the network operator's acceptance test documentation. These tests are scoped more broadly than the GCF- and PTCRB-validated conformance tests. Network operators, drawing on the experience gained over many years of providing 2G and 3G services, develop acceptance tests with the goal of ensuring that the devices promoted for their network will meet customer expectations. Thus an operator whose network is marketed as the fastest and most reliable will be motivated to ensure that the devices approved for use on that network perform as promised under many different network conditions.

7.5.2 Acceptance Tests

Acceptance tests are designed around specific characteristics and conditions of the wireless network and cover a multitude of scenarios that stress both network and UEs to ensure that problems are detected and resolved before they can affect end users. Network operators are particularly concerned about perceived quality of service, which can be greatly affected by the performance of network devices. New features and capabilities are being introduced on devices in ever-shorter life cycles. While new features or performance enhancements may strike a chord with customers, they add complexity to the user equipment as well as to network operation. Operators use acceptance tests to ensure that enhancements do not degrade the quality or usability of devices. Unlike conformance tests, which are publicly available, the details of an operator's acceptance tests are usually confidential. Network operator acceptance tests tend to differ in specifications tested, measurement limits, and implementation.

7.5.2.1 Usage Profiles and Battery Performance

Designers of user equipment are continually being pushed to add more power-consuming features to smartphones and tablets while also reducing the size and weight of these devices and extending their battery life. Consumers get frustrated when batteries perform poorly during real-world operation, making battery performance a priority for both network operators and UE vendors. Operators push the UE vendors for extended battery life not only to satisfy end users but also to encourage greater consumption of pay-for-use data services, which provides a critical revenue stream. A UE vendor who can supply a phone that operates longer between charges can gain an important competitive advantage in the crowded mobile device market.

Previous testing of battery performance focused primarily on how well the battery lasted during talk time, while the device was sending and receiving text messages, and while the device was in idle mode for an extended period of time. However, the devices available today are able to manage multiple, complex tasks simultaneously, so much more sophisticated testing is required in order to stress the phone to reflect actual customer use.

Evaluating battery performance in real-world situations must address a seemingly endless number of permutations and combinations of activities for which the phone is used. The only way to manage the potential complexity is to limit the number of usage scenarios to a few representative cases so that devices can be compared to each other as well as be tested against limits established by the operator.

One method that is employed is to create different usage profiles for specific customer groups. Once these profiles are defined, test scenarios are developed that enable the operator to test for adequate battery performance under the predetermined usage conditions.

Typical profiles reflect usage of the following features and services:
- Data consumption
- Email and web surfing
- Talk time
- SMS/MMS
- Backlight usage
- Camera
- Location-based services
- Amount and speed of travel between cells.

Operators develop their own unique sets of customer profiles based on the services and features listed above. Table 7.5-1 shows examples of typical usage by different groups.

Table 7.5-1 Example customer profiles

Feature	Description	Teenager	Soccer mom	Business user	Grandparent
Data consumption	Online gaming, video streaming, music and movie downloads	Very high	Low	Medium	Low
Email and web surfing	Social networking, web browsing, email applications	Medium	Low	High	Low
Talk time	Voice calls	High	Very high	Very high or high (depends on job)	High
SMS/MMS	Texting, sending photos to friends	Very high	Medium	Low	Low
Back light	Using phone other than voice or at night	Very high	Medium	High	Low
Camera	Taking photos or video	Very high	Low	Low	Low
Location-based services	Navigation, geo-tagging	Low	High	High	Low
Mobility	Using different modes of transport	Low	High	Very high	Low

7.5.2.2 Working With Existing Infrastructure

Today, services such as voice and SMS are still handled primarily by existing circuit-switched network elements. Until the LTE network is fully deployed, maintaining a satisfactory service will depend heavily on the ability of UEs to gracefully manage handovers from one LTE cell to another LTE cell as well as handovers from an LTE cell to a 2G/3G cell when the UE strays out of LTE coverage. Additionally, when a UE is on a data-only call and a voice page is received, the UE must be able to fall back to the circuit-switched network to complete the voice call while keeping the data connection alive. Testing handovers between different radio access technologies (RATs) is becoming increasingly important in the verification of LTE UEs. For a positive end-user experience, the UE needs to transition smoothly between different RATs and must also be able to handle multiple handovers during a single session.

Although conformance testing is necessary, it is not sufficient to determine the user experience during all operator-specific handover scenarios. As a result, developers and network operators are increasing their focus on testing the real-world handover performance of UEs before deploying them on live networks. Most network operators have developed their own sets of interRAT test plans, which are designed to test how well a candidate UE performs during such handovers. Test systems based on these test plans have already been deployed and are in service today. These inter-RAT systems provide not only pass or fail information related to individual test cases but also detailed log files that can be analyzed by the network operator and UE vendor for fault analysis and performance optimization.

7.5.2.3 Automated Testing

As more devices are designed and submitted to the network operator for acceptance testing, the strain on the operator's test facility may increase. Therefore test methods must be highly effective, efficient, reliable, and repeatable. Stress-testing a device in a lab under simulated environmental conditions is ideal to fully understand how a device is behaving and to analyze what happened if something goes wrong. However, executing complex stress tests manually can be very time consuming. The ultimate goal is to automate as many of the test cases in a given test plan as possible, which frees the operator to focus on identifying and troubleshooting any problems that arise. Additionally, complete test reports produced for each device at the end of an automated process make for a better dialogue between UE vendor and network operator should this be necessary.

7.5.3 Influence of Network Operators in the Cellular Ecosystem

It is both costly and time-consuming for network operators when multiple UE vendors submit new devices for acceptance testing every few months. Therefore it is expected that eventually the UE vendors, rather than submit their new devices to the network operator for acceptance testing, will themselves perform the operator-defined acceptance tests. Alternatively, the UE vendors may elect to outsource such testing to operator-approved third party labs. If a device does not pass acceptance testing at this stage, the UE vendor will be able to take immediate action to resolve the problem.

Due to the complexity and proprietary nature of acceptance tests, operators need to approve the exact implementation of the test systems used by UE vendors. An approved list of test systems ensures that any test carried out prior to formal acceptance testing will be done to the same standards as those used by the operator.

7.6 References

[1]   3GPP TS 36.508 V10.1.0 (2012-06) Common Test Environments for User Equipment (UE) Conformance Testing

[2]   3GPP TS 36.509 V10.0.0 (2012-09) Special Conformance Testing Functions for User Equipment (UE)

[3]   3GPP TS 36.124 V11.1.0 (2012-06) Electromagnetic Compatibility (EMC) Requirements for Mobile Terminals and Ancillary Equipmentt

[4]   3GPP TS 36.521-1 V10.2.0 (2012-06) User Equipment (UE) Conformance Specification; Radio Transmission and Reception Part 1: Conformance Testing (FDD/TDD)

[5]   3GPP TS 36.101 V11.1.0 (2012-06) UE Radio Transmission and Reception

[6]   3GPP TS 36.521-2 V10.2.0 (2012-06) User Equipment (UE) Conformance Specification; Radio Transmission and Reception Part 2: ICS

[7]   3GPP TS 36.521-3 V10.1.0 (2012-06) User Equipment (UE) Conformance Specification; Radio Transmission and Reception Part 3: Radio Resource Management Conformance Testing

[8]   3GPP TS 36.133 V11.1.0 (2012-06) Requirements for support of Radio Resource Management

[9]   3GPP TS 36.113 V11.1.0 (2012-06) Base Station (BS) and Repeater Electromagnetic Compatibility (EMC)

[10]   3GPP TS 36.141 V11.1.0 (2012-06) Base Station (BS) Conformance Testing

[11]   3GPP TS 36.104 V11.1.0 (2012-06) Base Station Radio Transmission and Reception

[12]   3GPP TS 36.143 V10.4.0 (2012-06) FDD Repeater Conformance Testing

[13]   3GPP TS 36.106 V10.4.0 (2012-06) Repeater Radio Transmission and Reception

[14]   ETR 273 (1998-02) Uncertainties in the Measurement of Mobile Radio Equipment Characteristics,
       Part 1, sub-part 2, subclause 6.5

[15]   ITU-R SM.329-10 (Feb 2003) Unwanted emissions in the spurious domain

[16]   3GPP TS 36.211 V10.5.0 (2012-06) Physical Channels and Modulation

[17]   3GPP TS 36.523-1 V10.1.0 (2012-07) User Equipment (UE) conformance specification; Part 1; Protocol conformance
       specification

[18]   3GPP TS 36.523-2 V10.1.1 (2012-07) Implementation Conformance Statement (ICS) Proforma Specification

[19]   3GPP TS 36.523-3 V10.0.0 (2012-07) User Equipment (UE) conformance specification; Part 3: Abstract Test Suites
       (ATS)

[20]   Testing and Test Control Notation Version 3 (TTCN-3). Available from http://www.ttcn-3.org/.

[21]   3GPP T 37.901 V11.4.0 (2012-09) User Equipment (UE) application layer data throughput performance

Links to all reference documents can be found at www.agilent.com/find/ltebook.

Chapter 8

Looking Towards 4G: LTE-Advanced

8.1 Summary of Release 8

The baseline LTE radio access network (RAN) and evolved packet core (EPC) network were defined in 3GPP Release 8, which was functionally frozen in December 2008. This provided the world with a comprehensive and highly capable new cellular communications standard that, according to a November 2012 Global Suppliers Association report, has been successfully launched in 113 commercial networks in 51 countries. The main attributes that differentiate this new standard from previous standards are the following:

- Single-channel peak data rates of up to 300 Mbps in the downlink and 75 Mbps in the uplink
- Improved spectral efficiency over legacy systems, particularly for the uplink
- Full integration of FDD and TDD access modes
- Packet-based EPC network to eliminate cost and complexity associated with legacy circuit-switched networks.

Some of the key technologies introduced in Release 8 that enable the new capabilities include:

- Adoption of OFDMA and SC-FDMA for the downlink and uplink air interfaces to enable narrowband scheduling and efficient support of spatial multiplexing
- Support for six channel bandwidths from 1.4 MHz to 20 MHz to enable high data rates and also efficient spectrum re-farming for narrowband legacy systems
- Baseline support for spatial multiplexing (MIMO) of up to four layers on the downlink
- Faster physical layer control mechanisms leading to lower latency.

Despite the substantial capabilities of LTE in Release 8, the 3GPP standard has continued to evolve through Releases 9, 10, 11, and now 12. The following sections summarize important additions to the LTE specifications that have been made since the first edition of this book. These include the most significant changes to the 3GPP standard, which occurred in Release 10 for the support of LTE-Advanced, 3GPP's submission to the ITU-R IMT-Advanced (4G) program. Full information about 3GPP releases can be found at www.3gpp.org/Releases. The most useful summary of the work items for each release can be found at ftp.3gpp.org/Information/WORK_PLAN/Description_Releases. This web page includes a short explanation of each work item. A complete list of all work items from Release 4 through Release 12 (nearly 4,000 items) can be found at ftp.3gpp.org/Specs/html-info/WI-List.htm. The list provides links to the specification documents impacted by each work item plus the specific change requests.

LTE and the Evolution to 4G Wireless: Design and Measurement Challenges, Second Edition, Edited by Moray Rumney.
Copyright Agilent Technologies, Inc. 2013. Published by John Wiley & Sons, Ltd.

8.2 Release 9

Release 9 was considered a "short" release in that it came between the major effort required to finish Release 8 and the definition of Release 10. The work on Release 9 was done with the knowledge that significant changes were due in Release 10 as part of the plans for LTE-Advanced. Some of the items in Release 9 were carryovers from Release 8 that had not yet been completed; others were new items not in the original Release 8 definition. At a formal level, Release 9 included over 80 identifiable features. Since it is not within the scope of this book to go through each Release 9 feature individually, a few of the key items have been selected here for further explanation, with a focus on the radio aspects.

8.2.1 New Frequency Bands

Every release introduces new frequency bands. As discussed in Section 2.1.1 and defined in 36.307 [1], new frequency bands are specified independent of release. This is a pragmatic approach to managing the evolving specifications such that bands defined in later releases can be applied to an earlier release without the need to modify that release. Within Release 9 four FDD bands were added as shown in Table 8.2-1.

Table 8.2-1. Frequency bands added during Release 9

Band number	Uplink		Downlink		Bandwidth	Duplex spacing	Gap	Duplex mode
	Low	High	Low	High				
18	815	830	860	875	15	45	30	FDD
19	830	845	875	890	15	45	30	FDD
20	832	862	791	821	30	-41	11	FDD
21	1447.9	1462.9	1495.9	1510.9	15	48	33	FDD

Bands 18 and 19 are referred to as the extended LTE 800 bands and were specified for use in Japan. The background can be found in 36.800 [2]. Band 20 was added for the so-called "digital dividend" spectrum within Europe made available through the switchover to digital television. Note that the uplink and downlink frequencies are reversed from the usual arrangement. The background to band 20 can be found in 36.810 [3]. The final new band is the extended LTE 1500 band in Japan. The background can be found in 36.821 [4].

8.2.2 Home Base Station

Work on femtocell inclusion in UMTS was ongoing during Release 8, and this work continued in Release 9 for the home base station (home BS, also known as home eNB or femtocell). The femtocell concept is not unique to LTE or LTE-Advanced, but there was an opportunity for LTE to incorporate this technology from the start rather than retrospectively designing it into legacy systems such as UMTS and GSM. Figure 8.2-1 shows the topology of a femtocell deployment.

From a radio deployment perspective the femtocell operates over a small area within a larger cell. The radio channel can be the same as that of the larger cell (known as co-channel deployment) or on a dedicated channel. The femtocell concept is fundamentally different from relaying (covered in 8.3.3.4) since the femtocell connection back into the core network is provided locally by an existing wired internet connection rather than over the air back to the macrocell. Most femtocell deployments will be indoors, which helps provide isolation between the femtocell and the macrocell. Also shown in Figure 8.2-1 is a femtocell outside the coverage area. This could be, for example, a way to provide local cellular coverage in rural areas where DSL exists but there is no cellular coverage provided by the operator.

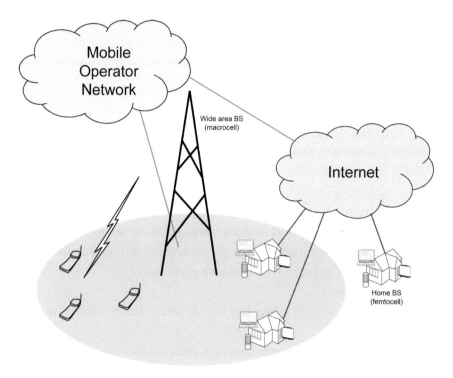

Figure 8.2-1. Femtocell (Home eNB)

Although the name femtocell suggests that the major difference from existing systems is one of coverage area, the defining attributes of femtocells are actually far more significant than coverage area alone. Table 8.2.2 compares the main attributes of traditional macro-, micro- and picocell systems with those of femtocells.

Table 8.2-2. Traditional cellular versus femtocellular technology

Attribute	Traditional cellular	Femtocellular
Infrastructure cost	$10,000–100,000	$100–200
Infrastructure finance	Operator	End user
Backhaul	Expensive leased E1/T1 lines	Existing end-user DSL or cable broadband
Planning	Operator	End-user (no central planning)
Deployment	Operator truck roll	End user one touch provisioning
Quality of Service (QoS)	Operator controlled	Best effort
Control	Operator via O&M	Operator via Internet
Mobility	Good/excellent	Nomadic/best effort
Data throughput	Limited	Excellent

The two main deployment scenarios for femtocells are as follows:

- In rural areas with poor or no coverage, probably using co-channel deployment
- In dense areas to provide high data rates and capacity (primarily indoors).

In both cases it must be decided whether the femtocell will be operated for closed subscriber group (CSG) UE or for open access. This along with other practical considerations such as pricing can be considered commercial issues, although in the co-channel CSG case (see 8.3.4.1), the probability that areas of dense femtocell deployment will block macrocells becomes an issue.

Although the potential gains from femtocells are substantial, many challenges remain:

- Need for cognitive methods to reduce interference to the macro network
- Need for radio resource management requirements
- Security concerns from making base station technology widely available—including backhaul protection, device authentication, and user authentication
- Verification of geographic location and roaming aspects
- Business models of open versus closed access
- No obvious solution yet for cross-network femtocells—any one femtocell can support only one network operator. Therefore, household members can only share a single femtocell if they all choose the same network operator.
- Net neutrality—who owns the backhaul? The answer will vary by country.
- Possible public safety concerns
- Need for optimized and balanced interworking between macro- and femtocells to minimize unnecessary handovers (ping pong)
- Potential bottleneck over fixed broadband backhaul (such as DSL or cable) connection, especially on the uplink for services requiring symmetric bandwidths, prioritization, and congestion management
- QoS control for real-time services and applications requiring guaranteed bit rates, such as voice, with all the other traffic types on the broadband access network
- Access control providing CSG local and roaming access
- Self-configuration (plug and play), self-organization, and self-optimization including fault management and failure recovery (self-healing).

Despite these issues, studies have shown that increases in average data rates and 100 times greater capacity are possible with femtocells over what can be achieved from the macro network. On the other hand, femtocells do not provide the mobility of macrocellular systems, and differences exist in the use models of these systems, as shown in Table 8.2-3.

Table 8.2-3. Comparison of macro- and microcellular with femtocellular use models

Macro-/microcellular	Femtocell/hotspot
Ubiquitous mobile data and voice	Opportunistic nomadic data
Mobility and continuous coverage	Hotspot coverage
Ability to control QoS	Limited QoS for lower value data
Limited capacity and data rates	Distributed cost (not low cost)
High costs, acceptable for high value traffic	Free or charged
Often outdoors and moving	Sitting down indoors

For these reasons, femtocellular and hotspot deployments should be considered as complementary rather than competitive with the macro-/microcellular systems.

The background to the home BS Release 9 work can be found in 36.921 [5] for FDD and 36.922 [6] for TDD. The work had two objectives: first, to complete the RF specifications in 36.104 [7] for the introduction of the home BS class and second, to introduce features in the home BS and network that enable control of the home BS output power, in order to mitigate interference to the macro network or other home BS. A number of relaxations to the RF specifications were introduced, not least in importance the maximum output power, which is limited to 20 dBm and lower in some scenarios. The expected low UE speeds in home BS deployments enabled a five times looser requirement for frequency error and there are various other relaxations for spurious emissions.

However, to enable effective interference mitigation, the home BS must be able to measure the signal strength of other base stations in the neighborhood. Downlink measurement is not an issue for TDD, but for FDD a downlink measurement function is required in the home BS although some measurements may also be gathered from UEs connected to the home BS.

The need for interference mitigation is most important when the home BS is deployed in a co-channel closed subscriber group. In this mode the home BS is deployed on the same frequency as the macro network and access to the home BS is restricted to a closed group of users. In this environment, UEs not part of the closed group would likely experience a loss of coverage when close to the home BS whereas UEs that are part of the closed group would hand over to the home BS. For this reason it is important to limit the potential for the home BS to interfere with the macro network when the home BS is operated in a co-channel CSG mode. The general term applied to this form of interference mitigation is inter-cell interference coordination (ICIC). Interference mitigation work continues in Release 10 with enhanced ICIC or eICIC and in Release 11 further enhanced ICIC (FeICIC) is introduced. These developments are described later.

8.2.3 Multimedia Broadcast Multicast Service (MBMS)

The MBMS television service was specified at the physical layer in Release 8 but was not functionally complete until Release 9. The features in Release 9 provide a basic MBMS service carried over an MBMS single frequency network (MBSFN).

In Release 9 only the guaranteed bit rate (GBR) bearers were specified, which means that the maximum bit rate (MBR) is always equal to the GBR. This is not good for variable bit rate services which, by exploiting statistical multiplexing, would otherwise allow the MBR to exceed the GBR.

Another limitation of the Release 9 definition was the lack of a feedback mechanism from the UEs that would inform the network if sufficient UEs were present in the target area to justify turning on the MBSFN locally.

In Release 11 further MBMS enhancements for service continuity were specified including support on multiple frequencies, reception during RRC idle and RRC connected states, and support to take UE positioning into account for further optimization of the received services.

8.2.4 Positioning Support

Positioning support work included specifications for support of the Assisted Global Navigation Satellite System (AGNSS) in 36.171 [8]. The GNSS incorporates the following satellite positioning systems:

- Galileo
- Global Positioning System (GPS) and modernized GPS
- GLObal'naya NAvigatsionnaya Sputnikovaya Sistema (GLONASS)
- Quazi-Zenith Satellite System
- Space Based Augmentation System (SBAS).

The LTE physical layer was augmented to support the observed time difference of arrival (OTDOA) positioning scheme with the introduction of the positioning reference signal (PRS). See Section 3.2.12.3. Network-based positioning for LTE was included in Release 11 with a further study item in Release 12 on positioning based on RF pattern matching.

8.2.5 RF Requirements for Multicarrier and Multi-RAT Base Station

This enhancement is better known as multi-standard radio (MSR). It is introduced in Section 2.1.10 with design and test aspects covered in Section 6.4.7. The work was continued in Release 10 with non-contiguous (inter-band) MSR and in Release 11 with the specification of a medium-range and local area MSR base station classes.

8.2.6 RF Requirements for Local Area Base Stations

The local area BS (picocell), along with the home BS (femtocell), is another important introduction to the LTE specifications. The local area BS enables the deployment of a heterogeneous network comprising macro- (wide area BS), pico-, and femtocells. The RF requirements for local area base stations are based on a reduced UE-to-BS coupling loss of 45 dB compared to the 70 dB used for macrocells. This allows for a lower maximum output power requirement of 24 dBm and other relaxations such as for frequency error and unwanted emissions consistent with small cell deployment.

8.2.7 Enhanced Dual-Layer Transmission

Release 8 specified seven downlink transmission modes (TMs). Transmission mode 7 (TM7) introduced the concept of UE-specific reference symbols (RS), described in Section 2.4.4.7, which enabled non-codebook precoding of the physical downlink shared channel (PDSCH) for single layer transmission. In Release 9 TM8 was introduced, which adds dual-layer transmission to TM7. See Section 2.4.4.8 for further details.

8.2.8 Self Organizing Networks (SON)

Today's cellular systems are very much centrally planned and the addition of new nodes to the network involves expensive and time-consuming work, site visits for optimization, etc. The background to SON can be found in 32.500 [9].

This technical report identified a number of use cases in which SON could be applied:

- Automation of neighbor relation lists in the E-UTRAN and UTRAN and between different 3GPP radio access technologies
- Self-establishment of a new eNB in the network
- Self-configuration and self-healing of the BS
- Automated coverage and capacity optimization
- Optimization of parameters affected by troubleshooting
- Continuous optimization to accommodate dynamic changes in the network
- Automated handover optimization
- Optimization of QoS-related radio parameters.

The use cases are further elaborated in 36.902 [10]. Release 8 introduced a basic version of SON including automatic neighbor relations (ANR) list management and self-establishment of new base stations.

In Release 9 SON was extended to include the following operation and maintenance features:

- Load balancing
- Handover parameter optimization.

The SON work was continued in Release 10 with specification of the management aspects for the following:

- Interference control
- Capacity and coverage optimization
- Random access channel (RACH) optimization.

The concept of self-healing was also developed in Release 10. This feature involves the detection and, analysis of network faults and identification of the corrective action required of the network to respond to disruptive events with minimal manual intervention.

8.3 Release 10 and LTE-Advanced

Release 10 developed the 3GPP proposal for the International Telecommunications Union Radiocommunication Sector (ITU-R) International Mobile Telecommunications Advanced (IMT-Advanced) program. This program is often referred to as "4G" although that term in not formally defined by the ITU or any other body. Due to the involvement of ITU-R in setting the requirements for Release 10, the specification process was more complicated than for previous or subsequent releases:

- ITU-R defined the requirements for IMT-Advanced in ITU-R M [IMT-TECH] [11].
- 3GPP defined requirements for LTE-Advanced in 36.913 [12] to meet or exceed the ITU-R requirements in [11].
- 3GPP undertook a feasibility study 36.912 [13] that proposed LTE-Advanced as an IMT-Advanced candidate technology.
- 3GPP then created work items to develop the many detailed specification in Release 10 to define LTE-Advanced.

To explain the evolution to IMT-Advanced it is helpful to remember what came before. The term "third generation" (3G) has been widely and consistently used to describe the ITU-R's IMT-2000 cellular communications project. The requirements for IMT-2000, defined in 1997, were quite simple, being expressed in terms of peak user data rates:

- 2048 kbps for indoor office
- 384 kbps for outdoor to indoor and pedestrian
- 144 kbps for vehicular
- 9.6 kbps for satellite.

Early 3G systems, of which there were five, did not immediately meet the high peak data rate targets in practical deployment although they did in theory. However, later improvements to the standards brought deployed systems closer to and even beyond the original 3G targets. From a 3GPP perspective, the addition of high speed downlink packet access (HSDPA) to UMTS ushered in the informally named 3.5G, and the subsequent addition of the enhanced dedicated channel (E-DCH), better known as high speed uplink packet access (HSUPA), completed 3.5G. The combination of HSDPA and HSUPA is now referred to as high speed packet access (HSPA).

At the time the IEEE 802.16e standard (Mobile WiMAX) was being developed, and later 3GPP's LTE/SAE, the ITU-R framework for IMT-Advanced (at the time informally referred to as 4G) was not in place. For this reason the term 3.9G was widely used to describe the first release of LTE and sometimes 802.16e with the expectation of their evolving towards official "4G" status in due course. However, 802.16e was also described by some as 4G and more recently 4G has been used to describe the evolution of UMTS HSPA. This inconsistent application of the term 4G has effectively devalued its meaning to that of a marketing term used to describe anything new and fast. It is therefore always more accurate to use the term IMT-Advanced when referring to the ITU-R "4G" program.

The formal definition of IMT-Advanced was developed by Working Party 5D of the ITU-R. A timeline of the program and the parallel 3GPP activities for LTE-Advanced is shown in Figure 8.3-1.

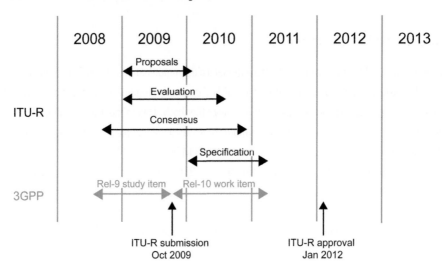

Figure 8.3-1. Overall IMT-Advanced and LTE-Advanced timeline

n naming its IMT-Advanced program, the ITU consciously reused the "IMT" (International Mobile Telecommunications) from he IMT-2000 program. This naming is significant because it was agreed that spectrum currently allocated for exclusive use y IMT-2000 technologies will now be known as just "IMT" spectrum and will be available to any approved IMT-Advanced echnology. At the 2007 World Radio Conference (WRC-07), new IMT spectrum was identified in the following bands: 50 MHz, 698–960 MHz, 2.3 GHz, and 3.4–4.2 GHz. Crucially, no plans for exclusive IMT-Advanced spectrum were proposed. This is pragmatic since spectrum is scarce and largely occupied. Also noteworthy was the addition in 2008 of 802.16e to the st of approved IMT-2000 technologies. This addition opened up the entire IMT spectrum to access by 802.16e prior to the ossibility of 802.16m gaining the same access via the IMT-Advanced route. 802.16m is a planned enhancement to 802.16e, and it was submitted to the ITU-R as a candidate IMT-Advanced technology known as Wireless Mobile Area Network Advanced (Wireless MAN-Advanced).

After considering the attributes of the candidate technologies, the ITU in January 2012 formally approved both LTE-Advanced and Wireless MAN Advanced as meeting the requirements of the IMT-Advanced program. In the period between formal submission and acceptance, 3GPP developed the specifications for LTE-Advanced in Release 10, which were functionally rozen in March 2011.

t is worth pointing out that both of the approved IMT-Advanced technologies are based heavily on pre-existing standards and the modifications that were required of these technologies to meet IMT-Advanced requirements are not considered major, in particular for LTE Release 8, which already met most of the IMT-Advanced requirements.

8.3.1 IMT-Advanced and LTE-Advanced High Level Requirements

The high level requirements for IMT-Advanced defined by ITU-R in [11] are the following:

- A high degree of common functionality worldwide while retaining the flexibility to support a wide range of local services and applications in a cost efficient manner
- Compatibility of services within IMT and with fixed networks
- Capability of interworking with other radio access systems
- High quality mobile services
- User equipment suitable for worldwide use
- User-friendly applications, services, and equipment
- Worldwide roaming capability
- Enhanced downlink peak data rates to support advanced services and applications (100 Mbps for high mobility and 1 Gbps for low mobility were established as targets for research).

The first seven of the eight requirements are "soft" and are largely being pursued by the industry already. However, the eighth requirement, for 100 Mbps high mobility and 1 Gbps low mobility, is quite a different matter and has fundamental repercussions on system design. The 1 Gbps peak target for IMT-Advanced is akin to the 2 Mbps target for IMT-2000 set some ten years earlier. Like its predecessor, the 1 Gbps peak figure is not without qualification since it applies only for low mobility in excellent radio conditions and may require up to 100 MHz of spectrum. Nevertheless, if publicity focuses on the peak rates without taking account of the caveats, the expectations for IMT-Advanced may outstrip practical reality for what could be a long time.

The work by 3GPP to define a candidate radio interface technology (RIT) started in Release 9 with a study phase for LTE-Advanced. The requirements for LTE-Advanced have been captured in 36.913 "Requirements for Further Advancements for E-UTRA (LTE-Advanced)" [12]. These requirements are defined based on the ITU-R requirements for IMT-Advanced as well as on 3GPP operators' own requirements for advancing LTE. Key elements include the following:

- Continual improvement to the LTE radio technology and architecture
- Scenarios and performance requirements for interworking with legacy radio access technologies (RATs)
- Backward compatibility of LTE-Advanced with LTE; i.e., an LTE terminal can work in an LTE-Advanced network, and an LTE-Advanced terminal can work in an LTE network. Any exceptions will be considered by 3GPP.
- Account to be taken of recent WRC-07 decisions for new IMT spectrum as well as existing frequency bands to ensure that LTE-Advanced accommodates geographically available spectrum for channel allocations above 20 MHz. Also, requirements must recognize those parts of the world in which wideband channels will not be available.

8.3.2 IMT-Advanced and LTE-Advanced Detailed Requirements

When IMT-2000 was defined, the only requirements were for peak data rates with no targets for latency or for the more important average or cell-edge performance, which defines the experience for the typical user. Fortunately, this requirement gap has been eliminated with IMT-Advanced, which specifies a much broader range of performance. The ITU-R requirements, specified in [11], were used by 3GPP along with operator requirements to develop TR 36.913 [12], which defines detailed requirements for LTE-Advanced in the following areas:

- Peak data rate: 1 Gbps downlink, 500 Mbps uplink
- Latency
  - Control plane: idle to connected < 50 ms, un-sync to in-sync < 10 ms (see Figure 8.3-2)
  - User plane: Improvements over Release 8 for with and without scheduling assignment
- Spectral efficiency
  - Peak spectral efficiency—see Table 8.3.1
  - Average spectral efficiency—see Table 8.3.1
  - Cell-edge user data throughput—see Table 8.3.1
  - VoIP capacity
- Mobility
  - Support for up to 350 km/h and for some frequency bands 500 km/h
  - Enhanced performance for 0–10 km/h over Release 8 with no degradation and preferred enhancement for higher speeds
- Further enhancements to MBMS—improved requirements for spectrum efficiency over Release 8.

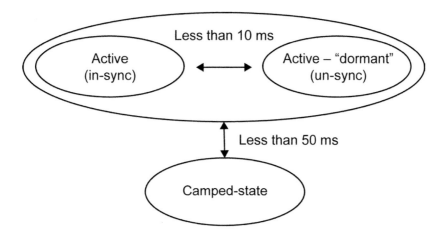

Figure 8.3-2. Requirements for state transitions (36.913 [12] Figure 7.1)

The link between LTE's targets and the retrospective performance derived for UMTS Release 6 is given in Chapter 1 of this book in Tables 1.4-3 and 1.4-4. Looking forward, the relationship between a few of the LTE targets and those for LTE-Advanced and IMT-Advanced is given in Table 8.3-1. The cell and cell-edge spectral efficiency figures are for inter-site distance (ISD) of 500 m.

Table 8.3-1. LTE, LTE-Advanced and IMT-Advanced spectral efficiency performance targets

Item	Sub-category	LTE (Release 8) target [14]	LTE-Advanced target [12]	IMT-Advanced requirement [11]
Peak spectral efficiency (b/s/Hz)	Downlink	16.3 (4x4 MIMO)	30 (8x8 MIMO or less)	15 (4x4 MIMO)
	Uplink	4.32 (64QAM SISO)	15 (4x4 MIMO or less)	6.75 (2x4 MIMO)
Downlink cell spectral efficiency b/s/Hz/user Microcellular 3 km/h, 500 m ISD	(2x2 MIMO)	1.69	2.4	2.6
	(4x2 MIMO)	1.87	2.6	
	(4x4 MIMO)	2.67	3.7	
Uplink cell spectral efficiency b/s/Hz/user Microcellular 3 km/h, 500 m ISD	(1x2 MIMO)		1.2	1.8
	(2x4 MIMO)		2.0	
Downlink cell-edge user spectral efficiency (b/s/Hz/user), (5 percentile, 10 users), 500m ISD	(2x2 MIMO)	0.05	0.07	0.075
	(4x2 MIMO)	0.06	0.09	
	(4x4 MIMO)	0.08	0.12	
Uplink cell-edge user spectral efficiency (b/s/Hz/user), (5 percentile, 10 users), 500m ISD	(1x2 MIMO)		0.04	0.05
	(2x4 MIMO)		0.07	

The first point of note is that the peak efficiency targets for LTE-Advanced are substantially higher than the IMT-Advanced requirements—thus the desire to drive up peak performance is maintained despite the average targets and requirements being very similar. However, 36.913 [12] states: "The target for average spectrum efficiency and the cell edge user throughput efficiency should be given a higher priority than the target for peak spectrum efficiency and VoIP capacity." Another point of note is that with the exception of uplink spectral efficiency, LTE Release 8 meets the requirements for IMT-Advanced. The next section will discuss the challenge of raising the average and cell edge performance.

Improving the Average and Cell Edge Spectral Efficiency

As discussed in Chapter 1, the LTE targets for average and cell-edge spectrum efficiency are based on 2x to 4x improvements to Release 6 HSPA. The reference HSPA configuration is receive diversity with no equalizer for the downlink and a single transmitter for the uplink (25.913 [15] Subclause 7.1). This reference configuration was analyzed during the LTE study phase to provide an average downlink cell spectral efficiency of around 0.53 b/s/Hz/cell using a 500 m ISD [14].

Increasing peak data rates by using more spectrum or higher-order modulation in a good radio environment is a well-understood process that has been in use for years. However, improving the targets for average and cell-edge performance is a much harder task due to radio propagation and interference issues, which are independent of the air interface technology. Many of the UMTS enhancements in Release 7 and Release 8 as well as Release 8 LTE have addressed the challenge of increasing average efficiency. LTE-Advanced takes this challenge to the next level.

The underlying problem of interference is illustrated in Figure 8.3-3, which shows a cumulative distribution function plot of the geometry factor within a typical urban cell.

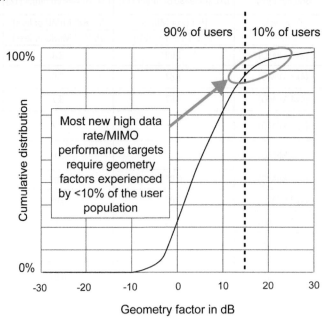

Figure 8.3-3. Geometry factor distribution in a typical urban cell with frequency reuse 1

The geometry factor is the term used in UMTS to indicate the ratio of the wanted signal to the interference plus noise. It is equivalent to the signal to interference plus noise ratio (SINR). From the figure it can be seen that 10% of users experience a better than 15 dB geometry factor but 50% of users experience worse than 5 dB. The exact shape of the curve varies significantly depending primarily on the frequency reuse factor followed by the cell size and cell loading. An isolated cell (e.g., a hotspot) would exhibit a shift to the right, indicating that most users are experiencing very good signal conditions. A cell in an urban area with significant co-channel inter-cell interference would shift to the left. Penetration loss through buildings, as experienced when indoor coverage is provided from an external cell, would also cause a shift to the left. However, on the assumption that the deployment in a particular area has resulted in a certain geometry factor distribution, the challenge then becomes how to deal with the interference to improve cell-average and cell-edge performance.

In 2G systems, performance was obtained by means of interference avoidance through the use of high frequency reuse factors of up to 21. In 3G systems the frequency reuse was optimized at 1 and methods such as scrambling and spreading were used to minimize the impact of interference with resulting gains in average spectral efficiency. Later systems employed receive diversity, equalizers, transmit diversity, and limited spatial multiplexing (MIMO). For LTE-Advanced, the planned performance enhancement techniques will take further steps by using more advanced MIMO and beamsteering, interference cancellation, fractional frequency reuse, and other advanced methods.

It is worth explaining how the cell-edge performance targets used by ITU-R and 3GPP were developed based on simulated geometry-factor distributions for the target deployment environments. Ten users were randomly distributed within each cell and the resulting geometry factor was calculated for each UE. This information was converted into a data throughput rate that was in turn used to plot a distribution of throughput. The process was repeated many times in a multi-drop simulation to create a smooth throughput distribution. The cell-edge performance was then defined as the fifth percentile of the throughput distribution. Because the simulation was carried out using groups of 10 UEs, the units are b/s/Hz/user and therefore appear to be 10 times lower than might be expected. It is not straightforward to take the cell-edge figure per user and multiply by 10 to predict the cell-edge average since the distribution is complicated by the type of scheduler used. The scheduler may have allocated more resources to the cell-edge users in a proportionally fair system. Leaving this complication aside, it can be seen that the cell-edge performance is about one third that of the average of the cell.

8.3.3 Release 10 LTE-Advanced Enhancements

As first noted in 8.3.1, the LTE-Advanced submission to ITU-R was made in September 2009 in 36.912 "Feasibility study for Further Advancements for E-UTRA (LTE-Advanced)" [13]. This document outlines those features identified for development in Release 10 relevant for the IMT-Advanced requirements. This subset of Release 10 is what is meant by the term LTE-Advanced. The original key features of LTE-Advanced proposals were the following:

- Support of wider bandwidths
- Uplink transmission scheme
- Downlink transmission scheme
- Coordinated multi-point transmission and reception (CoMP)
- Relaying.

Not all the above were essential to meet the IMT-Advanced requirements and not all aspects were subsequently developed in Release 10 (e.g., CoMP, which is a work item in Release 11 and will be covered later in this chapter). There were other areas for development also identified in 36.912 for which details were not elaborated. These included mobility enhancements, radio resource management enhancements, MBMS enhancements, and further work on SON. 36.912 concludes with a self-evaluation that reports how LTE-Advanced meets or exceeds the ITU-R IMT-Advanced requirements. The following sections outline the main functional areas that were developed in Release 10 for LTE-Advanced. These sections are followed by other work items in Release 10 that were not part of the ITU-R submission.

8.3.3.1 Support of Wider Bandwidths: Carrier Aggregation

Support of wider bandwidths is primarily aimed at addressing the IMT-Advanced requirements for peak single user data rates up to 1 Gbps, although there are additional system-level benefits in terms of deployment flexibility and associated trunking gains that come from the availability of a wider instantaneous transmission bandwidth. Today's spectrum allocations (frequency bands) offer almost no opportunity for finding 100 MHz of contiguous spectrum needed for 1 Gbps peak data rates. Some new IMT spectrum was identified at the World Radio Conference in 2007 (WRC-07), but there are still only a few places where continuous blocks of 100 MHz might be found (e.g., at 2.6 GHz or 3.5 GHz). One possible way of increasing available bandwidths would be to encourage network sharing, which reduces fragmentation caused by splitting one band between several operators. However, sharing the spectrum, as opposed to just the sites and towers, is a considerable step up in difficulty. The ITU-R recognizes the challenge that wide-bandwidth channels present and so expects that the required 100 MHz will be created by the aggregation of non-contiguous channels from different bands in a multi-transceiver mobile device.

The beginnings of such aggregation techniques are already showing up in established technologies—first with EDGE Evolution, for which aggregation of two non-adjacent 200 kHz channels was specified in Release 7, potentially doubling the single-user data rates that are possible with standard EDGE. Along similar lines, there were 3GPP specifications introduced in Release 8 for dual-carrier HSDPA that close the bandwidth gap between 5 MHz UMTS and 20 MHz LTE. Contiguous multicarrier cdma2000 (3xRTT) has also been defined, which avoids the need for multiple transceivers.

Carrier aggregation is clearly not a new idea; however, the proposal to extend aggregation up to 100 MHz in multiple bands presents numerous design challenges, particularly for the UE in terms of additional cost and complexity. At each of the layers in the radio protocol, from the physical layer up through radio resource control (RRC), changes are required for carrier aggregation. An overview of these can be found in 36.912 [13] Section 5. Although it is possible to conceive of applications for 1 Gbps data rates to a single mobile device, the commercial viability has yet to be understood. It should also be noted that carrier aggregation does not fundamentally increase spectral efficiency (or network capacity) per se, although wider channels do offer better trunking efficiency. Taking all these factors into account suggests that 100 MHz multi-transceiver carrier aggregation as a means of delivering extreme single-user peak data rates is not likely being pursued at this time. The scenarios that are being actively studied are two-carrier aggregation for intra-band and for inter-band. Section 2.1.11 provides more details about specific band combinations and Section 6.4.8 discusses test aspects.

8.3.3.2 Uplink Transmission Scheme

Several enhancements were introduced to the uplink for LTE-Advanced:

- Spatial multiplexing up to four layers
- Transmit diversity
- Clustered SC-FDMA
- Simultaneous PUCCH/PUSCH.

Spatial Multiplexing and Transmit Diversity

The introduction of spatial multiplexing and transmit diversity to the uplink makes a significant departure from the UE architecture of Release 8 since both enhancements require the support of more than one uplink transmitter. This has implications for cost, space, power handling, and many new spurious emission scenarios that need to be studied and will require new designs. The benefits of spatial multiplexing provide the necessary improvements in spectral efficiency over Release 8 for LTE-Advanced to meet the requirements for IMT-Advanced. This subject is covered more fully in Section 2.4.

Clustered SC-FDMA

The uplink multiple access scheme has been enhanced by adopting clustered discrete Fourier transform spread OFDM (DFT-S-OFDM). This scheme is similar to SC-FDMA but has the advantage that it allows non-contiguous (clustered) groups of subcarriers to be allocated for transmission by a single UE, thus increasing the flexibility available for frequency-selective scheduling. Clustered SC-FDMA was chosen in preference to pure OFDM in order to avoid a large increase in peak-to-average power ratio (PAPR).

Simultaneous PUCCH/PUSCH Transmission

In Release 8 the user data carried on the physical uplink shared channel (PUSCH) and the control data carried on the physical uplink control channel (PUCCH) are time-multiplexed as shown earlier in Figure 3.2.13. It is also possible to multiplex control data with user data on the PUSCH. LTE-Advanced introduces a new mechanism for simultaneous transmission of control and data by allowing the PUSCH and the PUCCH to be transmitted simultaneously. This mechanism has some latency and scheduling advantages over time-multiplexed approaches although it does generate a multicarrier signal within one component carrier of the uplink. Simultaneous PUCCH/PUSCH transmission should not be confused with carrier aggregation, which involves more than one component carrier. Simultaneous PUCCH/PUSCH transmission is known to increase PAPR, which makes it more likely that the power amplifier will create unwanted intermodulation products. This effect is similar to the one described for clustered SC-FDMA in Section 2.3.5.

8.3.3.3 Downlink Transmission Scheme

The enhancement to the downlink for LTE-Advanced include the following:

- Extension of spatial multiplexing from four to eight layers
- Enhancements to downlink reference signals.

Eight Layer Spatial Multiplexing

The increase in the number of spatial multiplexing layers on the downlink from four to eight may appear to be a symbolic extension to the standard, since performance requirements through Release 11 have existed only for two-layer transmission to a single UE, even though four-layer transmission has been defined since Release 8. The main drawback to the implementation of eight-layer single user spatial multiplexing is not so much at the base station end, where eight-antenna systems already exist, but at the UE receiver, which would require implementation of eight receive antennas per carrier. This proposition is not practical today due mainly to space constraints.

The potential for eight spatial layers does open up new possibilities for multi-user spatial multiplexing (MU-MIMO), which offers new combinations for the simultaneous support of more than one user sharing the eight layers. This is discussed further in Section 2.4.4.9 for transmission mode 9. Also, the potential for eight transmitters at the base station opens up the potential for enhanced transmission using beamforming; for example, in an 8x2 configuration. Eight-antenna beamforming in covered Section 6.9.

Downlink Reference Signals

The UE-specific reference signals are extended to support up to eight layers and a new class of channel state information reference signals (CSI-RS) are introduced. The purpose of the CSI-RS is limited to channel state information reporting of the channel quality indicator (CQI), precoding matrix indicator (PMI), and rank indication (RI). Specifically, the CSI-RS are not used in support of PDSCH demodulation, which is the task of the precoded UE-specific RS and the non-precoded cell RS. The CSI-RS are discussed further in Section 3.2.12.4.

8.3.3.4 Relaying

The concept of relaying is not new but the level of sophistication continues to grow. The most basic relay method is the use of a repeater, which receives, amplifies, and then retransmits the downlink and uplink signals to overcome areas of poor coverage. The repeater could be located at the cell edge or in some other area of poor coverage. Repeaters are relatively simple devices operating purely at the RF level. Typically they receive and retransmit an entire frequency band; therefore, care is needed when repeaters are sited. In general repeaters can improve coverage but do not substantially increase capacity.

More advanced relays can in principle decode transmissions before retransmitting them. This gives the ability to selectively forward traffic to and from the UE local to the relay station thus minimizing interference. Depending on the level at which the protocol stack is terminated in the relay node (RN), such types of relay may require the development of relay-specific standards. This can be largely avoided by extending the protocol stack of the RN up to Layer 3 to create a wireless router that operates in the same way as a normal eNB, using standard air interface protocols and performing its own resource allocation and scheduling. The distinguishing feature of such relays compared to normal eNBs is that the backhaul connecting the relays to the other eNBs operates as an in-band LTE radio link to the donor eNB. This link, called the Un interface, can be on the same frequency as the RN-to-UE link (inband) or on a different frequency (outband). The concept of the relay station can also be applied in low density deployments where a lack of suitable backhaul would otherwise preclude use of a cellular network. The use of in-band or in-channel backhaul can be optimized using narrow point-to-point connections to avoid creating unnecessary interference in the rest of the network. Multi-hop relaying is also possible as shown in Figure 8.3-4.

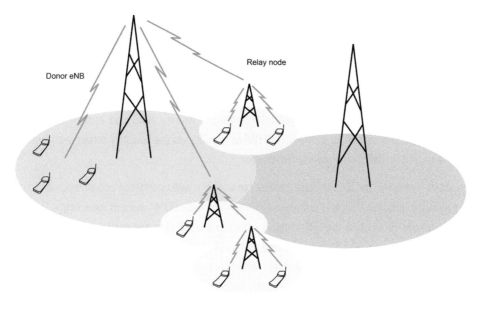

Figure 8.3-4. In-channel relay

Since the RN cannot simultaneously receive from the donor eNB and transmit to a local UE at the same time and frequency, downlink transmission gaps during which the eNB communicates with the RN can be created by configuring MBSFN subframes at the RN. This principle is shown in Figure 8.3-5.

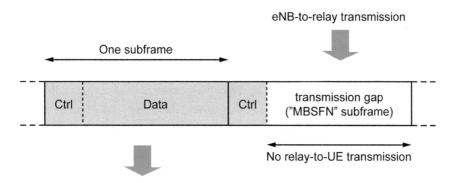

Figure 8.3-5: Relay-to-UE communication using normal subframes (left) and eNB-to-relay communication using MBSFN subframes (right). (36.912 [13] Figure 9.1)

The essential functionality to enable relaying is specified in Release 10 but the radio requirements for the RN transmitter and receiver performance are specified in Release 11. In Release 12 a study item has started to investigate mobile relaying as a solution for improving performance on high speed trains. Currently, the handover success rate from high speed trains is problematic due to the large number of UE attempting to handover at the same time. By using a mobile relay, possibly equipped with a group handover mechanism, the signaling load on the macro network could be substantially reduced.

8.3.4 Release 10 Other Enhancements

LTE-Advanced is a subset of Release 10 and the following sections describe further work items in Release 10 that were not originally identified to meet the ITU-R requirements for IMT-Advanced in 36.912.

8.3.4.1 Enhanced Inter-cell Interference Coordination (eICIC)

Basic support for ICIC started in Release 8 and is enhanced in Release 10 with the eICIC work item. Before discussing eICIC it is worthwhile reviewing the attributes of the CDMA and OFDM air interfaces to see how they behave with regard to inter-cell interference and the techniques that can be applied to mitigate it.

In the CDMA systems that dominate 3G, cell-edge interference is now a well-understood phenomenon and techniques for dealing with it continue to advance. This was not always the case and early CDMA systems were dogged with unexpected issues such as "cell breathing" in which the cell boundary moves as a result of power-control problems and excessive soft handover activity. Cell breathing can now be used with care as a tool for inter-cell load balancing. UMTS Release 7 introduced the HSDPA Type 3i receiver, which incorporated diversity reception, an equalizer, and dual-input interference cancellation capability. Due to the use of cell-specific scrambling codes and the presence of patterns within the signal caused by frequency selective fading, a cell-edge interferer in a CDMA system has considerably more structure than additive white Gaussian noise (AWGN). This structure can be used by an interference-cancelling receiver to remove significant portions of the co-channel interference.

The introduction of OFDMA to cellular systems—starting with 802.16e and continuing with LTE—has significantly changed the nature of cell-edge interference. In CDMA systems all the transmissions occupy the entire channel and are summed to create a signal with relatively stable dynamics. In OFDMA the potential for frequency-selective scheduling within the channel opens up new possibilities for optimizing intra-cell performance but also creates dynamic conditions in which inter-cell co-channel interference may occur. Work continues in 3GPP to better understand the effect of this interference on operational performance. In particular it has been noted that the narrowband and statistical (temporal) nature of the interference can influence the behavior of subband CQI and PMI reporting. While the presence of interference in CDMA systems is largely consistent across the channel bandwidth, the presence of interference in OFDMA systems using frequency-selective scheduling can change rapidly from the time of CQI reporting to its impact on the next scheduled transmission.

The interference protection between CDMA cells offered by the use of scrambling codes is not available in narrowband OFDMA transmissions, which leaves the narrowband signals vulnerable to narrowband interference. However, the ability of cells to coordinate their narrowband scheduling offers some potential for interference avoidance. Support for coordination of resource block (RB) allocation between cells in the downlink was introduced in Release 8 with the inclusion of the relative narrowband transmit power (RNTP) indicator. This support feature is a bitmap that can be shared between base stations over the X2 interface. It represents those RB for which the base station intends to limit its output power to a configurable upper limit for some period of agreed-upon time. This allows schedulers to agree on how cell-edge RB will be used so that, for instance, cell-edge users who cause the most interference can be restricted to certain parts of the channel. This coordination could be implemented using a semi-static agreement for partial frequency reuse at the cell edge or might involve more dynamic scheduling based on real-time network loading.

Two interference coordination mechanisms based on RB bitmaps are available for the uplink. The first is a bitmap called the overload indicator (OI), which can be provided by a base station to neighbor base stations indicating the level of uplink power plus noise as being "low," "medium," or "high." The second is more proactive and is the high interference indicator (HII). This is communicated to neighbor base stations prior the UE being scheduled, giving other base stations the chance to avoid the identified RB rather than allowing interference to occur and then having to deal with the consequences. These basic frequency domain approaches to ICIC are elaborated in Release 10 with the additional ability to coordinate inter-cell scheduling in the time domain.

Heterogeneous Networks

The original cellular deployment scenario in Release 8 was the traditional cellular pattern of adjacent cells sharing the same frequency. By Release 10 a variety of new base station types were introduced including the afore-mentioned local area BS (picocell), home BS (femtocell), and relay node. The inter-cell coexistence techniques that might be employed in a Release 8 network comprising wide area base stations are well understood; however, the introduction of the new base station types creates new coexistence scenarios. The issue is not that a network incorporating only one base station class might be deployed—in which case existing techniques might suffice—but that the network might include a mixture of different base station classes, all occupying the same frequency. This scenario has been termed the heterogeneous network or "het-net" for short. In the het-net environment new co-channel interference scenarios arise that require new inter-cell interference coordination solutions.

There are two forms of co-channel heterogeneous deployment, each requiring a different approach to interference avoidance. The first is the open subscriber group (OSG), a type of deployment that might be used by an operator with a macro network providing broad coverage overlaid with local area base stations in areas where coverage issues exist or where higher capacity is needed—for example, in a shopping mall. In this scenario a user is free to roam between the macro network and any local area BS deployed by the operator on the same frequency. For OSG deployment, the local area BS is located in the center of the area in the network where the increased capacity is required. At the perimeter of this area the strengths of the wide area and local area base stations are similar and performance may be significantly degraded. Closer to the local area BS the interference becomes less problematic. It is also possible to have an OSG scenario with a home BS, provided that the home BS is configured to be open to all users of that operator.

The second form of co-channel deployment is the closed subscriber group (CSG). This type of deployment is essentially limited to a home BS scenario in which access to the home BS is limited to a fixed group of subscribers, most likely the occupants of a dwelling or employees of an enterprise who have installed the home BS. This form of deployment provides good service for the closed subscriber group but creates a much more difficult interference situation for all other users since the problem area is no longer limited to a ring around the local area BS or home BS but extends to the entire coverage area of the home BS. This situation could be acceptable in low density rural areas but is likely to cause severe difficulties for macro network coverage in more densely populated areas. The obvious solution to home BS CSG is to assign different channels to the home BS and the macro network, thus reducing the interference to that which exists between adjacent home BS. This approach, however, is not available to operators with only a single channel. Some form of partial frequency reuse is also possible although this does not solve interference in the control channels, which always occupy the central 1.08 MHz of the channel. Given the difficulty of CSG, the initial work on eICIC in heterogeneous networks has been focused on the OSG case.

Almost Blank Subframes

The frequency domain ICIC techniques available in Release 8 and Release 9 are effective in managing interference that is caused by data traffic, but these techniques are not suited to minimizing interference between the control channels, which always occupy the same central 1.08 MHz of the channel regardless of channel bandwidth. To deal better with control-channel interference issues, Release 10 introduces the almost blank subframe (ABS) as the primary mechanism for eICIC. In this time-domain approach, the macro network chooses to minimize scheduled transmissions on certain subframes so that they can be used by the local area BS with minimal degradation of performance. These subframes are considered "almost blank" since minimal control traffic on the PDCCH may still be present in order to schedule macro uplink traffic and maintain HARQ ACK/NACK feedback to the macro UE. Backward compatibility to Release 8 and Release 9 UEs must also be maintained, which requires that the base station downlink still be measurable by legacy UEs. To do this, the downlink subframe must contain the cell RS, synchronization signals, and the paging channel. If the downlink subframe is designated as an MBSFN subframe, then fewer signals will be required.

As with the RNTP indicator introduced for frequency-domain ICIC, the use of ABS by the macro BS is indicated by an ABS pattern bitmap, but in this case we are not dealing with frequency domain RBs but with the time-domain subframe. There is also a secondary indicator known as the measurement subset, which indicates to the victim BS those subframes that the UE connected to the victim BS should use to assess the interference from the macro network when ABS is not configured. There is a great deal of flexibility in how ABS can be used and as such the standards specify the mechanisms for use in proprietary implementations but does not mandate specific solutions.

Further enhanced ICIC (FeICIC)

Some of the work on eICIC was not completed in Release 10 and so the FeICIC work item was created for Release 11. This includes specification of system performance requirements for scenarios involving a dominant downlink interferer.

Carrier-based Het-Net ICIC

The ICIC requirements developed through Release 10 are all based on co-channel (intra-frequency) scenarios. It was originally planned to develop ICIC further in Release 11 to take advantage of network-based carrier selection and this work has now been carried over to Release 12.

8.3.4.2 Minimization of Drive Test

It has long been the case that the planning and operational optimization of networks has been facilitated by the use of drive testing (see Section 6.14). Drive testing is a powerful technique; however, it is time-consuming and expensive to carry out. To alleviate some of the cost associated with drive testing a new set of UE measurement capabilities are introduced in Release 10 under the minimization of drive test (MDT) work item. The Release 10 work was focused on coverage and Release 11 added QoS verification. The MDT technical report is in 37.320 [16].

8.3.4.3 Machine-Type Communications (MTC)

For most of the history of cellular communications the goal has been to provide services between people. However, since the advent of data services there has been an increasing desire to support cellular communication between machines. These could e vending machines communicating with a corporate server to indicate sales activity and the need for restocking, or perhaps machines providing remote meter reading. The types and frequency of traffic in such scenarios is quite different from that for which LTE was originally developed. Machine-type communications often involve small amounts of data sent infrequently, preferably using very low cost infrastructure. These attributes are well-served by legacy systems such as GSM but are not well-suited to the footprint provided by LTE Release 8, whose lowest UE category mandates support for at least 10 Mbps in the downlink with two receivers and 5 Mbps in the uplink. The purpose of the MTC work item is therefore to develop additional UE categories more suited to the lower requirements of MTC.

The work on MTC started in Release 10 and has continued through Release 11 into Release 12. The scope has been clarified to indicate a target improvement in coverage over legacy systems of some 20 dB for very small data packets on the order of 100 bytes per message in the uplink and 20 bytes per message in the downlink. This may be achieved through drastically reduced latency of up to 10 seconds in the downlink and one hour in the uplink. High overall system efficiency can then be delivered through scheduling during quiet times. The MTC technical report is in 36.888 [17].

8.3.4.4 New Frequency Bands

The new frequency bands added in Release 10 are shown in Table 8.3-2.

Table 8.3-2. Frequency bands added during Release 10

Band number	Uplink		Downlink		Bandwidth	Duplex spacing	Gap	Duplex mode
	Low	High	Low	High				
22	3410	3490	3510	3590	80	100	20	FDD
23	2000	2020	2180	2200	20	180	160	FDD
24	1626.5	1660.5	1525	1559	34	-101.5	67.5	FDD
25	1850	1915	1930	1995	65	80	15	FDD
41	2496	2690	2496	2690	194	0	0	TDD
42	3400	3600	3400	3600	200	0	0	TDD
43	3600	3800	3600	3800	200	0	0	TDD

8.4 Release 11

Of the 48 radio-related work items identified for Release 11, some 31 relate to work in new frequency bands and various band combinations for carrier aggregation. These work items are covered in some detail in Section 2.1.11. Several other work items are continuations of work started in earlier releases and are referenced in Sections 8.2 and 8.3.

The remaining ten work items introduce new concepts to the LTE specifications:

- Verification of radiated multi-antenna reception performance of UEs in LTE/UMTS
- LTE RAN enhancements for diverse data applications
- Signaling and procedure for interference avoidance for in-device coexistence
- Coordinated multi-point operation for LTE
- Network energy saving for the E-UTRAN
- Enhanced downlink control channel(s) for LTE-Advanced

- Public Safety Broadband High Power UE for Band 14, Region 2
- Improved minimum performance requirements for E-UTRA: interference rejection
- Additional special subframe configuration for LTE TDD
- Carrier aggregation enhancements.

8.4.1 Verification of Radiated Multi-Antenna Reception Performance of UEs in LTE/ UMTS

This work item is the culmination of the study item introduced in March 2009 for MIMO over the air (OTA) performance verification. This subject is covered in detail in Section 6.10.

8.4.2 LTE RAN Enhancements for Diverse Data Applications

The diverse range of mobile data applications is now very extensive and includes short message service (SMS), instant messaging, web browsing, social networking, and a variety of push services. Modern devices such as tablets and smartphones will often activate some or all of these services in parallel, putting considerable strain on the radio network—not just due to the volume of data but also the substantial signaling overhead created by the "chatty" nature of many applications. In addition, the user expectation of an always-on mobile broadband experience puts great demands on battery consumption since the device may be prevented from reaching the idle state. Moreover, most modern applications were not developed with the unique characteristics of cellular networks in mind and so the use of network resources is often inefficient. How to balance the demands of user experience with battery consumption and network efficiency will depend on the characteristics of individual applications that may vary over time.

The outcome of the RAN enhancement work item is captured in 36.822 [16]. It has resulted in the specification of a power-preference feature that allows the UE to signal the network its preference for a configuration that reduces power consumption.

8.4.3 Signaling and Procedure for Interference Avoidance for In-device Coexistence

So that users can access various networks and services wherever they are, an increasing number of UEs are equipped with multiple radio transceivers for LTE, Wi-Fi, *Bluetooth*, GNSS, etc. As a result, UEs are challenged to avoid coexistence interference between those co-located radio transceivers. The studies done for this work item have shown that existing RRM mechanisms in some cases are not effective enough to handle the coexistence issues, and some enhanced signaling and other procedures are necessary to avoid or mitigate the coexistence interference in the identified usage scenarios.

As a result of this work item a new in-device coexistence (IDC) indication message has been defined. This message enables the UE to alert the network of an interference issue and provide information regarding the direction and nature of the interference, which may be identified in either the time or frequency domain. Upon receipt of the IDC message the network will take appropriate steps to alleviate the problem by reallocating radio resources. The stage 2 specification for this message is in 36.300 [17] Section 23.4 and the details are captured in 36.331 [18].

8.4.4 Coordinated Multi-Point Transmission (CoMP)

Section 2.4.7 briefly introduced the concept of co-operative MIMO, which also goes by the name of network MIMO or coordinated multi-point (CoMP). The goal of CoMP is to improve the coverage of high data rates and cell-edge throughput, and also to increase system throughput. Figure 8.4-1 compares standard MIMO with CoMP.

The primary difference between standard MIMO and CoMP is that for the latter, the transmitters are not physically co-located. In the case of downlink CoMP, however, there is the possibility of linking the transmitters at baseband (shown as the link between the transmitters on the right half of Figure 8.4-1) to enable sharing of payload data for the purposes of coordinated precoding. This sharing is not physically possible for the uplink, which limits the options for uplink CoMP. For the standard network topology in which the eNBs are physically distributed, provision of a high capacity, low latency baseband link is challenging and would probably require augmentation of the X2 inter-eNB interface using fiber. However, a cost-effective solution for inter-eNB connectivity is offered by the move towards a network architecture in which the baseband and RF transceivers are located at a central site with distribution of the RF to the remote radio heads via fiber.

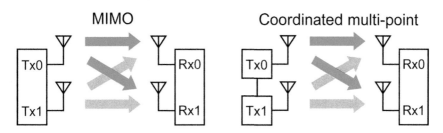

Figure 8.4-1. Standard MIMO versus coordinated multi-point

The physical layer framework for CoMP is described in the Release 11 feasibility study in 36.819 [19].

8.4.4.1 CoMP Deployment Scenarios

Four downlink scenarios were defined for the feasibility study:

Scenario 1 is a homogeneous network (all cells have the same coverage area) with intra-site CoMP. This is the least complex form of CoMP and is limited to eNBs sharing the same site.

Scenario 2 is also a homogeneous network but with high Tx-power RRHs. This is an extension of scenario 1 in which the six sites adjacent to the central site are connected via fiber optic links to enable baseband cooperation across a wider area than is possible with scenario 1.

Scenarios 3 and 4 are heterogeneous networks in which low power RRHs with limited coverage are located within the macrocell coverage area. In scenario 3 the transmission/reception points created by the RRHs have different cell identifications than does the macro cell and for scenario 4 the cell identifications are the same as that of the macro cell.

8.4.4.2 CoMP Categories

The introduction of CoMP enables several new categories of network operation, which are defined for the downlink as follows.

Joint processing (JP): Data for a UE is available at more than one point in the CoMP cooperating set (see Section 8.4.4.3) for the same time-frequency resource.

- **Joint transmission (JT):** This is a form of spatial multiplexing that takes advantage of decorrelated transmission from more than one point within the cooperating set. Data to a UE is simultaneously transmitted from multiple points; e.g., to coherently or non-coherently improve the received signal quality or data throughput.

- **Dynamic point selection (DPS)/muting:** The UE data is available at all points in the cooperating set but is only transmitted from one point based on dynamic selection in time and frequency. The DPS includes dynamic cell selection (DCS). DPS may be combined with JT, in which case multiple points can be selected for data transmission in the time-frequency resource.

Coordinated scheduling and beamforming (CS/CB): Data for a UE is only available at and transmitted from one point in the CoMP cooperating set but user scheduling and beamforming decisions are made across all points in the cooperating set. Semi-static point selection (SSPS) is used to make the transmission decisions. Dynamic or semi-static muting may be applied to both JP and CS/CB.

Hybrid JP and CS/CB: Data for a UE may be available in a subset of points in the CoMP cooperating set for a time-frequency resource but user scheduling and beamforming decisions are made with coordination among points corresponding to the CoMP cooperating set. For example, some points in the cooperating set may transmit data to the target UE according to JP while other points in the cooperating set may perform CS/CB.

New categories in the uplink are the following.

Joint reception (JR): The PUSCH transmitted by the UE is simultaneously (jointly) received at some or all of the points in the cooperating set. This simultaneous reception can be used with inter-point processing to improve the received signal quality.

Coordinated scheduling and beamforming (CS/CB): User scheduling and precoding selection decisions are made with coordination among points corresponding to the cooperating set. Data is intended for one point only.

8.4.4.3 CoMP Sets

Various sets of eNBs are identified for downlink CoMP purposes.

CoMP cooperating set: The set of eNB points within a geographic area that are directly or indirectly participating in data transmission to a UE. The UE may or may not know about this set. The direct participation points are those actually transmitting data and the indirect points are those involved in cooperative decision making for user scheduling and beamforming in the time and frequency domains.

CoMP transmission point(s): The point or set of points transmitting data to a UE. CoMP transmission points are a subset of the CoMP cooperating set.

- For JT, CoMP transmission points may include multiple points in the CoMP cooperating set at each subframe for a certain frequency resource.
- For CS/CB, DPS, and SSPS, a single point in the CoMP cooperating set is the CoMP transmission point at each subframe for a given frequency resource.
- For SSPS, the CoMP transmission point can change semi-statically within the CoMP cooperating set.

CoMP measurement set: The set of points about which channel state and statistical information related to the UE radio link is measured and reported.

RRM measurement set: The set of cells for which Release 8 radio resource management (RRM) measurements are performed. Additional RRM measurement methods may be developed; e.g., in order to separate different points belonging to the same logical cell entity or in order to select the CoMP measurement set.

For the uplink, the following sets are identified.

CoMP reception point(s): The point or set of points that is a subset of the cooperating set receiving data from a UE.

- For JR, CoMP reception points may include multiple points in the CoMP cooperating set at each subframe for a certain frequency resource.
- For CS/CB, a single point in the CoMP cooperating set is the CoMP reception point at each subframe for a certain frequency resource.

8.4.4.4 Radio Interface Aspects

To enable CoMP operation, changes to the radio interface will likely be needed in the areas of channel state information (CSI) feedback from the UE, preprocessing schemes for coordination of joint transmission, and possibly new reference signal designs and new control signaling mechanisms. Reuse of existing Release 8 CSI measurements extended for CoMP, called explicit feedback, is expected. Channel parameters (per point) include the channel matrix H, the transmit covariance matrix R, and possibly inter-point properties such as the inter-point phase relationship required for JT. Noise and interference parameters are also required. To take full advantage of CoMP, more advanced implicit feedback will be required based on UE hypotheses about different CoMP transmission and reception processing. The potential for CoMP becomes greater for TDD operation since UE transmission of the sounding reference signal (SRS) can be used by the eNB to precisely determine the downlink channel conditions on the assumption of TDD channel reciprocity.

8.4.4.5 Simulation Results

Extensive simulation of CoMP performance has been performed by multiple companies for the four deployment scenarios identified in Section 8.4.4.1 for uplink and downlink FDD and TDD. Both 3GPP and ITU channel models were used, and the impact of cell loading and inter-cell communication latency and bandwidth was also studied. Although the simulation criteria were specified, the results showed variations in performance that may be due to different assumptions being made for the channel estimation error modeling, channel reciprocity modeling, feedback and SRS mechanisms, the scheduler, and the receiver. The impact of CoMP on legacy UEs is not considered.

The results show that CoMP gains vary widely depending on the specific scenario and whether the focus is on average cell performance, mean user performance, or improving the performance of the worst 5% of users in the cell. Some scenarios provide no gain at all and others, particularly TDD with its channel reciprocity advantage, show gains of up to 80%. Typical gains fall in the range of 10% to 30%. As a result a work item to progress CoMP is defined in Release 11 and will focus on the following aspects:

- Joint transmission
- Dynamic point selection, including dynamic point blanking
- Coordinated scheduling and beamforming, including dynamic point blanking.

8.4.5 Network Energy Saving for E-UTRAN

With the growth in network capacity there is an increasing need to consider the energy costs of operating the network. In particular, opportunities exist to dynamically dimension the network based on traffic loading. The stage 2 definition of network energy saving is defined in 36.300 [17] Section 22.4.4. The basic mechanism is that an eNB containing one or more capacity booster cells in addition to basic coverage cells may choose to deactivate the booster cells based on a drop in the network load. The deactivation may also require communication with peer eNBs over the X2 interface to indicate that the booster cell is to be deactivated. Offload of users from the booster cell to the coverage cells may also be necessary through handovers.

8.4.6 Enhanced Downlink Control Channels for LTE-Advanced

The addition of carrier aggregation in combination with a new carrier type (see Section 8.5), CoMP, and DL MIMO has resulted in the need to enhance the capabilities of the physical downlink control channel (PDCCH). The enhanced PDCCH (EPDCCH) will be restricted to QPSK modulation. The EPDCCH will be compatible with legacy carriers, provide more signaling capacity, support frequency domain ICIC, improve the spatial reuse of the control channel, support beamforming and diversity schemes, and operate in MBSFN subframes. Frequency-selective scheduling for the EPDCCH is also desirable as is mitigation of inter-cell interference.

8.4.7 Public Safety Broadband High Power UE for Band 14, Region 2

The US Federal Communications Commission Public Safety and Homeland Security Bureau has selected LTE as the technology for public safety services in the 700 MHz public safety band (3GPP Band 14; see Table 2.1-1). Due to the coverage and uplink performance requirements for public safety broadband (PSBB) systems, the existing 23 dBm UE power class (class 3) is not considered sufficient.

Public safety "first responders" will rely on handheld UEs as well as vehicular mobile applications that have fewer constraints on size, weight, and power consumption than handheld UEs. A vehicular mobile application also has the possibility of incorporating very efficient vehicle-mounted antennas. Unlike commercial cellular systems, which often generally have a 95% population coverage target, PSBB systems target 99% coverage. Although this change may seem insignificant, to reach the additional 4% of the US population requires a 60% increase in the coverage area.

Providing such coverage using base stations alone would be very expensive, so a higher power UE (HPUE) power class 1 is being specified for a vehicular mobile form factor with vehicular-mounted antennas. The provisional requirements are captured in 36.837 [20].

In order to optimize reuse of the existing LTE UE ecosystem, the requirements will minimize change that might impact the design of the baseband and lower-power RF components of the UE. The bulk of the changes are expected in the RF front end containing the power amplifier (PA), filtering, and signal-combining components. The headline parameter driving the HPUE specification is the proposed 10 dB increase of maximum output power to 33 dBm. Although it is expected that few other transmitter and receiver requirements will be changed from those defined for the existing power class 3 UE (23 dBm), this increased maximum power has considerable design implications for both the transmitter and the receiver. For instance, the dynamic range of the transmitter increases 10 dB and all fixed-level unwanted emissions become 10 dB harder to meet.

For the receiver to maintain the existing RF sensitivity the duplex filter has to provide 10 dB more isolation from the transmitter. The tighter filtering requirements represent probably the biggest design change for the HPUE because existing miniature surface acoustic wave (SAW) filters measuring perhaps 5 mm^3 cannot handle the higher output power or provide the necessary filtering performance. Alternative technologies will be required—for example, ceramic or cavity filters, which are substantially larger at around 8000 mm^3. Fortunately, the form factor of the vehicular mobile has more relaxed constraints on size and power than does the standard handheld UE.

One of the few performance requirements likely to change for the HPUE is the ACLR requirement. Studies have shown that to maintain the existing co-existence performance of power class 3 UE, the HPUE will need to have an ACLR in the region of 28 dB, some 5 dB tighter than for power class 3.

In summary, the increase in maximum output power along with the potential for vehicular-mounted antennas means that the power class 1 HPUE will offer substantially better performance in areas of poor reception than was possible with the power class 3 UE. It's expected that the increased cost of the HPUE will be offset by substantial savings in the number of base stations needed to achieve 99% population coverage.

8.4.8 Improved Minimum Performance Requirements for E-UTRA: Interference Rejection

Existing LTE UE demodulation requirements are based on an assumption of a linear minimum mean squared error (LMMSE) dual receiver. This is a powerful receiver architecture capable of suppressing both inter- and intra-cell interference. However, existing demodulation requirements are based on additive white Gaussian noise (AWGN), which is decorrelated between the antennas. This simplified method of modeling interference has been widely used for many years and is suitable for measuring the performance of receivers without interference cancellation capabilities. However, to exploit the full potential an LMMSE receiver with interference rejection combining (IRC) capabilities and achieve a performance gain over standard receivers, it is necessary to more accurately model the interference. Studies carried out in Release 10 showed that enhanced receivers capable of RS-based interference covariance estimation to mitigate spatial domain interference could provide significant throughput gains in the high interference conditions of a heavily loaded network.

The scope of the Release 11 work includes a variety of deployment scenarios that take into account the number of interfering sources, their structure (including transmission rank and precoding), and their power ratio relative to the total interference from other cells. This ratio is known as the dominant interferer proportion (DIP) ratio. Both synchronized and asynchronous cases are considered since they can have a major impact on interference susceptibility. Also within the scope of Release 11 were definitions of cell RS and UE-specific RS in anticipation of future network deployment scenarios.

8.4.9 Additional Special Subframe Configuration for LTE TDD

The operation of TDD networks requires careful coordination between systems deployed on adjacent channels. Co-existence of LTE TDD with legacy UMTS TD-SCDMA systems is required and for this case, special subframe configuration number 5 is chosen for the normal cyclic prefix case and configuration number 4 for the extended cyclic prefix case (see Table 3.2-2). The special subframe lasts for one ms and always comes between the transition from downlink transmission to uplink transmission, although it is not required from the uplink back to the downlink (see Table 3.2-1). The special subframe comprises the downlink pilot timeslot (DwPTS), a gap period (GP), and an uplink pilot timeslot (UpPTS). The ratio between the DwPTS, GP, and UpPTS is configurable, and for TD-SCDMA co-existence, configuration 5 uses a ratio of 3:9:2 and configuration 4 uses 3:7:2. Although these configurations provide the necessary protection when LTE TDD and TD-SCDMA systems are in adjacent channels, the use of a relatively large GP in these configurations is seen as inefficient since no data can be transmitted during the gap period.

To address this shortcoming, two new special subframe configurations have been specified in Release 11. For the extended cyclic prefix case, a new option for special configuration number 7 has been defined for a ratio of 5:5:2, which provides an additional two symbols for data communication per special subframe. For the normal cyclic prefix case a new special subframe configuration number 9 provides a ratio of 6:6:2, which is three extra useful symbols per special subframe.

8.4.10 Carrier Aggregation Enhancements

Many new band combinations specific to carrier aggregation are being developed in Release 11 and are covered in Section 2.1.11 of this book. Also in Release 11 is the introduction of new CA capabilities. For example, some uplink CA scenarios will require the ability to define different timing advances for each carrier. This could occur in an inter-band case that uses repeaters for one band but not the other. To deal with the situation the UE is allowed to adjust the timing advance of the two carriers independently such that the time orthogonality of the uplink in the cell is preserved.

Another new CA feature introduced in Release 11 is the ability for TDD to support different uplink and downlink configurations for each band. This provides more flexibility than was possible in Release 10, which required that the format of each carrier be the same.

8.4.11 New Frequency Bands

The new frequency bands added in Release 11 are shown in Table 8.4-1.

Table 8.4-1. Frequency bands added during Release 11

Band number	Uplink		Downlink		Bandwidth	Duplex spacing	Gap	Duplex mode
	Low	High	Low	High				
26	814	849	859	894	35	45	10	FDD
27	807	824	852	869	17	45	28	FDD
28	703	748	758	803	45	55	10	FDD
44	703	803	703	803	100	0	0	TDD

Band 29 is also being specified. This is a downlink only band from 717 MHz to 728 MHz intended for use in CA scenarios which are likely to initially include bands 2 and 4.

8.5 Release 12

At the time of this writing Release 12 is still under discussion. A workshop to consider proposals was held in June 2012. The broad areas for future radio evolution were identified as follows:

- Energy saving
- Cost efficiency
- Support for diverse application and traffic types
- Backhaul enhancements.

The following proposals from the workshop were identified as most likely to be developed in Release 12:

- Interference coordination and management
- Dynamic TDD
- Enhanced discovery and mobility
- Frequency separation between macro and small cells, using higher frequency bands in small cells (e.g., 3.5 GHz)
- Inter-site carrier aggregation and macrocell-assisted small cells
- Wireless backhaul for small cells.

Other possible areas for study include the following:

- Support for diverse traffic types (control signaling reduction, etc.)
- Interworking with Wi-Fi
- Continuous enhancements for machine-type communications, SON, MDT, and advanced receivers
- Proximity services and device-to-device communications
- Further enhancements for HSPA including interworking with LTE.

The timeframe for Release 12 is likely to be 18 to 24 months beyond Release 11 with March 2013 being the date proposed for completion of the Stage 1 specifications. This would put the Stage 3 completion sometime in 2014.

As of RAN plenary #57 (Oct 2012), twenty new work items have been approved for Release 12. Most of these are spectrum related with three for new frequency bands and thirteen for new carrier aggregation scenarios. The remaining four are unrelated.

8.5.1 New Frequency Bands

Three new frequency bands will be defined:

- FDD downlink 1670 MHz–1675 MHz, uplink 1646.7 MHz–1651.7 MHz for ITU Region 2 (US)
- FDD downlink 461MHz–468 MHz, uplink 451–458 MHz for Brazil
- FDD downlink 2350–2360 MHz, uplink 2305–2315 MHz, US Wireless Communications Service (WCS) band.

Assuming that band 29 is defined in Release 11, the introduction of the bands listed above will require a break in the band numbering since the TDD bands start at band 32.

In addition to the work items, a study item in Release 12 exists for the 30 MHz of paired spectrum immediately above band 1; that is, uplink from 1980 MHz–2010 MHz and downlink from 2170 MHz– 2200 MHz. This band is currently widely allocated for satellite communications but its use for terrestrial communications is now being considered, particularly for ITU Region 3. The potential for a combined 110 MHz paired band including band 1 is also being considered.

8.5.2 Carrier Aggregation Scenarios

With 28 FDD and 12 TDD bands defined up to Release 11, the theoretical number of two-carrier CA combinations is enormous. Fortunately, only those scenarios relevant to specific geographic deployments or potential deployments are covered in the standardization process. In Release 11, work items were created for five intra-band scenarios and twenty inter-band scenarios (see Section 2.1.11).

In Release 12 work items exist for five intra-band scenarios:

- Band 1 (contiguous)
- Band 3 (non-contiguous), carried over from Release 11
- Band 3 (contiguous)
- Band 4 (non-contiguous)
- Band 25 (non-contiguous).

An additional eight inter-band scenarios are defined:

- Bands 3 and 5 with two uplink carriers
- Bands 2 and 4
- Bands 3 and 26
- Bands 3 and 28
- Bands 3 and 19
- Bands 38 and 39
- Bands 23 and 29*
- Bands 1 and 8.

Band 29 is being specified as part of Release11. It is a downlink-only band from 717 to 728 MHz and to be used only for the purposes of carrier aggregation with other bands.

Future evolution of carrier aggregation is likely to include inter-site aggregation and macrocell-assisted small cells. The goal is to enable the UE to remain connected at all times to the macro network on one carrier, which is likely to be at a lower (< 1 GHz) frequency for coverage reasons, while opportunistically connecting to the macro network on a second carrier provided by a small cell (probably not co-located) to provide higher capacity. The advantage of doing this using carrier aggregation rather than handover is that CA should provide much faster adaptation to the network conditions than handover-based approaches.

Also identified in the Release 12 workshop are opportunities to exploit LTE aggregation with other radio systems such as UMTS and Wi-Fi to optimize connectivity.

8.5.3 Carrier-based Het-Net ICIC for LTE

See Section 8.3.4.1.

8.5.4 New Carrier Type for LTE

Backward compatibility of LTE carriers has been a priority since the first LTE Release up to Release 11. However, it is also desirable to consider introducing a new carrier type (NCT) for LTE that would not be constrained by legacy requirements such as cell reference symbols (CRS) and signaling overhead that impact overall network efficiency at low-to-medium loads. This NCT could be useful in several deployment scenarios such as at the cell edge in a homogenous network, in the cell range expansion zone of heterogeneous networks, and in low power eNB deployments whether they be isolated or part of a wider macro deployment. There is also potential for energy savings by allowing the downlink signal to switch off when low demand occurs.

The use cases for the NCT also include new carrier aggregation scenarios such as the aggregation of a downlink-only carrier with a legacy carrier (see Section 8.5.2, bands 23 and 29).

8.5.5 Further Downlink MIMO Enhancement for LTE-Advanced

A study item in Release 11 considered several aspects of downlink MIMO performance including rank reporting, time misalignment and antenna calibration, and numerous aspects related to CSI feedback. The conclusions of the study are captured in 36.871 [21], which concluded that UE rank reporting is problematic in some conditions such as non-co-located antenna deployments. These issues are best addressed by specifying new performance requirements. The study also identified potential CSI performance improvements through the use of four transmit antennas.

The scope of the work item will cover the following topics:

- Four transmit antenna PMI feedback codebook enhancements to provide finer spatial domain granularity and to support different antenna configurations for macro and small cells, especially cross-polarized antennas, both closely and widely spaced, and non-co-located antennas with power imbalance
- New CSI feedback mode providing subband CQI and subband PMI

- Finer frequency-domain granularity
- Enhanced control of the reported rank and corresponding assumptions for CQI/PMI derivation to improve support for MU-MIMO.

8.5.6 Further Enhancements for H(e)NB Mobility-Part 3

Following a Release 11 study item, this work item has been introduced to further enhance the mobility between home (e)NBs, and from a home (e)NB to a wide area eNB. Both UMTS and LTE are considered. The aspects relevant to LTE will focus on RAN sharing for the scenario in which the UE reports the PLMN identities of the home eNB that are accessible and can pass a closed subscriber group check. The home eNB verifies the access check, selecting just one identity if more than one are identified, and finally the MME/SGSN verifies the CSG membership check.

8.5.7 Release 12 Study Items

Looking further ahead, several radio-related study items that introduce new concepts to LTE have been defined for Release 12.

8.5.7.1 RF and EMC Requirements for Active Antenna Array System

The multiple antenna base station techniques that have been deployed to date are largely proprietary in nature and have no formal specifications or performance requirements. With the increasing sophistication of multiple antenna techniques it has become apparent that the largely omnidirectional assumptions about base station RF and EMC performance are becoming less representative of actual system performance. The current reference point for base station requirements is the antenna connector and excludes the antenna behavior and any multi-antenna array affects such as beamforming. This study item will investigate defining a new point in the system, independent of implementation, that will better represent the true spatial performance of the base station. A natural conclusion would be some kind of radiated over-the-air test point similar to what is being developed for the UE (see Sections 8.4.1 and 6.10). The study item is being captured in 37.840 [22].

In order to progress the work the concept of an active antenna system (AAS) has been defined as a base station system that combines an antenna array with an active transceiver unit array. An AAS may also include a radio distribution network, which is a passive network that physically separates the active transceiver unit array form the antenna array. Figure 8.5-1 shows the general AAS architecture.

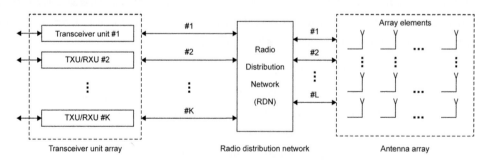

Figure 8.5-1. General AAS Radio Architecture (TR 37.840 [22] Figure 4.2-1)

focal point for the study has been spatial ACLR. It is understood that if the unwanted emissions from the antenna array are not fully correlated, then the ACLR performance will vary in space. The ACLR variation can be used to advantage or it can create interference conditions worse than what would be seen with an omnidirectional transmission scheme. For these reasons it is necessary to study the spatial aspects of ACLR in order that the use of multiple antenna systems benefit the network.

8.5.7.2 Passive Intermodulation Handling for UTRA and LTE Base Stations

During the development of the non-contiguous (inter-band) MSR work (see Section 8.2.4) it became apparent that the passive inter-modulation (PIM) performance of the physical elements in the base station transmitter had the potential to significantly degrade system performance. The mechanisms through which signal degradation occurs are based on the non-linear behavior of conducting elements in the vicinity of the transmitter when it is subjected to very high RF power. These elements can then re-radiate at intermodulation frequencies determined by the order of the non-linearity in the components. Initial work on resolving this issue will focus on identifying the sources of PIM while the longer term goal will be to measure and control PIM products. Given the diverse nature of PIM sources, this will not be an easy task and may require testing of the entire base station structure in an anechoic chamber.

8.5.7.3 Scenarios and Requirements of LTE Small Cell Enhancements

The contribution that small cells offer in providing increased network capacity through frequency reuse is well established. However, the propagation, mobility, interference, and backhaul assumptions for small cells can be very different from those of a network comprising the homogenous macro cells that were used to develop LTE and legacy systems. Therefore various features in support of small cells have been incorporated into the LTE specifications since Release 8, including the definition of the home area base station class and the ongoing work on such topics as ICIC and mobility in heterogeneous networks (see Sections 2.1.4.2, 8.2.1, 8.3.4.1, and 8.5.6). This Release 12 study item will look broadly at the unique challenges presented by small cell deployment to identify future areas where the LTE specifications can be further enhanced.

8.5.7.4 Feasibility Study for Proximity Services (LTE-Direct)

A study item for UE peer-to-peer communication has been carried out in TR 22.803 [25]. There are many potential applications including public safety and location-based services. The underlying goal is to develop a system whereby UEs can discover each other within a local area without the direct intervention of the network. This might be achieved by having the network allocate uplink subframes to be used by the UEs to broadcast their identity to the local area. Periodically each UE could be allocated a unique timeslot so that large numbers of UEs might have the opportunity to identify themselves over a period of several seconds. During its allocated timeslot a UE could broadcast information that may be of use to other UEs in the immediate area. For instance, a shop owner might use a UE to broadcast a web address for further information. For a UE to receive these transmissions, it would have to be augmented with a receiver in its uplink band, similar to the base station. The UE could then choose to listen to or (not to listen to) the local UEs that are broadcasting their identities and decide what, if any, further action to take. The detailed operation of such a system needs to be carefully considered from a radio interference perspective and the security aspects are also a concern. A further extension to the broadcast-only feature would be to enable peer-to-peer communication between UEs without relying on the network. This has obvious benefits for public safety applications although overlaying such a feature on top of an existing cellular network presents many challenges yet to be solved.

8.6 References

[1]   3GPP TS 36.307 V11.2.0 (2012-09) Requirements on User Equipments (UEs) Supporting a release-independent frequency band

[2]   3GPP TR 36.800 v9.0.0 (2009-09) Extended UMTS / LTE 800 Work Item Technical Report

[3]   3GPP TR 36.810 V9.0.0 (2010-03) UMTS / LTE in 800 MHz for Europe

[4]   3GPP TR 36.821 v9.1.0 {2010-03) Extended UMTS/LTE 1500 work item technical report

[5]   3GPP TR 36.921 V11.0.0 (2012-09) Home eNode B (HeNB) Radio Frequency (RF) requirements analysis

[6]   3GPP TR 36.922 v11.0.0 (2012-09) TDD Home eNode B (HeNB) Radio Frequency (RF)

[7]   3GPP TS 36.104 V11.1.0 (2012-06) Base Station Radio Transmission and Reception

[8]   3GPP TS 36.171 v11.0.0 (2012-09) Requirements for Support of Assisted Global Navigation Satellite System (A-GNSS) requirements analysis

[9]   3GPP TR 32.500 V11.1.0 (2011-12) Self-Organizing Networks (SON); Concepts and requirements

[10]   3GPP TR 36.902 V9.3.1 (2011-03) Self-configuring and self-optimizing network (SON) use cases and solutions

[11]   ITU-R M.[IMT-TECH] "Requirements related to technical performance for IMT-Advanced radio interface(s)," August 2008

[12]   3GPP TR 36.913 V11.0.0 (2012-09) Requirements for Further Advancements of E-UTRA (LTE-Advanced)

[13]   3GPP TR 36.912 V11.0.0 (2012-09) Feasibility study for Further Advancements for E-UTRA (LTE-Advanced)

[14]   3GPP TSG RAN Tdoc RP-070466

[15]   3GPP TR 25.913 V9.0.0 (2009-12) Requirements for E-UTRA and E-UTRAN

[16]   3GPP TR 37.320 V11.1.0 (2012-09) Radio measurement collection for Minimization of Drive Tests (MDT)

[17]   3GPP TR 36.888 V2.0.0 (2012-06) Study on provision of low-cost MTC UEs based on LTE

[18]   3GPP TR 36.822 V1.0.0 (2012-09)

[19]   3GPP TS 36.300 V11.3.0 (2012-09) Evolved Universal Terrestrial Radio Access (E-UTRA) and Evolved Universal Terrestrial Radio Access Network (E-UTRAN); Overall description; Stage 2

[20]   3GPP TS 36.331 V11.1.0 (2012-09) Evolved Universal Terrestrial Radio Access (E-UTRA); Radio Resource Control (RRC); Protocol specification

[21]   3GPP 36.819 V11.1.0 (2012-12) Coordinated multi-point operation for LTE physical layer aspects

[22]   3GPP TR 36.837 V0.3.0 (2012-08) Band 14 Public safety broadband high power User Equipment (UE) for Region 2

[23]   3GPP TR 36.871 V11.1.0 (2011-12) Evolved Universal Terrestrial Radio Access (E-UTRA); Downlink Multiple Input Multiple Output (MIMO) enhancement for LTE-Advanced

[24]   3GPP TR 37.840 V0.3.0 (2012-10) Study of AAS Base Station

[25]   3GPP TR 22.803 V1.0.0 (2012-06) Study on provision of low-cost MTC UEs based on LTE

Links to all reference documents can be found at www.agilent.com/find/ltebook.

List of Acronyms

1xRTT	Single-channel (1x) Radio Transmission Technology
1xEV-DO	1xRTT Evolution–Data Only
2G	2nd Generation
3G	3rd Generation
3GPP	3rd Generation Partnership Project
3GPP2	3rd Generation Partnership Project 2

A

AAS	Active Antenna System
ABS	Almost Blank Subframe
ACK	Acknowledgement
ACK/NACK	Acknowledgement/Negative Acknowledgement
ACLR	Adjacent Channel Leakage Ratio
ACS	Adjacent Channel Selectivity
A/D	Analog to Digital
ADC	Analog-to-Digital Converter
ADS	Advanced Design System
AF	Authentication Framework
AGC	Automatic Gain Control
AGNSS	Assisted Global Navigation Satellite System
AICH	Acquisition Indictor Channel
AIPN	All Internet Protocol Network
AKA	Authentication and Key Agreement
AM	Acknowledge Mode
	Amplitude Modulation

AMBR	Aggregate Maximum Bit Rate
AMC	Adaptive Modulation and Coding
A-MPR	Additional Maximum Power Reduction
AN	Access Network
ANR	Automatic Neighbour Relation
AoA	Angle of Arrival
AoD	Angle of Departure
AP	Application Part
	Application Protocol
API	Application Programming Interface
APN	Access Point Name
APN-AMBR	Access Point Name–Aggregate Maximum Bit Rate
ARIB	Association of Radio Industries and Businesses (Japan)
ARP	Allocation and Retention Priority
ARQ	Automatic Repeat Request
AS	Access Stratum
	Application Server
	Azimuth Spread
ASIC	Application Specific Integrated Circuit
ASN.1	Abstract Syntax Notation number 1
ASSP	Application Specific Standard Product
AT	Attention
ATC	Air Traffic Control

LTE and the Evolution to 4G Wireless: Design and Measurement Challenges, Second Edition, Edited by Moray Rumney.
Copyright Agilent Technologies, Inc. 2013. Published by John Wiley & Sons, Ltd.

ATIS	Alliance for Telecommunications Industry Solutions (USA)		CCDF	Complementary Cumulative Distribution Function
ATS	Abstract Test Suite		CCE	Control Channel Element
AUTN	Authentication Token		CCF	CDMA Certification Forum
AWGN	Additive White Gaussian Noise		CCO	Cell Change Order
			CCSA	China Communications Standards Association

B

BB	Baseband
BBIC	Baseband IC
BBIQ	Baseband IQ
BCCH	Broadcast Control Channel
BCH	Broadcast Channel
BER	Bit Error Ratio
BERT	Bit Error Ratio Tester
BLER	Block Error Ratio
BO	Buffer Occupancy
BP	Bandwidth Parts
BPSK	Binary Phase-Shift Keying
BS	Bearer Service
BSIC	Base Station Identity Code
BSC	Base Station Controller
BTS	Base Transceiver Station
BW	Bandwidth
BWRF	RF Bandwidth

C

CA	Carrier Aggregation
CACLR	Cumulative ACLR
CAG	Common Agreement Group
CB	Code Block
	Cooperative Beamforming
CBC	Cell Broadcast Center
CC	Call Control
	Component Carrier
CCCH	Common Control Channel

CDD	Cyclic Delay Diversity
CDMA	Code Division Multiple Access
CFI	Control Format Indicator
CFO	Carrier Frequency Offset
CFR	Crest Factor Reduction
CI	Codebook Index
CINR	Carrier to Interference Plus Noise Ratio
CIR	Channel Impulse Response
CK	Ciphering Key
CLI	Command Line Interface
CLR	Campaign Loader Files
CL-SM	Closed-loop Spatial Multiplexing
CM	Cubic Metric
CMAS	Commercial Mobile Alert System
CMOS	Complementary Metal Oxide Semiconductor
CoMP	Coordinated Multi-Point
CP	Cyclic Prefix
	Control Plane
CPICH	Common Pilot Channel
CPRI	Common Public Radio Interface
CPT	Control PDU Type
CPU	Central Processing Unit
CQI	Channel Quality Indicator
CR	Change Request
CRC	Cyclic Redundancy Check
C-RNTI	Cell Radio Network Temporary Identity
CRS	Cell Reference Symbol

CS Circuit Switched
Coordinated Scheduling

CS/CB Coordinated Scheduling and
Cooperative Beamforming

CSFB Circuit Switched Fall Back

CSG Closed Subscriber Group

CSI Camera Serial Interface
Channel State Information

CSI-RS Channel State Information–
Reference Signal

CSIT Channel State Information at the
eNB Transmitter

CTIA Cellular Telecommunications Industry
Association

CW Code Word
Continuous Wave

D

D/A Digital to Analog

DAB Digital Audio Broadcast

DAC Digital-to-Analog Converter

DAI Downlink Assignment Index

DC Direct Current

D/C Data or Control

DCCH Dedicated Control Channel

DCI Downlink Control Information

DCS Display Command Set

DeNB Donor evolved Node B

DFT Discrete Fourier Transform

DFT-S-OFDM Discrete Fourier Transform Spread
OFDM

DHCP Dynamic Host Configuration Protocol

DIP Dominant Interferer Proportion

DL Downlink (base station to subscriber
transmission)

DLA Direct Language Architecture

DL-MIMO Downlink Multiple Input Multiple
Output

DL-SCH Downlink Shared Channel

DMRS Demodulation Reference Signal

DNS Directory Name Service

DPCCH Downlink Physical Control Channel

DPD Digital Predistortion

D-PHY D-Physical Layer

DPS Dynamic Point Selection

DRAM Dynamic Random Access Memory

DRB Data Radio Bearer

DRX Discontinuous Reception

DSI Display Serial Interface

DSIM Digital Signal Interface Module

DSL Digital Subscriber Line

DSP Digital Signal Processing

DTCH Dedicated Traffic Channel

DTX Discontinuous Transmission

DUT Device Under Test

DVB-T Digital Video Broadcast-Terrestrial

DVM Digital Volt Meter

DwPTS Downlink Pilot Time Slot

E

E Extension

E2E End-to-End

E-ARFCN E-UTRA Absolute Radio Frequency
Channel Number

ECGI E-UTRAN Cell Global Id

ECM EPS Connection Management

ECN Explicit Congestion Notification

ECTEL European Conference on
Technology-Enhanced Learning

ED Error Detecting

EDA Electronic Design Automation

E-DCH Enhanced Dedicated Channel

EDGE	Enhanced Data rates for GSM Evolution		FDD	Frequency Division Duplex
EDL-MIMO	Enhanced Downlink-MIMO		FDMA	Frequency Division Multiple Access
eHRPD	Enhanced High Rate Packet Data		FEC	Forward Error Correction
eICIC	Enhanced Inter-Cell Interference Coordination		FeICIC	Further enhanced Inter-Cell Interference Coordination
EIR	Equipment Identity Register		FFT	Fast Fourier Transform
EM	Electromagnetic		FI	Framing Indicator
EMC	Electromagnetic Compatibility		FM	Frequency Modulation
EMM	EPS Mobility Management		FPGA	Field Programmable Gate Array
eNB	Evolved Node B		FRC	Fixed Reference Channel
EPA	Extended Pedestrian A		FS1	Frame Structure type 1
EPC	Evolved Packet Core		FS2	Frame Structure type 2
EPDCCH	Enhanced Physical Downlink Control Channel		FSS	Frequency Selective Scheduling
			FTAG	Field Trial Agreement Group
ePDG	Evolved Packet Data Gateway		FTP	File Transfer Protocol
EPS	Evolved Packet System			
E-RAB	Evolved Radio Access Bearer			

G

ESL	Electronic System Level		GAN	Generic Access Network
ESM	EPS Session Management		Gbps	Gigabit per second
ETC	Extreme Temperature Conditions		GBR	Guaranteed Bit Rate
E-TM	E-UTRA Test Model		GCF	Global Certification Forum
ETSI	European Telecommunications Standards Institute		GERAN	GSM Enhanced Radio Access Network
ETU	Extended Typical Urban		GG	Golden Gate
ETWS	Earthquake and Tsunami Warning System		GGSN	Gateway GPRS Support Node
			GLONASS	GLObal' naya NAvigatsionnaya Sputnikovaya Sistema
E-UTRA	Evolved Universal Terrestrial Radio Access			
			GNSS	Global Navigation Satellite System
E-UTRAN	Evolved Universal Terrestrial Radio Access Network		GP	Guard Period
			G-PDU	GTP-U non-signaling PDU
EVA	Extended Vehicular A		GPIB	General Purpose Interface Bus
EVM	Error Vector Magnitude		GPP	General Purpose Processor
			GPRS	General Packet Radio Service

F

			GPS	Global Positioning System
FACH	Forward Access Channel		GSM	Global System for Mobile Communication
FCC	Federal Communications Commission		GSMA	GSM Association

GTP	GPRS Tunneling Protocol
GTP-C	GTP Control
GTP-U	GTP User
GUI	Graphical User Interface
GUMMEI	Globally Unique MME Identity
GUTI	Globally Unique Temporary Identity
GW	Gateway
GWCN	Gateway Core Network

H

HARQ	Hybrid Automatic Repeat Request
HATS	Hand and Torso Simulator
HDL	Hardware Description Language
HeNB	Home eNB
HF	High Frequency
HFN	Hyper Frame Number
HI	HARQ Indicator
HII	High Interference Indicator
HLR	Home Location Register
HPLMN	Home Public Land Mobile Network
HPUE	High Power User Equipment
HRPD	High Rate Packet Data
HSDPA	High Speed Downlink Packet Access
HSPA	High Speed Packet Access
HSS	Home Subscriber Server
HST	High Speed Train
HSUPA	High Speed Uplink Packet Access
HTML	Hypertext Markup Language
HTTP	Hypertext Transfer Protocol

I

IC	Integrated Circuit
ICI	Inter-Carrier Interference
ICIC	Inter-Cell Interference Coordination
ICS	Implementation Conformance Statement

	In-Channel Selectivity
Id	Identity
IDC	In-Device Coexistence
IDFT	Inverse Discrete Fourier Transform
IDLA	Indirect Learning Architecture
IF	Intermediate Frequency
IFFT	Inverse FFT
IFOM	IP Flow Mobility
IFT	Interactive Functional Test
IK	Integrity Key
IMCS	Index Modulation and Coding Scheme
IMEI	International Mobile Equipment Identity
IMS	IP Multimedia Subsystem
IMN CN SS	IMS Core Network Subsystem
IMSI	International Mobile Subscriber Identity
IMT	International Mobile Telecommunications
Inter-RAT	Inter-Radio Access Technology
I/O	Input/Output
IODT	Interoperability Development Testing
IOT	Interoperability Testing
IP	Internet Protocol
IPv4	Internet Protocol version 4
IPv6	Internet Protocol version 6
IQ	In-phase Quadrature
IQoIP	IQ over Internet Protocol
ISD	Inter-Site Distance
ISI	Inter Symbol Interference
ITU	International Telecommunications Union
ITU-R	International Telecommunications Union Radiocommunication Sector
IWS	Interworking Solution

J

JP	Joint Processing
JR	Joint Reception
JT	Joint Transmission

K

K_{ASME}	Key Access Security Management Entity
K_{eNB}	eNB Key
K_{NASenc}	Non-Access Stratum Encryption Key
$K_{NASiint}$	Non-Access Stratum Integrity Protection Key
KPI	Key Performance Indicator
K_{RRCenc}	Radio Resource Control Encryption Key
K_{RRCint}	Radio Resource Control Integrity Protection Key
K_{UPenc}	Uplink User Data Encryption Key
KSI	Key Set Identifier
KSI_{ASME}	Key Set Identifier for Access Security Management Entity

L

L	Level
L1	Layer 1 (physical layer)
L2	Layer 2
L3	Layer 3
LAN	Local Access Network
LBI	Linked EPS Bearer Identity
LCR	Low Chip Rate
LCS	Location Service
LI	Length Indicator
LIPA	Local IP Access
LLI	Low Latency Interface
LMMSE	Linear Minimum Mean Square Error
LO	Local Oscillator
LOS	Line of Sight

LPP	LTE Positioning Protocol
LS	Large Scale
LSF	Last Segment Field
LSTI	LTE/SAE Trial Initiative
LTE	Long Term Evolution
LUT	Lookup Table

M

M2M	Machine to Machine
MAC	Medium Access Control
MAC-I	Message Authentication Code
MBMS	Multimedia Broadcast and Multicast Service
MBMS-RNTI	MBMS-Radio Network Temporary Identity (also, M-RNTI)
Mbps	Megabits per second
MBR	Maximum Bit Rate
MBSFN	Multimedia Broadcast Multicast Service Single Frequency Network
MCC	Mobile Country Code
MCCH	Multicast Control Channel
MCE	MAC Control Element
MCH	Multicast Channel
Mcps	Mega-chips per second
MCS	Modulation and Coding Scheme
MDT	Minimization of Drive Test
MIB	Master Information Block
MIMO	Multiple Input Multiple Output
MIPI	Mobile Industry Processor Interface
MISO	Multiple Input Single Output
MM	Mobility Management
MME	Mobility Management Entity
MMIC	Monolithic Microwave Integrated Circuit
MMS	Multimedia Service
MNC	Mobile Network Code
MO	Mobile Originated

MOCN	Multi-Operator Core Network
MOP	Maximum Output Power
M-PHY	M-Physical Layer
MPR	Maximum Power Reduction
MPS	Multimedia Priority Service
MRC	Maximum Ratio Combining
M-RNTI	MBMS-RNTI
MSC	Mobile Switching Center
Msps	Mega-samples per second
MSR	Multi-Standard Radio
MTC	Machine-Type Communication
MTCH	Multicast Traffic Channel
MU-MIMO	Multi-User MIMO

N

NACK	Negative Acknowledgement
NAS	Non-Access Stratum
NB	Node B
NCT	New Carrier Type
NDI	New Data Indication
NDS	Network Domain Security
NLOS	Non-Line of Sight
NS	Network Signaling
NW	Network
NRT	Non-Real Time

O

O&M	Operations and Maintenance
OBSAI	Open Base Station Architecture Interface
OBW	Occupied Bandwidth
OCNG	OFDMA Channel Noise Generator
OFCS	Offline Charging System
OFDM	Orthogonal Frequency Division Multiplexing

OFDMA	Orthogonal Frequency Division Multiple Access
OI	Overload Indicator
OMC	Operation and Maintenance Center
OSG	Open Subscriber Group
OTA	Over The Air
OTDOA	Observed Time Difference of Arrival

P

P	Polling
PA	Power Amplifier
PAG	Performance Agreement Group
PAPR	Peak-to-Average Power Ratio
PAS	Power Azimuth Spectrum
PBCH	Physical Broadcast Channel
PBR	Prioritized Bit Rate
PCCC	Parallel Concatenated Convolutional Code
PCCH	Paging Control Channel
P-CCPCH	Primary Common Control Physical Channel
PCEF	Policy and Charging Enforcement Function
PCell	Primary Cell
PCFICH	Physical Control Format Indicator Channel
PCH	Paging Channel
PCI	Physical Cell Identity
PCRF	Policy and Charging Rules Function
PCS	Personal Communication System
PDCCH	Physical Downlink Control Channel
PDCP	Packet Data Convergence Protocol
PDN	Packet Data Network
PDN-GW	Packet Data Network Gateway (also, P-GW)
PDSCH	Physical Downlink Shared Channel
PDU	Protocol Data Unit

PECF	Policy Enforcement and Charging Function
PESQ	Perceptual Evaluation of Speech Quality
P-GW	Packet Data Network Gateway (also, PDN-GW)
PHICH	Physical Hybrid ARQ Indicator Channel
PHY	Physical layer
PI	Paging Information
PICS	Protocol Implementation Conformance Standard
PIM	Passive Intermodulation
PIXIT	Protocol Implementation Extra Information for Test
PLMN	Public Land Mobile Network
PMCH	Physical Multicast Channel
PM	Phase Modulation
PMCH	Physical Multicast Channel
PMI	Precoding Matrix Indicator
PMIP	Proxy Mobile Internet Protocol
P-MPR	Power Management MPR
PRACH	Physical Random Access Channel
PRAT	Rated Output Power (base station)
PRB	Physical Resource Block
P-RNTI	Paging Radio Network Temporary Identity
PRS	Positioning Reference Signal
PRBS	Pseudo Random Binary Signal
PS	Packet Switched
PSHD	Packet Switched Handover
P-SS	Packet Switched Service
	Primary Synchronization Signal
PTCRB	PCS Type Certification Review Board
PTI	Precoding Type Indicator
PUCCH	Physical Uplink Control Channel
PUSCH	Physical Uplink Shared Channel

PVT	Power Versus Time
PWS	Public Warning System

Q

QAM	Quadrature Amplitude Modulation
QCI	QoS Class Identifier
QoE	Quality of Experience
QoS	Quality of Service
QPSK	Quadrature Phase-Shift Keying

R

R&TTE	Radio and Telecommunications Terminal Equipment
RAB	Radio Access Bearer
RACH	Random Access Channel
RAN	Radio Access Network
RAND	Random Number
RA-RNTI	Random Access Radio Network Temporary Identity
RAT	Radio Access Technology
RB	Resource Block
RBW	Resolution Bandwidth
RCLP	Relative Carrier Leakage Power
RDN	Radio Distribution Network
RDX	Radio Digital Cross Domain
RE	Resource Element
REFSENS	Reference Sensitivity
RES	Response
RF	Radio Frequency
	Re-segmentation Flag
RFFE	RF Front End
RFIC	Radio Frequency Integrated Circuit
RFSP	RAT/Frequency Selection Priority
RI	Rank Indication
RLC	Radio Link Control
RMC	Reference Measurement Channel

rms	Root Mean Square		SC-FDMA	Single Carrier Frequency Division Multiple Access
RN	Relay Node		SCH	Synchronization Channel
RNC	Radio Network Controller		SCME	Spatial Channel Model Extended
RNTI	Radio Network Temporary Identifier		SCTP	S Common Transport Protocol
RNTP	Relative Narrowband Transmit Power		SDF	Service Data Flow
RoHC	Robust Header Compression		SDMA	Space Division Multiple Access
RR	Radio Resource		SDO	Standards Development Organization
RRC	Radio Resource Control		SDU	Service Data Unit
	Root Raised Cosine		SE-BER	Single-Ended Bit Error Ratio
RRH	Remote Radio Head		SEM	Spectrum Emission Mask
RRM	Radio Resource Management		SFBC	Space-Frequency Block Coding
RS	Reference Signal		SFN	Single Frequency Network
RSCP	Received Signal Code Power		SFO	Sampling Frequency Offset
RSRP	Reference Signal Received Power		SGSN	Serving GPRS Support Node
RSRQ	Reference Signal Received Quality		S-GW	Serving Gateway
RSS	Root Sum Square		SI	System Information
RSSI	Received Signal Strength Indicator		SIB	System Information Block
RSTD	Reference Signal Time Difference		SIGCOMP	Signaling Compression Protocol
RT	Real Time		SIMO	Single Input Multiple Output
RTT	Radio Transmission Technology		SINR	Signal to Interference plus Noise Ratio
RV	Redundancy Version		SiP	System in Package
RX	Receiver		SIP	Session Initiation Protocol
			SIPTO	Selected IP Traffic Control
S			SI-RNTI	System Information Radio Network Temporary Identity
S	Subband		SISO	Single Input Single Output
SACM	Single Acquisition Combined Measurement		SMS	Short Message Service
SAE	System Architecture Evolution		SN	Sequence Number
SAM	Specific Anthropomorphic Mannequin			Serving Network
SAMM	Single Acquisition Multiple Measurement		SNid	Serving Network Identity
			SNR	Signal-to-Noise Ratio
SAP	Service Access Point		SO	Segment Offset
SAW	Surface Acoustic Wave		SoC	System on Chip
SBAS	Space Based Augmentation System		SON	Self Optimizing Network
SCell	Secondary Cell		SPID	Subscriber Profile ID

SPMI	System Power Management Interface
SPS	Semi-Persistent Scheduling
SR	Scheduling Request
SRB	Signalling Radio Bearer
SRS	Sounding Reference Signal
SRVCC	Single Radio Voice Call Continuity
SS	Subsystem
	Small Scale
	System Simulator
S-SS	Secondary Synchronization Signal
SSPS	Semi-Static Point Selection
STBC	Space Time Block Coding
SU-MIMO	Single User MIMO

T

TA	Timing Adjustment
	Timing Advance (also, TADV)
	Tracking Area
TAI	Tracking Area Identity
TAS	Time Accurate Strobe
TAU	Tracking Area Update
TB	Transport Block
TBS	Transport Block Size
TCP	Transmission Control Protocol
TDD	Time Division Duplex
TDMA	Time Division Multiple Access
TD-SCDMA	Time Division Synchronous Code Division Multiple Access
TEI	Terminal Endpoint Identifier
TEID	Tunnel Endpoint Identifier
TF	Transport Format
TFT	Traffic Flow Template
TIS	Total Isotropic Sensitivity
TM	Transmission Mode
	Transparent Mode

TP	Transmit Power
TPC	Transmit Power Control
TR	Technical Report
TRP	Total Radiated Power
TrCH	Transport Channel
TRS	Total Radiated Sensitivity
TS	Technical Specification
TSG	Technical Specification Group
TTA	Telecommunications Technology Association (Korea)
TTC	Telecommunications Technology Committee (Japan)
TTCN-2	Tree and Tabular Combined Notation 2
TTCN-3	Testing and Test Control Notation 3
TTI	Transmission Time Interval
TTL	Transistor to Transistor Logic
TX	Transmitter

U

UCI	Uplink Control Information
UDP	User Datagram Protocol
UE	User Equipment
UE-AMBR	UE Aggregate Maximum Bit Rate
UL	Uplink (subscriber to base station transmission)
ULA	Uniform Linear Array
UL-MIMO	Uplink Multiple Input Multiple Output
UL-SCH	Uplink Shared Channel
UM	Unacknowledged Mode
UMB	Ultra-Mobile Broadband
UMTS	Universal Mobile Telecommunications System
UniPro	Unified Protocol
UPE	User Plane Entity
UpPTS	Uplink Pilot Time Slot
URA	UTRAN Registration Area
USIM	Universal Subscriber Identity Module

UTRA   Universal Terrestrial Radio Access

UTRAN   Universal Terrestrial Radio Access
Network

V

VHDL   VHSIC Hardware Description
Language

VHSIC   Very High Speed Integrated Circuit

VLR   Visitor Location Register

VoIP   Voice over Internet Protocol

VoLGA   Voice over LTE Generic Access

VoLTE   Voice over LTE

VRB   Virtual Resource Block

VRC   Variable Reference Channel

VSA   Vector Signal Analyzer

VSG   Vector Signal Generator

W

WAP   Wireless Application Protocol

W-CDMA   Wideband Code Division Multiple
Access

WCS   Wireless Communication Service

WG   Working Group

WI   Work Item

WLAN   Wireless Local Area Network

WRC   World Radio Conference

X

X2-AP   X2–Application Part

X2-C   X2–Control

XML   Extensible Markup Language

XOR   Exclusive Or

XPR   Cross Polarization Ratio

XRES   Expected User Response

Index

2X2 MIMO, 5, **70**, 244, 362, 403, 409, 417, 423, 458, 577

3GPP, **see** Third Generation Partnership Project

4G, 54, 567, 573-574

4x4 MIMO, 5, 393, 395, 403-409, 577

16QAM, 55, 100, 321, 406

64QAM, 55, 100, 131, 317

8X8 MIMO, 85, 244, 577

A

Acceptance test (operator), 473, 485, 529, 560–561

Access gateway, 179

Access point name (APN), 193

Access stratum (AS), 159, 172–178, 200

ACK/NACK, 100, 104, 119, **128–135**, 162, 170, 237, 349, 366, 376, 381, 390, 484, 541, 546, 586

Acknowledged mode (AM), 169

Acknowledgement (ACK), see ACK/NACK

Active antenna system (AAS), 598

Active probe, 292

Adaptive modulation and coding (AMC), 7, 129, 131–136, 361, 562

Additional maximum power reduction (A-MPR), 20, 52

Additive white Gaussian noise (AWGN), 44, 298, 310, 345, 352–354, 366-367, 372-375, 543, 584, 593

Adjacent channel interferer and interference, 24, 46, 247, 263

Adjacent channel leakage ratio (ACLR), 33, **38–40**, 52, 272, 297, 306, 335, 518, 521, 539, 540, 545, 593, 598

Adjacent channel selectivity (ACS), 43, 45–46, 356, 541, 546

All IP network (AIPN), 5, 506, 178

Allocation and retention priority (ARP), 219–220

Almost blank subframe (ABS), 586

Amplitude flatness, 304, 341–344

Analog-to-digital converter (ADC), 346

Anechoic chamber, 244, 456–459, 461–465, 469–470, 599

Angle of arrival (AoA), 151, 244, 356, 415, 436, 457

Antenna configuration, 4–6, 72-73, 77, 102, 113, 362, 370, 418, 420–421, 425–426, 440, 597

Antenna techniques, 68, 87–88,100, 296, 430, 598

Antenna correlation matrices, 360

Antenna spacing, 393, 415–416, 422, 425–426, 460

Aperiodic reporting, 138–139

Application layer handover testing, 485

Application programming interfaces (APIs), 512

Application specific integrated circuit (ASIC), 274, 346

Assisted Global Navigation Satellite System (AGNSS), 572

Asynchronous Type-II HARQ, 130

Attach procedure, 181, 184–185, 190

Authentication, 171, 179, 182-185, 200, 216, 481, 551, 570

Authentication and key agreement (AKA), 182

E

M

S

T

Printed and bound by CPI Group (UK) Ltd, Croydon, CR0 4YY

16/04/2025

14658382-0005